Experimental Vibro-acoustics

Experimental Vibro-acoustics is the first comprehensive practical engineering guide for the effective use of measured vibro-acoustic data in a component-based approach to the analysis, simulation, and virtual prototyping of machines and mechanical systems.

The book provides practical techniques which cover measurement, data processing and uncertainties and includes many 'tricks of the trade.' It also includes a range of case studies and a detailed walk-through example in a tutorial style. Further, it focuses on the in situ blocked force method, now a full international standard, through which many of the developments in the component-based approach have been based.

This book is essential for design engineers in vibration, acoustics and structural dynamics diagnosing and troubleshooting vibro-acoustic problems in machines and mechanical systems, as well as simulation of existing and virtual assemblies. It extends beyond the core of the automotive industries, to new applications in air, rail and marine transport, as well as for domestic and industrial equipment and buildings and is relevant to both researchers and industrial engineers.

Experimental Vibro-acoustics

In situ and blocked force methods for component-based simulation and virtual prototyping

JWR Meggitt and AT Moorhouse

CRC Press
Taylor & Francis Group
Boca Raton London New York

CRC Press is an imprint of the
Taylor & Francis Group, an **informa** business

Designed cover image: Designed by Joshua Meggitt and Andrew Moorhouse

First edition published 2026
by CRC Press
4 Park Square, Milton Park, Abingdon, Oxon, OX14 4RN

and by CRC Press
2385 NW Executive Center Drive, Suite 320, Boca Raton FL 33431

CRC Press is an imprint of Informa UK Limited

British Library Cataloguing-in-Publication Data
A catalogue record for this book is available from the British Library

ISBN: 978-1-032-47971-2 (hbk)
ISBN: 978-1-032-50578-7 (pbk)
ISBN: 978-1-003-39913-1 (ebk)

DOI: 10.1201/9781003399131

Typeset in Nimbus font
by KnowledgeWorks Global Ltd.

To our colleagues and students, past, present and future, at the Acoustics Research Centre.

Contents

Part 2: Practical Implementation

Foreword

The paper 'Virtual Acoustic Prototypes: Listening to Machines That Don't Exist' by Professor Andy Moorhouse has had a profound impact on my career. Published two decades ago, it introduced groundbreaking concepts that revolutionised our understanding and characterisation of product sound. By enabling a clear separation of noise-emitting components from the passive structure itself, it has since transformed our approach to vibro-acoustics. Although at the time the experimental methods required for Virtual Acoustic Prototyping were still in their infancy, this paper marked the start of an intensive collaboration between Professor Moorhouse, an esteemed authority in acoustics from the University of Salford, and Robert Bosch GmbH, the world's largest automotive supplier.

A significant milestone in this journey was the development of the in situ blocked force method. Introduced at Bosch in the early 2000s through dedicated research projects and the efforts of several talented PhD students, this method placed us at the forefront of vibro-acoustic technology. It generated widespread industry interest, enabling revolutionary approaches like in situ Transfer Path Analysis (TPA) for diagnostic testing and component-based TPA for Virtual Acoustic Prototyping. These methods are now considered game-changers in the relationship between OEMs and suppliers, providing a clear distinction between source and path contributions to the overall acoustic experience of customers. It paved the way to shift the focus from traditional vehicle interior sound pressure levels to a more granular, component-level perspective. Component-based TPA allows, for example, interchangeability of active components in a vehicle to investigate its impact on customer perception, facilitating what-if analysis based on auralised data. In situ TPA provides invaluable advantages in finding root causes of unwanted noise, providing an essential tool for target-oriented product improvement.

The foundational research of Professor Moorhouse's team, now presented in this book, has had a profound influence on experimental vibro-acoustics. This impact extends beyond the automotive industry to sectors like public transport, household appliances, and aviation. The work also contributed to the development of the ISO 20270 standard in recent years, which has been instrumental in the widespread adoption of these methods across industries. From an industrial perspective, the development of the in situ blocked force method has enabled what was unimaginable 20 years ago. It has allowed for precise diagnostic testing and prototyping using digital twins for NVH, leading to significant improvements in product quality and customer satisfaction. The ability to separate source and path contributions has revolutionised our approach to noise and vibration challenges, making it easier to identify and address root causes. This has not only enhanced our product development process but also strengthened our partnerships with OEM's, ensuring a more efficient and effective workflow.

Looking to the future, I foresee a strong demand in the industry for acousticians proficient in vibro-acoustic engineering as described in this book, since NVH becomes increasingly central in shaping the overall perceived quality of products. Furthermore, I envision a need to simplify the methods and reduce the required expertise for test engineers, democratizing the approach for a wider range of professionals. Additionally, there is a need for procedures that enable a higher degree of automation in obtaining the necessary input data for Virtual Acoustic Prototyping. Finally, with the increasing level of electrification of the vehicle powertrain, a focus shift to higher frequencies is expected in both measurement and simulation, requiring advances in these fields as well.

This book serves as an invaluable resource, presenting knowledge in a clear and profound manner that effectively bridges the gap between theory and practice. It caters to the needs of students seeking to immerse themselves in the field of vibro-acoustics, as well as practitioners and experts. With its well-rounded exploration of theoretical and practical aspects, this book is a cornerstone for advancing the field and shaping the future of experimental vibro-acoustics in general and Vibro-Acoustic Virtual Prototyping (VAVP) in particular.

Prepare to embark on a transformative journey that will empower you with the knowledge and tools needed to excel in the dynamic field of Experimental Vibro-Acoustics!

Thomas Alber
Senior Expert NVH at Robert Bosch GmbH Filderstadt, December 2024

Preface

The material in this book has been collected over nearly 40 years since around 1985. At that time, virtual prototyping methods for vibro-acoustics were almost nonexistent, but things have moved on since then to the point where it now seems timely to summarise current theory and practice in one place - hence this book.

An important milestone in the development of vibro-acoustic virtual prototyping was the publication in 2019 of the International standard ISO 20270: 2019 *Characterization of sources of structure-borne sound and vibration — Indirect measurement of blocked forces*. Whilst numerical modelling and substructuring had been around for some time, the in situ blocked force method provided the last piece of the vibro-acoustic virtual prototype (VAVP) puzzle, finally enabling us to 'switch on' virtual assemblies with real excitations.

To make full use of blocked force data from the standard however, one needs to employ it within the wider VAVP framework alongside other techniques such as component characterisation and substructuring that are not described in the standard. While the relevant material to enable this is mostly available in journal papers, it has not at any point been brought together in one place and presented in a manner accessible for practitioners. In short, there has been no 'user manual' for use of blocked forces. This may have put VAVP technology beyond the reach of some otherwise interested practitioners. With this book, we aim to address the lack of a single source by providing, in one place, everything a practitioner might need to build a complete VAVP, including theory, practical guidance and worked examples. In this respect, it could be said that the book aims to provide a 'companion' to ISO 20270, while further extending coverage to the broader subject of Vibro-Acoustic Virtual Prototyping as a whole.

In aiming to fulfil this brief, we have tried to give a first-hand account as far as possible and have therefore concentrated on methods and techniques employed by the authors since 1985, primarily at the Universities of Liverpool and Salford in the UK. In some cases, this means that other valid approaches have not been reported. Some of the material is original, of most importance being the *in situ* blocked force method in Chapter 3, on which ISO 20270 was based. Further original contributions include *in situ TPA* (or *blocked force TPA*) in Chapter 5, the *Round Trip* method in Chapter 2, *Hybrid FEA-SEA-eXperimental analysis* in Chapter 5 and the treatment of *Uncertainties* in Chapter 8. Of course, numerous other researchers and research groups that have contributed significantly towards towards the VAVP framework as it stands. Many of these are cited throughout the book but we are also aware of some high quality research and industrial practice that has not been included due to lack of space.

Salford, October 2024

Acknowledgements

Many people have contributed to this book.

We would like to extend our immense gratitude to Profs. Barry Gibbs and Goran Pavic: these elder statesmen of Acoustics have been a constant source of support over several decades; the scale of their vision and their generosity of spirit are much appreciated.

Special acknowledgment is due to several people for their direct contributions to the methods presented: Andy Elliott, who through meticulous experimental work and understanding of structural dynamics was able to turn promising theory into practical methodologies; Thomas Alber, who had the vision to stay with a good idea for as long as it took; Michael Sturm and Kevin Wienen, who had the energy, enthusiasm and skill to take complex ideas and turn them into a reality.

We would like to acknowledge the contribution of research collaborators on numerous research projects, including: Robin Langley and Arnau Clot at the University of Cambridge; Guy Banwell and colleagues at Dyson Technology; Heng Yi (Kevin) Lai at Boing Corporation; Oliver Farrell and colleagues at Farrat Isolevel; John Smith and colleagues at DSTL; various colleagues at Bentley Motors and Hottinger Bruel and Kjaer; collaborators on the NABUCCO project, including Goran Pavic, Leon Gavric, Jean Tourret, Mats Abom and colleagues at KTH, Barry Gibbs and Gary Seiffert at the University of Liverpool's Acoustic Research Unit.

Most of the original work in this book was carried out as part of funded projects and this support is gratefully acknowledged: EPSRC projects: GR/S01979/01, EP/D002109/1, EP/G066582/1, EP/P005489/, EC Framework V, NABUCCO project; Innovate UK; Technology Strategy Board (UK); DEFRA (UK); Robert Bosch GMBH; Boeing Corporation; Bentley Motors; Dyson Technology.

A number of graduate students developed aspects of the work as part of their studies and their contributions are gratefully acknowledged: Su Jianxin, Andy Elliott, Tomos Evans, Marcus Bauer, Michael Hudelmaier, Michael Sturm, Nikos Zafeiropoulos, Alsdeg Abohnik, Nikhilesh Patil, Dennis Zabel, Kevin Wienen, James Evans, Daniela Filipe, Lucy Barton, Ramin McGee, Gabriel Whittle, Martin McNulty. The contributions of a number of project students - too many to mention individually - is also gratefully acknowledged.

We are grateful to friends who gave their time to review and comment on draft sections of the book: Barry Gibbs, Fabio Bianciardi, Thomas Alber, David Lennstrom, Nikos Zafeiropoulos, Andy Elliott, Arnau Clot, Tomos Evans, Gabriel Whittle, Lucy Barton, Martin McNulty.

Authors

Joshua Meggitt has been working in the area of blocked forces, component-based simulation and virtual prototyping since starting his PhD in 2013. In this time he has contributed to the field by: presenting one of the first experimental studies combining blocked forces with dynamics substructuring (2014), developing an in situ characterisation method for vibration isolators (2015), formulating a rigorous uncertainty framework for blocked forces, substructuring and TPA (2020), and establishing a metric for assessing interface completeness (2020). Joshua regularly works alongside industry, offering both contractual research and PhD supervision. In addition to this, he frequently delivers bespoke training on blocked force and virtual prototyping methods. In 2023 Josh received the prestigious Tyndall medal from the UK's Institute of Acoustics in recognition of his achievement and services in the field of acoustics.

Andy Moorhouse is best known as the lead author of the *In situ Blocked Force Method* which is of central importance to subject of this book. He has worked in the vibro-acoustics field since starting his PhD in 1985, most notably in pioneering *in situ* measurement methods in vibro-acoustics. He was Professor of Engineering Acoustics and Vibration, as well as Director of the Acoustics Laboratories at the University of Salford between 2007 and 2021 and Convener of ISO TC43 Working group 57 which led to the publication of the in situ blocked force method in ISO 20270 in 2019. He was principle investigator of a number of funded projects, supervised 16 PhDs to completion and produced numerous technical reports for the UK government. In 2018 he was awarded the prestigious Institute of Acoustics Engineering Medal for outstanding contributions in the field of acoustical engineering. He is a chartered engineer and a Fellow of the Institute of Acoustics. He is also a performer and composer jazz music and has had various compositions played on local and national radio.

Case Study Contributors

Nezam Sanei
Alstom
Saint-Ouen, France

Simon Tate
Ricardo UK
Shoreham-by-Sea, UK

Fabio Bianciardi
Siemens Industry Software NV
Leuven, Belgium

Dr. David Lennström
Volvo Car Group
Göteborg, Sweden

Martin McNulty
Ramboll
Manchester, UK

Dr. Andrew Elliott
Acoustics Laboratories
University of Salford, UK

Nomenclature

GENERAL REMARKS

Symbols and formulations are by the most part given in the frequency domain, unless otherwise specified. For clarity, explicit frequency dependence notations are typically omitted, except where confusion may arise. For the most part, variables are described by the symbols given below. In some cases a symbol might have multiple uses; its meaning should be clear from the surrounding context.

In general, matrix quantities are indicated by bold upper-case font whilst vector quantities are shown in bold lower-case. Matrix dimensions are indicated using set notation, with $\mathbf{A} \in \mathbb{R}^{M\times N}$ and $\mathbf{A} \in \mathbb{C}^{M\times N}$ denoting, respectively, real and complex matrices of size $M \times N$. Real and complex vector dimensions are similarly indicated as $\mathbf{a} \in \mathbb{R}^N$ and $\mathbf{a} \in \mathbb{C}^N$. Scalar elements are shown in standard typeface. Matrix and vector elements are adorned with subscripts ij and i, respectively, to denote their position within the matrix/vector.

Capitalised matrix subscripts are used to denote the component to which marked quantity (typically an FRF) belongs, whilst lower case subscripts denote the DoFs at, or between, which the quantity is defined. As an example, $\mathbf{A}_{Cbc}$ would be a matrix that belongs to the coupled assembly C and is defined between the DoFs b and c.

ACRONYMS

DoF(s)	Degree(s) of freedom
EMA	Experimental modal analysis
EMPC	Equivalent multi-point connection
FD	Finite difference
FE(A)	Finite element (analysis)
FRF	Frequency response function
LPU	Law of propagation of uncertainty
NVH	Noise, vibration and harshness
SEA	Statistical energy analysis
SVD	Singular value decomposition
TPA	Transfer path analysis
TSVD	Truncated singular value decomposition
VP	Virtual point

LOWER-CASE ROMAN LETTERS

c	The set of interface DoFs
c_i	The set of known interface DoFs
c_j	The set of unknown interface DoFs

b	The set of remote receiver DoFs (used for indirect estimations)
r	The set of target receiver DoFs for which a prediction is to be made
a	The set of remote source DoFs
i	The set of internal source DoFs (where forces are generated)
$f_\square$	External force applied to subscripted DoF
g_c	Coupling/contact force occurring at the interface DoFs c
$v_\square$	Operational velocity at the subscripted DoF
$p_\square$	Operational pressure at the subscripted DoF
Q	Complex power
$\tilde{v}_c$	Free velocity at interface DoFs c
$\bar{f}_c$	Blocked force at the interface DoFs c
$\check{f}_a$	Pseudo-force at the source DoFs a
$\mathbf{f}$	Vector of external forces
$\mathbf{g}_c$	Vector of coupling/contact forces
$\mathbf{v}$	Vector of operational velocities
$\mathbf{p}$	Vector of sound pressures
$\tilde{\mathbf{v}}_c$	Vector of free velocities
$\bar{\mathbf{f}}_c$	Vector of blocked forces
$\check{\mathbf{f}}_a$	Vector of pseudo-forces

ROMAN CAPITALS

S	Source component
R	Receiver component
I	Isolator or resilient coupling component
C	Coupled assembly
SI	Coupled source-isolator assembly
IR	Coupled isolator-receiver assembly
Y_{ij}	Mobility (ijth matrix entry)
Z_{ij}	Impedance (ijth matrix entry)
H_{ij}	Vibro-acoustic or acoustic FRF (ijth matrix entry)
T^f_{ij}	Force transmissibility FRF (ijth matrix entry)
T^r_{ij}	Response transmissibility FRF (ijth matrix entry)
$\mathbf{Y}_\square$	Mobility matrix of subscripted component/DoFs
$\mathbf{Z}_\square$	Impedance matrix of subscripted component/DoFs
$\mathbf{H}_\square$	Vibro-acoustic or acoustic FRF matrix of subscripted component/DoFs
$\mathbf{T}^f_\square$	Force transmissibility matrix of subscripted component/DoFs
$\mathbf{T}^r_\square$	Response transmissibility matrix of subscripted component/DoFs
$\mathbf{V}_\square$	Operational velocity matrix at the subscripted DoFs
$\mathbf{I}$	Identity matrix
$\mathbf{U}$	Left singular vector matrix
$\mathbf{V}$	Right singular vector matrix

$\mathbf{B}$	Signed Boolean matrix for satisfying continuity
$\mathbf{L}$	Unsigned Boolean matrix for satisfying equilibrium
$\mathbf{T}_{v,f}$	Interface transformation matrix for measured forces (f) and response (v)

LOWER-CASE GREEK LETTERS

ω	Radian frequency
σ_i	ith singular value
λ	Vector of interface force intensities
$\Delta_\square$	Finite difference force or response spacing along the subscripted axis
$\mu_\square$	Mean value of subscripted quantity
$\sigma^2_\square$	Variance of subscripted quantity
$\phi_\square$	Phase of subscripted quantity

GREEK CAPITALS

Σ	Singular value matrix (diagonal)
Λ	Diagonal matrix of natural frequencies squared
Φ	Mass normalised mode shape matrix
$\Sigma_\square$	Covariance matrix of subscripted quantity

ACCENTS AND OTHER SYMBOLS

$\square \in \mathbb{R}^N$	Denotes a real column vector of size $1 \times N$
$\square \in \mathbb{C}^N$	Denotes a complex column vector of size $1 \times N$
$\square \in \mathbb{R}^{M \times N}$	Denotes a real matrix of size $M \times N$
$\square \in \mathbb{C}^{M \times N}$	Denotes a complex matrix of size $M \times N$
$\mathrm{diag}(\square)$	Diagonal matrix with $\square$ along the diagonal
$\square^{-1}$	Matrix inverse
$\square^{+}$	Matrix pseudo-inverse
$\square^{*}$	Complex conjugate
$\square^{\mathrm{T}}$	Matrix transpose
$\square^{\mathrm{H}}$	Hermitian operator (conjugate transpose)
$\mathrm{Re}(\square)$	Real part of $\square$
$\mathrm{Im}(\square)$	Imaginary part of $\square$
$M(\square)$, $M_v(\square)$, $M_H(\square,\square)$	Functions to separate real and imaginary parts of complex quantities
$\lvert\square\rvert$	Magnitude of marked quantity
$\angle\square$	Phase of marked quantity

1 Introduction

1.1 AIMS AND SCOPE

The aim of this book is to provide a comprehensive theoretical and practical background for effective use of measured vibro-acoustic data in a component-based approach to analysis, simulation and virtual prototyping of machines and mechanical systems.

It is aimed at engineering graduates with some background in vibration, acoustics and/or structural dynamics. It has applications for diagnosis and troubleshooting of vibro-acoustic problems in machines and mechanical systems, simulation of existing and virtual assemblies and as a tool for design. Historically, the main application of these techniques has been in NVH (Noise, Vibration and Harshness) within the automotive industries, but there are increasing applications for other modes of transport including aerospace, rail and marine, as well as for satellites, domestic and industrial equipment and buildings.

In the component-based approach, machines and mechanical systems are viewed as assemblies of separable, active and passive components: active components generate the dynamic forces that excite the rest of the assembly; passive components do not generate forces themselves but transmit vibrational energy to and from connected components and into surrounding fluid as sound. In transmitting vibration, each component modifies the signature of the dynamic forces and hence imposes its characteristics on the output of the assembly as a whole. The customer or user hears the sound (or feels the vibration) of the whole machine in which every component plays its part. One cannot evaluate the sound of a fridge from the compressor alone or the acoustics of a vehicle solely from the sound of its engine – it is necessary to simulate the operating machine in its entirety.

The key assumption in the component-based approach therefore is that all components may be vibro-acoustically characterised in an independent way such that they may be interchanged and recombined by mathematically coupling and decoupling them. Virtual prototypes take this idea to its logical conclusion by combining existing or modelled components into not-yet-existing combinations. A virtual prototype is, in many respects, like the 'digital twin' of a physical prototype, assembled from digital rather than physical components. Hence, this book is concerned with facilitating the representation, or characterisation, of these digital components by means of vibro-acoustic measurements and their combination into virtual machines.

1.2 INDUSTRIAL CONTEXT

Today, whether we know it or not, we live at the dawn of a new era: the fourth industrial revolution is upon us! We are entering the age of *cyber-physical* systems.

DOI: 10.1201/9781003399131-1

Figure 1.1 In the vision of the fourth industrial revolution, machines will be accompanied by digital counterparts or 'digital twins'.

Part of the grand vision of so-called '*Industry 4.0*' is that all machines, as well as industrial processes, will be accompanied by a 'digital twin'; a purely digital representation of all aspects of the system's appearance and performance. Among the main advantages of working with these virtual systems is that it is easier, faster, cheaper, safer and less resource-intensive to experiment and make tests on the digital copy than on a real physical machine. The use of digital representations in this way is, we are told, the major industrial trend of our time.

Convenience, cost and speed, though, are not the only advantages of this paradigm. This digital representation is a model; engineers have known for centuries that the process of building a model forces us to think about what is really important for whatever it is we are modelling to function as it does. By formulating a representation of the real thing, we gain *understanding* into its operation. This is further enhanced when we use the model to investigate the effect of modifications or to test 'what-if' speculations. Arguably then, the really important benefit from digital representations is the *insight* that they bring.

In the area of prototyping, this major trend towards modelling as a replacement for physical prototyping has been underway for decades, and confidence has grown to such an extent that major vehicle manufacturers have announced their intention to launch future production runs entirely without (physical) prototypes. In this 'zero prototypes' environment, all the trials and testing usually carried out on physical prototypes will instead be conducted numerically using 'virtual prototypes'. Already, many aspects of vehicle appearance and performance are fully or partially tested in this way: visual designers can employ a wide range of tools varying from sketches, to CAD drawings to full 3D simulations. Ergonomics, structural design, control, dynamics, even crash tests of vehicles can nowadays be reliably modelled. But what about acoustics or vibro-acoustics?

1.2.1 VIBRO-ACOUSTIC DESIGN

Acoustics and vibration design, once often something of an afterthought, has steadily been increasing in importance over the last few decades. Nowhere is this trend more evident than in the automotive industry where vehicle attributes relating to NVH are nowadays often given the highest priority. There are complex reasons for this growth in status, but in general it all comes back to two major drivers: legislation and customer perception.

Regarding *legislation*, manufacturers of many products, vehicles, industrial, domestic and outdoor products, for example, are typically subject to mandatory limits on noise or vibration emission. In the European Union, for example, certain types of equipment may not be sold unless they meet limits for sound power [1].

Customer perception, on the other hand, is typically a more complex issue that may take several forms, the most important of which is 'brand image'. The iconic '*potato-potato-potato*' sound of the Harley Davidson engine and the satisfying '*clunk*' of the VW car door are classic examples where sounds have augmented brand value by making a product more desirable. Product sounds may also attract negative associations though, which can be damaging to the brand, for example if a product sounds 'cheap' or 'nasty'. Avoiding negative connotations of this type can, in the most extreme cases, be a matter of life or death for a product [18]. Another area where things can go wrong is with 'no fault returns'; here the buyer returns a product believing it to be faulty even though it is functioning correctly as designed. A common reason for such returns is that the product 'sounds wrong', whether or not it actually is wrong. Manufacturers report that the cost burden from such returns can be enormous because every returned item has to be individually checked and cannot be resold at full price.

To satisfy legislation or customer preference, manufacturers have adopted a range of specific acoustics and vibration design targets. Those derived from legislation are for the most part single number limits designed for simplicity and clarity. Designing to such limits typically involves application of classical engineering noise control techniques, often by blocking or absorbing the sound on its way to the receiver. This type of noise control is a specialist engineering discipline, though nowadays fairly well established.

Targets for customer perception on the other hand, are a far more nuanced and complex matter. What makes a product sound 'cool' or 'tinny' or 'wrong', and how do we improve it? We first need to quantify 'cool' or 'powerful' or whatever perceived attribute we are seeking. We then need to understand how positive or negative associations are triggered by combinations of features in the sound. These first steps bring us firmly into the realms of psycho-acoustics. Designers then need to modify the product to improve the perception, but quite how engineering modifications map to specific features in the sound and then to perception is a complex question. Altogether then, 'sound design' of engineered products is a far more difficult and advanced science than traditional noise control.

It is clear then that today's manufacturers cannot afford to get the acoustic design wrong and furthermore that getting it right may demand a sophisticated, possibly multidisciplinary approach. The task is made all the more difficult by constant pressure to reduce mass and cost. A measure of the importance of vibro-acoustics is that vehicle manufacturers (including rail) nowadays generally employ teams of acoustic specialists (often including psycho-acousticians), as do some producers of domestic and outdoor equipment.

To work effectively though, designers need tools; ideally they would like to listen to their designs and hear the effects of modifications for themselves or play them to a jury of listeners for quantitative or qualitative testing. Traditionally this was done with physical prototypes, but as already stated, to comply with the industrial trends of the times, it increasingly has to be achieved digitally. A need has therefore emerged for *Vibro-Acoustic Virtual Prototypes*.

1.2.2 VIBRO-ACOUSTIC VIRTUAL PROTOTYPING

The need for 'vibro-acoustic virtual prototypes' (VAVPs) has emerged from the desire to listen to designs and design modifications before machines physically exist. In today's zero-prototype context, this must be done virtually. In that case our digital copy must be capable of reproducing all the functions of the real prototype.

One such function is to evaluate the effect of modifying or replacing components for the purposes of design optimisation. This implies that our VAVP is also built from components, each with their own digital copies, and defined in such a way that they can be removed and replaced independently of the assembly and other connected components. This further implies that we need the means to couple these digital components to, and decouple them from, the assembly as a whole. The interchangeability of separable components in this way is at the heart of the component-based approach that is central to this book (and will be picked up again in 1.5.1).

Another use of prototypes is to display the functionality of a design for validating specifications and targets. In a vehicle, for example, the vibro-acoustic specifications may be at a whole vehicle level (such as drive-by noise) or relate to specific components (such as steering system noise). The prototype may be required to *verify* that specific targets are met or to *validate* that the design fulfils its intended purpose overall. Either way, our digital representation will need to facilitate reliable judgements based on entirely simulated sound and vibration. Moreover, in the vibro-acoustics field these judgements will often be subjective. In other words, to perform the role of a prototype, our VAVP, assembled entirely from digital components, will need to sound and feel exactly like the real thing.

1.3 SCIENTIFIC AND TECHNICAL CHALLENGES

The need for digital representations to produce realistic sounds raises immense scientific and technical challenges. In part, these are due to the complexity of the machines being represented and in part because, in striving for realism, we are up against a formidable opponent – the extraordinary sophistication of human hearing!

Consider a comparatively simple vibro-acoustic assembly as an example – a music box attached to a sounding board. From the sound of the assembled unit the

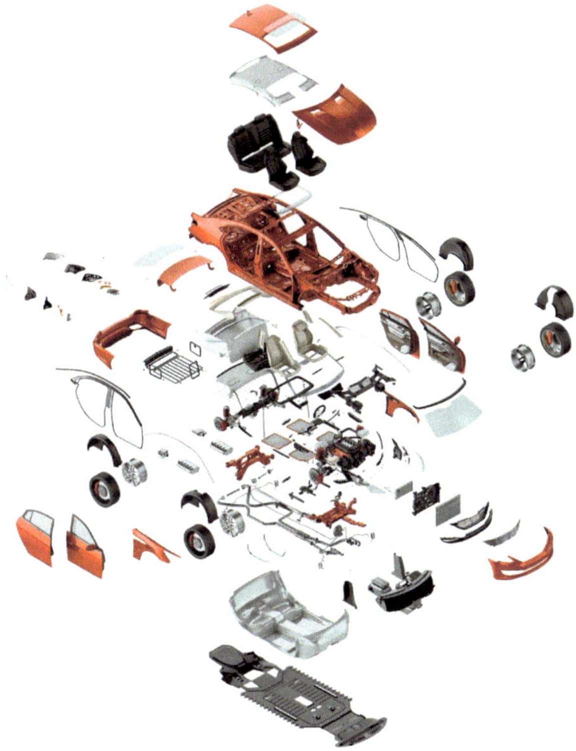

Figure 1.2 Machines are composed of multiple components each with its own vibro-acoustic characteristics.

average person is able to discern fine detail in the active mechanism itself and also certain properties of the (passive) sounding board, whether it is wood or metal for example. To produce a VAVP of this relatively simple system then, we need not only to model the active mechanism in significant detail but also the passive support structure. Neither component can be omitted. When we scale up this example to a complex machine like a vehicle, there may be multiple active components and thousands of passive components involved in the transmission path to the ear of the listener (Figure 1.2), none of which can necessarily be neglected *a priori*.

Human ability to discern fine detail in sound as mentioned above is related to the impressive frequency range over which our hearing system operates. Adults are normally able to hear up to at least 10 kHz, and nowadays specifications for electric vehicles often extend to the very top of this range. At the other end of the scale, most people can hear down to 20 Hz, and if we include tactile vibration in our VAVP, then we may need to model down to 1 Hz. Thus, the full working range of a VAVP could span four whole orders of magnitude (a factor of 10000!). This is an enormous range, compared with say our vision which operates between about 400 and 700 nanometres, a factor of only 1.75.

The range of structural and acoustic wavelengths corresponding to this frequency range is also very wide, usually extending from larger than the whole machine to smaller than individual components. As a consequence, no one method of modelling is capable of spanning the whole frequency range. 'Long wavelength' methods, based on modal decomposition such as finite element methods (FEM) are able, nowadays, to model the first few hundred modes of a structure or component but are not suitable for handling many short waves. Conversely, energy-based methods like Statistical Energy Analysis (SEA) [35], or more recently Dynamical Energy Analysis (DEA) [31], are suitable for short wavelengths, but not long. Thus, we have the so-called 'mid frequency problem' which is not yet fully solved [7], though some major advancements have been made, see for example [29].

Another major challenge is the modelling of *vibration-generating mechanisms* found inside every active component – fluctuating forces brought about by various mechanisms including aero- and fluid-dynamic forces, magnetic forces, impacts, friction forces, gear transmission error, forces at joints, etc. These mechanisms tend to be highly complex, the inputs are often difficult to determine and the mechanisms themselves are often non-linear, all of which makes modelling challenging. Perhaps for these reasons, while enormous effort has been devoted to vibro-acoustic modelling over the last few decades, only a tiny fraction of this effort has been directed towards source mechanisms. As a result the vast majority of models tell us how a structure responds to unit force input but do not attempt to 'switch on' the machine with realistic input. Today then, although there have been recent advances in modelling of magnetic forces in electric motors [36], reliable models of vibro-acoustic sources must be considered beyond state-of-the-art.

Further areas of difficulty include joints, complex geometry and material properties. The dynamic effects of *joints* are difficult to model because of the various non-linear mechanisms that contribute, e.g. friction, stick-slip, etc. Certain complex geometries, such as double curvature, are known to present particular challenges for modelling. Finally, the *material properties* of some materials, such as rubber compounds, display complex multi-dimensional temperature, load and frequency-dependent behaviour that is not easily incorporated into FEA models.

To complicate matters even further, a VAVP may need to represent a production run rather than an individual machine or vehicle. Each nominally identical vehicle, or machine, will have its own variations caused by manufacturing tolerances and the impossibility of sourcing identical components every time. In principle, we must assume that these variations will cause audible variations in sound output, particularly considering the fidelity of human hearing as mentioned above. A VAVP of a production run will therefore somehow need to incorporate models of *ensemble variance*. Understanding of 'uncertainty in structural dynamics' as a subject has moved on significantly over the past couple of decades and a wide field of literature has emerged [19]. Nevertheless, there are many unsolved questions, particularly on the acoustics side – for example how does one auralise an ensemble made up of many products?

1.4 WHY EXPERIMENTAL METHODS?

Why then, in this context of digital representations and virtual prototypes do we need a book on experimental methods?

One answer, hinted at above, is that modelling vibro-acoustics is simply too hard – at current state-of-the-art there are too many unsolved problems for us to move forward with the zero-prototype vision if we are to rely on numerical modelling alone. Measurements are therefore needed to 'fill in the gaps'.

Above, we have alluded to some specific areas where models do not yet provide what we need:

- source mechanisms,
- responses spanning the mid-frequency range,
- dynamic properties of some materials,
- some complex geometries.

Of course, if measurements are involved then physical components must exist and in that sense our VAVP will not be 'virtual' at all. Nevertheless, it will be seen later in this chapter that one can retain many of the advantages of the virtual realm, flexibility and insight for example, by integrating measured data with numerical models, or even building VAVPs entirely from measured data. In this way it is still possible to combine components in ways that have not been physically realised, so we can at least claim that the *assemblies* are virtual even if their constituent *components* are not.

Experiments, however, should not be seen simply as a backstop whose role it is to come to the rescue whenever modelling proves too difficult. Even where numerical models do exist, measurements are required for *validation*. Whereas good models are the bedrock of modern engineering, unreliable models do not provide useful insight and may obscure important behaviour. Ultimately, only by validation against measurements are we able to distinguish the useful from the unhelpful.

A related application is *model updating* where measurements are needed to refine model parameters to give a closer match with reality. Measurements are also needed for extracting material properties to serve as input to numerical models. Even in a world of fully virtual prototypes it is likely that measurements of these types are here to stay. Indeed, even in the full-blown Industry 4.0 vision, there is a constant conversation between the real process and its digital twin; for this to happen there must be a data stream based on physical sensors, in other words there must be measurements.

Experimental methods are not then about to disappear; on the contrary, if anything they are destined for an enduring role in ever-closer integration with numerical methods. However, their nature is changing and evolving to suit the needs of the day. In the early days of machinery acoustics, the most common use of measurements was in rating the overall sound power of equipment. For this purpose one or more external sensors suffice – one does not need to 'get inside' the machine at all. As numerical methods, particularly FEM, began to blossom in the 1970s and 80s we saw a new experimental paradigm emerge in the form of 'Experimental Modal Analysis' (EMA, see 1.6). EMA provides the wherewithal to identify system parameters, primarily natural modes, from measured data which represented a giant leap in sophistication from the earlier sound power measurements.

Today, the landscape has shifted yet again: in the component-based approach, we require the excitation by complex sources to be represented with realistic data, and we require every component to be described independently of its surroundings and with a sophistication compatible with that of human perception. New experimental methods are needed to meet these needs, methods able to provide the digital copies to build a VAVP. Part of the rationale of this book is to bring together some of the relevant experimental methods, particularly *in situ* and blocked force methods. We will describe later why these approaches deserve particular attention but first we need to understand more about what they will be used for, in other words what is required of a VAVP.

1.5 VIBRO-ACOUSTIC VIRTUAL PROTOTYPES

In this section, we lay out the basic conceptual scheme together with the main assumptions to be used in the component-based approach and for VAVPs. The scheme will provide the framework for all material in the book.

The general principles are illustrated in the scheme for a VAVP of a refrigerator (Figure 1.3), taken from a European project in 2002 [23]. Broadly similar schemes were in use in vehicle simulators around that time. In Figure 1.3, the active component, the compressor, is represented simultaneously as a source of airborne (AB), structure-borne (SB) and fluid-borne (FB) sound (the figure is a simplification in that it does not show that four modes, or degrees of freedom (DoFs), were required to

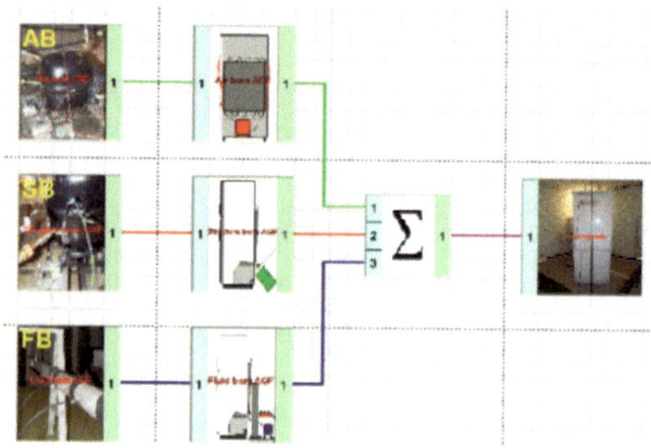

Figure 1.3 Scheme of a VAVP of a refrigerator showing airborne (AB), structure-borne (SB) and fluid-borne (FB) sound-generating mechanisms in the active component (compressor), which are filtered in different ways and summed together at the ear of the listener (EC Framework V project 'Nabucco' GRD1-1999-10785).

represent AB excitation and three for SB). The fluctuating forces generated by these mechanisms are coloured (filtered) in different ways by the passive structure of the refrigerator, as represented by the filter boxes in Figure 1.3. The filtered sounds then converge and are summed at the receiver point, taken to be at the ear position of a hypothetical listener. The digital representation of the sound is finally 'auralised', through headphones, loudspeakers or 3D sound rendering systems, so as to reproduce what would be heard by a listener at this position.

When considering the role of VAVPs, it is worthwhile to consider the steps in the process of using *physical* prototypes for vibro-acoustic design. Broadly speaking: (1) the (physical) *components* are constructed, then (2) connected together to form the *assembly*, (3) any active elements or *sources* are then switched on and (4) the vibro-acoustic *outputs* are monitored and compared with targets (potentially including perceptual targets). If design optimisation is required then (5) components are *modified* or *replaced* and we then repeat steps (3) and (4).

The same basic elements will be required for a *virtual* prototype except that all physical aspects are replaced by digital representations. Thus, the VAVP is expected to include the following elements:

- passive and active components
- means of assembly
- prediction of output including by auralisation
- means of replacing components.

Going through the above list, we therefore need to consider how the (digital) components are defined as well as the means by which they may be connected. Basic definitions will be considered in the following and more detailed discussion of these topics will form the main material in Chapters 2, 3 and 5 respectively. It will become evident that correct definition of components can only be achieved with accurate representation of the *interfaces* between them. This is of such crucial importance that it is given its own chapter: Chapter 4, 'Interfaces'. Regarding prediction of outputs, one of the main points for consideration is that predicted outputs must be in a form suitable for auralisation, which effectively means that we require data in the form of time series. Notes on processing are provided in Chapter 7 to this effect. Although auralised output is in many respects the final goal of a VAVP, the topic of Auralisation has been well covered elsewhere [32] and so will not be discussed at any length. Thus, for the purposes of the book, a complete VAVP will include everything up to the time series output but will stop short of the creation of an audible experience.

Last in the above list is the requirement to replace components. It will be seen that a natural consequence of the 'component-based' approach is that components can be readily interchanged. Therefore, before we go on to define components, sources, assemblies and interfaces we introduce some key concepts, namely the 'component-based approach' and the closely-related 'black-box' model.

1.5.1 THE COMPONENT-BASED APPROACH

We have seen above that an important aspect of *physical* prototypes is the ability to remove, replace and transfer components from one assembly to another. Provided the fixing locations are compatible between the components and the rest of the assembly (in other words, provided the bolt holes line up), then we expect to be able to interchange (physical) components with complete freedom.

Interchangeability is equally important in the *virtual* domain with a VAVP and the term 'component-based' has been coined to indicate just this: the component-based approach then can be defined as one in which all (virtual) components are *separable, interchangeable, transferable* or *independent* (of the assembly) – all these terms will be considered as equivalent in this context. We might also talk of component *properties* or *intrinsic* properties, meaning that they belong to the component alone and are free from any influence of the rest of the assembly. The terms *component-based* simulation and *VAVP* will similarly be assumed equivalent.

When representing components with *measured* data however we cannot take for granted that this all-important interchangeability will be achieved, in fact, in general this will not be the case unless specific measures are adopted. Briefly, when measurements are made *in situ*, that is, on an assembly, then one automatically measures properties of the *assembly* rather than of any individual components. In order to obtain independent properties of components, they must be separated from the assembly somehow, typically either physically or mathematically or a combination of both. When we talk of *in situ* measurement methods, we are really talking about how to organise measurements on an assembly such that one can achieve this separation and thereby extract independent (transferable, separable or interchangeable) properties of individual components.

In essence in the component-based approach, components must be characterised in a state with the interfaces either free or rigidly fixed. If one of these ideal boundary conditions can be achieved physically, then component-based properties can be measured directly. If not, as is often the case, then one may need an *in situ* measurement with some special measures to remove the effect of the rest of the assembly. The question of how, practically, to obtain independent component properties within the constraints of the real world is one of the central themes of this whole subject and will be discussed at length throughout the book. Further discussion will be deferred until Chapter 2 but for now we introduce a closely related concept needed for construction of the VAVP.

1.5.2 THE BLACK-BOX MODEL

In the VAVP scheme presented here, all components will be treated as 'black boxes'. The black box concept has nowadays expanded to describe systems as diverse as the workings of the human brain and governments, but its original application was for electrical networks where it is still used in a form very similar to that employed here. The essential black box concept in electrical networks is that a system is fully described by the behaviour at its terminals; the mechanics of what goes on inside the box are unknown and of no great importance. In other words in a black box, the inputs and outputs are related in a known and defined way but the workings in between are unknown.

In the case of vibro-acoustics, the 'systems' to be described by black boxes are the components of the VAVP. But what constitutes the 'terminals' of these black boxes? Generally, we need to consider 'interfaces' and 'remote', specifically 'target', points. Clearly, we need to capture all inputs (normally forces) and outputs (normally vibration or acoustic responses) to and from the component. *Interfaces*, as the regions where a component connects to another component, are where this transmission takes place and so all interface points must be included in our black box description. Additionally, we may be interested in the response at *target* locations; for example, the acoustic response at a driver's ear position might be needed even if no external forces are applied there. Target locations are therefore also included in the black box description. (For completeness, there is another category of remote points called *indicator* points that do not form part of the component characterisation but are required as an intermediate step in the indirect determination of interface properties. Indicator points will be revisited in Chapter 3 but will not be discussed further here.) The 'terminals' of the black box description therefore comprise *interface* and *target* locations and the component is defined solely by the properties at these locations.

This is quite different to what we are used to in modelling. A FEM model of a component, an isolator for example, will normally provide structural dynamic properties at every point of the structure, only limited by the resolution of the mesh employed. A description based on measurements on the other hand will typically only include the properties at the ends of the isolator where it connects to other components (the interfaces); we have no information about what goes on in between the inputs and outputs unless we specifically define a response location there.

In electrical networks, black boxes are defined so that they can be *connected* to form larger networks by applying coupling rules. The same is true of the black box components defined for a VAVP which can be connected to form larger assemblies (the main topic of Chapter 5). The reverse process is also possible in that an assembly of multiple black box components can be disconnected into individual components. It should be obvious that the ability to assemble and disassemble components in this way relates to the interchangeability property defined earlier (1.5.1) that was said to be essential to the component-based approach.

Whilst it can be connected to, and disconnected from others, a black box component cannot be subdivided because, without taking further measurements, we have no information about the inside of the box. Again, this highlights that digital representations based on measurements, unlike those from most models, are solely defined at measurement points.

1.5.3 THE ELEMENTS OF A VAVP

We are now ready to define the various elements in a VAVP in more detail.

Components

Components are any parts of the assembly that one may need to separate from the whole, for example to allow components from different manufacturers to be modified or replaced. It should be clear from the above that a component is the fundamental unit of the VAVP that is not further divisible without taking further measurements (in which case the components would be redefined). Typical examples of components are frames, housings, subframes, floors of buildings or even entire buildings, isolators and linkages as well as active components (see next section) like motors, pumps, compressors, fans etc.

Usually, a component of a VAVP coincides with a physically separable component, but this is not necessarily the case. In fact, in the virtual world there is considerable flexibility in defining the boundaries. For example, a subassembly of several physical components such as a suspension system may, if convenient, be considered as a single virtual component with interfaces at either end and no details about the intervening joints. Moreover as will be seen in Chapter 3, the boundaries of a component do not even have to coincide with physical boundary so that one may occasionally define a virtual component as a physically non-separable part of a (physical) component; for example the part of a housing containing a bearing could, in a VAVP, be considered as a separate component from the rest of the housing.

The first step in constructing a VAVP then is to define the components. In effect, since a black box is fully defined by the way it responds with the outside world, this means defining their *boundaries*, and in particular their *interfaces*, i.e. the regions of the boundary where they connect to neighbouring components in the VAVP. We also need to include any other points where the response needs to be known such as target locations. The component is therefore defined solely at the interfaces and target points.

What properties are needed to define the component at these points? This question will be discussed in more detail in Chapter 2 but for now we can get a fast-track impression by analogy with electrical circuits. Electrical components are defined by their dynamic impedance which defines how the input (current) is modified to become the output (voltage). The impedance is in effect a frequency-dependent, input-output filter.

The same is true of a VAVP component. For example, tapping any object with a fingernail and listening to the response usually reveals what it is made from because the object colours (filters) the sound in a characteristic way. In other words, the object acts as a vibro-acoustic filter. VAVP components then, like electrical components can be characterised by vibro-acoustic filter functions or Frequency Response Functions (FRFs). FRFs are measured by applying a known excitation (usually a force) at one point on the component and registering the response (usually an acoustic or kinematic response – displacement, velocity or acceleration) at the same or different points. The main assumptions inherent in this approach are that the component is linear and time-invariant. Nonlinearity in passive substructures as sometimes occurs in practice, for example with certain types of bolted joints, is beyond the scope of this book and the reader is referred to Allen et al. [2].

In some ways, VAVP components are more complicated than electrical components since there are typically multiple points (or degrees of freedom) at each interface and therefore multiple inputs and outputs to and from the same component. A matrix of FRFs is therefore required describing the input-output function of every point with every other point and with itself.

In summary then, the components in a VAVP are linear, time-invariant substructures, not restricted to individual machine components, that are characterised by a matrix of multiple-input multiple-output (MIMO) input-output filters (FRFs) connecting all points on the interfaces as well as any response-only target points.

Sources

The components considered above were implicitly *passive* and characterised by passive properties, i.e. FRFs. However, an important class of components such as motors, pumps, compressors, gears, switches, catches etc. are clearly *active* in that they contain moving parts that generate fluctuating forces and hence create vibration and sound in the VAVP. Active components of this type will be referred to as *sources*.

Sources are considered a subset of components because everything that has been said above regarding passive properties also applies to sources. However, sources incorporate active mechanisms of vibration generation so in addition to their passive properties as described above, they require an extra set of data to characterise their 'activity'.

Examples of active mechanisms include the fluctuating magnetic forces generated in the gap between rotor and stator of an electric motor, the fluctuating aerodynamic forces acting on the blades of fans, transmission error in gears and so on. These active mechanisms are usually (though not always) buried inside the component, away from its interface. However, in accordance with the black box concept (1.5.2) they are characterised at the interfaces rather than at the point of generation. Thus, when we characterise the 'activity' of a source like an electric motor, we do not directly characterise the magnetic forces themselves but rather their effect where the component connects to the outside world, i.e. at the interfaces. This requirement is also convenient practically because the point of generation itself is often hidden whereas interfaces are more accessible and in any case must be accessed for measurement of passive component properties.

The need for independent and transferable data has been emphasised in 1.5.1 above and of course this requirement applies equally to the active properties of components. The same situation arises with electrical networks and the neat solution there is to describe source by its source strength and internal impedance. The source strength, either open circuit voltage or short circuit current, is then defined independently of the rest of the circuit. By analogy, structure-borne sound source 'activity' is defined either by blocked force or free velocity which, as with electrical sources, is independent of the rest of the assembly (see Chapter 3). However, unlike many electrical sources, vibro-acoustic sources are inherently multi-output because multiple DoFs are typically required at every interface.

Although we are not yet discussing the detailed requirements for experiments it is worth noting that quite different test regimes are required to obtain active and passive properties: passive properties are measured in *system definition* tests with the source switched off; active properties on the other hand require *operational* tests with the source operating under whatever loading regimes are desired or specified. Operational tests of this type present various practical challenges which will be discussed at various points throughout the book.

As mentioned earlier, we have so far made the assumption that our components are linear and time-invariant (LTI). However when it comes to source mechanisms, non-linearity often plays an important role, in impact for example. So too does time-variance, such as in gear mesh forces where the time-varying contact stiffness throughout each cycle is an important cause of dynamic forces. It is important that these details are correctly represented in a VAVP since the time structure of the dynamic forces is of great importance in determining the overall character of sound. Fortunately, it turns out that, while non-linearity and time-variance are important factors in the *generation* of the dynamic forces, it is not necessary to depart from the LTI assumptions to represent them in a VAVP. This is because, as mentioned in 1.5.2, sources are characterised at the *interfaces* rather than at the point of generation. So, even when there is significant non-linearity and time-variance in the source region itself, provided the behaviour is approximately linear at the interfaces the LTI representation remains valid; the details of non-LTI behaviour in the source region are by then represented in the source activity at the interfaces, particularly in its time structure. See Chapter 3 for further discussion.

In summary, sources are characterised by a set of passive properties (FRFs) as for any other component, plus a set of active properties (either blocked forces or free velocities), all these properties being defined at the interfaces. Sources will be discussed further in Chapter 3.

Receivers and coupling elements

It is convenient to categorise passive components as 'receivers' or 'coupling elements'.

- Receivers typically include at least one interface and one or more response-only points, such as a target location.
- Coupling elements are components that connect to two or more other components, such as vibration isolators. Typically, they have an interface at either end and no internal response-only points.

Interfaces

Correct characterisation of a component's interface(s), as will become evident particularly in Chapter 4, is one of the most critical aspects of building a VAVP – *everything goes right or wrong at the interface!*

Interfaces have already been partially defined as those regions of the boundary where a component connects to other components, but there is more to interfaces than this. Determining where there is contact between two components is normally relatively straightforward, but defining the *interface* requires that all dynamic forces are properly accounted for. This a significantly more difficult task for two main reasons: the nature of the dynamic forces themselves and the geometry of the contact as discussed in the following.

First, dynamic forces do not only act perpendicular to the boundary but may occur simultaneously in different directions or degrees of freedom (DoFs). In general, at a single point up to six DoFs are required for a full description, three perpendicular forces and three moments (rotational forces) about perpendicular axes. Thus, even a single point interface may require up to six DoFs and a 6×6 matrix of FRFs. Failure to include the important DoFs can amount to a disastrous misrepresentation. However, successful characterisation is not simply a question of 'playing safe' and including all six DoFs every time since not all DoFs are important on every occasion; if the digital representation of a component is filled with unimportant details there is a price to be paid not just in increased measurement effort but also in terms of reduced accuracy (this point will be explained further in 7.4). Thus, both under- and over-specification are undesirable and there is no alternative but to correctly identify the important DoFs. This demands a thorough understanding of the dynamics of the specific interface in question and often requires preliminary testing.

Secondly, there is the question of how to represent a contact patch by discreet points. Recall that the FRFs used to characterise components are point-to-point quantities but in reality, a 'point' is an idealisation that does not exist. A contact patch may approximate a point when the structural wavelength is 'long' compared with the dimension of the contact, but even here, the exact location of the equivalent point may require some judgement. Moreover, when this assumption breaks down some other form of discretisation is required that captures the dynamics of the interaction over an extended interface.

Thus, the art of defining an interface consists of selecting the points and the DoFs that fully capture the dynamic interaction between connected components. These important points will be discussed further in Chapter 4.

1.5.4 THE VAVP SCHEME

Having described the elements of a VAVP we now present examples of the VAVP scheme which provides a framework for all the material in the book.

To summarise so far, a VAVP is a digital representation of a machine whose (digital) components can be interchanged and modified and which sounds and feels just like the real thing. To achieve this we have seen we will need to include digital representations of both active components, which generate the initial disturbance, and passive components which transmit and filter the sound (Figure 1.3 for example). All components must be represented independently so that they can be decoupled from the assembly and replaced. We also need a way to couple and decouple components and to auralise the resulting sound. The ensemble uncertainty of a production run may also need to be incorporated (not shown in Figure 1.3).

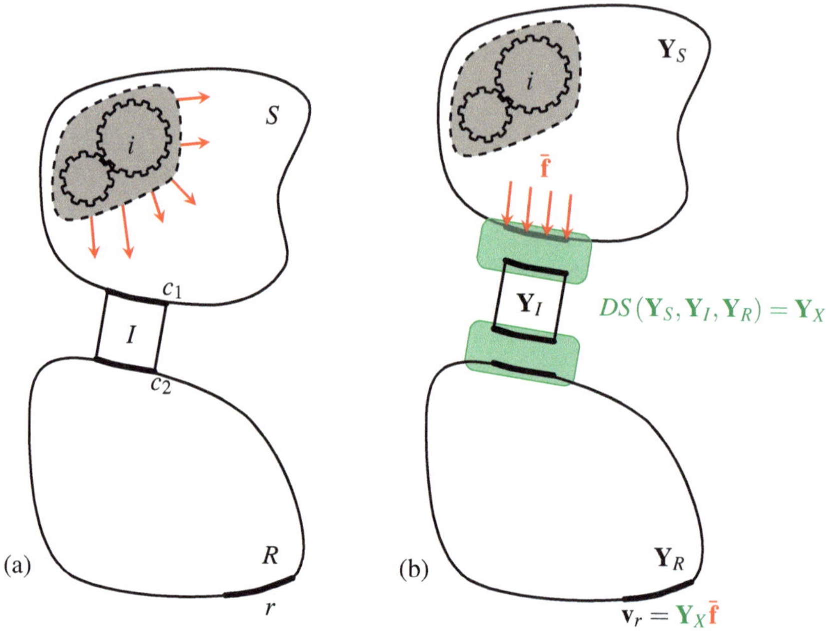

Figure 1.4 Schematic of an operational source resiliently connected to a passive receiver (a) and its VAVP representation (b) in terms of component-level FRFs ($\mathbf{Y}_S$, $\mathbf{Y}_I$, $\mathbf{Y}_R$) and an operational blocked force ($\bar{\mathbf{f}}$).

The VAVP scheme is a highly flexible and general framework that applies to a very broad range of assemblies including machines, vehicles, ships, buildings, any system effectively that is approximately linear and time-invariant. Moreover, provided the requirements for the component-based approach are met, both modelled and measured data can in theory be combined within the same framework. Given that modelling alone is not yet able to deliver the virtual prototypes that industry demands, the ability to *hybridise* data in this way within a single self-consistent framework is potentially highly significant.

Shown in Figure 1.4 is a simple arrangement with two substructures, an active source and a passive receiver, separated by the coupling element I, representing any intermediate structure, including, but not limited to, resilient isolators, bushings, boots, linkages, etc. The *source*, S, is excited by internal dynamic forces occurring within the region marked o which are assumed not to be accessible for direct measurement. The *coupling element*, I, and *receiver*, R, contain no such forces but are excited through the *interfaces*, c_1 and c_2, causing a response at r.

This simple assembly contains all the elements to enable more complex assemblies to be devised with multiple sources, receivers and coupling elements. Also possible are nested schemes where we model 'sources within a source' as will be discussed further in Chapter 3.

As done above, throughout this book we will adopt the convention that substructures are referred to by uppercase letters, S, R, I, for source, receiver and connecting elements respectively and occasionally by A and B. Response and excitation locations are referred to by lowercase letters: i, for the location of internal excitation by operational forces (assumed inaccessible), a, b, c etc. for input or output locations.

1.5.5 MAIN EQUATIONS FOR A VAVP

The mathematical descriptions used in building a VAVP give us another way of looking at the process. From this perspective, the main steps are:

- Characterise passive properties
- Characterise active properties
- Couple components
- Make response prediction and auralise.

Step 1. Characterise passive properties

As discussed above, passive properties of each component are represented by their FRF matrices,

$$\mathbf{Y}_S, \quad \mathbf{Y}_R, \quad \mathbf{Y}_I \tag{1.1}$$

where, $\mathbf{Y}$ indicates a matrix of mobility-type FRFs. In the frequency domain representation, which is used throughout the book, all properties are presented as functions of frequency; matrices then, apply to a single frequency and there is a similar matrix for every frequency in the spectrum. The upper case S, R or I in the subscript indicates the component to which the FRFs belong, source, receiver or coupling element, respectively. For more detailed definition of FRFs and their measurement see 2.2.

Step 2. Characterise active components

For sources we require a measure of their *activity* in addition to their passive properties. Here we quantify activity in terms of *blocked forces*, although free velocity can also be used (see Chapter 3). The main equation for obtaining blocked forces is,

$$\mathbf{v}_c = \mathbf{Y}_{Ccc}\bar{\mathbf{f}}_c \tag{1.2}$$

where $\bar{\mathbf{f}}_c$ is a vector of blocked forces with one entry for every DoF on the interface; the subscript c indicates that it applies on the interface labelled c. $\mathbf{Y}_{Ccc}$ is the matrix of mobility FRFs pertaining to the assembly of the source mounted on whatever receiver substructure is being used for the test – a test bench or a real installation for example. In other words $\mathbf{Y}_{Ccc}$ is the mobility of the *coupled* source and a receiver, the assembly of the two being denoted with subscript C. The first lower case c indicates the set of response (output) points and the second the set of excitation (input) points which in this formulation are co-located on interface c. $\mathbf{v}_c$ represents the vector of velocity, again with one entry for every point on interface c.

Two separate tests are required here, a passive system identification test to obtain the FRF matrix $\mathbf{Y}_{Ccc}$ and an *operational* test with the machine running to obtain $\mathbf{v}_c$. Having obtained the data for $\mathbf{Y}_{Ccc}$ and $\mathbf{v}_c$, one then needs to solve Equation 1.2 to obtain the unknown blocked forces. This step is known as solving the inverse problem. It is not intuitively obvious why these forces should be the blocked forces but this question is addressed in more detail in Chapter 3. There are other forms of Equation 1.2 but this will suffice for illustration. Obtaining blocked forces, including solving the inverse problem, is one of the most critical stages in building a VAVP and will be described in some detail in Chapters 3 and 6.

Step 3. Couple components

Having obtained FRF data describing each component in Step 1 they then need to be coupled together to build a virtual assembly. The rules for coupling are well known and directly analogous to those in electrical systems. Essentially, one simply needs to satisfy continuity and equilibrium conditions across the connecting interfaces, a process which has become known as frequency-domain substructuring [37]. There are three main formulations, the direct, primal and dual methods. Without making a preference at this stage we will simply indicate the coupling step as,

$$\mathbf{Y}_X = DS(\mathbf{Y}_S, \mathbf{Y}_I, \mathbf{Y}_R), \tag{1.3}$$

in which $DS(\square)$ indicates any formulation of dynamic substructuring, that is combining the properties of the source S, coupling element I, and receiver R on the right-hand side to form the FRFs of the assembly X.

Note that whilst this substructuring step is typically used to predict the FRFs of a *virtual* assembly (i.e. one that does not exist), it can also be applied to an existing assembly for example to investigate structural modification/redesign.

Step 4. Forward prediction and auralisation

Having built the assembly in Step 3, the machine must then be 'switched on' so that the resulting sound and vibration can be simulated. The switching on is done by applying the blocked forces obtained in Step 2 to the assembly built in Step 3. The acoustic or vibration response at target locations is then given by,

$$\begin{pmatrix} \mathbf{v}_{r_1} \\ \mathbf{p}_{r_2} \end{pmatrix} = \begin{bmatrix} \mathbf{Y}_{Cr_1c} \\ \mathbf{H}_{Cr_2c} \end{bmatrix} \bar{\mathbf{f}}_c \tag{1.4}$$

where $\mathbf{v}_{r_1}$ is the vector of predicted velocity at receiver locations r_1, one entry per DoF. $\mathbf{p}_{r_2}$ is the vector of predicted sound pressures, again with one entry for each location in r_2. $\mathbf{Y}_{Cr_1c}$ is the mobility matrix for excitation at locations on interface c and velocity response at r_1 and $\mathbf{H}_{Cr_2c}$ is the FRF connecting acoustic responses at r_2 to force excitation at interface c. $\bar{\mathbf{f}}_c$ is the blocked force vector at interface c as obtained in Step 2.

The subscript C indicates a property of the assembly as a whole. Note that this may or may not be the same assembly as in Step 2 from which blocked forces were obtained; for example the blocked forces could be obtained with the source mounted on a test bench and the responses in Step 4 with the same source mounted in a virtual assembly whose FRFs are obtained by substructuring as in Step (i.e. $\mathbf{Y}_X$).

Equation 1.2, and hence also Equation 1.4, can be solved for multiple time snapshots. This allows the target response prediction to be transformed back into the time domain and auralised to provide a subjective impression of the product in operation – a component-based simulation.

1.5.6 PRECISION VS. ENGINEERING APPROACHES

There are, broadly speaking, two different approaches to the characterisation of components (including sources) which we will term 'precision' and 'engineering' approaches. This terminology is in many respects a misnomer since it can hardly be claimed, despite significant improvements in accuracy in recent years, that the digital representation of vibro-acoustic phenomena achieves 'precision' comparable to other branches of metrology! However, the terminology has been adopted in some international standards and it seems appropriate to retain it here.

In 'precision' methods, one attempts to retain all details of the vibro-acoustic behaviour, in particular by representing all important DoFs at interfaces. In 'engineering' methods on the other hand, one typically simplifies the interface to a single equivalent point. This has the advantage of simplifying measurements; the disadvantage is that information is lost in reducing the real multi-DoF behaviour to a single value and hence the characterisation is inherently approximate. These two approaches have much in common with respectively 'deterministic' and 'statistical' methods of analysis [29].

For the purposes of building VAVPs, precision approaches are clearly advantageous. Indeed in some cases the data from engineering methods is insufficiently detailed and cannot be used. Nevertheless, engineering methods do have a role to play and are discussed at appropriate points in the text.

1.6 BACKGROUND TO LITERATURE

What are the main areas of literature relevant to the component-based paradigm? The review below is intended to provide an overview of the context rather than a detailed analysis and hence only books and a few main papers are included at this stage. More detailed references to relevant research papers are included throughout the text at the appropriate points.

We start with the main areas relevant to the scope, namely Experimental Modal Analysis (EMA), Noise, Vibration and Harshness (NVH) and Substructuring before going on to mention other areas of relevance: Auralisation, Model Updating, Digital Signal Processing (DSP), Condition Monitoring, Structural Health Monitoring and Uncertainty in Structural Dynamics. Before starting the review however, it is necessary to acknowledge the seminal importance of Cremer and Heckl's classic text on structure-borne sound [5] first published in the 1970s (updated by Petersson [6]) which still underpins the entire field of vibro-acoustics. (See also Nilssen and Liu [25], [24]).

The component-based approach is built on the legacy of the experimental modal analysis (EMA) approaches pioneered by Ewins [10] and more recently by He and Fu [13] and Silva and Maia [8], among many others, from the 1970s to the present day. In EMA, the vibration response of a structure is considered as the sum of contributions from its natural modes. EMA therefore aims to define the structure through its *modal* properties: natural frequencies and damping ratios. The basic measurements and assumptions in the component-based approach do not differ to any great extent from EMA but the demands of industry have evolved over time with a corresponding evolution in experimental methods.

First, there is a difference in the purpose of the measurements. The development of EMA was largely complementary to that of numerical methods, particularly Finite Element Analysis for which EMA provided much needed experimental validation. As such, EMA was largely used to describe single structures or assemblies with only limited attention given to the problem of coupling substructures together [10] [8]. However, today's industry is gearing up for full 'digitalisation' of the design process with as little reliance on physical prototypes as possible. In this context, there is more emphasis on combining and recombining substructures into new assemblies. Experiments are now required not only to verify and validate numerical models but also to provide data for components not amenable to modelling, particularly active components (sources). The component-based approach has evolved to meet this demand by allowing components to be characterised independently so that they can be interchanged. Furthermore, since some components cannot be removed from their assemblies for testing, it has proved useful to develop *in situ* measurement methods in which the components' properties are mathematically decoupled from those of their surroundings.

Secondly, there is the question of frequency range. The early Finite Element Analysis, which EMA supported, was initially aimed at vibration problems with a maximum frequency of a few hundred Hz, encompassing the first few tens of modes of the structure or so. Vibro-acoustic problems on the other hand, are defined over a significantly wider frequency range, these days up to 10 kHz or so. As the frequency increases, the role of individual structural modes becomes both less important and more difficult to extract from measurements. The component-based approach therefore tends to favour a description based on frequency response functions (FRFs) rather than on modes.

A more recent body of literature relevant to the current topic is the field of *NVH* (Noise, Vibration and Harshness) [33], [15], [14]. NVH grew in response to demands, both from customers and legislation, for improved refinement of vehicles. NVH deals with all aspects of vehicle refinement from generation, transmission and perception of noise and vibration through to mitigation of its effects. It also considers diagnosis, rating and target setting. Component-based simulation does not attempt to tackle all these issues but provides practical tools for the NVH engineer in several ways. First, blocked force Transfer Path Analysis (TPA) [9] provides a new diagnostic method for determining the dominant transmission routes from a vibration source to some target location (driver's ear, external microphone etc.). It is usually easier to perform than classical TPA and has the added advantage that the data is transferable as will be discussed in Chapters 3 and 5.

Secondly, blocked forces, which will be seen to be a central concept in component-based simulation, allow vibration sources to be characterised independently of their surroundings. This opens up the possibility of specifying 'source strength' for active components even when there is no possibility of testing in the vehicle, for example at the design stage. This has knock-on implications for target setting and specification as well as facilitating communication between suppliers and purchasers of active components that would otherwise be impossible. Thirdly, the component-based approach provides the data for simulation and prediction of what-if scenarios including for full vibro-acoustic virtual prototypes. It allows simulation in both time and frequency domains enabling comparison with targets, including subjective targets if the results are auralised [32].

The component-based approach can therefore be viewed as providing important tools for the NVH engineer and hence complements the field of NVH. However, its application is not restricted to vehicles or to vehicle refinement but extends to any mechanical systems (ships, buildings, domestic products etc) for which noise and vibration is important, and then not merely for the purposes of refinement.

Regarding the field of *substructuring*, it will be seen in Chapter 5 that methods for coupling and decoupling components form an essential element of the component-based approach. Substructuring theory was initially developed for connecting FEM models of subassemblies (usually credited to Craig and Bampton [4] but now recognised to have owed much to Hurty [16]), but was later adopted into the experimental field. While simple in theory being based simply on the assumptions of compatibility and equilibrium, obtaining reliable results for experimental substructuring has proved remarkably challenging and at the time of writing the problem can hardly be considered fully solved. Much of the most important work in the field has been reported through the IMAC (*International Modal Analysis Conference*) organised by the *Society of Experimental Mechanics* over several decades and much of this has now been condensed into the book by Allen et al. [2]. Since the aim of the current book is to provide the reader with a full set of tools for component-based simulation, a working summary of the essential material on substructuring is included in Chapter 5. The treatment focuses on the practical application of the theory to experimental problems. Readers requiring more detail on theoretical aspects of substructuring, particularly for use in numerical methods, are referred to [2].

Other areas of relevance include: Auralisation, Model Updating, Digital Signal Processing (DSP), Condition Monitoring, Structural Health Monitoring and Uncertainty in Structural Dynamics.

Auralisation, is *'the technique of creating audible sound files from numerical (simulated, measured or synthesised) data'* [32]. This is relevant to the current work because creating audible sounds for subjective evaluation is one of the main reasons for producing a VAVP. The subject has been well covered elsewhere [32] and so we do not go into any great detail in the book. The field of Psychoacoustics [20, 21, 34] is also relevant here in understanding how auralised product sounds are perceived but we will not discuss this aspect further.

Model updating [12] provides a method by which the parameters of a numerical model may be optimised (updated) so as to yield a closer fit with physical measurements. The use of numerical models of components is of course entirely compatible with the component-based approach and model updating therefore provides an important area of supporting literature to the current work. A specific application, '*in situ* model updating', will be discussed in Chapter 2 in relation to the problem of how to separately characterise components that cannot be physically separated. Furthermore, some of the experimental methods described in relation to VAVPs, particularly *in situ* methods, may potentially be of relevance in model updating.

DSP is a vast area of literature from which we adopt most of the data processing techniques used throughout the book, particularly in Chapter 7 (see for example [3, 26, 27]). Mostly, these are well-known and mature techniques for which there is no need for derivation from first principles and therefore we focus on their specific application to component-based analysis.

Condition Monitoring and Structural Health Monitoring [11, 28] are, again, huge fields of literature of relevance but not core to the current work. The reason for inclusion here is that both rely on measured data and thus there is potential synergy between the fields. For example, the *in situ* blocked force method may have potential advantages in Condition Monitoring [22] due to the independence of the data from the rest of the assembly.

Finally, we mention another field with an extensive literature, namely Uncertainty in Structural Dynamics. Our specific interest lies in a narrow niche of the field and is outlined in some detail in Chapter 8. Readers with a broader interest may wish to refer to the published reviews [17, 19, 30].

1.7 LAYOUT OF THE BOOK

Chapter 1 states the aims and scope of the book, defines essential terms and lays out the basic scheme and terminology for a vibro-acoustic virtual prototype around which the rest of the book is structured. Having laid out the context and defined our general theme, the remainder of the book is in three parts: Part 1 presents theory and concepts, Part 2, considers practical implementation and Part 3, case studies.

In more detail, Part 1 comprises Chapters 2 to 5. All components in a working machine respond to forces in a characteristic manner: representation of these passive properties is the subject of Chapter 2, 'Components'. Active components also include internal source mechanisms, the characterisation of which is considered in Chapter 3, 'Sources', particularly through the *in situ* blocked force method. In Chapter 4. 'Interfaces', we consider how to represent the important dynamic interactions at the interfaces where components are connected. Having defined passive and active properties and the interfaces that connect them, Chapter 5, 'Assemblies', presents concepts related to connecting components, including substructuring and TPA methods.

Part 2 comprises chapters focused on Measurement, Processing and Uncertainties as well as a worked example of the whole process. Chapter 6, 'Measurement', focuses on implementation with a particular emphasis on checking data and validation of the VAVP and its components. Chapter 7, 'Processing', recounts some familiar material for processing and organising raw data and goes on to address the important process of solving the inverse problem. Chapter 8, 'Uncertainties', categorises the sources of error in a VAVP and provides tools to quantify the completeness of the interface description and to propagate uncertainty from measurements onto the predicted output of the VAVP. Finally in Part 2, Chapter 9 provides a 'walk-through' example illustrating, in detail, the practical application of the techniques described throughout the book including providing relevant code.

The final section of the book is Part 3, Chapter 10, which presents a series of case studies provided by industrial partners illustrating the principles applied to real industrial problems in a range of industries. The book finishes with Chapter 11 summarising the main findings and speculating on what the future may hold.

Bibliography

1. Directive 2000/14/EC of the European Parliament and of the Council of 8 May 2000 on the approximation of the laws of the member states relating to the noise emission in the environment by equipment for use outdoors. *OJ*, L 162:1–78, 2000-07-03. `http://http://data.europa.eu/eli/dir/2000/14/oj`.
2. MS Allen, D Rixen, M van der Seijs, P Tiso, T Abrahamsson, and RL Mayes. *Substructuring in Engineering Dynamics*. Springer, 2020.
3. JS Bendat and AG Piersol. *Random data: analysis and measurement procedures*, volume 729. John Wiley & Sons, 2011.
4. RR Craig Jr and MCC Bampton. Coupling of substructures for dynamic analyses. *AIAA journal*, 6(7):1313–1319, 1968.
5. L Cremer and M Heckl. *Structure-borne sound: structural vibrations and sound radiation at audio frequencies*. Springer Science & Business Media, 2013.
6. L Cremer, M Heckl, and BAT Petersson. *Structure-borne sound: structural vibrations and sound radiation at audio frequencies*. Springer Science & Business Media, 2005.
7. W Desmet. Mid-frequency vibro-acoustic modelling: challenges and potential solutions. In *Proceedings of ISMA*, volume 2, page 12. Citeseer, 2002.
8. JM Montalvão e Silva and NMM Maia. *Modal analysis and testing*, volume 363. Springer Science & Business Media, 2012.
9. AS Elliott, AT Moorhouse, T Huntley, and S Tate. In-situ source path contribution analysis of structure borne road noise. *Journal of Sound and Vibration*, 332(24):6276–6295, 2013.
10. DJ Ewins. *Modal testing: theory and practice*, volume 15. Research studies press Letchworth, 1984.
11. CR Farrar and K Worden. *Structural health monitoring: a machine learning perspective*. John Wiley & Sons, 2012.
12. M Friswell and JE Mottershead. *Finite element model updating in structural dynamics*, volume 38. Springer Science & Business Media, 2013.
13. ZF Fu and J He. *Modal analysis*. Elsevier, 2001.
14. A Fuchs, E Nijman, and HH Priebsch. *Automotive NVH technology*. Springer, 2016.
15. M Harrison. *Vehicle refinement: controlling noise and vibration in road vehicles*. Elsevier, 2004.
16. WC Hurty. Dynamic analysis of structural systems using component modes. *AIAA Journal*, 3(4):678–685, 1965.
17. RA Ibrahim. Structural dynamics with parameter uncertainties. *Applied Mechanics Reviews*, 40(3): 309-328. 1987.
18. M Lindstrom. *Buyology: How everything we believe about why we buy is wrong*. Random House, 2012.

19. BR Mace, K Worden, and G Manson. Uncertainty in structural dynamics. *Journal of Sound and Vibration*, 288(3):423–429, 2005.
20. BCJ Moore. *An introduction to the psychology of hearing*. Brill, 2012.
21. BCJ Moore. Psychoacoustics. *Springer handbook of acoustics*, pages 475–517, 2014.
22. AT Moorhouse. Vibro-acoustic measurement methods in condition monitoring. *CM/MFPT. London*, 2012.
23. AT Moorhouse. Virtual acoustic prototypes: Listening to machines that don't exist. *Acoustics Australia*, 33(3):97, 2005.
24. AC Nilsson and B Liu. *Vibro-acoustics, Volume 2*. Springer, 2016.
25. AC Nilsson and B Liu. *Vibro-acoustics*, Volume 1, Springer 2016.
26. AV Oppenheim and RW Schafer. *Digital Signal Processing*, Prentice-Hall Englewood Cliffss N.J., 1975.
27. RB Randall. *Frequency analysis*. Naerum (DK): Bruel & Kjaer, 1987.
28. RB Randall. *Vibration-based condition monitoring: industrial, aerospace and automotive applications*. John Wiley & Sons, 2011.
29. PJ Shorter and RS Langley. Vibro-acoustic analysis of complex systems. *Journal of Sound and Vibration*, 288(3):669–699, 2005.
30. C Soize. Stochastic modeling of uncertainties in computational structural dynamics—recent theoretical advances. *Journal of Sound and Vibration*, 332(10):2379–2395, 2013.
31. G Tanner. Dynamical energy analysis—determining wave energy distributions in vibro-acoustical structures in the high-frequency regime. *Journal of Sound and Vibration*, 320(4-5):1023–1038, 2009.
32. M Vorländer. *Auralization: fundamentals of acoustics, modelling, simulation, algorithms and acoustic virtual reality*. Springer Science & Business Media, 2007.
33. X Wang. *Vehicle noise and vibration refinement*. Elsevier, 2010.
34. E Zwicker and H Fastl. *Psychoacoustics: Facts and models*, 2nd edition. Springer Science & Business Media, 2013.
35. RH Lyon. *Theory and application of statistical energy analysis*. Elsevier, 2014.
36. R Pile and J Le Besnerais. Comparison of main magnetic force computation methods for noise and vibration assessment in electrical machines. *IEEE Transactions on Magnetics*, 54(7):1–13. IEEE, 2018.
37. D de Kerk, DJ Rixen and SN Voormeeren. General framework for dynamic substructuring: history, review and classification of techniques. *AIAA journal*, 46(5) 1169–1181, 2008.

Part 1: Theory and Concepts

2 Components

2.1 INTRODUCTION

Part 1 of the book is where we aim to lay out the theory and concepts needed for component-based simulation with measured data. The first two chapters consider characterisation of *passive* and then *active* components. We then move on to consider the *interfaces* between components and then the *assembly* of our digital components into a VAVP.

The aim of this first chapter is to provide the reader with the means to characterise passive components in manner suitable for inclusion in a VAVP. We describe *what* properties are needed, *where* to measure and *how* to mount the component for testing. Practical description of the measurements themselves is discussed further in Part 2 of the book. Passive components have no capacity to generate vibration, but in a real working assembly every component is excited in some way and will respond, as well as potentially transmitting excitation to other connected components. Characterising such components therefore amounts to finding digital representations of the component's ability to *respond* to, and to *transmit*, dynamic excitation. After a brief review (2.2) we conclude that the most suitable system definition approach is to use *Frequency Response Functions* (FRFs) in the frequency domain. To a large extent then, this chapter is about FRFs.

Regarding where to measure, recall from Chapter 1 that with a black box representation, a component is represented only at interfaces and target (remote) locations. Thus, the digital representation of passive components consists of a set of independent FRFs connecting all points on its interfaces and remote points of interest. The first step therefore, in characterising a component is to identify the interfaces and remote locations: other locations are not represented in a black box model. One must also recognise that forces may be transmitted across the interfaces from one component to another in different directions, or degrees of freedom (DoFs). The interfaces defining passive components are therefore inherently *multi-DoF*.

A requirement with important implications for testing is that the components in a VAVP must be *interchangeable*. Theoretically, the required independent properties can be measured with either *free* or *fixed-type* mounting conditions. Free conditions can sometimes be adequately approximated with soft supports (2.3); coupling elements and resilient isolators (2.3.2) on the other hand often require fixed-interface conditions which may or may not be achievable practically. Inevitably, in the real world we encounter non-ideal boundary conditions which leads to a major challenge of the component-based approach, namely how to extract independent component properties from in situ measurements, i.e. those made on a coupled assembly. The focus of the chapter therefore starts to move from conventional measurement towards a range of indirect approaches. Such techniques include implementing ideal boundary conditions mathematically rather than physically (2.2.8), decoupling techniques

DOI: 10.1201/9781003399131-2

to remove the dynamic influence of the rest of the assembly (2.3.1) and use of numerical models (2.6) alongside measurements.

Indirect methods often make use of *transmissibilities*; although not required for the final description of a passive component, they form an intermediate step in several indirect methods and are hence discussed in 2.4. One of these is the Round Trip method which enables FRFs to be obtained indirectly even at locations where external forces cannot be applied and is discussed in 2.5.

Finally, although this book is focused on experimental methods, within the context of virtual prototyping it will increasingly be necessary or desirable to characterise components with the output of numerical models. In 2.6 therefore we present, first, techniques for synthesising suitable FRFs from the output of a pre-existing FE model and secondly, an in situ model updating procedure whereby model parameters can be optimised based on in situ measurements.

2.2 FREQUENCY RESPONSE FUNCTIONS

This section is primarily concerned with definitions relating to FRFs. We describe the various types of FRF, discuss the boundary conditions required to achieve independence for each type and provide formulations for multi-point, multi-DoF and vibro-acoustic interfaces. We also describe the principle of reciprocity. Before we deal with FRFs themselves however, we provide some context by briefly laying out the various alternative approaches to 'system definition', i.e. the representation of passive structures.

2.2.1 REPRESENTATIONS OF PASSIVE COMPONENTS

The invariant properties of a passive component can be described in various 'domains', each providing a different perspective on essentially the same dynamic behaviour. The physical, state-space, modal, time and frequency domains are the most commonly encountered in vibro-acoustics and will be briefly reviewed below before we adopt the FRF domain as the default for the component-based approach.

In the *physical domain*, a component's inertial and elastic properties are discretised by N DoFs, for example by the Finite Element method, and represented by the mass and stiffness matrices, $\mathbf{M} \in \mathbb{R}^{N\times N}$ and $\mathbf{K} \in \mathbb{R}^{N\times N}$, respectively. Together with an appropriate matrix representation of its damping, $\mathbf{C} \in \mathbb{R}^{N\times N}$, these matrices characterise the passive properties of a component. They can be used to formulate a second order equation of motion that relates the external force $\mathbf{f}(t)$ applied to the component, to its resulting kinematics,

$$\mathbf{M}\ddot{\mathbf{x}}(t) + \mathbf{C}\dot{\mathbf{x}}(t) + \mathbf{K}\mathbf{x}(t) = \mathbf{f}(t) \tag{2.1}$$

where $\mathbf{x}(t)$, $\dot{\mathbf{x}}(t)$ and $\ddot{\mathbf{x}}(t)$ are vectors of displacement, velocity and acceleration, respectively at each of the N DoFs.

In the *state-space domain*, a component is represented by a set of first order differential equations that relate the state vector $\mathbf{y}(t) = [\mathbf{x}^{\mathrm{T}}(t) \quad \dot{\mathbf{x}}^{\mathrm{T}}(t)]^{\mathrm{T}}$ to its temporal derivative, $\dot{\mathbf{y}}(t) = [\dot{\mathbf{x}}^{\mathrm{T}}(t) \quad \ddot{\mathbf{x}}^{\mathrm{T}}(t)]^{\mathrm{T}}$. With reference to Equation 2.1, we have,

$$\dot{\mathbf{y}}(t) = \begin{bmatrix} \dot{\mathbf{x}}(t) \\ \ddot{\mathbf{x}}(t) \end{bmatrix} = \begin{bmatrix} \mathbf{0} & \mathbf{I} \\ -\mathbf{M}^{-1}\mathbf{K} & -\mathbf{M}^{-1}\mathbf{C} \end{bmatrix} \begin{bmatrix} \mathbf{x}(t) \\ \dot{\mathbf{x}}(t) \end{bmatrix} + \begin{bmatrix} \mathbf{0} \\ \mathbf{M}^{-1} \end{bmatrix} \mathbf{f}(t) \tag{2.2}$$

or equivalently,

$$\dot{\mathbf{y}}(t) = \mathbf{A}\mathbf{y}(t) + \mathbf{B}\mathbf{f}(t) \tag{2.3}$$

where $\mathbf{A}$ is the so-called state matrix, and $\mathbf{B}$ is the input matrix. Together, $\mathbf{A}$ and $\mathbf{B}$ characterise the passive properties of a component in state-space.

In the *modal domain* a component is represented by a set of mode shapes $\Phi = [\phi_1 \quad \phi_2 \quad \cdots \quad \phi_N]$ and natural frequencies $\Lambda = \mathrm{diag}[\omega_1^2 \quad \omega_2^2 \quad \cdots \quad \omega_N^2]$ which individually satisfy the homogenous undamped equation of motion,

$$\mathbf{K}\mathbf{x} - \omega^2\mathbf{M}\mathbf{x} = \mathbf{0} \quad \rightarrow \quad \left[\mathbf{K} - \omega_i^2\mathbf{M}\right]\phi_i = 0. \tag{2.4}$$

Together Φ and Λ, with an appropriate representation of damping (e.g. modal loss factors), characterise the passive properties of the component.

In the *time domain* a component's kinematic response $\mathbf{x}(t)$, $\dot{\mathbf{x}}(t)$ or $\ddot{\mathbf{x}}(t)$ to a time varying force $\mathbf{f}(t)$ satisfies the convolution integral (shown only for displacement),

$$\mathbf{x}(t) = \int_0^{\mathrm{T}} \mathbf{G}(t-\tau)\mathbf{f}(\tau)d\tau \tag{2.5}$$

where $\mathbf{G}(t)$ is a matrix of displacement impulse response functions (IRFs) that describe the component response to a unit (dirac) force impulse. The IRF matrix $\mathbf{G}(t)$ characterises the passive properties of a component.

In the *frequency domain* a component is characterised by its frequency response function (FRF) matrix, which can take several forms. Assuming time harmonic motion, Equation 2.1 can be Fourier transformed into the frequency domain,

$$\left(i\omega\mathbf{M} + \mathbf{C} + \frac{1}{i\omega}\mathbf{K}\right)\dot{\mathbf{x}}(\omega) = \mathbf{Z}(\omega)\mathbf{v}(\omega) = \mathbf{f}(\omega) \tag{2.6}$$

where $\mathbf{Z}(\omega)$ is the component impedance FRF matrix. Equivalently, we can write,

$$\mathbf{v}(\omega) = \mathbf{Y}(\omega)\mathbf{f}(\omega) \tag{2.7}$$

where $\mathbf{Y}(\omega) = \mathbf{Z}^{-1}(\omega)$ is the component mobility FRF matrix. Both $\mathbf{Y}(\omega)$ and $\mathbf{Z}(\omega)$ characterise the passive properties of a component. Similar FRF definitions exist for displacement and acceleration, as discussed in 2.2.2.

Remaining within the frequency domain, a component can also be viewed in terms of transmission *paths* connecting the various points of interest [30].

The choice of domain depends, amongst other things, on whether a physical structure is available for test. If no experimental characterisation is possible, the conventional starting point is a finite element model, i.e. in the physical domain. In this case, reliable estimates of the mass, stiffness and damping properties of the structure are required (not always easily found) for input to Equation 2.1. From here, state-space, modal, IRF, or FRF representations can be determined. If a physical component *is* available for testing, then representation by measured FRFs is often preferred. These do not require estimates of the component's physical properties because the FRF is directly obtained from measured forces and velocities (Equation 2.7). From measured FRFs, IRF and modal representations can be determined with relative ease. However, physical and state-space domains are typically limited to numerical models because identifying system matrices (e.g. mass, stiffness, state, etc) from experimental data is a difficult problem requiring advanced system identification methods. Being able to represent both experimentally and numerically sourced components, the FRF domain is the obvious choice for VAVPs. Added to the almost universal adoption of the frequency domain in the literature relating to VAVPs, we will adopt FRFs as our default representation from now on.

2.2.2 TYPES OF FRF

Having chosen the FRF as our preferred representation, it still remains to choose the dynamic quantities involved: FRFs simply represent an excitation-response (input-output) relation and it is to be decided whether the inputs and outputs are forces, kinematic or acoustic quantities. If a kinematic quantity, is this in the form of acceleration, velocity or displacement? In addition, are rotational quantities, moments or angular velocity for example, involved?

Our first choice will be to consider excitation in terms of applied force and response in terms of resultant velocity giving a force-velocity type of FRF known as 'mobility'. Much literature now uses acceleration rather than velocity, and 'accelerance' rather than mobility since acceleration is most often the directly measured quantity. However, we have opted here for velocity (and mobility) for consistency with measurement standards [11] and Ewins' classic book [3]. With that said, the same underlying equations apply to displacement, velocity and acceleration-based FRFs and provided the response type is *self-consistent* through all stages of simulation the choice of response variable is largely a matter of personal preference. Only when dealing with power, which is defined specifically in terms of velocity, is the choice important. Hence, we often refer to 'response' and 'FRF' rather than to specific types. Conversion between the various forms in the frequency domain, if required, is simply a matter of multiplication or division by $i\omega$ as described shortly. Note that FRFs with the same input and output types, such as force-force or velocity-velocity, are known as transmissibilities and are not included in this section but are discussed separately in 2.4.

The force-velocity FRF is known as the mobility and is typically denoted by $Y(\omega) \in \mathbb{C}$. For a single input-output system, it is a complex frequency dependent scalar quantity that describes the velocity response to a unit force excitation at a point, defined mathematically as,

$$v(\omega) = Y(\omega)f(\omega), \qquad Y(\omega) = \frac{v(\omega)}{f(\omega)} \tag{2.8}$$

where $v(\omega)$ and $f(\omega)$ are the complex frequency domain spectra of the velocity and force, respectively.

Being complex, the mobility has a real and imaginary part. Physically, the real part is associated with damping mechanisms and losses, whilst the imaginary part is related to energy storage (see 3.3). When visualising or plotting a mobility FRF is it rare that we look directly at its real or imaginary parts. Rather, we consider its polar representation in terms of magnitude $|Y|$ and phase $\angle Y$,

$$\begin{aligned} v(\omega,t) =&\text{Re}(v(\omega,t)) + i\text{Im}(v(\omega,t)) \\ =&|v(\omega)|e^{i(\omega t+\angle Y)} = |Y(\omega)|e^{i\angle Y} f_o(\omega)e^{i\omega t} \end{aligned} \tag{2.9}$$

where the magnitude of the mobility (or any FRF) describes the change in amplitude between a harmonic input (in the case of mobility, force) and output signal (velocity), whilst the phase relates to the time difference between the input and output introduced by the system at a particular frequency. In most applications it is the magnitude of the FRF that allows easiest comparison between two or more results and for this reason, the phase response is rarely presented throughout the book – this does not mean phase is not important however, it is!

FRFs in terms of displacement and acceleration, rather than velocity, are frequently found in the literature. It is straightforward to convert between the various kinematic quantities. By integration and differentiation with respect to time (assuming harmonic motion), Equation 2.8 can be transformed into a pair of force-displacement and force-acceleration relations, respectively,

$$v = \frac{dx}{dt} = i\omega x = Yf, \qquad v = \int a dt = \frac{a}{i\omega} = Yf \tag{2.10}$$

where for simplicity we have omitted explicit frequency dependence (ω). From the above we have the FRF relations,

$$x = \frac{Y}{i\omega} f = Rf, \qquad a = i\omega Y f = Af \tag{2.11}$$

where $R \in \mathbb{C}$ is called the (dynamic) compliance and $A \in \mathbb{C}$ the accelerance. Thus in the frequency domain, multiplication or division by $i\omega$ is all that is required to convert one from one form to another.

Accompanying each of the FRFs above is an *inverse* FRF in which the excitation is an imposed motion and the response a reaction force,

$$f = Dx, \qquad f = Zv, \qquad f = \hat{M}a. \tag{2.12}$$

where,

$$D = \frac{1}{R}, \qquad Z = \frac{1}{Y}, \qquad \hat{M} = \frac{1}{A}. \tag{2.13}$$

The FRFs $D \in \mathbb{C}$, $Z \in \mathbb{C}$ and $\hat{M} \in \mathbb{C}$ are called, respectively, the dynamic stiffness, impedance, and apparent mass. Table 2.1 presents a summary of the standard force-response FRFs encountered in the literature, alongside their inverse relations and conversions. For reasons described later in 2.2.8, we term FRFs whose input is an applied force as *free*-type FRFs, and those for which force is the response as *fixed*-type FRFs.

Table 2.1

A list of commonly used frequency response functions along with their relations for single-input single-output systems. E.g. $D = i\omega Z = -\omega^2\hat{M}$ **or** $i\omega R = Y = A/i\omega$

	Name	Def.	Sym.			
Free	Dynamic compliance	x/f	$R =$	R	$Y/i\omega$	$-A/\omega^2$
	Mobility	v/f	$Y =$	$i\omega R$	Y	$A/i\omega$
	Accelerance	a/f	$A =$	$-\omega^2 R$	$i\omega Y$	A
	Name	**Def.**	**Sym.**			
Fixed	Dynamic stiffness	f/x	$D =$	D	$i\omega Z$	$-\omega^2\hat{M}$
	Mechanical impedance	f/v	$Z =$	$D/i\omega$	Z	$i\omega\hat{M}$
	Effective mass	f/a	$\hat{M} =$	$-D/\omega^2$	$Z/i\omega$	$\hat{M}$

Importantly, the above definitions concern single-input single-output systems in which all FRFs are (complex) *scalar* quantities. When dealing with multi-input multi-output systems, as is necessary for virtual prototyping, a more general FRF definition is required that allows for simultaneous excitation and response at multiple points.

2.2.3 MULTI-INPUT MULTI-OUTPUT FRFS

Very rarely are structural components excited at, or connected through, a single point. In almost all scenarios, the FRFs needed for component-level simulation are *multi-point*. In the following we will specifically consider multi-point mobility and impedance FRFs, though the definitions apply equally to other types of FRF. As is common practice, we will also omit specific reference to frequency dependence from now on; all formulations throughout the remainder of this book are assumed to be in the frequency domain unless stated otherwise.

For a component with N inputs and outputs, its *mobility* is defined by the vector-matrix equation,

$$\mathbf{v} = \mathbf{Y}\mathbf{f} \tag{2.14}$$

where $\mathbf{v} \in \mathbb{C}^N$ is an N dimensional vector of resultant velocities, $\mathbf{f} \in \mathbb{C}^N$ is an N dimensional vector of applied forces, and $\mathbf{Y} \in \mathbb{C}^{N\times N}$ is the $N \times N$ dimensional mobility matrix that relates the two [41]. Written out in full Equation 2.14 reads,

$$\begin{pmatrix} v_1 \\ v_2 \\ \vdots \\ v_N \end{pmatrix} = \begin{bmatrix} Y_{11} & Y_{12} & \cdots & Y_{1N} \\ Y_{21} & Y_{22} & \cdots & Y_{2N} \\ \vdots & \vdots & \cdots & \ddots \\ Y_{N1} & Y_{N2} & \cdots & Y_{NN} \end{bmatrix} \begin{pmatrix} f_1 \\ f_2 \\ \vdots \\ f_N \end{pmatrix} \tag{2.15}$$

where given the matrix entry Y_{ij}, the first subscripted index i denotes the response position (the row of $\mathbf{Y}$), and the second index j the force excitation position (the column of $\mathbf{Y}$).

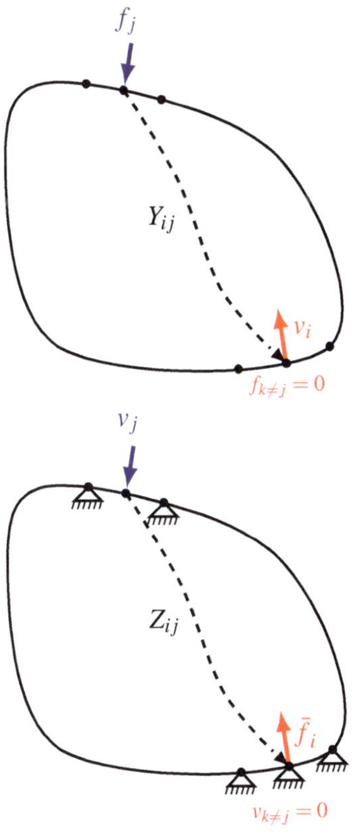

Figure 2.1 Mobility (free-type) and impedance (fixed-type) FRFs.

Hereafter, we will often refer to $\mathbf{Y}$ and $\mathbf{v}$ more generally as an FRF and response, respectively. The reader can readily substitute these for acceleration ($\mathbf{A}$ and $\mathbf{a}$) or displacement-based ($\mathbf{R}$ and $\mathbf{x}$) quantities provided self-consistency is achieved. As mentioned earlier, the only time this substitution cannot be made directly, is when dealing with power which is defined specifically with velocity.

Note that Equation 2.15 is simply a statement that the velocity response at any given position i, is the linear combination, or superposition, of the response caused by each of the N forces individually,

$$v_i = Y_{i1}f_1 + Y_{i2}f_2 + \cdots + Y_{1N}f_N. \tag{2.16}$$

With reference to Equation 2.15, we can distinguish between two *types* of mobility; point and transfer. Point mobilities describe the velocity at a point when the force excitation is at the same position, i.e. co-located, $v_i = Y_{ii}f_i$. Transfer mobilities describe the velocity at a point when the excitation force is located elsewhere, $v_i = Y_{ij}f_j, j \neq i$.

When excitations and responses are considered across the same set of positions, $\mathbf{Y}$ is a square *symmetric* matrix ($Y_{ij} = Y_{ji}$ or $\mathbf{Y} = \mathbf{Y}^{\mathrm{T}}$). Its diagonal elements contain the point mobilities at each position, whilst the off-diagonal elements contain the transfer mobilities between each pair of positions. The symmetry of $\mathbf{Y}$ arises due to an important feature of FRFs; they are *reciprocal*, meaning that when the input and output positions are interchanged the FRF remains the same, i.e. $Y_{ij} = Y_{ji}$ [4, 52, 53]. The same is true for any other FRF defined in Table 2.1. Reciprocity is discussed further in 2.2.7.

In the more general case whereby the number of excitations and responses differ, and their positions are not necessarily co-located, say $\mathbf{v} \in \mathbb{C}^N$ and $\mathbf{f} \in \mathbb{C}^M$, the mobility matrix $\mathbf{Y} \in \mathbb{C}^{N\times M}$ is rectangular, only some elements of which are paired with a reciprocal element. See for example in 2.2.5.

Equation 2.15 provides a component-level definition of mobility for the multi-input multi-output case. To obtain an element-wise definition for Y_{ij}, we must isolate a single entry of the mobility matrix. We do this by considering the response at an arbitrary position i, due to *a single applied force* at position j; that is, we set all forces other than that at j to zero, $f_{k\neq j} = 0$. Equation 2.15 then reduces to the element wise mobility definition,

$$Y_{ij} = \left.\frac{v_i}{f_j}\right|_{f_{k\neq j}=0}. \tag{2.17}$$

Based on Equation 2.17, the measurement of a mobility matrix should be done in such a way that (a) forces are applied one at a time to each point of interest; (b) external forces at all other points are zero (the structure is allowed to respond freely); (c) the individual elements of the matrix are then measured as the complex ratio of the velocity response at i to the excitation force at j. In practice, this calculation is often done using auto and cross-spectra rather than a simple ratio so as to allow averaging, but this detail is picked up in Chapter 7 and need not concern us now.

As implied by Equation 2.17, for the measurement of mobility it is important that no external forces are applied to the test object other than at the excited point, in other words the test structure must be *free* from physical constraints.

Considering the same component with N inputs and outputs, its *impedance* is defined by the vector-matrix equation,

$$\mathbf{f} = \mathbf{Z}\mathbf{v} \tag{2.18}$$

where $\mathbf{f} \in \mathbb{C}^N$ is an N dimensional vector of resultant forces, $\mathbf{v} \in \mathbb{C}^N$ is an N dimensional vector of applied velocities and $\mathbf{Z} \in \mathbb{C}^{N\times N}$ is the $N \times N$ dimensional impedance matrix that relates the two. Written out in full,

$$\begin{pmatrix} f_1 \\ f_2 \\ \vdots \\ f_N \end{pmatrix} = \begin{bmatrix} Z_{11} & Z_{12} & \cdots & Z_{1N} \\ Z_{21} & Z_{22} & \cdots & Z_{2N} \\ \vdots & \vdots & \cdots & \ddots \\ Z_{N1} & Z_{N2} & \cdots & Z_{NN} \end{bmatrix} \begin{pmatrix} v_1 \\ v_2 \\ \vdots \\ v_N \end{pmatrix}. \tag{2.19}$$

As above for the mobility, the first subscripted index denotes the response position (now a resultant force) and the second index the excitation position (now an applied velocity).

Like the mobility matrix, when excitations and responses are considered across the same set of positions, $\mathbf{Z}$ is a square *symmetric* (i.e. reciprocal) matrix ($Z_{ij} = Z_{ji}$ or $\mathbf{Z} = \mathbf{Z}^{\mathrm{T}}$), with the so-called point impedances along the diagonal, and transfer impedances elsewhere. Similarly, in the more general case where say $\mathbf{v} \in \mathbb{C}^M$ and $\mathbf{f} \in \mathbb{C}^N$, the impedance matrix $\mathbf{Z} \in \mathbb{C}^{N\times M}$ is rectangular, only some elements of which have reciprocal partners.

To obtain an element-wise definition of the impedance Z_{ij}, we must isolate a single element of the matrix $\mathbf{Z}$. Following the same procedure as used for the mobility, we consider an arbitrary force location i and set all velocities, bar that applied to position j, to zero, $v_{k\neq j} = 0$. We are then left with the element-wise definition,

$$Z_{ij} = \frac{\bar{f}_i}{v_j}\bigg|_{v_{k\neq j}=0} \tag{2.20}$$

where it is noted that in order to satisfy the constraint $v_{k\neq j} = 0$, an external *blocking force* $\bar{f}$ is required at all positions $k \neq j$.

Based on Equation 2.20 an impedance matrix would be measured in such a way that (1) velocities are applied one at a time to each point of interest; (2) the velocities at all other points are constrained to zero (the structure is *not* allowed to move); (3) the individual elements of the matrix are then measured as the complex ratio of the force response to the single excitation velocity.

In contrast to mobility, for the measurement of impedance (dynamic stiffness or apparent mass) all points other than the excited point must be completely prevented from moving, in other words they must be *fixed*. These different constraints for mobility and impedance, as established in Equations 2.17 and 2.20 respectively, have major practical implications for measurement which will be discussed in the following.

Example 2.1: FRFs of simple structures

The aim of this example is to show some measured FRFs from both idealised and representative components, and highlight some of their important features, in particular reciprocity, the effect of damping and inversion.

Shown in Figure 2.2 are the point and transfer mobilities measured at and between two points on a perspex plate (910 × 910 × 10mm, red) and a steel beam (1515 × 30 × 10mm, blue). In both cases the measurement points are positioned 30cm apart.

Some general trends can be observed from the mobility magnitudes in Figure 2.2 and can understood with reference to Cremer and Heckl [61] (some corresponding phase plots are shown in Figure 2.3):

1) The low damping of the steel beam leads to well-defined resonances with sharp peaks across the entire frequency range. In contrast, the well-damped perspex plate yields less pronounced resonances that are only well-defined in the low frequency range.
2) Modal spacing is regular for the 1D beam and irregular for the 2D plate.
3) Modal density of the 2D plate increases with frequency more rapidly than that of the 1D beam.

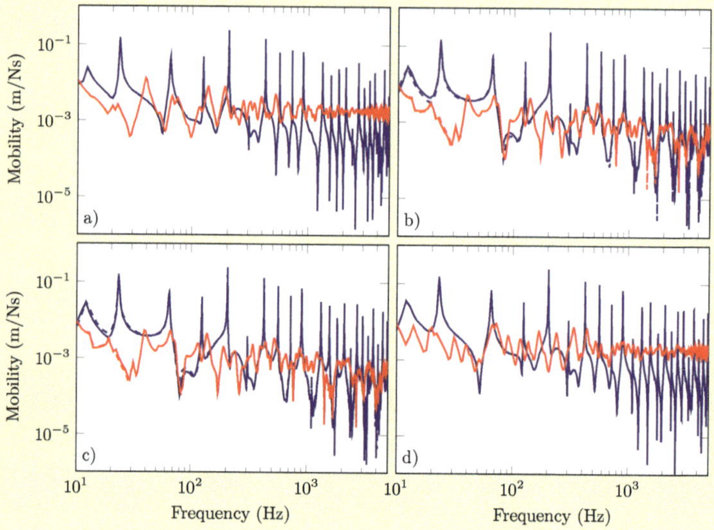

Figure 2.2 Point and transfer mobility magnitude for plate (red) and beam (blue); a) Y_{11}, b) Y_{12}, c) Y_{21}, d) Y_{22}. Dashed plots in b) and c) show the reciprocal transfer mobilities $Y_{12} = Y_{21}$.

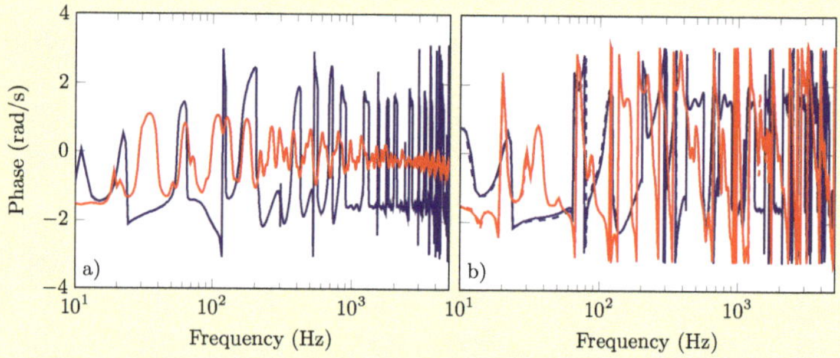

Figure 2.3 Point and transfer mobility phase for plate (red) and beam (blue); a) Y_{11}, b) Y_{12}. Dashed plots in b) show the reciprocal transfer mobilities $Y_{12} = Y_{21}$.

4) The frequency-averaged plate and beam point mobilities follow their characteristic trends (i.e. the mobility of an infinite beam or plate); the plate mobility tends to a constant value, whilst the beam falls with a slope proportional to $1/\sqrt{\omega}$ [6].

5) For the point mobilities, resonances are always separated by an anti-resonance, whilst for the transfer mobilities resonances can be separated by an anti-resonance or simply a minimum (a consequence of Foster's reactance theorem [13]).
6) The transfer mobilities satisfy reciprocity to a high degree of accuracy, illustrated by the dashed plots in Figures 2.2 b) and c).

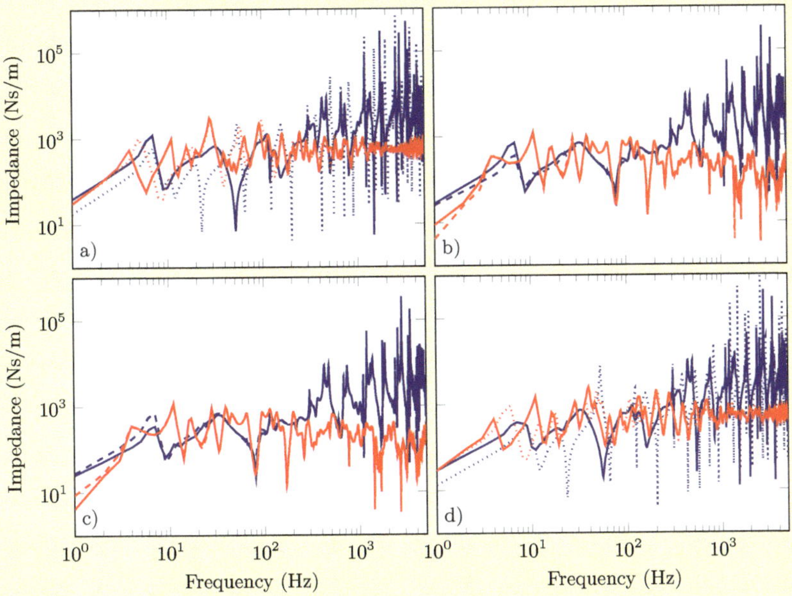

Figure 2.4 Point and transfer impedances for plate (red) and beam (blue); a) Z_{11}, b) Z_{12}, c) Z_{21}, d) Z_{22}. Dashed plots in b) and c) show the reciprocal transfer impedances $Z_{12} = Z_{21}$. Dotted plots in a) and d) show the point impedance obtained as the inverse of the point mobility.

Shown in Figure 2.4 are the corresponding point and transfer *impedances* of the plate and beam, obtained by inverting their respective mobility matrices as per,

$$\begin{bmatrix} Z_{11} & Z_{12} \\ Z_{21} & Z_{22} \end{bmatrix} = \begin{bmatrix} Y_{11} & Y_{12} \\ Y_{21} & Y_{22} \end{bmatrix}^{-1}. \tag{2.21}$$

Also shown in subplots a) and d) are the point impedances obtained by considering only a single DoF, i.e. by taking the reciprocal of the point mobility,

$$Z_{11} = \frac{1}{Y_{11}}, \qquad Z_{22} = \frac{1}{Y_{22}}. \tag{2.22}$$

Figure 2.5 Photo of steering system measured.

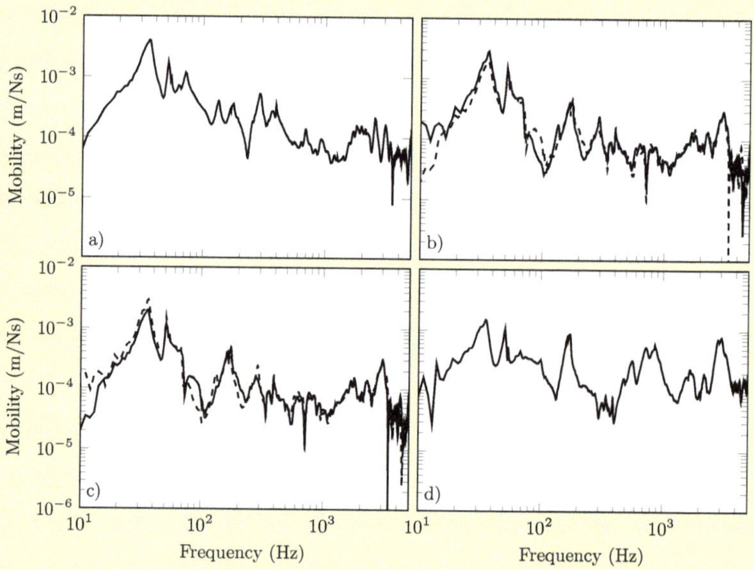

Figure 2.6 Point and transfer mobility for steering system; a) Y_{11}, b) Y_{12}, c) Y_{21}, d) Y_{22}. Dashed plots in b) and c) show the reciprocal transfer mobilities $Y_{12} = Y_{21}$.

The difference between the point impedance obtained by matrix and scalar inversion highlights the influence of the velocity constraint in the definition of impedance (see Equation 2.20).

Shown in Figure 2.6 and 2.7 are the point and transfer mobilities and impedances for the electric steering system shown in Figure 2.5.

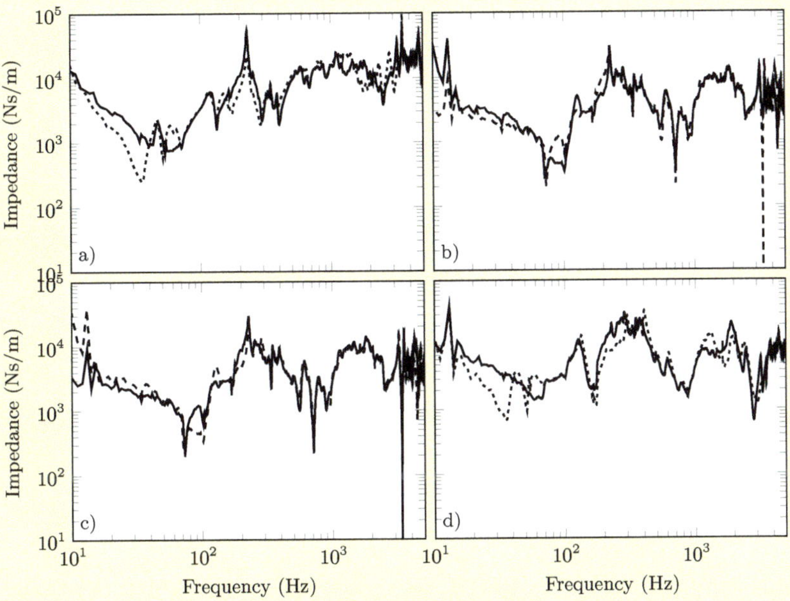

Figure 2.7 Point and transfer impedances for steering system; a) Z_{11}, b) Z_{12}, c) Z_{21}, d) Z_{22}. Dashed plots in b) and c) show the reciprocal transfer impedances $Z_{12} = Z_{21}$. Dotted plots in a) and d) show the point impedance obtained as the inverse of the point mobility.

Many of the same trends are observed as with the idealised beam and plate structures: for example reciprocal transfer mobilities, resonant behaviour at low frequency, a beam-like overall trend in the mobility $\propto 1/\sqrt{\omega}$, the inverse trends of the impedance and its dependence on the velocity constraint.

2.2.4 MEASUREMENT POINTS AND DEGREES OF FREEDOM

By introducing the multi-point FRF we have dealt with the fact that excitation and coupling rarely occurs through a single point. What is left to consider is the fact that at each point forces may be transmitted, and the structure may respond, in different *directions* or degrees of freedom (DoFs). Indeed, the motion of any 'point' on a structural component is in general described by 6 DoFs – 3 translational motions in the x, y and z coordinates, and a rotation about each of these axes, respectively, α, β and γ. Of course, point-like behaviour is an idealisation, nevertheless, if locally the component dynamics behave rigidly then the point-like assumption is valid and can be adopted in the measurement of FRFs. This will be the case when deformations of the structure across the point of interest are negligible or equivalently, when the structural wavelength is long compared with the size of the point (see Chapter 4). Hence, a component with N points of interest is characterised by $6N$ DoFs such that its complete mobility matrix $\mathbf{Y} \in \mathbb{C}^{6N \times 6N}$, is given by,

$$\mathbf{Y} = \begin{bmatrix} \mathbf{Y}_{11} & \mathbf{Y}_{12} & \cdots & \mathbf{Y}_{1N} \\ \mathbf{Y}_{11} & \mathbf{Y}_{22} & \cdots & \mathbf{Y}_{2N} \\ \vdots & \vdots & \ddots & \vdots \\ \mathbf{Y}_{N1} & \mathbf{Y}_{N2} & \cdots & \mathbf{Y}_{NN} \end{bmatrix} \tag{2.23}$$

where in general $\mathbf{Y}_{ij} \in \mathbb{C}^{6\times 6}$ is the mobility submatrix relating the translational and rotational forces at position j to the corresponding velocities at position i. Explicitly,

$$\begin{pmatrix} v_{x_i} \\ v_{y_i} \\ v_{z_i} \\ v_{\alpha_i} \\ v_{\beta_i} \\ v_{\gamma_i} \end{pmatrix} = \begin{bmatrix} Y_{x_i x_j} & Y_{x_i y_j} & Y_{x_i z_j} & Y_{x_i \alpha_j} & Y_{x_i \beta_j} & Y_{x_i \gamma_j} \\ Y_{y_i x_j} & Y_{y_i y_j} & Y_{y_i z_j} & Y_{y_i \alpha_j} & Y_{y_i \beta_j} & Y_{y_i \gamma_j} \\ Y_{z_i x_j} & Y_{z_i y_j} & Y_{z_i z_j} & Y_{z_i \alpha_j} & Y_{z_i \beta_j} & Y_{z_i \gamma_j} \\ Y_{\alpha_i x_j} & Y_{\alpha_i y_j} & Y_{\alpha_i z_j} & Y_{\alpha_i \alpha_j} & Y_{\alpha_i \beta_j} & Y_{\alpha_i \gamma_j} \\ Y_{\beta_i x_j} & Y_{\beta_i y_j} & Y_{\beta_i z_j} & Y_{\beta_i \alpha_j} & Y_{\beta_i \beta_j} & Y_{\beta_i \gamma_j} \\ Y_{\gamma_i x_j} & Y_{\gamma_i y_j} & Y_{\gamma_i z_j} & Y_{\gamma_i \alpha_j} & Y_{\gamma_i \beta_j} & Y_{\gamma_i \gamma_j} \end{bmatrix} \begin{pmatrix} f_{x_j} \\ f_{y_j} \\ f_{z_j} \\ f_{\alpha_j} \\ f_{\beta_j} \\ f_{\gamma_j} \end{pmatrix} \tag{2.24}$$

where, for example, $Y_{\alpha_i y_j}$ is the mobility relating a translational force in the y direction at position j to a rotational velocity about the x-axis at position i.

In practice, not all DoFs are necessary at all points. For example when considering an acoustic response, only the out-of-plane velocity response v_z contributes to sound radiation. In contrast, at the interface connection point between two components all DoFs are often important.

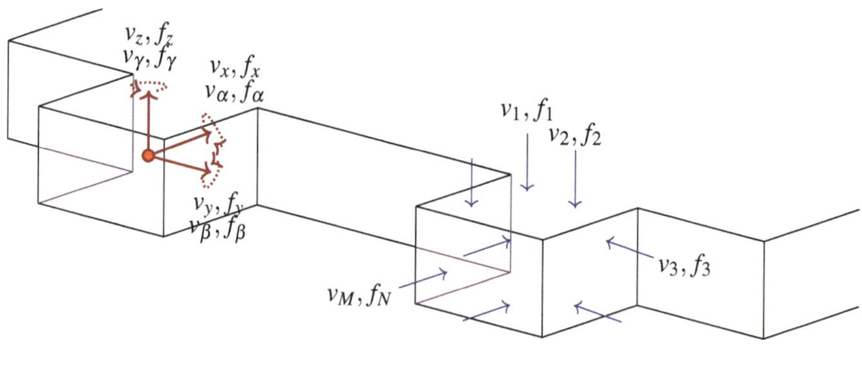

Figure 2.8 Comparison of co-located (left) vs. non-co-located (right) forces and responses, as per Equation 2.24 and 2.25.

At this point it is important to recognise that whilst the full $6N$ DoF matrix described above is theoretically important, mobility matrices are rarely measured in this form. The $6N$ DoFs above describe a component's motion in terms of the translations and rotations about a series of idealised point-like locations. In practice, it is not usually possible to directly measure translational and rotational motions for co-located excitation and response positions, nor is it straightforward to measure rotational motions directly. Rather, we measure a set of translational FRFs between N excitation positions surrounding point j, and M response positions surrounding point i, in the form,

$$\begin{pmatrix} v_1 \\ v_2 \\ \vdots \\ v_M \end{pmatrix}_i = \begin{bmatrix} Y_{11} & Y_{12} & \cdots & Y_{1N} \\ Y_{11} & Y_{22} & \cdots & Y_{2N} \\ \vdots & \vdots & \ddots & \vdots \\ Y_{M1} & Y_{M2} & \cdots & Y_{MN} \end{bmatrix}_{ij} \begin{pmatrix} f_1 \\ f_2 \\ \vdots \\ f_N \end{pmatrix}_j . \tag{2.25}$$

One of many possible examples is illustrated in Figure 2.26, where for clarity we consider co-located points, $i = j$.

Whilst directly measured FRF matrices in the form of Equation 2.25 can be used for some forms of analysis, due to their non-co-located excitation and response positions and lack of rotations, they are limited in terms of VAVP development, as described in Chapter 4.

From Equation 2.25, provided the measured DoFs span the vector space of the full 6 DoF matrix (Equation 2.24), the latter can be determined by applying an appropriate rigid body transformation to the measured FRFs,

$$\begin{bmatrix} Y_{x_ix_j} & Y_{x_iy_j} & Y_{x_iz_j} & Y_{x_i\alpha_j} & Y_{x_i\beta_j} & Y_{x_i\gamma_j} \\ Y_{y_ix_j} & Y_{y_iy_j} & Y_{y_iz_j} & Y_{y_i\alpha_j} & Y_{y_i\beta_j} & Y_{y_i\gamma_j} \\ Y_{z_ix_j} & Y_{z_iy_j} & Y_{z_iz_j} & Y_{z_i\alpha_j} & Y_{z_i\beta_j} & Y_{z_i\gamma_j} \\ Y_{\alpha_ix_j} & Y_{\alpha_iy_j} & Y_{\alpha_iz_j} & Y_{\alpha_i\alpha_j} & Y_{\alpha_i\beta_j} & Y_{\alpha_i\gamma_j} \\ Y_{\beta_ix_j} & Y_{\beta_iy_j} & Y_{\beta_iz_j} & Y_{\beta_i\alpha_j} & Y_{\beta_i\beta_j} & Y_{\beta_i\gamma_j} \\ Y_{\gamma_ix_j} & Y_{\gamma_iy_j} & Y_{\gamma_iz_j} & Y_{\gamma_i\alpha_j} & Y_{\gamma_i\beta_j} & Y_{\gamma_i\gamma_j} \end{bmatrix} = \mathbf{T}_v \begin{bmatrix} Y_{11} & Y_{12} & \cdots & Y_{1N} \\ Y_{11} & Y_{22} & \cdots & Y_{2N} \\ \vdots & \vdots & \ddots & \vdots \\ Y_{M1} & Y_{M2} & \cdots & Y_{MN} \end{bmatrix}_{ij} \mathbf{T}_f^{\mathrm{T}} \tag{2.26}$$

where $\mathbf{T}_{v,f}$ are the transformation matrices for response and force positions, respectively. Further details on the construction of $\mathbf{T}_{v,f}$ are presented in Chapter 4. Hereafter, we will assume (unless otherwise stated) that any FRF matrices introduced have already been transformed as described above.

2.2.5 INTERFACES AND REMOTE POINTS

We are now able to represent multi-point components by FRF matrices that describe the full 6 DoF motion at each point of interest. The next question to consider is *which points should we include in our FRF matrix to enable the construction of a VAVP?*

As discussed through Chapter 1, the black-box approach to system modelling requires individual components to be characterised only at the points that they connect to other components, i.e. their *interfaces*. This is true for any component type (vibration sources, isolators, receivers, etc.). Indeed, in order to mathematically couple two components, equilibrium and continuity must be enforced at their respective interface DoFs (see Chapter 5). In addition to the interface DoFs, we are often interested in predicting the response at some additional *remote* points. These DoFs are not required for coupling, rather they provide a means of predicting the response at a target location. Interfaces and remote (target) points will be denoted by c and r, respectively.

With the above in mind, the mobility matrix of a generic source S or receiver R component, considering only its interface DoFs, is denoted by,

$$\mathbf{Y}_S = \mathbf{Y}_{Scc}, \qquad \mathbf{Y}_R = \mathbf{Y}_{Rcc} \tag{2.27}$$

where subscript c denotes the entire set of interface DoFs, which may include several connection points each with up to 6 DoFs. Which DoFs are considered 'important' enough to be included in $\mathbf{Y}_{Scc}$ and/or $\mathbf{Y}_{Rcc}$ is of crucial importance, but this question will be deferred for later discussion.

Often, it is necessary to refer to a subset of the interface DoFs, for example a specific connection point, for which we introduce the subsets $c_1, c_2, \cdots$, which together make up the interface set c. Considering the source component S as in Figure 2.9, its interface mobility matrix is given by,

$$\mathbf{Y}_{Scc} = \begin{bmatrix} \mathbf{Y}_{Sc_1c_1} & \mathbf{Y}_{Sc_1c_2} \\ \mathbf{Y}_{Sc_2c_1} & \mathbf{Y}_{Sc_2c_2} \end{bmatrix} \tag{2.28}$$

where $\mathbf{Y}_{Sc_1c_1}$ and $\mathbf{Y}_{Sc_2c_2}$ represent the point mobility matrices of interface DoF subsets c_1 and c_2, respectively, and $\mathbf{Y}_{Sc_1c_2} = \mathbf{Y}^{\mathrm{T}}_{Sc_2c_1}$ contains the transfer mobilities between them. As an example, if the subsets c_1 and c_2 each represent a discrete connection with 6 DoFs, $\mathbf{Y}_{Scc}$ will be of dimensions 12×12.

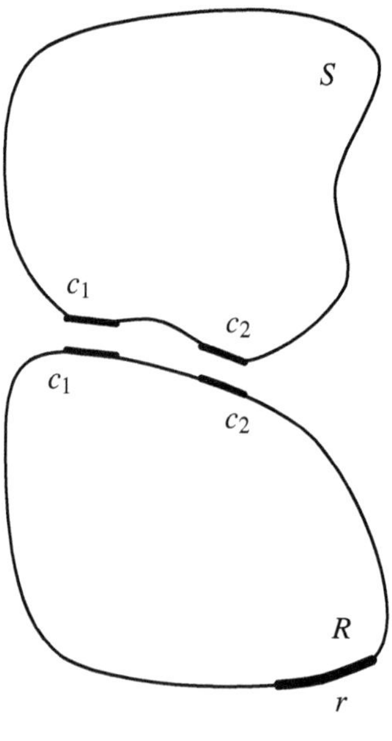

Figure 2.9 Example components with interface (c_1 and c_2) and remote (r) positions.

Components whose passive properties may be suitably characterised by their interface DoFs alone, as in Equation 2.27, are typically either *vibration sources* or *coupling elements*. With these component types there is usually no interest in their response away from the interface. For *receiver* components on the other hand, response predictions are often sought at remote positions away from the interface, and so DoFs r must be included. Consequently, a full receiver mobility matrix is given by,

$$\mathbf{Y}_R = \begin{bmatrix} \mathbf{Y}_{Rcc} & \mathbf{Y}_{Rcr} \\ \mathbf{Y}_{Rrc} & \mathbf{Y}_{Rrr} \end{bmatrix} \quad \text{or} \quad \mathbf{Y}_R = \begin{bmatrix} \mathbf{Y}_{Rc_1c_1} & \mathbf{Y}_{Rc_1c_2} & \mathbf{Y}_{Rc_1r} \\ \mathbf{Y}_{Rc_2c_1} & \mathbf{Y}_{Rc_2c_2} & \mathbf{Y}_{Rc_2r} \\ \mathbf{Y}_{Rrc_1} & \mathbf{Y}_{Rrc_2} & \mathbf{Y}_{Rrr} \end{bmatrix}. \tag{2.29}$$

The purpose of including r is typically to enable the response to be predicted at those locations. For this purpose it is not necessary to measure the point mobility $\mathbf{Y}_{Rrr}$ – this is only required if the point is also subject to an external force which is not usually the case for target locations. As such, for nearly all remote points there is scope to simplify the mobility measurement by omitting any excitation at r and recording only transfer mobility between the interface and remote DoFs. The DoFs r are then 'response-only' and the component is fully characterised by the reduced mobility matrix,

$$\mathbf{Y}_R = \begin{bmatrix} \mathbf{Y}_{Rcc} \\ \mathbf{Y}_{Rrc} \end{bmatrix} \tag{2.30}$$

or by reciprocity,

$$\mathbf{Y}_R = \begin{bmatrix} \mathbf{Y}_{Rcc} & \mathbf{Y}_{Rcr} \end{bmatrix}^{\mathrm{T}}. \tag{2.31}$$

Note that when including response-only positions as above, the resulting matrix is no longer square which can have implications for some processing, particularly matrix inversion, which will be discussed in 7.4.

2.2.6 VIBRO-ACOUSTIC COMPONENTS

For an acoustic component, the notion of an applied force and resulting velocity needs to be reconsidered. Applying a known force to the air in a particular direction is not practical. It is also difficult, though not impossible, to accurately measure particle velocity, the acoustic equivalent of structural velocity. Furthermore, the roles of force and velocity as excitation and response, respectively, are somewhat interchanged within an acoustic domain. The fundamental acoustic source is the monopole, which is described in terms of its *volume velocity* (related to particle velocity by a factor of surface area). Similarly, the natural response variable in air is pressure (related to force also by a factor of area). Consequently, an acoustic FRF, which relates an acoustic excitation to an acoustic response, is typically defined as,

$$p_i = H_{p_i q_j} q_j \tag{2.32}$$

where q_i is the volume velocity applied at position i, and p_j is the pressure response at position j. Note that, whereas structural FRFs have up to six DoF at each point, acoustic FRFs have only a single DoF since pressure and volume velocity are scalar quantities. For multiple points, Equation 2.32 can readily be extended to matrix form, from which the inverse FRF (acoustic impedance) can also be considered.

Having defined the acoustic FRF, we can now define vibro-acoustic FRFs in a similar way in terms of the same acoustic quantities. A vibro-acoustic FRF relates a structural excitation to an acoustic response, or alternatively, an acoustic excitation to a structural response. As before, structural excitations and responses are represented by forces and velocities, respectively. Here, an acoustic response is represented by a pressure p, and an acoustic excitation by a volume velocity q. Hence, we can define the vibro-acoustic FRFs as,

$$p_i = \mathbf{H}_{ij}\mathbf{f}_j = \begin{bmatrix} H_{p_i x_j} & H_{p_i y_j} & H_{p_i z_j} & H_{p_i \alpha_j} & H_{p_i \beta_j} & H_{p_i \gamma_j} \end{bmatrix} \begin{pmatrix} f_{x_j} \\ f_{y_j} \\ f_{z_j} \\ f_{\alpha_j} \\ f_{\beta_j} \\ f_{\gamma_j} \end{pmatrix} \tag{2.33}$$

and

$$\begin{pmatrix} v_{x_i} \\ v_{y_i} \\ v_{z_i} \\ v_{\alpha_i} \\ v_{\beta_i} \\ v_{\gamma_i} \end{pmatrix} = \mathbf{H}_{ij} q_j = \begin{bmatrix} H_{x_i q_j} \\ H_{y_i q_j} \\ H_{z_i q_j} \\ H_{\alpha_i q_j} \\ H_{\beta_i q_j} \\ H_{\gamma_i q_j} \end{bmatrix} q_j \tag{2.34}$$

where for example $H_{p_i \alpha_j}$ relates an applied moment about the x axis at position j, to the resultant pressure level at position i, and $H_{z_i q_j}$ relates the applied volume velocity at position j to the resultant structural velocity in the z axis at position i. Note we use $\mathbf{H}$ to denote both vibro-acoustic and acoustic FRFs. Which one is being used to should be clear from the context.

The complete FRF matrix for a vibro-acoustic component whose remote response positions r are entirely acoustic is given by,

$$\mathbf{H}_R = \begin{bmatrix} \mathbf{Y}_{Rcc} & \mathbf{H}_{Rcr} \\ \mathbf{H}_{Rrc} & \mathbf{H}_{Rrr} \end{bmatrix}. \tag{2.35}$$

To measure the remote point and transfer FRFs, $\mathbf{H}_{Rrr}$ and $\mathbf{H}_{Rcr}$, a calibrated volume velocity source is required to excite the component acoustically. Practically, this can be quite challenging. However, if points r in Equation 2.35 are response-only, then the component R can be represented by the reduced FRF matrix,

$$\mathbf{H}_R = \begin{bmatrix} \mathbf{Y}_{Rcc} \\ \mathbf{H}_{Rrc} \end{bmatrix} \tag{2.36}$$

for which the only acoustic measurements are of sound pressure at the remote DoFs due to force excitation at c, thereby avoiding the need for controlled acoustic excitation.

2.2.7 RECIPROCITY RELATIONS

The principle of reciprocity often finds applications in situations like that above where there is difficulty in applying excitation at a given location. Hence, a brief review is provided in what follows. In simple terms, the principle states that an identical FRF is obtained when excitation and response locations are interchanged. The reciprocity principle extends beyond structural systems to mechanical, acoustic and electro-magnetic systems, as well as composite systems, such as vibro-acoustic and electro-mechanical systems. Indeed, the principle applies to most stable and linear passive systems. An example of a system that does not satisfy the above would be an acoustic system with non-negligible mean flow.

Whilst the principle of reciprocity was known as early as the 1860s, with the work of Helmholtz and Rayleigh [46], it was not until the late 1970/80's that its practical benefits began to be realised. A detailed summary of its early development can be found in the excellent reviews by Fahy [4] and Ten Wolde [54].

From an experimental perspective, the interchange of excitation and response locations, can offer numerous practical advantages including:

- The reciprocal measurement of an FRF is often more practical than direct measurement when access is restricted because of the smaller size of sensors (accelerometers, microphones) compared with that of the corresponding structural and acoustic exciters (impact hammers, shakers and volume velocity sources).
- Reciprocal techniques can be used to avoid the measurement of difficult quantities (e.g. moments, volume velocity) by instead using their paired reciprocal quantities (e.g. rotational velocities, sound pressure).
- In challenging environments, or for the measurement of low level signals, reciprocal transfer functions can offer improved signal to noise levels, e.g. by using a more sensitive sensor, or powerful source, in the reciprocal measurement.

Some key *structural reciprocity* relations are illustrated in Figure 2.10. It is seen that the forward and reciprocal quantities at a point form a 'power pair', that is, their product would yield power if they were measured simultaneously. For example, force-velocity or moment angluar velocity in the same direction both yield power. Thus, in the third figure, the FRF between a rotational force at DoF i and translational response at DoF j is equal to the FRF between a translational force at DoF j and rotational response at DoF i. This arrangement can be used to avoid having to apply moments.

Based on mobility-type FRFs, no constraints are required to satisfy the above reciprocity relations. This is not the case for impedance FRFs. Shown in Figure 2.10 is a single reciprocity relation for the impedance FRF, which states that the FRF between an applied velocity at DoF i and the blocking force at DoF j is equal to the FRF between an applied velocity at DoF j and the blocking force at DoF i. Similar rotational and cross relations exist, as for the mobility examples. An important distinction from the mobility-based reciprocity relation is that, when dealing with multiple DoFs, reciprocity is only satisfied if appropriate constraints are applied to the relevant DoFs. Given that impedance is rarely measured directly this is of no great practical concern and so we will not discuss it further.

A key *acoustic reciprocity* relation is illustrated in Figure 2.12. It shows the interchange of a volume velocity source and a microphone. An important requirement in this case is that the directivity of the source matches that of the microphone, in this

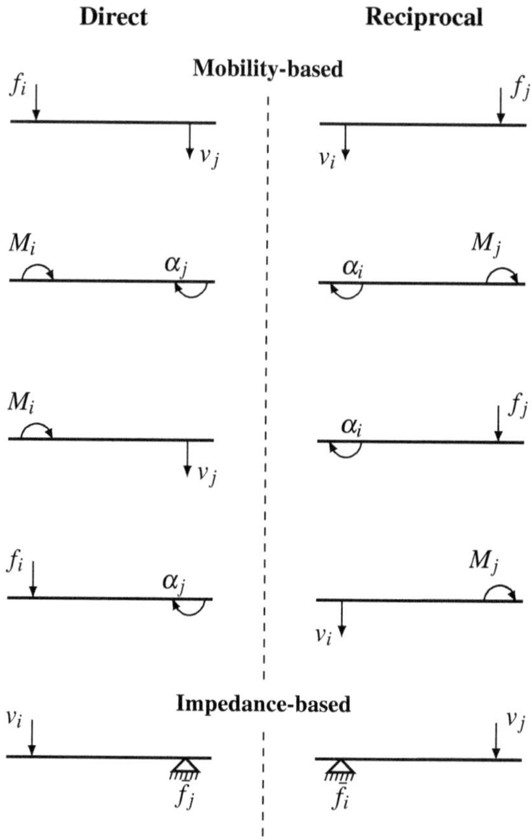

Figure 2.10 Mechanical reciprocity relations. For direct FRF, DoFs i and j are excitation and response, respectively. For the reciprocal FRF these roles are interchanged. For impedance-based reciprocity, only translational relations are shown for brevity.

case omni-directional. Similar reciprocity relations hold for high order directivities, for example, the FRF between a dipole excitation at DoF i and particle velocity at DoF j is equal to the FRF between a dipole excitation at DoF j and particle velocity at DoF i. In this directional case, it is important that when interchanged, the source and receiver are aligned correctly.

The most widely used *vibro-acoustic reciprocity* relation is illustrated in Figure 2.12. Again, the forward and reciprocal quantities at a point form a power-pair, e.g. pressure-volume velocity. An example application of vibro-acoustic reciprocity would be for measurement of vibro-acoustic FRFs from say an engine block to a passenger position. In this case it might be easier to place an accelerometer on the engine block and a volume velocity source at the passenger location, as opposed to exciting the engine block with a force hammer (difficult due to limited access) and measuring the pressure response at the passenger location.

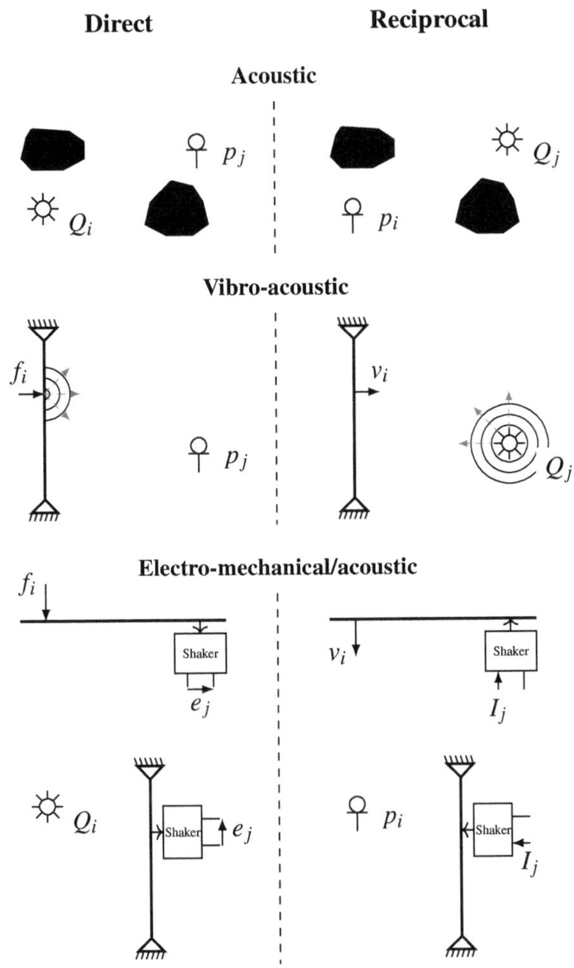

Figure 2.11 Acoustic, vibro-acoustic, and electro-mechanical/acoustic reciprocity relations. For direct FRF DoF i and j are excitation and response, respectively. For the reciprocal FRF these roles are interchanged.

Finally, some key *electro-mechanical/acoustic reciprocity* relations are illustrated in Figure 2.12. Again, the reciprocal quantities at a point form power-pairs, e.g. voltage-current.

2.2.8 FREE VS. FIXED FRFS

In 2.2.3 we introduced two forms of FRF; mobility and impedance. Whilst either one can be used to characterise the passive properties of a component or assembly, in practice it is usually only the mobility that is directly measurable.

In the definition of mobility (see Equation 2.17), we saw that there exists a zero force constraint which must be satisfied at all measurement positions with the exception of the one being excited. This force constraint simply means that a single excitation is applied at a time, and that the component or assembly is allowed to respond freely. This is relatively easy to acheive in practice. With the above in mind, we might describe the mobility (or equivalently the accelerance or compliance) as a *free*-type FRF, with reference to its zero force constraint.

In contrast, we saw in the definition of the impedance (see Equation 2.20) that there exists a zero velocity constraint, which must similarly be satisfied at all measurement positions with the exception of the one being excited. We might thus describe the impedance (or equivalently the effective mass or dynamic stiffness) as a *fixed*-type FRF, with reference to its zero response constraint.

Practically, the zero response constraint has important consequences regarding the direct measurement of fixed-type FRFs. Unlike the zero force constraint (which is often easily achieved in practice), the requirement to prevent all movement at non-excited points is practically impossible for most test objects. Even if these conditions can be achieved within a particular frequency region this is unlikely to extend to the whole working range of the component. To make matters worse, measurement of the reaction forces at all constrained points would involve inserting force transducers between the test component and a fixed boundary. This can rarely be achieved in practice and, even if it can, the presence of the transducers is liable to modify the component locally, thus altering the very properties sought from the measurement. Adding to the difficulty, the constraints must be applied, and forces measured, not just at every point of interest but simultaneously in up to six DOF at every location (see 3.1.2 and for further discussion on the difficulties of direct force measurement).

An additional complication of the fixed constraint is that the response will change when new measurement points are added because of the additional constraints required at that these new points. Thus, enlarging a fixed-type FRF matrix will alter all of its elements. In other words, the elements of a fixed-type FRF matrix are dependent on the number of DoFs considered, and their locations. This is not the case for free-type FRFs where existing elements are unchanged by the addition of new DoFs since the zero force constraints at the extra DoFs have no effect on the component's response.

It is clear then that there are significant practical disadvantages to the fixed-type FRF. For this reason, the free-type FRF is the preferred option for most components, particularly source and receiver components. There are two main exceptions: first, coupling elements and isolators as will be discussed later (2.3.2), and secondly, vibro-acoustic interfaces since fixed constraints are well approximated by most fairly rigid structures in contact with air.

Importantly, fixed-type FRFs *are* available indirectly by inversion of a measured free-type FRF matrix. Pre-multiplying both sides of Equation 2.14 by the inverse FRF matrix $\mathbf{Y}^{-1}$, we obtain,

$$\mathbf{f} = \mathbf{Y}^{-1}\mathbf{v}, \qquad \mathbf{Y}^{-1} = \mathbf{Z}. \tag{2.37}$$

The implications from this simple relationship will prove to be of great practical significance. First, Equation 2.37 implies that by simply inverting a measured free-type FRF matrix we are able to mathematically enforce the fixed constraints that are typically impossible to achieve physically! Secondly, the blocking forces that are not generally amenable to measurement can be obtained by calculation if the fixed-type FRF is known, since,

$$\bar{f}_i = Z_{i1}v_1 + Z_{i2}v_2 + \cdots + Z_{ij}v_j \tag{2.38}$$

where $\bar{f}_i$ is the blocking force at DoF i, i.e. the force required to achieve the zero velocity constraint in Equation 2.20. This mathematical trick will find application in many places throughout the book, most notably for the in situ measurement of isolator properties (2.3.2), the Round Trip method (2.5), the interface completeness criterion (6.4.1) and the *in situ* blocked force method (3.2).

2.3 MEASURING INDEPENDENT COMPONENT PROPERTIES

In this section we further consider the question of how to obtain the *independent* component properties that allow interchangeability. If the component can be physically separated from any connected structures for the test, then all measurements will automatically yield intrinsic component data. In theory, this separation can be achieved in two ways corresponding to the free and fixed boundary conditions discussed in the previous section. The most obvious approach would be to freely suspend the component from 'sky hooks', or in practice from resilient bungees or supports and measure the free FRF directly. Less commonly, it may be appropriate to fix the component to a rigid test bench and measure its fixed-interface FRF (although physically connected to the support in this case, the component is *dynamically* separated by virtue of its blocked interface). These test types will be considered in 2.3.1 and 2.3.2 respectively.

Before we move onto looking at arrangements for direct testing however, we note that there are situations where neither of the ideal free or fixed conditions can be approximated practically. If so, then the component cannot be physically separated and testing will have to be conducted with it installed in an assembly of some sort. Such measurements are known as *in situ* measurements. A major advantage of in situ testing is that one can be sure that the component is mounted in a manner representative of real conditions. The main disadvantage is that any measured data relates to the *assembly* rather than to the component alone. Special techniques are then required to extract intrinsic component properties from the measured assembly properties, including: implementing ideal boundary conditions by mathematical means (2.2.8), decoupling techniques to remove the dynamic influence of the attached components (2.3.1), and use of a numerical models alongside measurements (2.6), perhaps with in situ updating of model parameters. Later in the chapter we will increasingly turn our attention to indirect methods such as these.

2.3.1 FREE-INTERFACE FRFS

Free suspension, as shown in Figure 2.12a, is the preferred test arrangement for the majority of components, particularly for vibration sources which usually have only one interface. Some structures though, particularly larger ones, are designed to be mounted on a solid support and in this case it may be more practical to mount them as such for the test rather to try to suspend them (Figure 2.12b). In this arrangement, the ground is effectively included as part of the component. Note that in both cases the *interfaces* are free, hence we can refer to their properties as *free-interface* FRFs.

It should be understood that the free condition is an *idealised* state that can only be approximated in practice. For example, for suspended components (Figure 2.12a) the suspension is typically made as soft as possible but inevitably in practice has a finite stiffness. Thus, we need to consider the question of how well the free condition is approximated, or 'how soft is soft?'

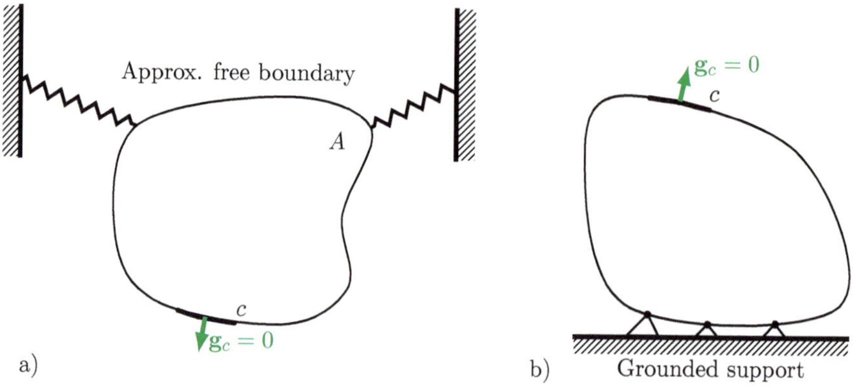

Figure 2.12 Illustration of (a) free boundary and (b) grounded support for the independent measurement of component free-interface FRFs.

With the above in mind, consider an arbitrary component A whose characterisation requires a free suspension. In the presence of a resilient support at the component's interface DoFs c, the measured mobility represents that of the coupled component-support (AI) assembly C, which can be expressed in the form,

$$\begin{bmatrix} \mathbf{Y}_{Ccc} & \mathbf{Y}_{Ccr} \\ \mathbf{Y}_{Crc} & \mathbf{Y}_{Crr} \end{bmatrix} = \begin{bmatrix} \mathbf{Z}_{Acc} + \mathbf{Z}_{Icc} & \mathbf{Z}_{Acr} \\ \mathbf{Z}_{Acr} & \mathbf{Z}_{Arr} \end{bmatrix}^{-1} \tag{2.39}$$

where $\mathbf{Z}_{Icc}$ represents the point impedance of the support system at its connection points and the $\mathbf{Z}_A$ terms are the independent impedances of the component in question. Note that the coupled impedance is given by the sum of the impedances of the joined elements $\mathbf{Z}_{Ccc} = \mathbf{Z}_{Acc} + \mathbf{Z}_{Icc}$ as will be shown explicitly in Chapter 5.

Typically, a support system is designed such that $\mathbf{Z}_{Icc}$ is as small as possible (while providing adequate structural support for the component) so that the measured coupled mobility approximates that of the component. Consider the simplified case where the suspension system can be represented by N ideal spring-damper elements,

$$\mathbf{Z}_{Icc} = \begin{bmatrix} \frac{K_1}{i\omega}+C_1 & 0 & 0 \\ 0 & \ddots & 0 \\ 0 & 0 & \frac{K_N}{i\omega}+C_N \end{bmatrix}. \tag{2.40}$$

The added mounting stiffness will ultimately lead to low frequency (rigid body) mass-spring-like resonances and a suitable suspension scheme should see these resonant frequencies positioned well below the lowest frequency of interest. ISO 9611:1996 for the measurement of free velocity [5], describes a method for evaluating the lower frequency limit f_l for which a resilient suspension can be regarded as a good approximation of a free condition. It states that, assuming the component to behave as a rigid mass below f_l, f_l is assumed to be three times f_0, where f_0 is the highest of the 6 natural (rigid body) frequencies of the mounted component. ISO 7626-2:2015 for the experimental determination of mechanical mobility provides an alternative requirement for component support [12]. It states that the magnitude of the driving point mobility of the suspension should be at least ten times greater than that of the structure at its attachment point, within the frequency range of interest. It further states that as a minimum requirement, all rigid body resonant frequencies should be less than half the lowest frequency of interest.

Even with the above measures in place, resilient supports might still influence the component's dynamics unacceptably. In this case it may be necessary to correct for the influence of said supports to obtain a more accurate set of free-interface FRFs. If the support impedance is known, its effect can be removed by simply subtracting it from the measured component's interface impedance,

$$\begin{bmatrix} \mathbf{Y}_{Acc} & \mathbf{Y}_{Acc} \\ \mathbf{Y}_{Arc} & \mathbf{Y}_{Arr} \end{bmatrix} = \left(\begin{bmatrix} \mathbf{Y}_{Ccc} & \mathbf{Y}_{Ccr} \\ \mathbf{Y}_{Crc} & \mathbf{Y}_{Crr} \end{bmatrix}^{-1} - \begin{bmatrix} \mathbf{Z}_{Icc} & \mathbf{0} \\ \mathbf{0} & \mathbf{0} \end{bmatrix} \right)^{-1}. \tag{2.41}$$

This is an example of a more general procedure known as substructure decoupling, whereby the dynamics of a residual component are subtracted, or removed, from those of a coupled system to obtain the free-interface dynamics of a target component.

In practice it is best to position the mounts away from the interface so as to minimise their influence on the interface dynamics (which are the most important when it comes to coupling components). If required the same decoupling procedure can be applied if the supports are positioned at remote DoFs r,

$$\begin{bmatrix} \mathbf{Y}_{Acc} & \mathbf{Y}_{Acr} \\ \mathbf{Y}_{Arc} & \mathbf{Y}_{Arr} \end{bmatrix} = \left(\begin{bmatrix} \mathbf{Y}_{Ccc} & \mathbf{Y}_{Ccr} \\ \mathbf{Y}_{Crc} & \mathbf{Y}_{Crr} \end{bmatrix}^{-1} - \begin{bmatrix} \mathbf{0} & \mathbf{0} \\ \mathbf{0} & \mathbf{Z}_{Irr} \end{bmatrix} \right)^{-1}. \tag{2.42}$$

If the suspension impedance is unknown, the in situ method for characterising coupling elements (see 2.3.2) can be used to obtain an estimate.

In some cases, a resilient suspension is not possible and the component must be mounted to a more supportive structure to operate, such as a test bench. In this scenario, the decoupling procedure described above remains valid, the isolation impedance $\mathbf{Z}_{Icc}$ simply being replaced by the more general test-bench impedance $\mathbf{Z}_{Bcc}$ (which can also be determined experimentally). It is worth remarking however, the closer the dynamics are between the component and test bench, the more difficult it is to obtain a successful decoupling. A more general framework for substructure decoupling exists, a brief description of which can be found in 5.2.3.

2.3.2 COUPLING ELEMENTS AND VIBRATION ISOLATORS

We now consider coupling elements, i.e. those that connect components by two or more sets of interface DoFs. Such components include, of course, vibration isolators but also linkages, pipes, resilient bellows, shaft couplings, power supply cables, hoses and pipe hangers. For fairly stiff and sufficiently large couplings, a direct measurement of its FRF might be possible using the free suspension approach described earlier. However, for most coupling elements it turns out to be more appropriate to use a fixed-interface approach by blocking one end of the component. This is because: (1) the component's dynamic properties may depend on installation effects, such as pre-load, which are not replicated by free-interface conditions, (2) the response of the freely suspended component tends to be dominated by a large rigid body response (the component 'flies off' when hit) and (3) their compact size and/or geometry make their instrumentation for measurement difficult. As we will see, these difficulties can largely be overcome by testing *in situ*, i.e. with the component installed in an assembly. The challenge then becomes how to extract the component's intrinsic properties from measurements made on an assembly.

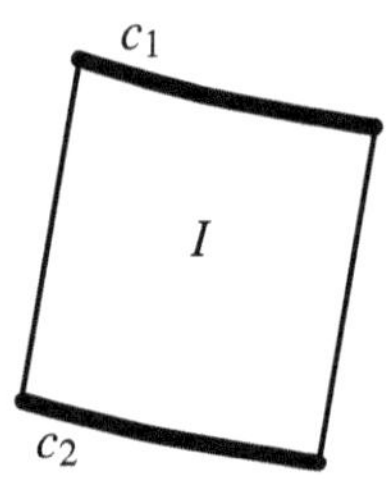

Figure 2.13 Isolator schematic with interface DoFs c_1 and c_2.

Coupling element impedance

Transmissibility is the most commonly used quantity to describe vibration isolation performance. However, as will be shown in 2.4, transmissibility is not an intrinsic property of an isolator, it depends also on the dynamics of the supported structure or foundation. As such, it does not satisfy the independence requirement and cannot be used for component-based simulation. Like other components, the complete and independent characterisation of a coupling element is provided by its free or fixed-interface FRF matrix,

$$\mathbf{Y}_I = \begin{bmatrix} \mathbf{Y}_{Ic_1c_1} & \mathbf{Y}_{Ic_1c_2} \\ \mathbf{Y}_{Ic_2c_1} & \mathbf{Y}_{Ic_2c_2} \end{bmatrix} \quad \text{or} \quad \mathbf{Z}_I = \begin{bmatrix} \mathbf{Z}_{Ic_1c_1} & \mathbf{Z}_{Ic_1c_2} \\ \mathbf{Z}_{Ic_2c_1} & \mathbf{Z}_{Ic_2c_2} \end{bmatrix} \tag{2.43}$$

where c_1 and c_2 represent the DoFs at the two ends of the element, and the subscript I that the FRFs are independent properties of the coupling element.

As already discussed, for coupling elements free-interface FRFs are challenging to measure, and not necessarily representative. For this reason, a much greater emphasis has been placed on their fixed-interface dynamics. Indeed, historically the preferred quantity to characterise a resilient element has been the *dynamic transfer stiffness* [57] (a fixed-FRF type), denoted $\mathbf{D}_{Iij}$ (or $\mathbf{D}_{Iji}$), which describes the relation between an applied displacement at one side of the element to the resultant blocked force at the other,

$$\bar{\mathbf{f}}_i = \mathbf{D}_{Iij}\mathbf{x}_j\Big|_{\mathbf{x}_i=0, i\neq j}. \tag{2.44}$$

The term *dynamic* indicates that, in contrast to static stiffness which is constant, the dynamic stiffness is frequency-dependent and indeed for isolators, often displays significant variation with frequency. In addition, isolator properties may be sensitive to static pre-load, dynamic excitation level, temperature and relative humidity [49]. These non-linearities make the complete characterisation of a resilient element a complicated and involved task. Nevertheless, for most practical situations (excluding extreme environmental conditions) a reasonable characterisation can be obtained by considering an element's linear (small excitation) behaviour under an appropriate pre-load and at a given temperature.

To be consistent with the remainder of this book, we will continue to use impedance rather than dynamic stiffness from now, recalling that $D = i\omega Z$.

Assuming a single point-like connection at each end (thus excluding distributed elements such as resilient matting), the point and transfer impedances of a coupling element are completely described by 6×6 matrices of the form,

$$\mathbf{Z}_{Ic_ic_j} = \begin{bmatrix} Z_{Ix_ix_j} & Z_{Ix_iy_j} & Z_{Ix_iz_j} & Z_{Ix_i\alpha_j} & Z_{Ix_i\beta_j} & Z_{Ix_i\gamma_j} \\ Z_{Iy_ix_j} & Z_{Iy_iy_j} & Z_{Iy_iz_j} & Z_{Iy_i\alpha_j} & Z_{Iy_i\beta_j} & Z_{Iy_i\gamma_j} \\ Z_{Iz_ix_j} & Z_{Iz_iy_j} & Z_{Iz_iz_j} & Z_{Iz_i\alpha_j} & Z_{Iz_i\beta_j} & Z_{Iz_i\gamma_j} \\ Z_{I\alpha_ix_j} & Z_{I\alpha_iy_j} & Z_{I\alpha_iz_j} & Z_{I\alpha_i\alpha_j} & Z_{I\alpha_i\beta_j} & Z_{I\alpha_i\gamma_j} \\ Z_{I\beta_ix_j} & Z_{I\beta_iy_j} & Z_{I\beta_iz_j} & Z_{I\beta_i\alpha_j} & Z_{I\beta_i\beta_j} & Z_{I\beta_i\gamma_j} \\ Z_{I\gamma_ix_j} & Z_{I\gamma_iy_j} & Z_{I\gamma_iz_j} & Z_{I\gamma_i\alpha_j} & Z_{I\gamma_i\beta_j} & Z_{I\gamma_i\gamma_j} \end{bmatrix}, \tag{2.45}$$

where, for example, $Z_{Iz_i\beta_j}$ represents the transfer impedance relating an applied angular velocity about the y axes at interface j, to the resultant translational blocked force in the z direction at interface i. The complete isolator impedance matrix $\mathbf{Z}_I$ is thus 12×12.

Whilst the entirety of $\mathbf{Z}_I$ is required for a complete isolator characterisation, the transfer impedance $\mathbf{Z}_{Ic_1c_2}$ is often sufficient to predict vibration transmission. Indeed, several ISO standards have been developed to obtain specifically the transfer characteristics of vibration isolators, with little guidance regarding the point impedance.

With regards to characterisation methods for resilient elements, broadly speaking, they can be categorised as either experimental or model-based, with the former being further categorised as either direct or indirect. Direct methods involve the *direct* measurement of force for the transfer impedance, whereas with indirect methods this is inferred from other, usually more easily measured quantities. Model-based methods typically involve establishing a numerical or analytical model that describes the underlying physics of the element, followed by a model updating procedure to determine the model inputs. Some examples of model-based approaches may be found in [1, 14, 18, 26, 35]. However, the remainder of this section will focus primarily on experimental methods.

Blocking mass methods

Blocking masses are widely used in the characterisation of coupling elements and can be applied with either a direct or indirect approach. Their purpose is to approximate the rigid constraints required by the definition of the fixed-interface FRFs.

Direct

The classical direct method determines the transfer impedance from the direct measurement of a receiver side blocking force $\bar{f}_{c_2}$ and source side velocity v_{c_1}, due to an applied harmonic excitation, as illustrated in Figure 2.14a. The blocking force is measured by placing a force transducer between the coupling element and a large blocking mass. Providing this blocking mass is made large enough, the measured force f_{c_2} approximates the necessary blocking force $\bar{f}_{c_2}$. The transfer impedance $Z_{Ic_2c_1}$ can then be approximated by,

$$Z_{Ic_2c_1} = \frac{\bar{f}_{c2}}{v_{c_1}} \approx \frac{f_{c2}}{v_{c_1}}. \tag{2.46}$$

The size of the upper mass can then be chosen to achieve an appropriate pre-load for the element. Often this pre-load is provided together with the excitation by a hydraulic actuator of some sort, although this type of excitation is restricted to low frequencies, typically below a few hundred Hz.

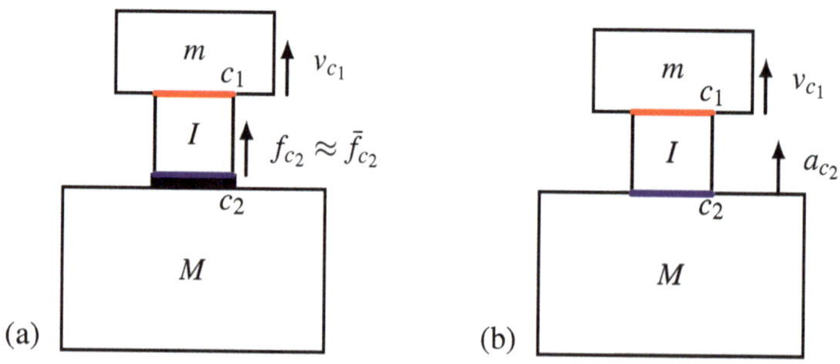

Figure 2.14 Test assemblies for (a) direct and (b) indirect characterisation of dynamic transfer stiffness.

Whilst a simple set-up on paper, by definition, the transfer impedance requires all DoFs bar the excitation DoF to be blocked. Hence, the test rig should be designed to minimise motion in all other DoFs. Furthermore, to enable characterisation in other

DoFs (e.g. in-plane and rotational) the test rig must be modified to allow motion in these DoFs whilst constraining all others. Clearly, the complete characterisation of a resilient element by means of the direct approach requires considerable effort, with multiple test rig configurations and repeated testing for each.

In-direct

One of the challenges with the direct method presented above is the measurement of the force in directions other than vertical. An extension to the direct method, termed the in-direct method, estimates the blocking force using Newton's second law as per,

$$\bar{f}_{c_2} \approx f_{c_2} = M a_{c_2} \tag{2.47}$$

where M is the mass of the blocking mass, and a_{c_2} its acceleration. Through the use of finite difference approximations (see Chapter 4), estimations of moment forces can also be obtained. Whilst the installation of a force transducer is avoided, the indirect method still requires the presence of a large blocking mass to achieve the rigid constraint. This has the downside that the acceleration response of the blocking mass is quite small and therefore subject to noise which can restrict the upper frequency limit of the indirect approach. Corrections for a non-rigid mass by Thompson et al. [55], and further extensions by Kari [25] have been proposed to improve high frequency performance.

Though limited in several respects, the direct and indirect methods described above are established methods and have gone on to form parts 2 and 3 of the International Standard series ISO 10846 for the characterisation of resilient elements [6, 8, 9]. Part 4 of the aforementioned standard demonstrates the application of both the direct and indirect approaches to the characterisation of resilient elements that are not simple resilient supports, i.e. flexible hoses, pipe hangers, etc [7]. Last in the series, part 5 [10] outlines a method for determining the low frequency dynamic point impedance using a rigid blocking mass with source side force and displacement measurements. Throughout the ISO series a number of test rig designs are given so as to apply the above methods to in-plane DoFs (i.e. x and y), although, with its focus largely on translational characterisation, only limited guidelines are given with regards to rotational impedance components.

Though well established, a fundamental assumption underpins all in-direct approaches whereby a blocking mass is used: theoretically, the blocking mass should be perfectly rigid if its acceleration is to be truly constrained. Since the in-direct method relies upon a measurable acceleration on the blocking mass to infer a receiver side force, this cannot be the true blocking force. Although likely a reasonable approximation at most frequencies, a method that avoids the need for a blocking mass would clearly be beneficial.

***In situ* characterisation**

The blocking mass methods described above determine the transfer impedance elements Z_{ij} of a resilient element by *physically* constraining one end with a blocking mass. *in situ* methods achieve these same constraints, instead, by *mathematical* means [33, 34, 37]. As well as avoiding the need for a blocking mass this opens up the potential for characterising elements while in real installations which by definition provides realistic operating conditions such as pre-load (see also e.g. [42, 58]).

The in situ characterisation of a resilient element involves the extraction of invariant component properties from measurements made on a coupled assembly, i.e. with the element installed between two arbitrary components, A and B (Figure 2.15). To see how this is possible we construct the impedance matrix for an AIB assembly, i.e. an isolator I between substructures A and B, noting that the impedance matrix of the assembly is given by the sum of those of the individual substructures (see 5.2 for more details on impedance coupling),

$$\begin{array}{ccccccc} \mathbf{Z}_{Ccc} & = & \mathbf{Z}_{Acc} & + & \mathbf{Z}_{Icc} & + & \mathbf{Z}_{Bcc} \\ \begin{bmatrix} \mathbf{Z}_{Cc_1c_1} & \mathbf{Z}_{Cc_1c_2} \\ \mathbf{Z}_{Cc_2c_1} & \mathbf{Z}_{Cc_2c_2} \end{bmatrix} & = & \begin{bmatrix} \mathbf{Z}_{Ac_1c_1} & \mathbf{0} \\ \mathbf{0} & \mathbf{0} \end{bmatrix} & + & \begin{bmatrix} \mathbf{Z}_{Ic_1c_1} & \mathbf{Z}_{Ic_1c_2} \\ \mathbf{Z}_{Ic_2c_1} & \mathbf{Z}_{Ic_2c_2} \end{bmatrix} & + & \begin{bmatrix} \mathbf{0} & \mathbf{0} \\ \mathbf{0} & \mathbf{Z}_{Bc_2c_2} \end{bmatrix} \end{array} \tag{2.48}$$

where as usual C represents assembly properties and A, B and I those of the substructures. Rewriting gives,

$$\mathbf{Z}_{Ccc} = \begin{bmatrix} \mathbf{Z}_{Ac_1c_1} + \mathbf{Z}_{Ic_1c_1} & \mathbf{Z}_{Ic_1c_2} \\ \mathbf{Z}_{Ic_2c_1} & \mathbf{Z}_{Bc_2c_2} + \mathbf{Z}_{Ic_2c_2} \end{bmatrix} = \mathbf{Y}_{Ccc}^{-1} \tag{2.49}$$

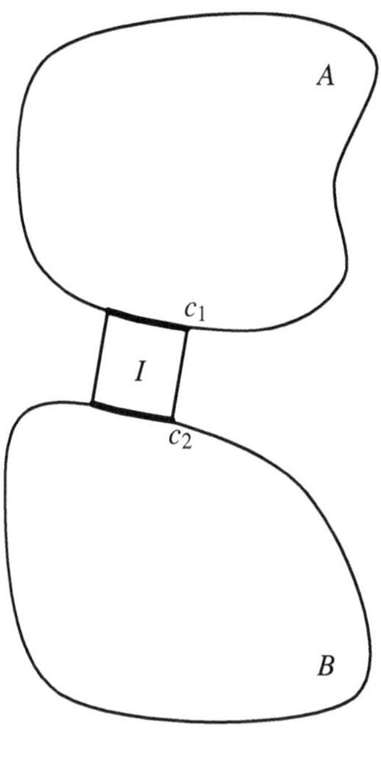

Figure 2.15 Isolator installed within arbitrary assembly with attached components A and B.

where it is recalled that the impedance matrix $\mathbf{Z}_{Ccc}$ is simply the inverse of the mobility matrix $\mathbf{Y}_{Ccc}$.

The remarkable conclusion from Equation 2.49 above is that the off-diagonal elements of the assembly's impedance matrix, obtained by inverting the measurable coupled mobility, are identically equal to those of the isolator alone ($\mathbf{Z}_{Ic_2c_1}$ or $\mathbf{Z}_{Ic_1c_2}$) and completely independent of the connected substructures A and B.

This important result can be explained from the definition of impedance: for the transfer impedance $\mathbf{Z}_{Ic_2c_1}$, the blocking force constraint at c_2 essentially removes the dynamic influence of component B, whilst at the same time, the velocity at c_1 is applied irrespective of the dynamics of A, and so neither A or B contribute to $\mathbf{Z}_{Ic_2c_1}$. Obviously, a similar argument applies in the other direction.

An important consequence of Equation 2.49 is that it provides a means of independently characterising a coupling element *in situ*, i.e. whilst installed within an arbitrary assembly. Furthermore, recalling that the interface DoFs c_1 and c_2 are assumed to include all 6 translational and rotational DoFs, the transfer impedance $\mathbf{Z}_{Ic_1c_2}$, by definition, constrains all the DoFs included in $\mathbf{Y}_C$. This enables the simultaneous characterisation of both in-plane and rotational transfer impedance terms. In essence, as was introduced in 2.2.8, the act of numerically inverting the mobility matrix $\mathbf{Y}_C$ constrains the DoFs that are included within it. If a particular DoF is omitted, it will not be constrained by the matrix inversion, and so the resulting impedance terms will not be subject to the correct set of constraints.

In summary, the in situ characterisation of a resilient element is performed as follows:

1) Install the element within a representative assembly. This might be a bespoke test rig that applies an appropriate pre-load, or an actual installation.
2) Instrument the interfaces above and below the isolator such that the DoFs of interest can be resolved.
3) Measure the coupled mobility by exciting above and below the element, and recording their simultaneous responses.
4) Apply appropriate interface transformation to obtain $\mathbf{Y}_C$ and invert numerically.
5) Extract the off-diagonal transfer impedance $\mathbf{Z}_{Ic_1c_2}$ or $\mathbf{Z}_{Ic_2c_1}$, which independently characterises the element I.

It is important to note that the in situ method, like the blocking mass methods described above, provides an estimate of only the *transfer impedance*, and that a complete characterisation also requires the *point impedance* matrices $\mathbf{Z}_{Ic_1c_1}$ and $\mathbf{Z}_{Ic_2c_2}$ to be known. As shown in Equation 2.49, the point impedance determined by the above procedure is that of the coupled *AI* or *IB* assemblies so a full characterisation cannot be obtained from *in situ* measurements alone. Methods to estimate the missing point impedances will be presented shortly, but first we present a brief example of *in situ* characterisation.

Example 2.2: In situ isolator characterisation

In this example the *in situ* method is used to characterise two simple cylindrical-type rubber mounts installed between various components.

The test set-up is as shown in Figure 2.16, where substructures A and B comprise masses, plates, and/ or beams. Readings from spaced accelerometers and forces applied with a hammer are averaged using the finite difference approximation (4.2.1) to yield the central translation-only mobility matrix $\mathbf{Y}_C$.

Shown in Figure 2.17 are the *transfer mobilities* across the first isolator whilst it is installed between two different sets of mass-like substructures (specific details of test the set-up can be found in [34]). It is clear that the measured transfer mobilities are dependent on the masses. On the other hand, the dynamic *transfer stiffness* (Figure 2.18), obtained from the same two assemblies by inversion of the mobility matrix, does not vary between the two assemblies; it is an independent isolator property as was proven in Equation 2.49 (note that we use dynamic stiffness rather than impedance here because it is easier to see the expected trends).

The stiffness itself displays many of the trends expected from an isolator of this type. In the low frequency range we see a relatively constant (frequency independent) stiffness value followed by a gradual upwards inflection towards a peak, representing an internal mount resonance. Above around 2 kHz the isolator attenuates the transmitted vibration (as is intended) such that the transfer mobilities $Y_{C c_1 c_2}$ and $Y_{C c_2 c_1}$ fall into the noise floor, and consequently, the dynamic stiffness becomes increasingly corrupted by noise.

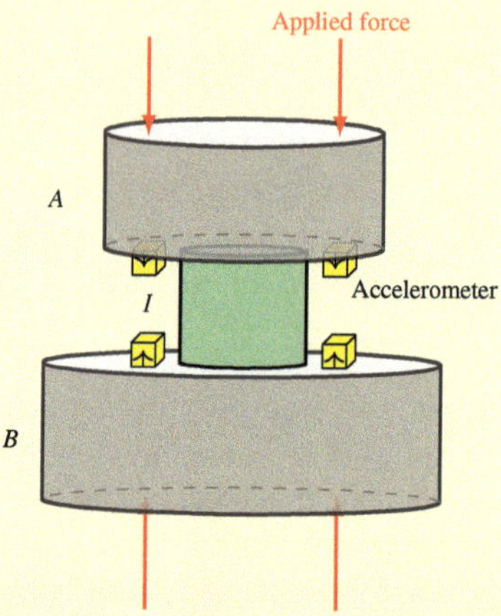

Figure 2.16 Test set-up for in situ characterisation with mass-like substructures.

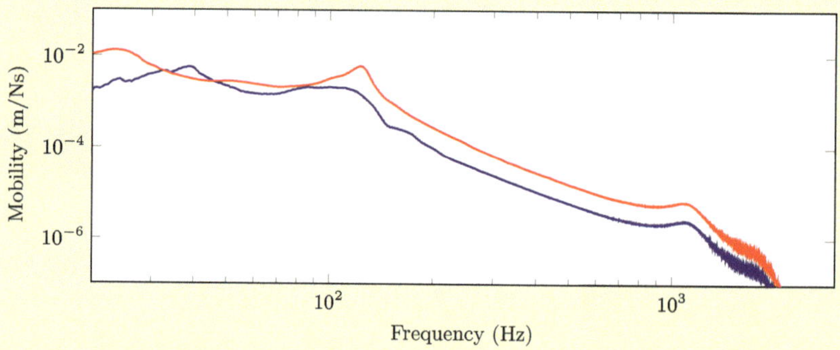

Figure 2.17 Transfer mobility across two MIM assemblies comprising different mass elements.

Shown in Figure 2.19 are the transfer mobilities across a similar rubber mount (slightly different geometry), whilst installed between: a) two masses (MIM, in blue), b) a mass and a perspex plate (MIP, in red), and c) a steel beam and a perspex plate (BIP, in green). As expected the measured mobilities display the characteristics of each assembly as a whole.

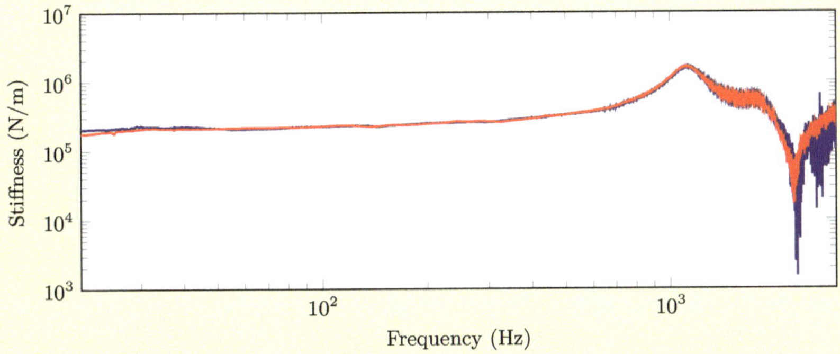

Figure 2.18 Dynamic transfer stiffnesses obtained from the same two MIM assemblies using the in situ method.

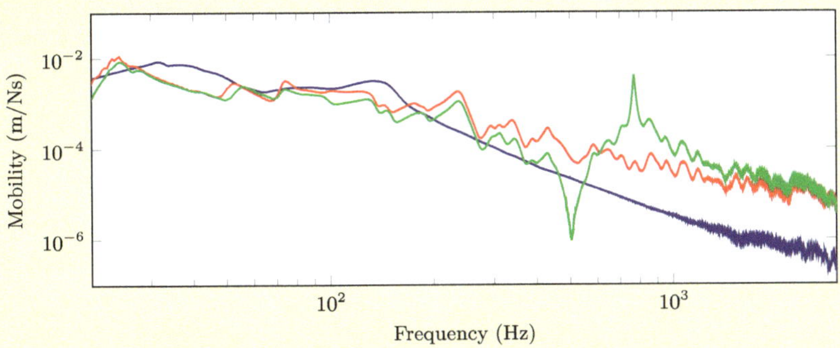

Figure 2.19 Transfer mobility across MIM (blue), MIP (red) and BIP (green) assemblies.

As shown in Figure 2.20 however, the dynamic transfer stiffness estimates from the same three set ups, obtained with the in situ method, are in agreement. It is clear though, that the lowly damped resonance of the steel beam has not been fully extracted by the inverse method, leaving some artefacts in the estimated stiffness. This, and other fluctuations in the stiffness can be traced to poor reciprocity in the measured mobility matrix [34].

Large artefacts, like those present in the BIP case, can potentially be removed using ad-hoc interpolation methods, or ameliorated by using different formulations of the in situ method, as will be illustrated in 2.5.

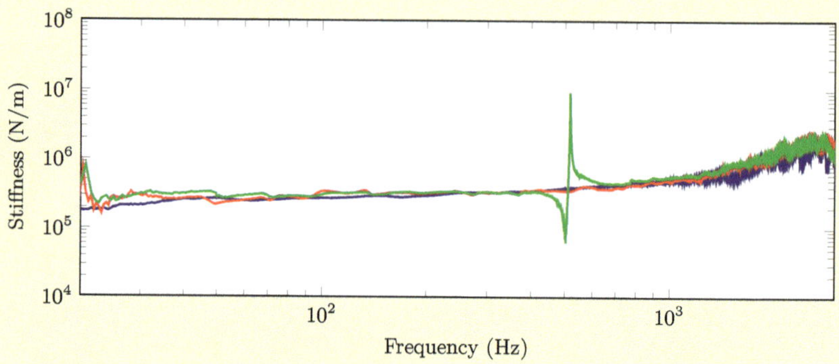

Figure 2.20 Dynamic transfer stiffnesses obtained from MIM (blue), MIP (red) and BIP (green) assemblies using in situ method.

As stated earlier, a complete isolator characterisation requires point impedances which are not available from the *in situ* method. In this case, three options are available: 1) the assumption of force conservation across the element, 2) substructure decoupling, or 3) in situ model updating, as discussed below.

Force conservation

If the coupling element can be assumed massless, i.e. it behaves like an ideal spring-damper, the interface forces ($\mathbf{g}$) acting on the element are conserved across its length such that,

$$\mathbf{g}_{c_1} = -\mathbf{g}_{c_2}. \tag{2.50}$$

Based on this assumption, the transfer impedance and point impedance are related as,

$$\mathbf{Z}_{Ic_1c_1} \approx -\mathbf{Z}_{Ic_1c_2}. \tag{2.51}$$

The entire coupling element impedance matrix can then be approximated by,

$$\begin{bmatrix} \mathbf{Z}_{Ic_1c_1} & \mathbf{Z}_{Ic_1c_2} \\ \mathbf{Z}_{Ic_2c_1} & \mathbf{Z}_{Ic_2c_2} \end{bmatrix} \approx \begin{bmatrix} -\mathbf{Z}_{Ic_1c_2} & \mathbf{Z}_{Ic_1c_2} \\ \mathbf{Z}_{Ic_2c_1} & -\mathbf{Z}_{Ic_2c_1} \end{bmatrix}. \tag{2.52}$$

Note that the above approximation is only valid where the coupling element satisfies the assumption of force conservation. This is largely the case for conventional vibration isolators at low frequencies when considering translational motions. A more detailed discussion can be found in [21].

Substructure decoupling

In cases where force conservation does not apply, for example at higher frequencies where internal isolator resonances or DoF cross-coupling become significant, the point impedance cannot be inferred from the transfer mobility. However, if the free interface mobility matrices of components A and B are known, then a decoupling procedure can be used, essentially by solving Equation 2.48 for the unknown isolator point mobilities (see also 2.3.1),

$$\begin{bmatrix} \mathbf{Z}_{Ic_1c_1} & \mathbf{Z}_{Ic_1c_2} \\ \mathbf{Z}_{Ic_2c_1} & \mathbf{Z}_{Ic_2c_2} \end{bmatrix} = \begin{bmatrix} \mathbf{Y}_{Cc_1c_1} & \mathbf{Y}_{Cc_1c_2} \\ \mathbf{Y}_{Cc_2c_1} & \mathbf{Y}_{Cc_2c_2} \end{bmatrix}^{-1} - \begin{bmatrix} \mathbf{Y}_{Ac_1c_1} & \mathbf{0} \\ \mathbf{0} & \mathbf{Y}_{Bc_2c_2} \end{bmatrix}^{-1}. \tag{2.53}$$

In practice, this method is most likely to be applied using specially designed components for A and B, see for example the cross-like fixtures used in [21]. Note that interface transformations may similarly be incorporated within this decoupling procedure (applied to both coupled C and uncoupled A and B measurements) to obtain the translation/rotational point impedance matrices.

In situ model updating

The in situ method is able extract the transfer characteristics of a coupling element from measurements made on the installed assembly. The missing point characteristics can be obtained by assuming force conservation, or by employing substructure decoupling. An alternative approach is to incorporate a numerical model and use so-called 'in situ model updating' [36].

In conventional model updating (see 2.6.2), experimental testing is performed on the uncoupled target component, and its free-interface FRF matrix is measured. This FRF matrix is used to update a numerical model directly, or to estimate modal parameters (natural frequencies and mode shapes) which are themselves used to update the numerical model. Key to this updating strategy is obtaining independent component FRFs by free suspension. For many components, resilient elements being an example, free suspension is simply not possible. The *in situ* method offers an alternative means of obtaining independent component properties without the need for free suspension, and so can provide an alternative route to model updating.

For a coupling element, the in situ model updating strategy is illustrated in Figure 2.21. A numerical model (labelled 'FE') of the coupling element (I) is built and its transfer impedance is computed (note that A and B are not modelled). The physical element (labelled 'exp') is installed in an appropriate manner, measured, and its transfer impedance obtained by the in situ method described above. The numerical and experimental transfer impedances are then used to formulate an appropriate cost function,

$$J_D(\theta) = J_D\left(\mathbf{Z}^{FE}_{Ic_1c_2}(\theta), \mathbf{Z}^{exp}_{Ic_1c_2}\right) \tag{2.54}$$

where θ is the vector of numerical model input parameters (for example, geometry, Young's modulus, density, etc.). This cost function describes the similarity between the FE and experimental transfer impedances and is minimised using an appropriate optimisation algorithm. An example cost function might be, for example, the sum of squared errors between the transfer impedance magnitudes,

$$J_D\left(\mathbf{Z}^{FE}_{Ic_1c_2}(\theta), \mathbf{Z}^{exp}_{Ic_1c_2}\right) = \sum_{ij}\left(|\mathbf{Z}^{FE}_{Ic_1c_2}| - |\mathbf{Z}^{exp}_{Ic_1c_2}|\right)^2_{ij} \tag{2.55}$$

where the summation is over all elements of the respective matrices.

From the optimised parameter vector θ, the numerical point impedances $\mathbf{Z}_{Ic_1c_1}^{FE}$ and $\mathbf{Z}_{Ic_2c_2}^{FE}$ can be computed and combined with the in situ experimental transfer impedances to obtain a complete component characterisation,

$$\mathbf{Z}_I = \begin{bmatrix} \mathbf{Z}_{Ic_1c_1}^{FE} & \mathbf{Z}_{Ic_1c_2}^{exp} \\ \mathbf{Z}_{Ic_2c_1}^{exp} & \mathbf{Z}_{Ic_2c_2}^{FE} \end{bmatrix}. \tag{2.56}$$

Alternatively, the numerically updated impedance matrix $\mathbf{Z}_I^{FE}$ can be used in its entirety to characterise the component.

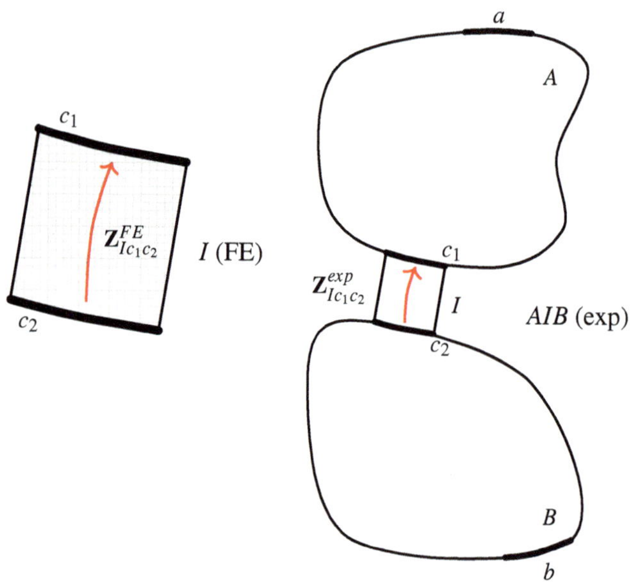

Figure 2.21 In situ model updating for a vibration isolator.

Having now considered characterisation of all component types, in the following sections we look further into indirect methods for obtaining the necessary data, starting with a discussion of transmissibility.

2.4 TRANSMISSIBILITY

In 2.3.1 and 2.3.2 we considered, respectively, the FRF-based characterisation of passive source/receiver-like components and resilient coupling elements. In this section, we introduce the closely related concept of transmissibility. Strictly speaking, transmissibility is a form of FRF, however, we will reserve this term for the type of cases encountered so far where input and output are different quantities, i.e. force-response or response-force. The term transmissibility will be used wherever input and output are of the same quantity, i.e. force-force or response-response. We have already seen that passive components can be fully characterised for our purposes by FRFs and therefore transmissibility will not generally be required as a means of representing components in a VAVP. However, it is an important intermediate quantity that will appear frequently throughout the book and our discussion of passive components would not be complete without introducing the concept and some of its multiple uses.

Among the advantages of transmissibility is that it can be obtained with 'output-only' measurements, i.e. with no need to apply a known excitation. This will be exploited for example in 5.3.5 in relation to 'operational TPA'. A further important property, valid only for the *generalised* transmissibility (see 2.4.2) is that, by careful definition of the DoFs involved, the transmissibility can be used to enforce constraints on a structure (in a manner similar to that described in 2.2.8). This can be exploited to extract independent component properties from *in situ* measurements (see also 2.6.2). Indeed, a transmissibility can be defined that is an independent component property and obtained in situ from output-only measurements. The transmissibility concept also appears in the derivation of the 'Round Trip' method (2.5) and, in the next chapter, in our discussion of embedded sources (3.2.2). However, we begin this section by briefly recounting the conventional transmissibility and its use in modelling and assessing the performance of isolators.

2.4.1 CONVENTIONAL TRANSMISSIBILITY

In general, a transmissibility is defined as the ratio of two like quantities, typically force or velocity, obtained at different locations. Noting that the velocity transmissibility is identical to that of displacement and acceleration, we refer to this form as simply the 'response' transmissibility. Consider the typical mass-spring-damper system in Figure 2.22, where the spring-dashpot represents a simple isolator. Excitation of the base (Figure 2.22a) represents the case where some sensitive object is isolated from vibration in the support; excitation of the mass (Figure 2.22b) represents the case where the support is isolated from a vibrating machine.

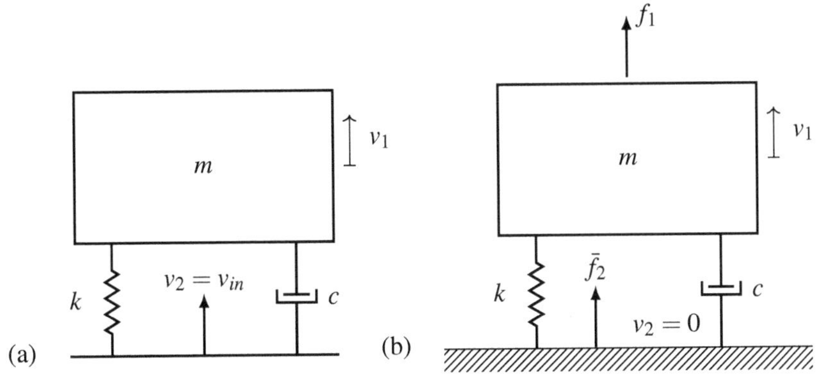

Figure 2.22 Single DoF mass-spring damper systems with velocity $v_1 = \dot{x}_1$. (a) Base-excited. (b) Mass-excited.

To better understand the notion of transmissibility, let us consider the response and force transmissibilites of the base and mass excited systems in Figure 2.22. These mass-spring systems are described by the equations of motion,

$$\begin{pmatrix} f_1 \\ f_2 \end{pmatrix} = \begin{bmatrix} m & 0 \\ 0 & 0 \end{bmatrix} \begin{pmatrix} \ddot{x}_1 \\ \ddot{x}_2 \end{pmatrix} + \begin{bmatrix} c & -c \\ -c & c \end{bmatrix} \begin{pmatrix} \dot{x}_1 \\ \dot{x}_2 \end{pmatrix} + \begin{bmatrix} k & -k \\ -k & k \end{bmatrix} \begin{pmatrix} x_1 \\ x_2 \end{pmatrix}. \tag{2.57}$$

Assuming periodic motion, Equation 2.57 can be rewritten in the frequency domain as,

$$\begin{pmatrix} f_1 \\ f_2 \end{pmatrix} = \left(i\omega \begin{bmatrix} m & 0 \\ 0 & 0 \end{bmatrix} + \begin{bmatrix} c & -c \\ -c & c \end{bmatrix} + \frac{1}{i\omega} \begin{bmatrix} k & -k \\ -k & k \end{bmatrix} \right) \begin{pmatrix} \dot{x}_1 \\ \dot{x}_2 \end{pmatrix} \tag{2.58}$$

or more compactly as,

$$\begin{pmatrix} f_1 \\ f_2 \end{pmatrix} = \begin{bmatrix} i\omega m + c + \frac{k}{i\omega} & -c - \frac{k}{i\omega} \\ -c - \frac{k}{i\omega} & c + \frac{k}{i\omega} \end{bmatrix} \begin{pmatrix} \dot{x}_1 \\ \dot{x}_2 \end{pmatrix} = \begin{bmatrix} Z_{11} & Z_{12} \\ Z_{21} & Z_{22} \end{bmatrix} \begin{pmatrix} \dot{x}_1 \\ \dot{x}_2 \end{pmatrix} \tag{2.59}$$

where Z_{11} and Z_{22} denote the point impedance of the mass and base, respectively, and $Z_{12} = Z_{21}$ is the transfer impedance across the spring-damper isolation system.

Considering first the *response* transmissibility of the base-excited system (Figure 2.22a). No external forces are applied to the mass, so $f_1 = 0$, and,

$$\begin{pmatrix} 0 \\ f_2 \end{pmatrix} = \begin{bmatrix} Z_{11} & Z_{12} \\ Z_{21} & Z_{22} \end{bmatrix} \begin{pmatrix} \dot{x}_1 \\ \dot{x}_2 \end{pmatrix}. \tag{2.60}$$

From the top row of Equation 2.60 we have,

$$Z_{11}\dot{x}_1 = -Z_{12}\dot{x}_2 \tag{2.61}$$

from which we can identify the response transmissibility as,

$$T^r_{12} = \frac{\dot{x}_1}{\dot{x}_2} = -\frac{Z_{12}}{Z_{11}}. \tag{2.62}$$

The transmissibility T^r_{12} (where r indicates a response transmissibility) can be interpreted as an FRF that relates the base response to that of the isolated mass,

$$\dot{x}_1 = T^r_{12}\dot{x}_2. \tag{2.63}$$

Substituting the relevant mass, stiffness and damping parameters into Equation 2.63, and evaluating its magnitude, we obtain the classical transmissibility curves as shown in Figure 2.23. At low frequencies (below resonance $\omega < \omega_c$) we see the transmissibility tend to 1; the isolated mass moves in unison with the base excitation. At high frequencies (above resonance $\omega > \omega_c$) we see the transmissibility decreasing with frequency; the mass response is smaller in magnitude than the base excitation. In this region, $T^r_{12} < 1$, the isolation system is seen to attenuate the base excitation. Between the low and high frequency limits (in the vicinity of resonance ω_c) we see a peak in the transmissibility whose maximum amplitude is dependent on the level of damping. Around this peak $T^r_{12} > 1$, which indicates an amplification of vibration across the isolation system; a small base excitation can yield a large mass response. When designing an isolation system it is important that the system is designed such that the operating frequency, or range thereof, remains within the attenuation regime of the transmissibility curve.

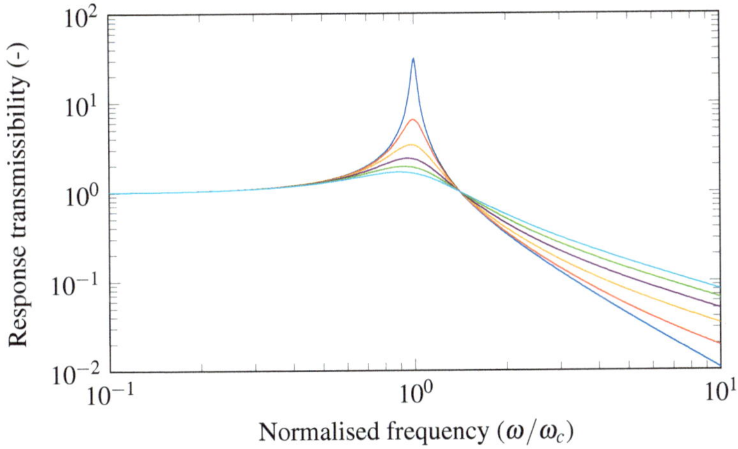

Figure 2.23 Response transmissibility magnitude for a base excited mass-spring-damper system with increased damping ($c = 1, 5, 10, 15, 20, 25$).

Let us now consider the *force* transmissibility of the mass-excited case (Figure 2.22b). To do so we consider the force transmitted to a stationary, or fixed, base. The system's equations of motions then become,

$$\begin{pmatrix} f_1 \\ \bar{f}_2 \end{pmatrix} = \begin{bmatrix} Z_{11} & Z_{12} \\ Z_{21} & Z_{22} \end{bmatrix} \begin{pmatrix} \dot{x}_1 \\ 0 \end{pmatrix} \tag{2.64}$$

where $\bar{f}_2$ is the blocking force required to constrain the base such that $\dot{x}_2 = 0$. Combining the top and bottom rows from Equation 2.64 we obtain,

$$f_1 = Z_{11}\dot{x}_1 = Z_{11}\frac{\bar{f}_2}{Z_{21}} \tag{2.65}$$

from which we can identify the force transmissibility as,

$$T^f_{21} = \frac{\bar{f}_2}{f_1} = \frac{Z_{21}}{Z_{11}}. \tag{2.66}$$

The force transmissibility T^f_{21} can similarly be interpreted as that relating the external force on the mass to the blocking force on the fixed base,

$$\bar{f}_2 = T^f_{21} f_1. \tag{2.67}$$

Comparing Equations 2.62 and 2.66 we see that $T^f_{21} = T^r_{12}$, so substituting the relevant mass, stiffness and damping parameters into Equation 2.66 yields the same set of transmissibility curves as in Figure 2.23; the response transmissibility is identical to the force transmissibility in the 'opposite direction' (i.e. base to mass vs. mass to base).

Transmissibility as an isolator characterisation

Whilst transmissibility provides a convenient means of assessing the performance of an isolation system, it is important to note that it is not an independent property of the isolator itself (i.e. the spring-damper) but includes properties of the isolated component (often a mass). A change in the mass of this component will thus cause the transmissibility peak to shift in frequency. The transmissibility is, however, independent of the dynamic properties of the base which, in this simple analysis, is either considered fixed or assigned a constant velocity.

2.4.2 GENERALISED TRANSMISSIBILITY

Having introduced the force and response transmissibility of a single DoF system, we are ready to generalise these concepts to the multi-DoF case [47].

The generalised *force* transmissibility T^f_{ij} between two points is defined here as the relation between a single force f_j applied at the DoF $j \in M$, and the blocking force $\bar{f}_i$ at a response DoF $i \in N$ with all other external forces set to zero $f_{k \neq j} = 0$,

$$\begin{pmatrix} \bar{f}_1 \\ \vdots \\ \bar{f}_N \end{pmatrix} = \begin{bmatrix} T^f_{11} & \cdots & T^f_{1M} \\ \vdots & \ddots & \vdots \\ T^f_{N1} & \cdots & T^f_{NM} \end{bmatrix} \begin{pmatrix} f_1 \\ \vdots \\ f_M \end{pmatrix} \tag{2.68}$$

with the element-wise definition,

$$T^f_{ij} = \left. \frac{\bar{f}_i}{f_j} \right|_{f_{k \neq j} = 0} . \tag{2.69}$$

Note that the excitation and blocking force DoFs belong to different sets ($j \in M$ and $i \in N$), and that if several blocking DoFs are considered, a constraint is applied to all, $v_{i \in N} = 0$. Equation 2.68 is simply an extension of Equation 2.66 to the multi-DoF case.

The generalised *response* transmissibility T^r_{ij} describes the relation between the applied velocity v_j at DoF $j \in N$, and the response v_i at DoF $i \in M$, whilst all other $k \neq j$ excitation DoFs are subject to a rigid constraint $v_{k \neq j} = 0$,

$$\begin{pmatrix} v_1 \\ \vdots \\ v_M \end{pmatrix} = \begin{bmatrix} T^r_{11} & \cdots & T^r_{1N} \\ \vdots & \ddots & \vdots \\ T^r_{M1} & \cdots & T^r_{MN} \end{bmatrix} \begin{pmatrix} v_1 \\ \vdots \\ v_N \end{pmatrix} \tag{2.70}$$

with the element-wise definition,

$$T^r_{ij} = \left.\frac{v_i}{v_j}\right|_{v_{k\neq j}=0}. \tag{2.71}$$

Again, the two response DoFs belong to different sets ($j \in N$ and $i \in M$).

These generalised force and response transmissibilities are illustrated in Figure 2.24.

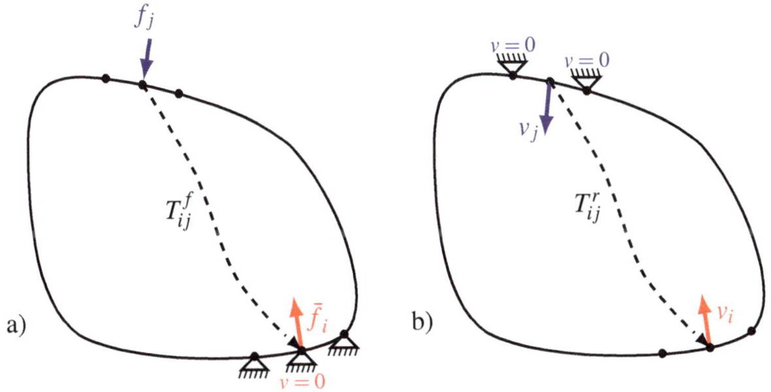

Figure 2.24 Generalised force (a) and response (b) transmissibility constraints.

Note that there is a clear symmetry between the force and response transmissibility. Defined in terms of a blocking force, implicit to the force transmissibility are a set of rigid constraints. Similarly, by definition the response transmissibility requires the rigid constraint of all DoFs $k \neq j$. Indeed, it has been shown that for multi-DoF structures the force and response transmissibilities are related through simple matrix manipulations, as will be described shortly [28].

An important consequence of the blocking constraints present in the definitions above is that, if defined appropriately, the transmissibility can become an independent component property. Specifically, if the blocking constraints are applied to the DoFs that separate the component of interest from the remainder of its assembly, the dynamics of neighbouring components are blocked and so do not the effect the transmissibility.

Force transmissibility

To derive some useful expressions for the generalised transmissibility we begin by considering component A of the coupled assembly in Figure 2.25. Three sets of DoFs are considered; the remote set a with N_a DoFs located internal to component A, the remote set b with N_b DoFs located internal to component B, and the interface set c with N_c DoFs. The equations of motion that govern the coupled AB assembly, in impedance form, are given by,

$$\begin{pmatrix} \mathbf{f}_a \\ \mathbf{f}_c \\ \mathbf{f}_b \end{pmatrix} = \begin{bmatrix} \mathbf{Z}_{Aaa} & \mathbf{Z}_{Aac} & \mathbf{0} \\ \mathbf{Z}_{Aca} & \mathbf{Z}_{Acc}+\mathbf{Z}_{Bcc} & \mathbf{Z}_{Acb} \\ \mathbf{0} & \mathbf{Z}_{Acb} & \mathbf{Z}_{Abb} \end{bmatrix} \begin{pmatrix} \mathbf{v}_a \\ \mathbf{v}_c \\ \mathbf{v}_b \end{pmatrix}. \tag{2.72}$$

We consider the case where an external force is applied at the remote DoFs a, whilst the interface c is constrained by a blocking force such that $\mathbf{v}_c = \mathbf{0}$ (and consequently $\mathbf{v}_b = \mathbf{0}$). The equation of motion then becomes,

$$\begin{pmatrix} \mathbf{f}_a \\ \bar{\mathbf{f}}_c \\ \mathbf{0} \end{pmatrix} = \begin{bmatrix} \mathbf{Z}_{Aaa} & \mathbf{Z}_{Aac} & \mathbf{0} \\ \mathbf{Z}_{Aca} & \mathbf{Z}_{Acc}+\mathbf{Z}_{Bcc} & \mathbf{Z}_{Acb} \\ \mathbf{0} & \mathbf{Z}_{Acb} & \mathbf{Z}_{Abb} \end{bmatrix} \begin{pmatrix} \mathbf{v}_a \\ \mathbf{0} \\ \mathbf{0} \end{pmatrix}. \tag{2.73}$$

Note that this blocking force effectively removes the influence of component B from any quantities derived henceforth.

From the top row of Equation 2.73 we obtain,

$$\mathbf{v}_a = \mathbf{Z}_{Aaa}^{-1}\mathbf{f}_a \tag{2.74}$$

which upon substitution into the second row yields,

$$\bar{\mathbf{f}}_c = \mathbf{Z}_{Aca}\mathbf{Z}_{Aaa}^{-1}\mathbf{f}_a. \tag{2.75}$$

The matrix product in Equation 2.75 may be interpreted as a generalised force transmissibility. It relates an applied force $\mathbf{f}_a$ at the remote DoFs a, to the blocked force necessary to constrain the interface DoFs c. We define,

$$\mathbf{T}_{Aca}^{f,a} = \mathbf{Z}_{Aca}\mathbf{Z}_{Aaa}^{-1} \tag{2.76}$$

such that,

$$\bar{\mathbf{f}}_c = \mathbf{T}_{Aca}^{f,a}\mathbf{f}_a \tag{2.77}$$

where $\mathbf{T}_{Aca}^{f,a} \in \mathbb{C}^{N_c \times N_a}$ is the force transmissibility, with the superscript a indicating it has been defined in terms of a shared response at a. Note that Equation 2.76 reduces to Equation 2.66 in the single DoF case, and so represents its multi-DoF generalisation.

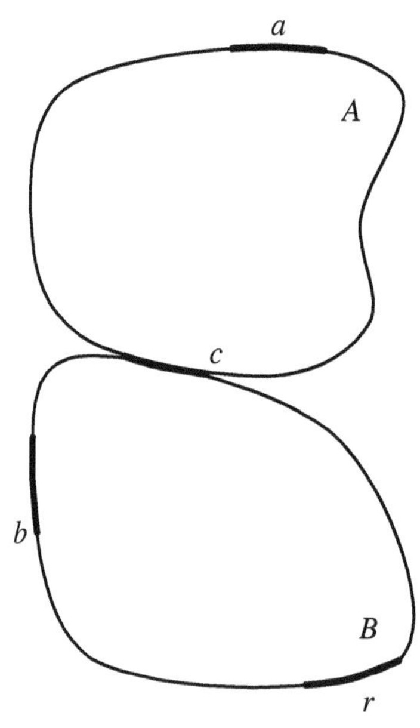

Figure 2.25 *AB* assembly.

It is clear from Equation 2.76 that the transmissibility is an independent property of component A, being defined as the product of its transfer and (inverse) point impedance. An equivalent transmissibility relation can be obtained for component B by considering an external force applied at b instead,

$$\mathbf{T}_{Bcb}^{f,b} = \mathbf{Z}_{Bcb}\mathbf{Z}_{Bbb}^{-1} \tag{2.78}$$

such that,

$$\bar{\mathbf{f}}_c = \mathbf{T}^{f,b}_{Bcb}\mathbf{f}_b \tag{2.79}$$

with $\mathbf{T}^{f,b}_{Bcb} \in \mathbb{C}^{N_c \times N_b}$.

Whilst informative, Equations 2.76 and 2.78 are of limited practical value as impedance matrices are not directly available by measurement. It is therefore convenient to derive the force transmissibility in terms of mobilities instead.

It is shown in 3.2.1 that the response field within a component (say B) can be reproduced identically by applying the (negative) blocking force along its connecting interface, in place of the original excitation. That is, given an external (or internal) force applied to A, $\mathbf{f}_a$, we can express the velocity at b and c directly as,

$$\mathbf{v}_b = \mathbf{Y}_{Cba}\mathbf{f}_a \tag{2.80}$$

$$\mathbf{v}_c = \mathbf{Y}_{Cca}\mathbf{f}_a \tag{2.81}$$

or equivalently using the blocked force in the following form,

$$\mathbf{v}_b = -\mathbf{Y}_{Cbc}\bar{\mathbf{f}}_c \tag{2.82}$$

$$\mathbf{v}_c = -\mathbf{Y}_{Ccc}\bar{\mathbf{f}}_c \tag{2.83}$$

where $\bar{\mathbf{f}}_c$ is the force required to restrain the interface motion generated by the external force $\mathbf{f}_a$. Equating Equations 2.80 and 2.82 we obtain,

$$\mathbf{v}_b = -\mathbf{Y}_{Cbc}\bar{\mathbf{f}}_c = \mathbf{Y}_{Cba}\mathbf{f}_a \tag{2.84}$$

which rearranges to,

$$\bar{\mathbf{f}}_c = -\mathbf{Y}^{-1}_{Cbc}\mathbf{Y}_{Cba}\mathbf{f}_a = \mathbf{T}^{f,b}_{Aca}\mathbf{f}_a \tag{2.85}$$

where $\mathbf{T}^{f,b}_{Aca}$ is the same force transmissibility as obtained from Equation 2.76. Note that for the force transmissibility the superscript b indicates that its definition is through a shared response at b.

Similarly, equating Equations 2.81 and 2.83 we obtain,

$$\mathbf{v}_c = -\mathbf{Y}_{Ccc}\bar{\mathbf{f}}_c = \mathbf{Y}_{Cca}\mathbf{f}_a \tag{2.86}$$

which rearranges to,

$$\bar{\mathbf{f}}_c = -\mathbf{Y}^{-1}_{Ccc}\mathbf{Y}_{Cca}\mathbf{f}_a = \mathbf{T}^{f,c}_{Aca}\mathbf{f}_a. \tag{2.87}$$

Note that Equation 2.85 and 2.87, though based on different mobility terms, yield the same transmissibility, that is $\mathbf{T}^{f,b}_{Aca} = \mathbf{T}^{f,c}_{Aca}$. Furthermore, recalling that a blocking constraint is imposed on the interface c, an identical result would be obtained from the uncoupled component A,

$$\bar{\mathbf{f}}_c = -\mathbf{Y}^{-1}_{Acc}\mathbf{Y}_{Aca}\mathbf{f}_a = \mathbf{T}^{f,c}_{Aca}\mathbf{f}_a. \tag{2.88}$$

And so for the force transmissibility of component A we have that,

$$\mathbf{T}^{f}_{Aca} = -\mathbf{Y}^{-1}_{Cbc}\mathbf{Y}_{Cba} = -\mathbf{Y}^{-1}_{Ccc}\mathbf{Y}_{Cca} = -\mathbf{Y}^{-1}_{Acc}\mathbf{Y}_{Aca}. \tag{2.89}$$

Considering an external force applied within B at the DoFs b, a similar set of transmissibility relations can be obtained,

$$\bar{\mathbf{f}}_c = -\mathbf{Y}_{Cac}^{-1}\mathbf{Y}_{Cab}\mathbf{f}_b = \mathbf{T}_{Bcb}^{f,a}\mathbf{f}_b \tag{2.90}$$

$$\bar{\mathbf{f}}_c = -\mathbf{Y}_{Ccc}^{-1}\mathbf{Y}_{Ccb}\mathbf{f}_b = -\mathbf{Y}_{Bcc}^{-1}\mathbf{Y}_{Bcb}\mathbf{f}_b = \mathbf{T}_{Bcb}^{f,c}\mathbf{f}_b. \tag{2.91}$$

Note that the blocked force in Equation 2.90 and 2.91 is different to that of Equation 2.87 and 2.88, since it is the result of a different external force (i.e. $\mathbf{f}_a$, not $\mathbf{f}_b$).

Response transmissibility

With reference to Equation 2.73, the force transmissibility can be interpreted as the relation between forces $\mathbf{f}_a$ and $\bar{\mathbf{f}}_c$, due to the applied velocity $\mathbf{v}_a$. We can similarly consider a response transmissibility due to an applied force. Consider an external force applied within B, at DoFs b. The responses at DoFs a and c due to this applied force are defined as,

$$\mathbf{v}_a = \mathbf{Y}_{Cab}\mathbf{f}_b, \qquad \mathbf{v}_c = \mathbf{Y}_{Ccb}\mathbf{f}_b. \tag{2.92}$$

Equating the forces $\mathbf{f}_b$ between $\mathbf{v}_a$ and $\mathbf{v}_c$, we obtain,

$$\mathbf{v}_a = \mathbf{Y}_{Cab}\mathbf{Y}_{Ccb}^{-1}\mathbf{v}_c = \mathbf{T}_{Aac}^{r,b}\mathbf{v}_c \tag{2.93}$$

where $\mathbf{T}_{Aac}^{r,b} \in \mathbb{C}^{N_a \times N_c}$ is the response transmissibility between c and a. Here we assume that $N_c = N_b$ such that the matrix inverse exists, though the above can easily be extended to cases where $N_c \neq N_b$ using the generalised pseudo inverse. Note that for the response transmissibility the superscript b indicates a definition through a shared force at b.

Considering an external force applied instead to the interface c, we can compute the velocity at a, c or b,

$$\mathbf{v}_a = \mathbf{Y}_{Cac}\mathbf{f}_c, \qquad \mathbf{v}_c = \mathbf{Y}_{Ccc}\mathbf{f}_c, \qquad \mathbf{v}_b = \mathbf{Y}_{Cbc}\mathbf{f}_c. \tag{2.94}$$

Rearranging the above we obtain the following transmissibility relations,

$$\mathbf{v}_a = \mathbf{Y}_{Cac}\mathbf{Y}_{Ccc}^{-1}\mathbf{v}_c = \mathbf{T}_{Aac}^{r,c}\mathbf{v}_c \tag{2.95}$$

$$\mathbf{v}_b = \mathbf{Y}_{Cbc}\mathbf{Y}_{Ccc}^{-1}\mathbf{v}_c = \mathbf{T}_{Bbc}^{r,c}\mathbf{v}_c. \tag{2.96}$$

If one instead considers an external force within A at DoFs a,

$$\mathbf{v}_c = \mathbf{Y}_{Cca}\mathbf{f}_a, \qquad \mathbf{v}_b = \mathbf{Y}_{Cba}\mathbf{f}_a \tag{2.97}$$

we obtain the transmissibility relation,

$$\mathbf{v}_b = \mathbf{Y}_{Cba}\mathbf{Y}_{Cca}^{-1}\mathbf{v}_c = \mathbf{T}_{Bbc}^{r,a}\mathbf{v}_c. \tag{2.98}$$

Invariance of response transmissibility

Being based on the blocked force, it is straightforward to see how the force transmissibility becomes an invariant component property. It is less clear that the response transmissibility can also be made an independent component property. This is made clearer by considering the nature of the external force employed in its definition. Taking Equations 2.97 and 2.98 as an example, the external forcing on A can be represented equivalently by the (negative) blocked force or contact force at the separating interface,

$$\mathbf{v}_b = \mathbf{Y}_{Cba}\mathbf{f}_a = -\mathbf{Y}_{Cbc}\bar{\mathbf{f}}_c = \mathbf{Y}_{Bbc}\mathbf{g}_c \quad (2.99)$$

$$\mathbf{v}_c = \mathbf{Y}_{Cca}\mathbf{f}_a = -\mathbf{Y}_{Ccc}\bar{\mathbf{f}}_c = \mathbf{Y}_{Bcc}\mathbf{g}_c \quad (2.100)$$

where $\bar{\mathbf{f}}_c$ and $\mathbf{g}_c$ are, respectively, the blocked and contact forces at the interface c due to the forcing $\mathbf{f}_a$ (see Chapter 5).

Use of the blocking force clearly introduces a rigid constraint at the interface c between A and B, effectively removing the dynamics of A. For the contact force, we see that the mobility terms are of B alone, A does not contribute. Hence, from Equation 2.99 and 2.100 we see that the same transmissibility is obtained from,

$$\mathbf{T}^{r,a/c}_{Bbc} = \mathbf{Y}_{Cba}\mathbf{Y}^{-1}_{Cca} = \mathbf{Y}_{Cbc}\mathbf{Y}^{-1}_{Ccc} = \mathbf{Y}_{Bbc}\mathbf{Y}^{-1}_{Bcc}. \quad (2.101)$$

The invariance of the response transmissibility to the exact position of the excitation (providing that it resides within component A) will form the basis of an output-only transmissibility measurement, as discussed shortly.

Perhaps surprisingly, we see that the position of the *response* DoFs in B is also arbitrary. Choosing b and r instead of b and c as our response locations,

$$\mathbf{v}_b = \mathbf{Y}_{Cba}\mathbf{f}_a = -\mathbf{Y}_{Cbc}\bar{\mathbf{f}}_c = \mathbf{Y}_{Bbc}\mathbf{g}_c \quad (2.102)$$

$$\mathbf{v}_r = \mathbf{Y}_{Cra}\mathbf{f}_a = -\mathbf{Y}_{Crc}\bar{\mathbf{f}}_c = \mathbf{Y}_{Brc}\mathbf{g}_c \quad (2.103)$$

we still obtain an independent transmissibility for component B,

$$\mathbf{v}_b = \mathbf{T}^{r,a}_{Bbr}\mathbf{v}_r = \mathbf{Y}_{Cba}\mathbf{Y}^{-1}_{Cra}\mathbf{v}_r = \mathbf{Y}_{Cbc}\mathbf{Y}^{-1}_{Crc}\mathbf{v}_r = \mathbf{Y}_{Bbc}\mathbf{Y}^{-1}_{Brc}\mathbf{v}_r. \quad (2.104)$$

In summary, we have seen that if a component B is excited entirely through the interface DoFs c, the response transmissibility between any two sets of DoFs becomes an independent property of that component. That is, the same transmissibility would be obtained if the component were isolated (A being removed) and excited externally at its interface. Clearly this has implications for obtaining independent component properties from in situ measurements.

It is worth noting that for Equation 2.104, the above independence relies on $N_r \geq N_c$, such that the interface is completely observable from the DoFs r.

Relation between force and response transmissibility

Inspecting the transmissibilities derived above, it can be seen that the force and response transmissibilities are closely related. Indeed, we have the following relations,

$$\mathbf{T}^{f,a/c}_{Bcb} = -\left(\mathbf{T}^{r,a/c}_{Bbc}\right)^{\mathrm{T}} \quad \begin{cases} \mathbf{T}^{r,a/c}_{Bbc} = \mathbf{Y}_{Cba}\mathbf{Y}^{-1}_{Cca} = \mathbf{Y}_{Cbc}\mathbf{Y}^{-1}_{Ccc} = \mathbf{Y}_{Bbc}\mathbf{Y}^{-1}_{Bcc} \\ \mathbf{T}^{f,a/c}_{Bcb} = -\mathbf{Y}^{-1}_{Cac}\mathbf{Y}_{Cab} = -\mathbf{Y}^{-1}_{Ccc}\mathbf{Y}_{Ccb} = -\mathbf{Y}^{-1}_{Bcc}\mathbf{Y}_{Bcb} \end{cases} \tag{2.105}$$

$$\mathbf{T}^{f,b/c}_{Aca} = -\left(\mathbf{T}^{r,b/c}_{Aac}\right)^{\mathrm{T}} \quad \begin{cases} \mathbf{T}^{r,b/c}_{Aac} = \mathbf{Y}_{Cab}\mathbf{Y}^{-1}_{Ccb} = \mathbf{Y}_{Cac}\mathbf{Y}^{-1}_{Ccc} = \mathbf{Y}_{A_{ac}}\mathbf{Y}^{-1}_{A_{cc}} \\ \mathbf{T}^{f,b/c}_{Aca} = -\mathbf{Y}^{-1}_{Cbc}\mathbf{Y}_{Cba} = -\mathbf{Y}^{-1}_{Ccc}\mathbf{Y}_{Cca} = -\mathbf{Y}^{-1}_{Acc}\mathbf{Y}_{Aca}. \end{cases} \tag{2.106}$$

The force are response transmissibilities, defined in opposite directions, are related by negative transposition. The response transmissibility $\mathbf{T}^{r,b/c}_{Aac}$ has the constraint on its input DoFs c, whilst the force transmissibility $\mathbf{T}^{f,b/c}_{Aca}$ has its constraint on the output DoFs, also c (see Figure 2.24). Transposing has the effect of swapping the force/response variable, reversing the direction of the transmissibility, and moving the constraint from output to input (or visa versa).

Output-only measurement

We now consider the indirect measurement of transmissibility using *output-only* measurements. We will also show how this opens up the interesting possibility that under some conditions transmissibility can be measured using *operational excitation* i.e. without any externally applied forces.

To extend the definition of transmissibility to output-only quantities it is sufficient to consider N linearly independent operational states of the assembly AB, with component A containing the set of unknown forcing mechanisms. From the perspective of component B, each operational state, i, can be represented by an equivalent blocked force vector $\bar{\mathbf{f}}^{(i)}_c$. Arranging the N_i blocked force vectors as the columns of a matrix we arrive at the blocked force matrix $\bar{\mathbf{F}}_c = \left[\begin{array}{ccc} \bar{\mathbf{f}}^{(1)}_c & \dots & \bar{\mathbf{f}}^{(N_i)}_c \end{array}\right] \in \mathbb{C}^{N_c \times N_i}$. Equations 2.99 and 2.100 can thus be rewritten as,

$$\mathbf{V}_b = \mathbf{Y}_{Cbc}\bar{\mathbf{F}}_c \tag{2.107}$$

and

$$\mathbf{V}_c = \mathbf{Y}_{Ccc}\bar{\mathbf{F}}_c \tag{2.108}$$

and Equation 2.96 as,

$$\mathbf{V}_b = \mathbf{Y}_{Cbc}\mathbf{Y}^{-1}_{Ccc}\mathbf{V}_c, \tag{2.109}$$

where $\mathbf{V}_c \in \mathbb{C}^{N_c \times N_i}$ and $\mathbf{V}_b \in \mathbb{C}^{N_b \times N_i}$ represent operational response matrices at, respectively, the interface c and remote b DoFs of component B and whose columns correspond to the independent force distributions in $\bar{\mathbf{F}}_c$. From Equation 2.109, assuming for now that $N_c = N_i$, it is straightforward to identify the response transmissibility matrix,

$$\mathbf{T}^{r,c}_{Bbc} = \mathbf{Y}_{Cbc}\mathbf{Y}^{-1}_{Ccc} = \mathbf{V}_b\mathbf{V}_c^{-1}. \tag{2.110}$$

Similarly, using Equation 2.105 we can define the force transmissibility matrix as,

$$\mathbf{T}^{f,c}_{Bcb} = -\left(\mathbf{T}^{r,c}_{Bbc}\right)^{\mathrm{T}} = -\left(\mathbf{V}_b\mathbf{V}_c^{-1}\right)^{\mathrm{T}}. \tag{2.111}$$

Equations 2.110 and 2.111 provide output only definitions of the generalised transmissibilities $\mathbf{T}^{r,c}_{Bbc}$ and $\mathbf{T}^{f,c}_{Bcb}$, respectively, requiring measurement of operational responses only. Recall that this transmissibility is in fact an invariant property of the component B, i.e. the passive component. If A were to be replaced by an entirely different component, the same transmissibility would be obtained.

It is important to note that the implementation of Equation 2.110 or 2.111 requires that rank$[\mathbf{V}_c] = N_c$, that is the interface response matrix $\mathbf{V}_c$ should be full rank. If $\mathbf{V}_c$ is rank deficient, or close to, its inverse cannot be computed or will be highly sensitive to small errors (i.e. ill-conditioned). A full rank response will only be achieved if each operational state produces a linearly independent response. For this reason, it is often beneficial to over-determine the problem by measuring $N_i > N_c$ operational states. The matrix inverse is then replaced by the generalised inverse $\mathbf{V}_c^+$, leading to a least-squares estimate of the transmissibility (see Chapter 7 for further discussion).

Whilst in the above we considered blocked forces at the interface c, these represent, in an equivalent manner, an arbitrary forcing within the component A. Clearly, this opens up the use of internal *operational forces* as the means of excitation. These forces might be generated by some controllable mechanism within the component, for example if A were a vibration source, or by some ambient vibration over which there is limited control. In either case, the columns of the response matrices would represent measurements made at different times and under different operational conditions. As before, this comes with the proviso that the operational excitation must be able to generate an interface response matrix of full rank. In practice, the more linearly independent source mechanisms there are inside the source then the more likely this condition is to be met. Running a source at different speeds and loads will tend to increase the variation in the operational response, but may not be sufficient to obtain full rank. Of course, external excitations can be used in conjunction with operational loading to help achieve full rank response matrices. For more details on the use of output only transmissibilities the reader is referred to [62].

One interesting advantage of the output only approach described above is that operational forces can be much larger in magnitude than those generated during a standard FRF test. For example, consider the forces generated by ship engine, compared to those with an impact hammer. In this sense, vibration sources can be used as heavy duty *exciters* for the purpose of measuring long distance transmissibility functions.

As a final comment, it is worth noting that the output only identification of transmissibility is a special case of a more general Multi-Input Multi-Output (MIMO) FRF identification, as will be discussed in 7.3.1.

2.5 ROUND TRIP METHOD

The Round Trip method is an indirect way to obtain FRFs that avoids the need to excite the structure at certain 'difficult' DoFs in return for applying additional excitations at other more convenient locations. It is particularly useful where access to measurement points is restricted. In such cases, while response sensors are relatively small, this is not the case for impact hammers or shakers so getting access to apply a reliable controlled force for FRF measurement can be problematic. For remote points, one may avoid this issue by invoking reciprocity but, for interfaces, excitation at every DoF is usually required and the Round Trip method may be the only option for measurement. A second important application is where some DoFs on an interface are difficult to excite for other reasons, for example, in-plane forces can never really be applied to plate-like structures in a reliable way to obtain the corresponding FRFs no matter how good the access. In both these cases one must look to some indirect method to obtain the necessary data. Possible indirect methods mentioned so far include substructure decoupling (2.3.2 and Chapter 5) or supplementing measurement with a model as described in 2.6. The Round Trip method provides a powerful indirect approach, entirely based on measurement, that has been shown to yield excellent results in practice. Since the theory is rather general for linear, time-invariant systems there is a wide range of potential applications, but we will focus here on its application to the indirect measurement of interface FRFs.

In the following we first derive the mathematical identity at the heart of the method (2.5.1), before going on to discuss its implementation in 2.5.2, with an example application to isolator characterisation. Finally, we discuss some finer details of the derivation and its implementation in 2.5.4.

2.5.1 THE ROUND TRIP IDENTITY

The core of the Round Trip method is a mathematical identity which has been derived in various places mostly by drawing on physical arguments [38, 39, 59]. Here we provide a more general mathematical proof not currently available in the literature.

Consider the coupled assembly in Figure 2.25. Three sets of DoFs are of interest: the remote set a with N_a DoFs located internal to component A, the remote set b internal to component B with N_b DoFs and the interface set c with N_c DoFs.

As stated earlier in this chapter, the impedance matrix for an assembly such as this can be constructed from the sum of the impedance matrices of the two components,

$$\begin{bmatrix} \mathbf{Z}_{Caa} & \mathbf{Z}_{Cac} & \mathbf{0} \\ \mathbf{Z}_{Cca} & \mathbf{Z}_{Ccc} & \mathbf{Z}_{Ccb} \\ \mathbf{0} & \mathbf{Z}_{Cbc} & \mathbf{Z}_{Cbb} \end{bmatrix} = \begin{bmatrix} \mathbf{Z}_{Aaa} & \mathbf{Z}_{Aac} & \mathbf{0} \\ \mathbf{Z}_{Aca} & \mathbf{Z}_{Acc} & \mathbf{0} \\ \mathbf{0} & \mathbf{0} & \mathbf{0} \end{bmatrix} + \begin{bmatrix} \mathbf{0} & \mathbf{0} & \mathbf{0} \\ \mathbf{0} & \mathbf{Z}_{Bcc} & \mathbf{Z}_{Bcb} \\ \mathbf{0} & \mathbf{Z}_{Bbc} & \mathbf{Z}_{Bbb} \end{bmatrix}$$
$$= \begin{bmatrix} \mathbf{Z}_{Aaa} & \mathbf{Z}_{Aac} & \mathbf{0} \\ \mathbf{Z}_{Aca} & \mathbf{Z}_{Acc}+\mathbf{Z}_{Bcc} & \mathbf{Z}_{Bcb} \\ \mathbf{0} & \mathbf{Z}_{Bbc} & \mathbf{Z}_{Bbb} \end{bmatrix} \tag{2.112}$$

where the subscripts A and B indicate component properties and C those of the assembly. We note in passing that the off-diagonal impedances of the assembly are equal to those of the components, a fact that was used in the *in situ* measurement of isolator properties in 2.3.2. Of particular interest here however, are the zero partitions on the upper right and the lower left. Physically, these zeros occur for this particular arrangement of components because there is no connection between the remote points a and b when the interface, c, is blocked. The free-type FRF $\mathbf{Y}$ (e.g. mobility) for this assembly is of course the inverse of the fixed-type FRF $\mathbf{Z}$ (e.g. impedance),

$$\begin{bmatrix} \mathbf{Y}_{Caa} & \mathbf{Y}_{Cac} & \mathbf{Y}_{Cab} \\ \mathbf{Y}_{Cca} & \mathbf{Y}_{Ccc} & \mathbf{Y}_{Ccb} \\ \mathbf{Y}_{Cba} & \mathbf{Y}_{Cbc} & \mathbf{Y}_{Cbb} \end{bmatrix} = \begin{bmatrix} \mathbf{Z}_{Caa} & \mathbf{Z}_{Cac} & \mathbf{0} \\ \mathbf{Z}_{Cca} & \mathbf{Z}_{Ccc} & \mathbf{Z}_{Ccb} \\ \mathbf{0} & \mathbf{Z}_{Cbc} & \mathbf{Z}_{Cbb} \end{bmatrix}^{-1} \tag{2.113}$$

where in general $Z_{ij} \neq Y_{ij}^{-1}$. Considering Equation 2.113 we see that the FRF matrix $\mathbf{Y}$ contains nine partitions which reduces to six when the reciprocity of symmetric blocks is taken into account $\mathbf{Y}_{ij} = \mathbf{Y}_{ji}^{\mathrm{T}}|_{i \neq j}$. On the other hand, the inverse FRF matrix $\mathbf{Z}$, which also fully describes the system, contains only seven non-zero elements, or five allowing for symmetry. This comparison indicates some redundancy in $\mathbf{Y}$, specifically, that we should expect one constraint equation between the partitions. This is derived as follows.

Equation 2.113 can be written equivalently as,

$$\begin{bmatrix} \mathbf{Z}_{Caa} & \mathbf{Z}_{Cac} & \mathbf{0} \\ \mathbf{Z}_{Cca} & \mathbf{Z}_{Ccc} & \mathbf{Z}_{Ccb} \\ \mathbf{0} & \mathbf{Z}_{Cbc} & \mathbf{Z}_{Cbb} \end{bmatrix} \begin{bmatrix} \mathbf{Y}_{Caa} & \mathbf{Y}_{Cac} & \mathbf{Y}_{Cab} \\ \mathbf{Y}_{Cca} & \mathbf{Y}_{Ccc} & \mathbf{Y}_{Ccb} \\ \mathbf{Y}_{Cba} & \mathbf{Y}_{Cbc} & \mathbf{Y}_{Cbb} \end{bmatrix} = \begin{bmatrix} \mathbf{I} & \mathbf{0} & \mathbf{0} \\ \mathbf{0} & \mathbf{I} & \mathbf{0} \\ \mathbf{0} & \mathbf{0} & \mathbf{I} \end{bmatrix}. \tag{2.114}$$

Expanding the two rightmost equations in the top row of Equation 2.114 we get,

$$\mathbf{Z}_{Caa}\mathbf{Y}_{Cac} + \mathbf{Z}_{Cac}\mathbf{Y}_{Ccc} = \mathbf{0} \tag{2.115}$$

$$\mathbf{Z}_{Caa}\mathbf{Y}_{Cab} + \mathbf{Z}_{Cac}\mathbf{Y}_{Ccb} = \mathbf{0}. \tag{2.116}$$

Post-multiplying Equation 2.115 by $\mathbf{Y}_{Ccc}^{-1}$,

$$\mathbf{Z}_{Cac} = -\mathbf{Z}_{Caa}\mathbf{Y}_{Cac}\mathbf{Y}_{Ccc}^{-1} \tag{2.117}$$

and substituting this into Equation 2.116 gives,

$$\mathbf{Z}_{Caa}\mathbf{Y}_{Cab} - \mathbf{Z}_{Caa}\mathbf{Y}_{Cac}\mathbf{Y}_{Ccc}^{-1}\mathbf{Y}_{Ccb} = \mathbf{0} \tag{2.118}$$

which implies,

$$\mathbf{Y}_{Cab} = \mathbf{Y}_{Cac}\mathbf{Y}_{Ccc}^{-1}\mathbf{Y}_{Ccb}. \tag{2.119}$$

Equation 2.119 is the constraint equation sought. It is an (exact) mathematical identity relating the partitions of the FRF matrix $\mathbf{Y}$.

This important result has practical application in measuring *long-distance* FRFs as presented later [60], as well as in the assessment of *interface completeness* as discussed in Chapter 8. However, for our purposes the main application of the Round

Trip theory is for the indirect measurement of interface FRFs and this requires some further steps of derivation. First, Equation 2.119 is inverted and rearranged to give,

$$\mathbf{Y}_{Ccb}\mathbf{Y}^{+}_{Cab}\mathbf{Y}_{Cac} = \mathbf{Y}_{Ccc}. \tag{2.120}$$

Some important implications arise from this operation which will be discussed in more detail in 2.5.4 but for now, we continue with the final step required to unlock the main practical benefits of the theory which is to use the reciprocity relationship $\mathbf{Y}_{Cac} = \mathbf{Y}^{\mathrm{T}}_{Cca}$ to give,

$$\mathbf{Y}_{Ccc} = \mathbf{Y}_{Ccb}\mathbf{Y}^{+}_{Cab}\mathbf{Y}^{\mathrm{T}}_{Cca}. \tag{2.121}$$

The remarkable conclusion of Equation 2.121 is that we can reconstruct the interface FRF matrix $\mathbf{Y}_{Ccc}$ indirectly from the three transfer FRF matrices on the right-hand side *none of which require excitation at the interface c.*

Equation 2.121 is the main result in the Round Trip theory as originally published in [39]. Equations 2.119 and 2.121 have become known as Round Trip identities in reference to the round trip path formed by the individual transfer FRFs (see Figures 2.26). For example, in Equation 2.121: the rightmost FRF $\mathbf{Y}^{\mathrm{T}}_{Cca}$ goes from c to a; the central FRF, $\mathbf{Y}_{Cab}$ (when inverted) goes from a to b and the leftmost one, $\mathbf{Y}_{Ccb}$, from b to c, hence completing the round trip.

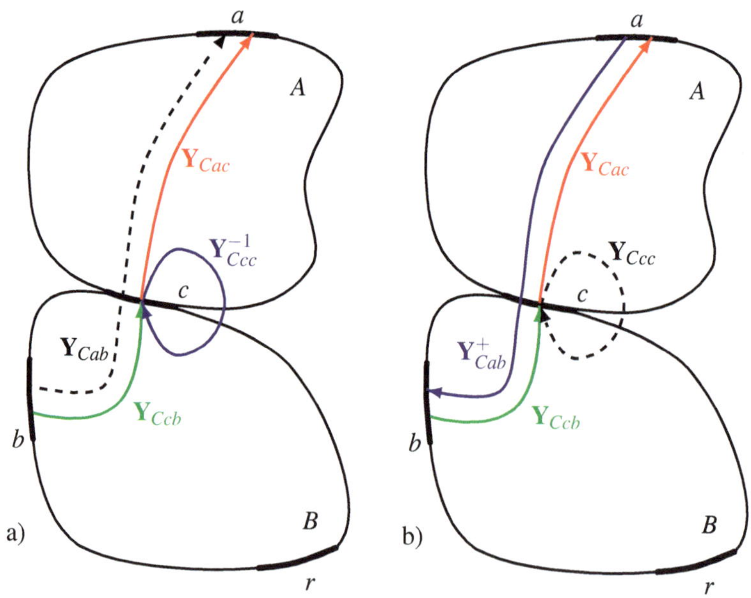

Figure 2.26 Transfer function 'paths' for round trip identities in Equations 2.119 (a) and 2.121 (b). Dashed line represents the FRF to be reconstructed, coloured lines represent the legs of the equivalent round trip.

For completeness we note that traversing the round trip in the reverse direction gives the transpose of Equation 2.121 which is mathematically identical to $\mathbf{Y}_{Ccc}$ by symmetry,

$$\mathbf{Y}_{Ccc}^{\mathrm{T}} = \mathbf{Y}_{Ccc} = \mathbf{Y}_{Cca}\mathbf{Y}_{Cba}^{+}\mathbf{Y}_{Ccb}^{\mathrm{T}}, \tag{2.122}$$

although the forward (Equation 2.121) and reverse (Equation 2.122) round trips generally yield slightly different results from experimental data. The Round Trip relations were applied to continuous interfaces in [38] but point-like connections will be assumed in the following.

2.5.2 IMPLEMENTATION

Implementation of Equation 2.121 requires a two-step measurement. First, excitation is applied at remote DoFs b and the responses measured simultaneously at c and a giving $\mathbf{Y}_{Ccb}$ and $\mathbf{Y}_{Cab}$. In the second stage, excitation is applied at a and the response measured at c giving $\mathbf{Y}_{Cca}$. In this arrangement, the sets of points are of three types:

- remote points at b are *excitation-only* points – they require excitation but no response measurements are needed;
- interface DoFs at c are *response-only* points – they require response measurements but no excitation;
- remote points at a are *excitation-response* points – they require both excitation and response.

Clearly, one can interchange a and b so that the a points become the excitation-only points if more convenient.

Some example results comparing directly measured FRFs with those obtained indirectly from the Round Trip method are shown later in this section. In addition, a number of examples in the literature show that, with appropriate care, the method can yield excellent agreement with directly measured FRFs; published examples include a laboratory beam, plate and mass structures [22], isolators [33, 34], plate structures [38], and automotive components [59].

To obtain good results from the Round Trip method it is important that the remote points are both sufficient in number and correctly located. These issues will be discussed in more detail shortly (2.5.4). For now, we simply remark that there must be at least as many remote points as there are interface DoFs and in practice some degree of *over-determination* is preferable so that $N_a \geq N_c$ and $N_b \geq N_c$ where N refers to the number of DoFs. Furthermore, the remote locations need to be located so as to have a good 'view' of the interface, as will be discussed in 2.5.4.

Subset of interface DoFs

We now look at the case where some of the interface DoFs can be measured directly but some others are inaccessible or difficult to excite reliably. In these cases we may ask if it is possible to characterise only the 'difficult' DoFs with the Round Trip method. To proceed, we partition the interface DoFs c into two sets, say c_u which are of interest, and c_v which are not. Equation 2.121 then becomes,

$$\mathbf{Y}_{Ccc} = \begin{bmatrix} \mathbf{Y}_{Cc_uc_u} & \mathbf{Y}_{Cc_uc_v} \\ \mathbf{Y}_{Cc_vc_u} & \mathbf{Y}_{Cc_vc_v} \end{bmatrix} = \begin{pmatrix} \mathbf{Y}_{Cc_ub} \\ \mathbf{Y}_{Cc_vb} \end{pmatrix} \mathbf{Y}_{Cab}^{+} \begin{pmatrix} \mathbf{Y}_{Cc_ua} \\ \mathbf{Y}_{Cc_va} \end{pmatrix}^{\mathrm{T}}. \tag{2.123}$$

Multiplying out the brackets we obtain a matrix of Round Trip identities for the partitioned interface,

$$\mathbf{Y}_{Ccc} = \begin{bmatrix} \mathbf{Y}_{Cc_uc_u} & \mathbf{Y}_{Cc_uc_v} \\ \mathbf{Y}_{Cc_vc_u} & \mathbf{Y}_{Cc_vc_v} \end{bmatrix} = \begin{bmatrix} \mathbf{Y}_{Cc_ub}\mathbf{Y}_{Cab}^{+}\mathbf{Y}_{Cc_ua}^{\mathrm{T}} & \mathbf{Y}_{Cc_ub}\mathbf{Y}_{Cab}^{+}\mathbf{Y}_{Cc_va}^{\mathrm{T}} \\ \mathbf{Y}_{Cc_va}\mathbf{Y}_{Cba}^{+}\mathbf{Y}_{Cc_ub}^{\mathrm{T}} & \mathbf{Y}_{Cc_vb}\mathbf{Y}_{Cab}^{+}\mathbf{Y}_{Cc_va}^{\mathrm{T}} \end{bmatrix} \tag{2.124}$$

the one of interest being the top left partition, i.e.

$$\mathbf{Y}_{Cc_uc_u} = \mathbf{Y}_{Cc_ub}\mathbf{Y}_{Cab}^{+}\mathbf{Y}_{Cc_ua}^{\mathrm{T}}. \tag{2.125}$$

Equation 2.125 is the same as the usual Round Trip identity (Equation 2.121) but based on a *subset* of interface DoFs. Since c_v does not appear in this equation, it is not necessary to instrument the other DoFs on the interface or even to know which they are! A note of caution is needed here though – it might be tempting to think that since we are dealing with a reduced set of interface DoFs the number of remote points can be reduced correspondingly. However, this is not the case, indeed the requirement still remains that the number of a and b points must exceed the *total* number of interface DoFs i.e. *not* restricted to those of particular interest. This statement can be justified by recalling that the point FRF Round Trip relation (Equation 2.121) is based on the inversion of the transfer FRF identity (Equation 2.119), which itself assumes that the interface c is completely described, i.e. all DoFs are accounted for. As such, the partitioned equations similarly require that the number of remote points at a and b are equal to or greater than the number of total interface DoFs, i.e. c. The question of *how many* remote points are needed is related to the topics of *interface matrix rank* and *interface completeness* which are discussed in Chapter 6.

In summary, FRFs can be obtained for certain interface DoFs of interest when other interface DoFs are not included or are even unknown, provided that the remote points are sufficient in number and suitably placed.

Transmissibility and output-only measurements

We now consider the role of transmissibility in the Round Trip method and the implications for measurement. Inspecting Equation 2.121, we see a pair of familiar matrix products. Indeed, two of the three FRF terms in each Round Trip relation form a transmissibility. Using Equation 2.93,

$$\mathbf{Y}_{Ccc} = \mathbf{Y}_{Ccb}\mathbf{Y}_{Cab}^{-1}\mathbf{Y}_{Cac} \tag{2.126}$$

$$= \left(\mathbf{T}_{Cac}^{r,b}\right)^{-1}\mathbf{Y}_{Cac} \tag{2.127}$$

where $\mathbf{T}_{Cac}^{r,b}$ is the response transmissibility of the assembly for excitation at b and response at a and c.

The response transmissibility in Equation 2.126 can be obtained with *output-only* measurements and thus there is no need to register the forces applied at the excitation-only points b. Furthermore, there is no requirement even to know where these excitation points are, provided they are sufficient in number. These facts suggest the idea of using naturally occurring excitation, for example from operational forces located within substructure B, instead of a force hammer to supply the excitation for the first part of the test. In [56] the authors demonstrated the combined use of operational transmissibilities and the Round Trip identity to obtain FRFs using partially operational data. This is an interesting prospect that could widen the application of the Round Trip method and seems likely to attract more attention in the future.

Application to long-distance FRFs

The same analysis as above can be applied to the *transfer* FRF version of the Round Trip identity, Equation 2.119. In this case, we have using Equations 2.95,

$$\mathbf{Y}_{Cab} = \mathbf{Y}_{Cac}\mathbf{Y}_{Ccc}^{-1}\mathbf{Y}_{Ccb} \tag{2.128}$$

$$= \mathbf{T}_{Cac}^{r,c}\mathbf{Y}_{Ccb}. \tag{2.129}$$

This has potential application in the measurement of *long-distance* FRFs. Essentially, if the points in a transfer FRF (a and b) are distant enough to cause problems with SNR, then the transfer FRF can be measured in shorter stages from a to c, then c to b. This approach was used in [59] and, despite the additional measurements, was found to significantly improve the estimation of a noisy FRF by measuring in stages.

With this method, as above, there is the prospect of using natural excitation, which might be stronger than can be applied manually, for the first part of the test. For example, after measuring $\mathbf{Y}_{Ccb}$ in Equation 2.128, the response transmissibility $\mathbf{T}_{Cac}^{r,c}$ could be obtained operationally by running machinery housed within component B which would cause an excitation at c (see 2.4.2). This transmissibility could potentially be measured over a large distance if the source mechanisms are sufficiently strong.

2.5.3 APPLICATION TO COUPLING ELEMENT CHARACTERISATION

We now consider the practical application of the Round Trip theory to the problem of isolator characterisation. The setup is as shown in Figure 2.27. Recall from 2.3.2

that, using the *in situ* isolator method one can extract independent isolator properties from the assembly's interface FRF matrix, i.e. from,

$$\mathbf{Y}_{Ccc} = \begin{bmatrix} \mathbf{Y}_{Cc_1c_1} & \mathbf{Y}_{Cc_1c_2} \\ \mathbf{Y}_{Cc_2c_1} & \mathbf{Y}_{Cc_2c_2} \end{bmatrix}. \tag{2.130}$$

However, access to the interfaces will often be restricted in practice making it difficult to measure this matrix directly. Hence, we now consider the possibility of obtaining the required FRFs *indirectly* using the Round Trip method.

First, by inspection of Figure 2.27 we note that if the isolator is considered part of substructure B, then Figure 2.27 is of the same form as Figure 2.26. Therefore, with remote excitation points at a and b we use the Round Trip relation Equation 2.121 to obtain the assembly FRF matrix at the *upper* interface,

$$\mathbf{Y}_{Cc_1c_1} = \mathbf{Y}_{Cc_1b}\mathbf{Y}^+_{Cab}\mathbf{Y}^{\mathrm{T}}_{Cc_1a}. \tag{2.131}$$

Note that this avoids excitation at either interface.

By a similar argument, if we consider the isolator as part of substructure A we may use the same (or different) a and b points to obtain the assembly FRF matrix at the *lower* interface,

$$\mathbf{Y}_{Cc_2c_2} = \mathbf{Y}_{Cc_2b}\mathbf{Y}^+_{Cab}\mathbf{Y}^{\mathrm{T}}_{Cc_2a}. \tag{2.132}$$

Again, no excitation is required at either interface.

Thus, we can obtain both diagonal elements of Equation 2.130 from the Round Trip method. It remains to obtain the off-diagonal elements. We cannot apply the same logic as above, however, by inspection of Equations 2.131 and 2.134 we might guess that these terms will be given as follows,

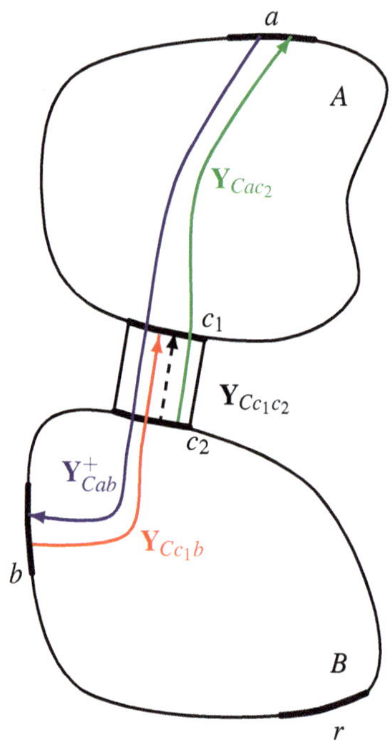

Figure 2.27 Dual interface round trip paths for Equation 2.140.

$$\mathbf{Y}_{Cc_1c_2} = \mathbf{Y}_{Cc_1b}\mathbf{Y}^+_{Cab}\mathbf{Y}^{\mathrm{T}}_{Cc_2a} \tag{2.133}$$

$$\mathbf{Y}_{Cc_2c_1} = \mathbf{Y}_{Cc_2b}\mathbf{Y}^+_{Cab}\mathbf{Y}^{\mathrm{T}}_{Cc_1a} \tag{2.134}$$

and indeed this turns out to be the case. In the following, a derivation will be given based on physical arguments in a similar approach to earlier derivations of the Round Trip identity in order to aid physical insight.

Following the original derivation in [39], we consider an external force applied at DoFs b and the resulting responses at a and c_1,

$$\mathbf{v}_a = \mathbf{Y}_{Cab}\mathbf{f}_b \qquad \mathbf{v}_{c_1} = \mathbf{Y}_{Cc_1b}\mathbf{f}_b. \tag{2.135}$$

Equating the above through the shared force $\mathbf{f}_b$ we have,

$$\mathbf{Y}_{Cab}^{-1}\mathbf{v}_a = \mathbf{Y}_{Cc_1b}^{-1}\mathbf{v}_{c_1}. \tag{2.136}$$

The responses at a and c_1 can alternatively be expressed using the *blocking* force at c_2 due to the external force $\mathbf{f}_b$,

$$\mathbf{v}_a = \mathbf{Y}_{Cac_2}\bar{\mathbf{f}}_{c_2} \qquad \mathbf{v}_{c_1} = \mathbf{Y}_{Cc_1c_2}\bar{\mathbf{f}}_{c_2}. \tag{2.137}$$

Substituting Equations 2.137 into Equation 2.136 gives,

$$\mathbf{Y}_{Cab}^{-1}\mathbf{Y}_{Cac_2}\bar{\mathbf{f}}_{c_2} = \mathbf{Y}_{Cc_1b}^{-1}\mathbf{Y}_{Cc_1c_2}\bar{\mathbf{f}}_{c_2}. \tag{2.138}$$

Noting that the external force and resulting blocked force $\bar{\mathbf{f}}_{c_2}$ are completely arbitrary, it must be true that,

$$\mathbf{Y}_{Cab}^{-1}\mathbf{Y}_{Cac_2} = \mathbf{Y}_{Cc_1b}^{-1}\mathbf{Y}_{Cc_1c_2}. \tag{2.139}$$

Finally, pre-multiplying both sides by $\mathbf{Y}_{Cc_1b}$ we obtain,

$$\mathbf{Y}_{Cc_1c_2} = \mathbf{Y}_{Cc_1b}\mathbf{Y}_{Cab}^{-1}\mathbf{Y}_{Cac_2} \tag{2.140}$$

or by reciprocity

$$\mathbf{Y}_{Cc_2c_1} = \mathbf{Y}_{Cc_2a}\mathbf{Y}_{Cba}^{-1}\mathbf{Y}_{Cbc_1}. \tag{2.141}$$

The above 'transfer' round trip is illustrated in Figure 2.27 and is seen to visit both interfaces en-route.

With these expressions for the off-diagonals, the full interface FRF matrix can now be constructed *without* applying forces at either c_1 or c_2 but instead applying forces at remote points a and b,

$$\mathbf{Y}_{Ccc} = \begin{bmatrix} \mathbf{Y}_{Cc_1c_1} & \mathbf{Y}_{Cc_1c_2} \\ \mathbf{Y}_{Cc_2c_1} & \mathbf{Y}_{Cc_2c_2} \end{bmatrix} \tag{2.142}$$

$$= \begin{bmatrix} \mathbf{Y}_{Cc_1b}\mathbf{Y}_{Cab}^{+}\mathbf{Y}_{Cc_1a}^{\mathrm{T}} & \mathbf{Y}_{Cc_1b}\mathbf{Y}_{Cab}^{+}\mathbf{Y}_{Cc_2a}^{\mathrm{T}} \\ \mathbf{Y}_{Cc_2a}\mathbf{Y}_{Cba}^{+}\mathbf{Y}_{Cc_1b}^{\mathrm{T}} & \mathbf{Y}_{Cc_2b}\mathbf{Y}_{Cab}^{+}\mathbf{Y}_{Cc_2a}^{\mathrm{T}} \end{bmatrix}. \tag{2.143}$$

Having obtained the interface FRF matrix by indirect means, the independent transfer impedance of the isolator can be obtained according to the *in situ* isolator method (2.3.2) by simply inverting Equation 2.142 and extracting the off diagonals of the resulting impedance matrix. This process is illustrated in the following example.

Example 2.3: Round-trip isolator characterisation

In cases where access to the interface is restricted, the Round Trip method provides a means of 'relocating' the forces applied for FRF measurement from the interface to remote and easily accessible locations. We demonstrate this application by returning to the isolator characterisation Example 2.2. There, the *in situ* method was used to extract an intrinsic characterisation of the isolator whilst installed between a range of different components. To demonstrate application of the Round Trip method, we revisit the beam-isolator-plate (BIP) example, as illustrated in Figure 2.28.

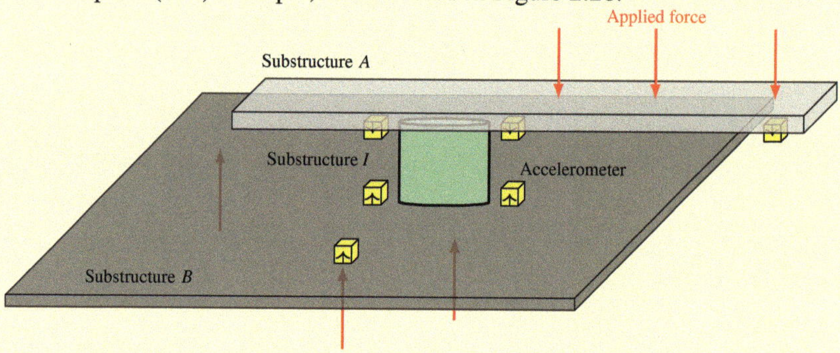

Figure 2.28 Test set-up for using the Round Trip method in the in situ characterisation of an isolator between beam and plate components. No forces are applied at the interface. Additional beam and plate excitations are included to provide over-determination of the Round Trip.

Recall that in the *in situ* isolator method (2.3.2) we first measure the FRF matrix of the assembly above and below the mount $\mathbf{Y}_{Ccc}$. This is then inverted to obtain the assembly impedance matrix $\mathbf{Z}_{Ccc}$ from which the off-diagonal elements are extracted and are identically equal to those of the isolator alone. As with the *in situ* method, the Round Trip implementation requires accelerometers to be positioned at the interfaces above and below the mount. However, *no forces* are applied at the interfaces in the Round Trip implementation. Rather, they are applied at additional remote positions on the beam or plate (both are included here for comparative purposes). As mentioned above, the problem can be over-determined by applying more forces than there are interface DoFs.

The first step is to obtain the FRFs of the assembly across the isolator. Results from direct measurement and via the Round Trip identity are compared in Figure 2.29 (for translational DoF only). Three estimates are presented from the round trip: a determined case (one excitation on beam) and two over-determined cases (three excitations on the beam and plate, respectively). Good agreement is obtained compared with the direct measurement up to about 800 Hz in this case but *without applying any forces* to the interface.

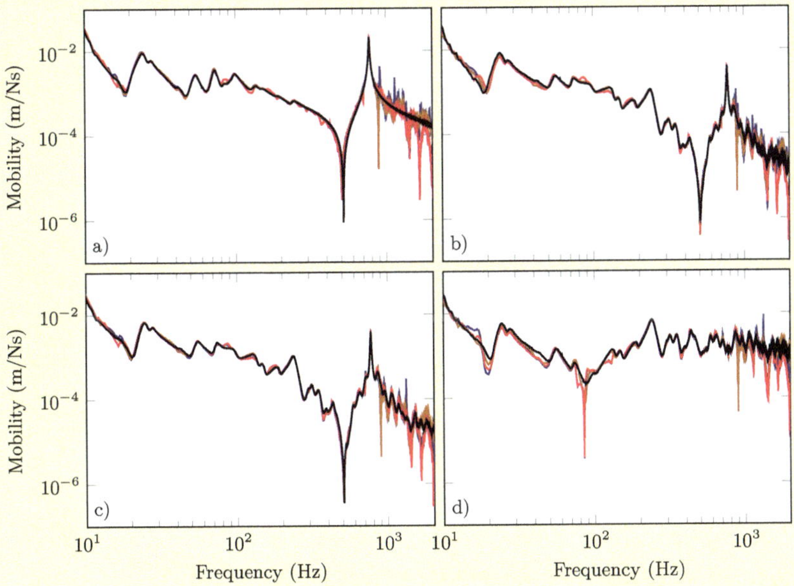

Figure 2.29 Interface FRF matrix (a – $Y_{Cc_1c_1}$, b – $Y_{Cc_1c_2}$, c – $Y_{Cc_2c_1}$, d – $Y_{Cc_2c_2}$) obtained by direct measurement (black) and Round Trip estimates: determined (blue), over-determined on beam (pink), and over-determined on plate (brown).

The next step is to invert this FRF matrix to obtain the isolator transfer impedance. Shown in Figure 2.30 are the results of the (determined) Round Trip characterisation (i.e. inversion of the FRF matrix shown in Figure 2.29) compared with those of the standard *in situ* test applied to the same BIP assembly. The *in situ* estimate from a simpler MIM assembly is also shown for comparison. Good overall agreement is obtained, albeit with some additional artefacts at higher frequencies. Indeed, the large artefact observed from the *in situ* method at 500 Hz in Figure 2.30 is no longer evident when using the Round Trip method, possibly as a benefit of relocating the excitation forces.

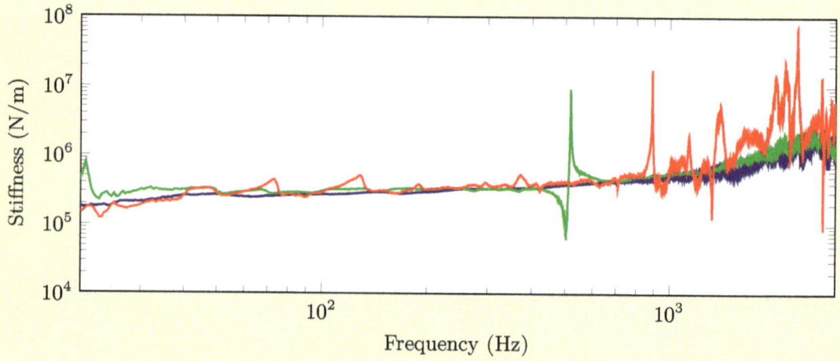

Figure 2.30 Dynamic transfer stiffness (magnitude only) obtained using standard in situ method applied to MIM (blue) and BIP (green) assemblies compared with Round Trip estimate (determined) applied to BIP assembly (red).

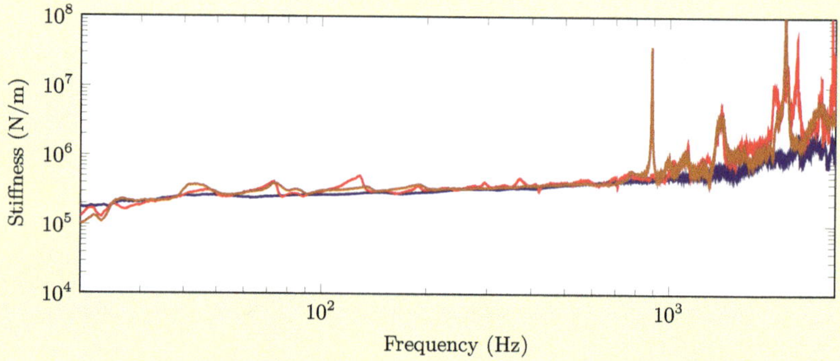

Figure 2.31 Dynamic transfer stiffness (magnitude only) obtained using standard *in situ* method (MIM, blue) and Round Trip on BIP assembly with over-determination on beam (pink) and plate (brown) substructures.

The effect of over-determination is shown in Figure 2.31. Here, three excitations were applied to the beam (pink) and plate (brown). It can be see that over-determination improves the overall stiffness estimate, reducing the severity of the observed artefacts, e.g. for the beam over-determination case the sharp artefact at approx 1 kHz is avoided, whilst for the plate over-determination the less pronounced artefacts below 1 kHz are reduced. Similar results have been demonstrated on more complex assemblies [34].

Example 2.2 illustrated the remarkable result that independent characteristics of isolators can be obtained *in situ*, i.e. without removing the isolator from its assembly. In this example it has been further shown that, using the Round Trip method, this can be achieved without applying any excitation to either of its interfaces.

2.5.4 THE BOTTLENECK EFFECT

A critically important stage in the component-based approach is to define the *interface DoFs* that transmit significant forces from one component to another. This topic, to which we will return throughout the book, has implications for measurement using the Round Trip method, particularly in relation to selecting the remote points. It is also relevant for the derivation of the Round Trip relation by inversion as introduced above. This will be discussed in the following where we will bring in the concept of *matrix rank* as well as considering how the interface acts as a *bottleneck*. We also introduce related concepts from control theory, namely: *observability* and *controllability*.

We start with the exact identity for the transfer FRF matrix across the interface, $\mathbf{Y}_{Cba}$, i.e. Equation 2.119, repeated here for convenience,

$$\mathbf{Y}_{Cab} = \mathbf{Y}_{Cac}\mathbf{Y}_{Ccc}^{-1}\mathbf{Y}_{Ccb}. \tag{2.144}$$

In this relation, the transfer FRF is written in terms of the interface FRF matrix, $\mathbf{Y}_{Ccc}$ which fully defines the interface DoFs. The *rank* of $\mathbf{Y}_{Ccc}$ therefore defines the *number* of DoFs at the interface and is related to the rank of $\mathbf{Y}_{Cab}$ as follows,

$$\mathrm{rank}(\mathbf{Y}_{Cab}) = \mathrm{rank}(\mathbf{Y}_{Cac}\mathbf{Y}_{Ccc}^{-1}\mathbf{Y}_{Ccb}) \leq \mathrm{rank}(\mathbf{Y}_{Ccc}). \tag{2.145}$$

In words, the rank of the transfer FRF, $\mathbf{Y}_{Cab}$, cannot exceed that of the interface FRF, $\mathbf{Y}_{Ccc}$. Equation 2.145 can therefore be seen as a mathematical expression of the *bottleneck* effect of the interface – more independent degrees of freedom cannot be observed on the far side of the interface than the interface will allow through. This important relationship will be picked up again in 4.5 and 6.4.1.

We now look at the inversion of Equation 2.119 (and Equation 2.144) in more detail. We begin by post-multiplying by the (right) pseudo-inverse of $\mathbf{Y}_{Ccb}$,

$$\mathbf{Y}_{Cab}\mathbf{Y}_{Ccb}^{+} = \mathbf{Y}_{Cac}\mathbf{Y}_{Ccc}^{-1}\mathbf{Y}_{Ccb}\mathbf{Y}_{Ccb}^{+}. \tag{2.146}$$

Now, if the rank of $\mathbf{Y}_{Ccb}$ is equal to the number of rows, N_c, then the matrix product of $\mathbf{Y}_{Ccb}\mathbf{Y}_{Ccb}^{+}$ on the right-hand side yields the identity matrix $\mathbf{I}$, i.e. they will cancel one another. However, if $\mathbf{Y}_{Ccb}$ is *rank deficient* then the cancellation will not be complete. Hence, it is important to understand the conditions under which rank deficiency may occur.

A fundamental result from linear algebra is that the number of linearly independent rows in a matrix is equal to the number of independent columns (the row rank equals the column rank). Consequently, in order to achieve a rank of N_c we must have at least as many columns as rows, i.e. there must be at least as many remote points b as there are interface DoFs c, or $N_b \geq N_c$. Physically, $\mathbf{Y}_{Ccb}$ being full rank means that the remote DoFs at b are able to fully *control* all possible dynamics displayed by the interface DoFs c and this can only occur if there are a sufficient number of excitation points at b.

Not only must the *number* of remote points be sufficient, but they must also produce a sufficient number of *linearly independent* responses. For example if two excitation points are too close together then we have essentially the same information in two columns of the matrix so they do not provide different ways to control the interface i.e. they are linearly dependent and would not both contribute to matrix rank. In this case, the number of linearly *independent* columns will be less than the total number of columns, N_b and if this is also less than N_c then the matrix will be rank deficient.

Assuming the conditions for full rank *are* achieved then Equation 2.146 reduces to,

$$\mathbf{Y}_{Cab}\mathbf{Y}_{Ccb}^{+} = \mathbf{Y}_{Cac}\mathbf{Y}_{Ccc}^{-1}. \tag{2.147}$$

We proceed by taking the pseudo-inverse of each side,

$$(\mathbf{Y}_{Cab}\mathbf{Y}_{Ccb}^{+})^{+} = (\mathbf{Y}_{Cac}\mathbf{Y}_{Ccc}^{-1})^{+} \tag{2.148}$$

leading to,

$$\mathbf{Y}_{Ccb}\mathbf{Y}_{Cab}^{+} = \mathbf{Y}_{Ccc}\mathbf{Y}_{Cac}^{+}. \tag{2.149}$$

Post-multiplying both sides by $\mathbf{Y}_{Cac}$ gives

$$\mathbf{Y}_{Ccb}\mathbf{Y}_{Cab}^{+}\mathbf{Y}_{Cac} = \mathbf{Y}_{Ccc}\mathbf{Y}_{Cac}^{+}\mathbf{Y}_{Cac}. \tag{2.150}$$

As discussed earlier, the product of $\mathbf{Y}_{Cac}$ with its own pseudo inverse on the right-hand side will equal the identity matrix provided it is not rank deficient. In this case, the condition for full rank is that $N_a \geq N_c$ in other words, the number of remote points at a must at least equal the number of interface DoFs. Whereas we previously described the requirement in terms of full *controllability* of the interface, in this case we say that the remote points at a must provide full *observability* (although this distinction is semantic rather than practical since the a and b points can readily be exchanged). If this condition is met then the last two terms on the right hand side cancel and we arrive at Equation 2.121, thereby completing the derivation.

We are now able to summarise the conditions relating to Equation 2.121: the number of remote points on each side of the interface must at least equal the number of interface DoFs, i.e. $N_a \geq N_c$ and $N_b \geq N_c$. Furthermore, these points must provide at least this many linearly independent rows and columns. In practice, one will typically *over-determine* the problem with additional remote points to ensure sufficient linear independence (for further discussion of over-determination see Chapter 7). Note that no such conditions were required relating to Equation 2.119 which is an exact mathematical identity. The conditions arise in relation to Equation 2.121 because it requires the *pseudo-inverse* of potentially non-square matrices which are not unique.

The above conditions can be interpreted physically as follows: when a set of forces act on the interface DoFs c, if the remote DoFs at a are able to completely *observe* c then they possess sufficient information to reconstruct an arbitrary interface forcing applied to c. If, in addition, the remote set of DoFs at b are able to fully *control* the interface then there exists a set of forces at b that can reproduce any response field across c. With a and b separated by the interface DoFs c, the number of independent contributions that can reach a from b (or *vice versa*) is limited to N_c because of the bottleneck effect.

For the sake of additional insight, note that Equation 2.149 equates two generalised transmissibilities. On the right-hand side, we have the 'ratio' of responses at c and a when excited at the interface, c. On the left-hand side, we have the same 'ratio' of responses but due to excitation further back from the interface at the remote points, b. In the first case, the excitation of the interface is direct from external forces and in the second case, it is indirect via substructure B. The equality suggests that substructure A does not 'care' whether its interface is excited directly or indirectly (as was discussed in relation to transmissibility earlier). This is to be expected given that the b locations are able to fully control the interface. This logic was the starting point for the original derivation of the Round Trip identity in [39] and follows from Ribeiro et al. [47].

2.6 NUMERICAL MODELLING AND UPDATING

Whilst the focus of this book is largely on experimental methods for component-based simulation, the component-based paradigm is not restricted to components characterised by measurement. In-fact, one of the principle advantages of the component-based approach is its ability to combine component models from measurement and simulation to form so-called 'hybrid' assemblies. This is particularly useful when, for example, a physical component is not available, in which case its experimental characterisation is impossible and numerical modelling is the only viable approach. Even if a physical component *is* available for testing, there are situations where it is necessary or desirable to work with a model alongside the physical component, for example to facilitate design optimisation. In these cases it makes sense to make use of any measured data to improve the reliability of the model by *model updating*. Therefore, in this section, we discuss both numerical modelling and model updating. Since these are large fields with extensive literature, we restrict our scope to the aspects specifically related to component simulation. Hence, the discussion of modelling is restricted to the synthesis of component FRFs from existing numerical models rather than the modelling itself. Similarly, the treatment of model updating emphasises *in situ* methods which are not widely discussed elsewhere but are of particular relevance in the context of component-based simulation as an aid to extracting intrinsic component properties.

2.6.1 SYNTHESISING FRFS FROM NUMERICAL MODELS

In this section, we introduce the key equations for synthesising FRFs from pre-existing numerical models, specifically physical domain representations where mass and stiffness matrices are known (e.g. by the Finite Element method). The reader is referred to the many text books in the field for further detail on the modelling itself, see for example [17, 45].

In the physical domain, a component is characterised by its mass, stiffness and damping characteristics, which together form the following equation of motion,

$$\mathbf{M}\ddot{\mathbf{x}}(t) + \mathbf{C}\dot{\mathbf{x}}(t) + \mathbf{K}\mathbf{x}(t) = \mathbf{f}(t). \tag{2.151}$$

To obtain a component mobility matrix from Equation 2.151, two approaches will be discussed; the direct (matrix inversion) method and modal synthesis.

In either case, we begin by assuming a harmonic forcing such that $\mathbf{f}(t) = \mathbf{f}_0 e^{i\omega t}$. Owing to linearity, the displacement, velocity and acceleration response will also be periodic with the same frequency. Equation 2.151 can thus be transformed into the frequency domain, and expressed in terms of displacement $\mathbf{x}(\omega)$ as,

$$\left[-\omega^2\mathbf{M} + i\omega\mathbf{C} + \mathbf{K}\right]\mathbf{x}(\omega) = \mathbf{D}(\omega)\mathbf{x}(\omega) = \mathbf{f}(\omega) \tag{2.152}$$

where $\mathbf{D}(\omega)$ represents the dynamic stiffness of the component (see Table 2.1).

Direct method

The direct determination of the mobility FRF involves computing the matrix inverse of $\mathbf{D}$ and pre-multiplying both sides of Equation 2.152 by it, giving,

$$\mathbf{x}(\omega) = \mathbf{R}(\omega)\mathbf{f}(\omega) \tag{2.153}$$

where,

$$\mathbf{R}(\omega) = \mathbf{D}^{-1}(\omega) = \left[-\omega^2\mathbf{M} + i\omega\mathbf{C} + \mathbf{K}\right]^{-1} \tag{2.154}$$

is the dynamic compliance of the component. The mobility is then obtained by differentiation, i.e. scaling by $i\omega$,

$$\mathbf{Y}(\omega) = i\omega\mathbf{R}(\omega). \tag{2.155}$$

This direct method is appropriate when working with small models containing not too many DoFs. However, as the matrix inversion has to be performed at each frequency, for large systems this becomes computationally expensive and is not recommended. Furthermore, this direct approach provides the full mobility matrix that includes every DoF within the component model. For virtual prototyping applications, this is usually not necessary since the black box models adopted for components only require interface and target DoFs to be described in full.

A reduced mobility matrix can be obtained by considering Equation 2.152 in a partitioned form (omitting frequency dependence for clarity),

$$\begin{pmatrix} \mathbf{f}_r \\ \mathbf{f}_d \end{pmatrix} = \begin{bmatrix} \mathbf{D}_{rr} & \mathbf{D}_{rd} \\ \mathbf{D}_{dr} & \mathbf{D}_{dd} \end{bmatrix} \begin{pmatrix} \mathbf{x}_r \\ \mathbf{x}_d \end{pmatrix} \tag{2.156}$$

where r denotes the set of DoFs we wish to retain, and d those that are required for computation but are not needed in the dynamic description of the component. Assuming no external forces are applied to any of the d DoFs, $\mathbf{f}_d = \mathbf{0}$, Equation 2.156 can be rearranged to yield,

$$\mathbf{f}_r = \left(\mathbf{D}_{rr} - \mathbf{D}_{rd}\mathbf{D}_{dd}^{-1}\mathbf{D}_{dr}\right)\mathbf{x}_r = \hat{\mathbf{D}}_{rr}\mathbf{x}_r \tag{2.157}$$

where $\hat{\mathbf{D}}_{rr}$ is the reduced dynamic stiffness matrix, from which the reduced mobility matrix is obtained by inversion and scaling,

$$\hat{\mathbf{Y}}_{rr} = i\omega\hat{\mathbf{D}}_{rr}^{-1}. \tag{2.158}$$

Note that although reduced in size to include only the retained DoFs r, Equation 2.157 still involves the inversion of the stiffness submatrix $\mathbf{D}_{dd}^{-1}$, which for large systems is still computationally demanding. More efficient reduction schemes are available at the cost of introducing some approximation in the estimation of the component dynamics, see for example [20].

Modal synthesis

The modal synthesis approach involves determining the mode shapes and natural frequencies of a component, and expressing its FRF as a truncated summation of modal contributions.

To formulate the modal synthesis of a mobility matrix, we begin by considering the case of an undamped and unforced system, whose equation of motion reads,

$$\mathbf{K}\mathbf{x}(\omega) = \omega^2\mathbf{M}\mathbf{x}(\omega). \tag{2.159}$$

Physically, we are interested in a special type of solution to Equation 2.159, where all coordinates execute synchronous motion, i.e. they have the same time dependence so that the ratio between the motion of any two points remains constant. For an N dimensional system, there exist N such solutions to Equation 2.159 in the form,

$$\mathbf{K}\tilde{\phi}_i = \lambda_i\mathbf{M}\tilde{\phi}_i \tag{2.160}$$

where $\lambda_i = \omega_i^2$ are the so-called system eigenvalues (representing the squared natural frequencies of the component), and $\tilde{\phi}_i$ are the corresponding eigenvectors (representing the relative deformations of the component at frequency ω_i that satisfy Equation 2.159). The eigenvectors $\tilde{\phi}_i$ are unique only in the sense that the ratio between the motion at any two points is constant. In other words, it is only the shape of the component's deflection that is well defined, not its amplitude. To render the eigenvectors unique a normalisation process is required. The most common normalisation scheme is that of mass normalisation, whereby the normalised eigenvectors are obtained as $\phi_i = \tilde{\phi}_i\left(\tilde{\phi}_i^{\mathrm{T}}\sqrt{\mathbf{M}}\tilde{\phi}_i\right)^{-1}$, or in matrix form $\Phi = \tilde{\Phi}\left(\tilde{\Phi}^{\mathrm{T}}\sqrt{\mathbf{M}}\tilde{\Phi}\right)^{-1}$. With this mass normalisation we have that,

$$\phi_i^{\mathrm{T}}\mathbf{M}\phi_i = 1 \tag{2.161}$$

and

$$\phi_i^{\mathrm{T}}\mathbf{K}\phi_i = \omega_i^2. \tag{2.162}$$

This latter relation can be easily proven by pre-multiplying both sides of Equation 2.160 by ϕ_i^{T}. Similarly, from Equation 2.160 and the uniqueness of eigenvalues, it is straightforward to show that the component eigenvectors/mode shapes are orthogonal, such that,

$$\phi_r^{\mathrm{T}}\mathbf{M}\phi_s = 0 \tag{2.163}$$

and

$$\phi_r^{\mathrm{T}}\mathbf{K}\phi_s = 0, \quad r \neq s. \tag{2.164}$$

The normalised eigenvectors described above are referred to as the component's normal modes, or simply its modes.

Noting that any N dimensional vector can be represented as a linear combination of N linearly independent vectors, and that normal modes are themselves orthogonal and linearly independent, the displacement $\mathbf{x}$ can be expressed in the form,

$$\mathbf{x} = \Sigma_i^N \phi_i q_i = \mathbf{\Phi}\mathbf{q} \tag{2.165}$$

where $\mathbf{\Phi} = [\phi_1, \phi_2, \cdots, \phi_N]$ is a modal matrix whose columns represent the normal modes and $\mathbf{q} = \mathbf{q}_0 e^{i\omega t}$ is a vector of modal amplitudes describing the contribution of the each mode to the response $\mathbf{x}$. With the modal matrix $\mathbf{\Phi}$ being constant, we have also that $\dot{\mathbf{x}} = \mathbf{\Phi}\dot{\mathbf{q}}$ and $\ddot{\mathbf{x}} = \mathbf{\Phi}\ddot{\mathbf{q}}$.

Returning now to the damped and forced equation of motion,

$$\mathbf{M}\ddot{\mathbf{x}}(t) + \mathbf{C}\dot{\mathbf{x}}(t) + \mathbf{K}\mathbf{x}(t) = \mathbf{f}(t) \tag{2.166}$$

substituting the displacement, velocity and acceleration for their modal representations from Equation 2.165 and its time derivatives, whilst pre-multiplying Equation 2.166 by the transposed mode shape matrix $\mathbf{\Phi}^{\mathrm{T}}$ yields,

$$\bar{\mathbf{M}}\ddot{\mathbf{q}}(t) + \bar{\mathbf{C}}\dot{\mathbf{q}}(t) + \bar{\mathbf{K}}\mathbf{q}(t) = \mathbf{Q}(t) \tag{2.167}$$

with,

$$\bar{\mathbf{M}} = \mathbf{\Phi}^{\mathrm{T}}\mathbf{M}\mathbf{\Phi}, \quad \bar{\mathbf{C}} = \mathbf{\Phi}^{\mathrm{T}}\mathbf{C}\mathbf{\Phi}, \quad \bar{\mathbf{K}} = \mathbf{\Phi}^{\mathrm{T}}\mathbf{K}\mathbf{\Phi}, \quad \mathbf{Q} = \mathbf{\Phi}^{\mathrm{T}}\mathbf{f}. \tag{2.168}$$

$\bar{\mathbf{M}}$, $\bar{\mathbf{K}}$ and $\bar{\mathbf{C}}$ are called the modal mass, stiffness and damping matrices, and $\mathbf{Q}$ the modal force vector. Recalling the mass normalised and orthogonality conditions in Equations 2.161-2.164,

$$\bar{\mathbf{M}} = \mathbf{I}, \quad \bar{\mathbf{K}} = \mathrm{diag}\left[\omega_1^2, \omega_2^2, \cdots, \omega_N^2\right] = \Lambda \tag{2.169}$$

Equation 2.167 becomes,

$$\left[-\omega^2\mathbf{I} + i\omega\bar{\mathbf{C}} + \Lambda\right]\mathbf{q} = \mathbf{Q}(t). \tag{2.170}$$

Inversion and pre-multiplication of both sides by the bracketed term yields a solution for the modal amplitudes,

$$\mathbf{q} = \left[-\omega^2\mathbf{I} + i\omega\bar{\mathbf{C}} + \Lambda\right]^{-1}\mathbf{Q}(t). \tag{2.171}$$

Substituting the modal amplitudes into Equation 2.165, whilst recalling $\mathbf{Q} = \Phi^{\mathrm{T}}\mathbf{f}$,

$$\mathbf{x} = \Phi\left[-\omega^2\mathbf{I} + i\omega\bar{\mathbf{C}} + \Lambda\right]^{-1}\Phi^{\mathrm{T}}\mathbf{f} \tag{2.172}$$

from which we can identify the component mobility as,

$$\mathbf{Y}(\omega) = i\omega\Phi\left[\Lambda - \omega^2\mathbf{I} + i\omega\bar{\mathbf{C}}\right]^{-1}\Phi^{\mathrm{T}}. \tag{2.173}$$

Equation 2.173 provides a means of synthesising the mobility matrix of a component based on its natural frequencies, mode shapes and modal damping characteristics. To provide an efficient synthesis, there are two important considerations.

First, regarding the number of modes, whilst an N DoF system will have N natural frequencies and associated mode shapes, often only a reduced subset of these are required to provide a good estimate of the mobility matrix. This is achieved by truncating the modal matrix Φ to include only the first n columns $\Phi = [\phi_1, \phi_2, \cdots, \phi_n]$ (i.e. the mode shapes of the first n natural frequencies). The matrix Λ should similarly be truncated. It is generally suggested to include modes up to twice the highest frequency (f_{max}) of interest, however this general guidance may not be sufficient for VAVP applications as discussed in Example 2.4. Going back to the modal matrix, the rows of Φ correspond to the component's physical DoFs and unwanted rows can readily be discarded to obtain a reduced FRF matrix including only the DoFs of interest.

Second, we require the square bracketed term to be diagonal, such that its matrix inversion becomes computationally trivial. Whilst the $-\omega^2\mathbf{I}$ and Λ terms satisfy this requirement, $\bar{\mathbf{C}}$ is generally fully populated. Hence, we are interested in simplified models of damping that yield diagonal modal damping matrices.

One such model is *structural damping* (also known as hysteric damping) which is represented in the physical domain as,

$$\left[-\omega^2\mathbf{M} + (\mathbf{K} + i\mathbf{H})\right]\mathbf{x}(\omega) = \mathbf{f}(\omega). \tag{2.174}$$

The complex stiffness matrix $(\mathbf{K} + i\mathbf{H})$ is most easily obtained by replacing Young's modulus E by a complex one of the form $E(1 + i\eta)$ when deriving the component's stiffness matrix, where η is the loss factor. Equivalently, Equation 2.173 can be written as,

$$\mathbf{Y}(\omega) = i\omega\Phi\left[\Lambda(1 + i\eta) - \omega^2\mathbf{I}\right]^{-1}\Phi^{\mathrm{T}}. \tag{2.175}$$

The loss factor η takes into account all losses, including material and coupling losses. Typical values are between 1% and 5% and tend to be higher where there are connections which allow sliding such as bolted rather than welded joints. In the absence of a physical component to determine damping characteristics, structural damping with an estimated loss factor is the simplest option.

Other damping models, such as viscous and modal damping are also available but will not be discussed here.

Example 2.4: Modal convergence for VAVP FRFs

When truncating a modal matrix it is often considered sufficient to include modes up to twice the frequency of interest. However, whilst convergence may be rapid close to resonance frequencies, at anti-resonances this is not the case. For VAVP applications, substructuring, and therefore matrix inversion, is required and anti-resonances may feature strongly in the inverted matrix. Thus, the normal guidelines may not be sufficient. The aim of this example is to demonstrate this effect.

To illustrate the issue of anti-resonance convergence, we synthesis the FRF matrix between two points, each with 3 DoFs (vertical z translation and x/y rotations, $\mathbf{Y} \in \mathbb{C}^{6\times 6}$), on a simply supported plate for different numbers of modes. The maximum frequency of interest is 1 kHz (indicated by the vertical dashed line). A baseline result is computed that includes 50^2 modes (50 modes along the x and y direction, along with all cross modes), we take this result to be the correct FRF. In Figure 2.32 we compare the baseline result to three other FRFs that use 3^2, 4^2 and 5^2 modes, corresponding to maximum modal frequencies of 1.8 kHz (approx. twice f_{max}), 2.8 kHz and 4.1 kHz, respectively. We see that whilst the resonant frequencies converge rapidly, away from isolated resonance peaks convergence is slower and at anti-resonance frequencies there is a lack of convergence even when including modes up to 4 times the f_{max}.

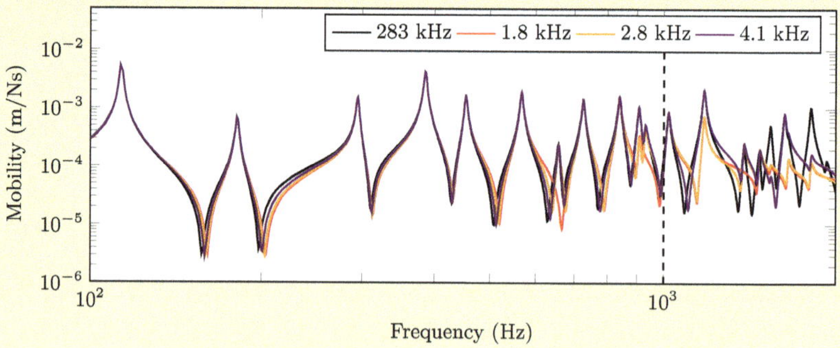

Figure 2.32 Point mobility FRFs obtained from modal summation model of a simply support plate with increasing number of modes. Legend specifies the largest modal frequency included in the sum. Number of modes included are: 50^2, 3^2, 4^2, and 5^2.

To highlight the importance of anti-resonance convergence Figure 2.33 shows the impedance FRFs, obtained from the inverse $\mathbf{Z} = \mathbf{Y}^{-1}$, for each of the simulations. It is clear that whilst the mobility appears to have converged reasonably well, this is not the case for the impedance.

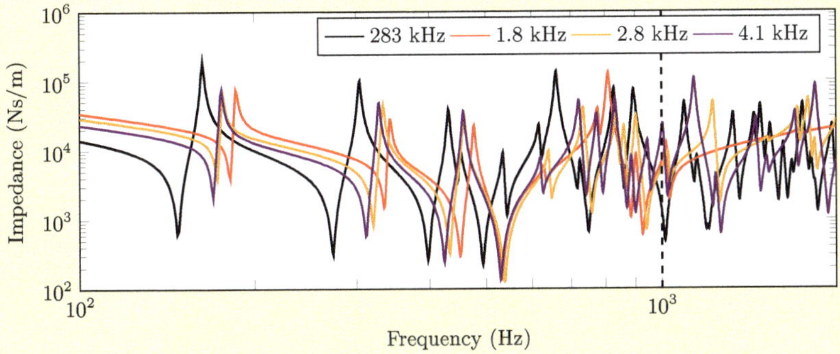

Figure 2.33 Point impedance FRFs obtained by inversion of the 6×6 mobility matrix computed by modal summation. Legend specifies the largest modal frequency included in the mobility sum. Number of modes included are: 50^2, 3^2, 4^2, and 5^2.

In conclusion, if a modal solution is used to compute an FRF matrix for a VAVP application, one cannot automatically assume that the usually adopted factor of 2 will be sufficient. Instead, it is important to conduct a preliminary convergence study to determine the number of modes to be included.

Modal parameter estimation

In the previous section we discussed how, given a set of modal parameters (mode shapes and natural frequencies), one can synthesise the FRFs of a component for use in the construction of a VAVP. Initially, we considered the modal parameters as being obtained from a numerical FE model. Of course, they could also be obtained from experiment, a process referred to as *modal identification* or *modal parameter estimation*.

With our intention being to construct a VAVP, this might seem somewhat counter intuitive at first. Why would we want to extract modal parameters from measured FRFs, just to re-synthesise the FRFs that we already had in the first place? Well, as will be discussed in more detail in Chapters 8 and 10, experimentally obtained FRFs are of course subject to experimental error. Individual FRFs might be corrupted by noise, whilst the entire FRF matrix might suffer from inconsistencies such as a lack of reciprocity or passivity. Such inconsistencies can cause large errors when, for example, substructuring is performed. By extracting modal parameters and re-synthesising the FRFs, we are effectively fitting a modal model. This ensures a consistent, noise free, set of FRFs, implicitly enforcing both reciprocity and passivity. This technique, called modal filtering, can be used to improve the performance of substructuring calculations. In addition to this modal filtering application, modal parameters are often used to update numerical models, which can then be used directly within the VAVP framework.

The subject of modal identification is a vast one and, though tangentially related to the construction of VAVPs, is largely beyond the scope of this book. The interested reader is referred to the classic textbook of Ewins [3], and also those by Maia and e Silvia [31] and He and Fu [16]. With regards to modal identification algorithms, various options are available. These include simple Single Input Single Output (SISO) methods such as peak picking and circle fitting [3], more advanced Single Input Multiple Output (SIMO) methods including Least Square Complex Exponential Method [2], and Multiple Input Multiple Output (MIMO) methods including Eigensystem Realisation Algorithm [24] and PolyMAX [19, 43, 44].

2.6.2 MODEL UPDATING

Even if a physical component is available for testing there are several scenarios in which a numerical model may be advantageous alongside the test results, for example, to provide improved insight or to facilitate design optimisation. Such applications require that the numerical model offers a sufficiently accurate representation of the physical component. This is unlikely to be achieved in the first instance, and so some tweaking of the model parameters will inevitably be required to bring it into alignment with experimental data. This 'tweaking' is known as *model updating*.

Model updating describes a broad class of methods that use experimental data to optimise the parameters of a numerical model [15, 40]. Its primary purpose is to improve test-model correlation, i.e. the ability of a model to reliably represent the dynamics of a target structure. Existing methods for model updating can be broadly categorised by the type of experimental data that is used to quantify model correctness. Common choices include modal parameter estimates (natural frequencies and mode shapes) [15, 40, 50], measured frequency response functions [23, 27, 29] and transmissibilities [36, 51]. Together, measured and modelled data are used to formulate an objective function (representing test-model error) which is minimised by an appropriate optimisation algorithm. A typical objective function would be the mean squared error between measured and modelled natural frequencies, for example.

Model updating has been the subject of extensive research for several decades. We will not attempt to provide a detailed survey on the variety of methods and their implementations, for this the reader is referred to the various review papers and text books found in the literature [15, 32, 40, 48]. Rather, we will briefly discuss some *in situ* methods which do not form part of the mainstream updating literature but have particular relevance for the component-based approach.

In situ model updating

To successfully update a numerical model based on experimental measurement, the acquired data, whether modal, FRF or transmissibility-based, should characterise the target component independently. This is most often achieved with the substructure uncoupled and suspended freely to achieve free-interface conditions. Alternatively, a fixed-interface boundary condition can be used. As described earlier in the chapter though, achieving these idealised boundary conditions in practise can often be challenging. Moreover, in some cases one requires representative mounting conditions, such as the isolators mentioned above whose characteristics might depend on pre-load. These issues can be avoided by using in situ methods to extract invariant component-level data from coupled (in situ) experimental tests, and using this data to update a component-level model. This approach has become known as *in situ model updating*. A brief example was provided earlier (2.3.2) in relation to isolator characterisation and we will now describe the method in more general terms.

Recall that for an arbitrary component A coupled to a second component B at the interface DoFs c, inversion of its mobility matrix yields,

$$\begin{bmatrix} \mathbf{Y}_{Caa} & \mathbf{Y}_{Cac} \\ \mathbf{Y}_{Cca} & \mathbf{Y}_{Ccc} \end{bmatrix}^{-1} = \begin{bmatrix} \mathbf{Z}_{Aaa} & \mathbf{Z}_{Aac} \\ \mathbf{Z}_{Aca} & \mathbf{Z}_{Acc} + \mathbf{Z}_{Bcc} \end{bmatrix} \tag{2.176}$$

where a denotes a set of remote DoFs on A. We see that by inverting the coupled mobility matrix, we obtain an impedance matrix, three quarters of which is a property of A alone, and independent of B. Alternatively, as shown in 2.4.2, the force transmissibility

$$\mathbf{T}^{f}_{Aca} = \mathbf{Y}^{-1}_{Ccc}\mathbf{Y}_{Cca} = \mathbf{Z}_{Aca}\mathbf{Z}^{-1}_{Aaa} \tag{2.177}$$

can be determined, which is also an independent property of A, as is the response transmissibility.

The quantities $\mathbf{Z}_{Aaa}$, $\mathbf{Z}_{Aac}$, $\mathbf{Z}_{Aca}$ and $\mathbf{T}^{f,r}_{Aca}$ are all indpendent component-level quantities, available from in situ testing. It is the use of these quantities in a model updating scheme that we refer to as *in situ* model updating [35].

Unlike a conventional model updating where both A and B would have to be modelled and updated together, by extracting and using invariant component-level quantities, the need to consider B in the updating has been avoided entirely, and so only A needs to be modelled.

Implementation of an in situ updating scheme follows the conventional approach. A cost function is set based on the in situ component quantities,

$$J_D(\theta) = J_D\left(\mathbf{Z}^{FE}_{Aaa}(\theta), \mathbf{Z}^{FE}_{Aac}(\theta), \mathbf{Z}^{FE}_{Aca}(\theta), \mathbf{T}^{f,r,FE}_{Aca}(\theta), \mathbf{Z}^{exp}_{Aaa}, \mathbf{Z}^{exp}_{Aac}, \mathbf{Z}^{exp}_{Aca}, \mathbf{T}^{f,r,exp}_{Aca}\right) \tag{2.178}$$

where θ describes the set of model parameters to be updated. This cost function is then minimised through some optimisation routine. A typical cost function might consist of the averaged squared magnitude error between the in situ and modelled quantities, as illustrated earlier in Equation 2.55. The cost function may be selected and solved using conventional approaches as described for example in [62]. To reiterate, the novelty here is the use of *in situ* measured component-level properties rather than the updating procedure itself.

Note that, as described in 2.3.2, the in situ updating of a coupling element I is slightly more restrictive, since the only independent quantity available is the transfer impedance,

$$\mathbf{Y}_{Ccc}^{-1} = \begin{bmatrix} \mathbf{Z}_{Ac_1c_1} + \mathbf{Z}_{Ic_1c_1} & \mathbf{Z}_{Ic_1c_2} \\ \mathbf{Z}_{Ic_2c_1} & \mathbf{Z}_{Bc_2c_2} + \mathbf{Z}_{Ic_2c_2} \end{bmatrix}. \tag{2.179}$$

Though straightforward in theory, it is important to recall that the extraction of component-level data from coupled in situ tests requires a complete interface representation (see Chapter 4). In some scenarios this might be impractical or at least quite challenging.

2.7 SUMMARY

This chapter has addressed the question of how to characterise passive components for use with the component-based approach. This is most conveniently done with FRFs in the frequency domain. A common form of FRF, mobility, quantifies the velocity response at a given point and in a given direction to a unit force excitation at the same or another point and in the same or another direction. The inverse of mobility is impedance. There are other forms based on the response measured as acceleration or displacement rather than velocity; converting between these types is straightforward. There are also vibro-acoustic and acoustic FRFs which introduce sound pressure or volume velocity at a point.

To characterise a component, FRFs only need to be defined at some specific points, namely the interface DoFs and any remote target DoFs. Since structures are usually connected at multiple points and transmit forces in up to six directions per point, FRFs are usually organised into matrices. The impedance matrix is therefore a matrix inverse of a mobility matrix and *vice versa*. An important property is reciprocity which allows a transfer FRF to be measured with excitation and response positions reversed, often simplifying the measurement procedure.

The key requirement for component-based application is that the component characteristics are *transferable*. This requires that it is mounted for testing with boundaries either *free* or *fixed*. Free conditions, being practically easier to achieve than fixed, are more often used for source and receiver components by taking measurements with the component suspended or hung from resilient supports. Coupling elements like resilient isolators on the other hand, often require fixed mounting. Generally however, these idealised mounting conditions are often difficult or impossible to achieve and so a range of techniques has evolved to simulate the conditions indirectly. The latter part of the chapter therefore considers indirect measurement approaches.

One method is to invert a measured mobility matrix, the effect of which is effectively to apply a full set of perfectly rigid constraints mathematically rather than physically. *In situ* measurements may also be conducted with the component installed in an assembly. In this case, in order to achieve independence, the dynamic influence of the rest of the assembly must be removed either by decoupling methods or by using a model alongside the measurements.

Transmissibility, while not itself used to characterise components, is an important form of FRF that will be used at various points throughout the book. It is sometimes claimed that it characterises isolators, but is shown to be a property of the installation rather than an independent property of the isolator itself. Perhaps surprisingly, some transmissibilities describe independent component properties even when measured with the component installed in an assembly. This fact has potential application for *in situ* measurements and model updating.

The Round Trip Method is an important indirect approach that allows indirect measurement of FRFs while avoiding the need to excite directly at the point of interest.

Finally, it is recognised that models will often need to be employed alongside measurements. Thus, we have described how to adapt data from FE models into a form suitable for the component-based framework. Additionally, a method of *in situ* model updating was introduced for updating model parameters to achieve a better fit with measurement.

In the following chapter we will consider the characterisation of active components – the *sources* of sound and vibration in machinery and equipment.

Bibliography

1. C Beijers, B Noordman, and A de Boer. Numerical modelling of rubber vibration isolators: identification of material parameters. In *Proceedings of the 11th International Congress on Sound and Vibration*, pages 3193–3200. Citeseer, 2004.
2. DL Brown, RJ Allemang, R Zimmerman, and M Mergeay. Parameter estimation techniques for modal analysis. *SAE transactions*, pages 828–846, 1979.
3. DJ Ewins. *Modal testing: theory, practice and application*. John Wiley & Sons, 2009.
4. FJ Fahy. The vibro-acoustic reciprocity principle and applications to noise control. *Acta Acustica United with Acustica*, 81(6):544–558, 1995.
5. BS ISO 9611:1996 Acoustics - characterisation of sources of structure-borne sound with respect to sound radiation from connected structures - measurement of velocity at the contact points of machinery when resiliently mounted. 1996.
6. BS EN ISO 10846-3:2002 Acoustics and vibration - laboratory measurement of vibro-acoustic transfer properties of resilient elements - part 3: Indirect method for determination of the dynamic stiffness of resilient supports for translatory motion. 2002.
7. BS EN ISO 10846-4:2002 Acoustics and vibration - laboratory measurement of vibro-acoustic transfer properties of resilient elements - part 4: Dynamic stiffness of elements other than resilient supports for translatory motion. 2002.
8. BS EN ISO 10846-1:2008 Acoustics and vibration - laboratory measurement of vibroacoustic transfer properties of resilient elements, part 1: Principles and guidelines. 2008.
9. BS EN ISO 10846-2:2008 Acoustics and vibration - laboratory measurement of vibro-acoustic transfer properties of resilient elements - part 2: Direct method for determination of the dynamic stiffness of resilient supports for translatory motion. 2008.
10. BS EN ISO 10846-5:2009 Acoustics and vibration - laboratory measurement of vibro-acoustic transfer properties of resilient elements - part 5: Driving point method for determination of the low-frequency transfer stiffness of resilient supports for translatory motion. 2009.

11. BS EN ISO 7626-1:2011 Mechanical vibration and shock - experimental determination of mechanical mobility - part 1: Basic terms and definitions, and transducer specifications. 2011.
12. BS ISO 7626-2:2015 Mechanical vibration and shock - experimental determination of mechanical mobility. part 2: Measurement using single point translation excitation with an attached vibration exciter. 2015.
13. RM Foster. A reactance theorem. *Bell System technical journal*, 3(2):259–267, 1924.
14. L Fredette and R Singh. High frequency, multi-axis dynamic stiffness analysis of a fractionally damped elastomeric isolator using continuous system theory. *Journal of Sound and Vibration*, 389:468–483, 2017.
15. M Friswell and JE Mottershead. *Finite element model updating in structural dynamics*, volume 38. Springer Science & Business Media, 1995.
16. ZF Fu and J He. *Modal analysis*. Elsevier, 2001.
17. M Géradin and DJ Rixen. *Mechanical vibrations: theory and application to structural dynamics*. John Wiley & Sons, 2014.
18. N Gil-Negrete, J Vinolas, and L Kari. A simplified methodology to predict the dynamic stiffness of carbon-black filled rubber isolators using a finite element code. *Journal of Sound and Vibration*, 296(4-5):757–776, 2006.
19. P Guillaume, P Verboven, S Vanlanduit, H Van Der Auweraer, and B Peeters. A poly-reference implementation of the least-squares complex frequency-domain estimator. In *Proceedings of IMAC*, 2003.
20. RJ Guyan. Reduction of stiffness and mass matrices. *AIAA Journal*, 3(2):380–380, 1965.
21. M Häußler, SWB Klaassen, and DJ Rixen. Experimental twelve degree of freedom rubber isolator models for use in substructuring assemblies. *Journal of Sound and Vibration*, 474:115253, 2020.
22. C Höller and BM Gibbs. Indirect determination of the mobility of structure-borne sound sources. *Journal of Sound and Vibration*, 344:38–58, 2015.
23. M Imregun, WJ Visser, and DJ Ewins. Finite element model updating using frequency response function data I. Theory and initial investigation. *Mechanical Systems and Signal Processing*, 9(2):187–202, 1995.
24. JN Juang and RS Pappa. An eigensystem realization algorithm for modal parameter identification and model reduction. *Journal of Guidance, Control, and Dynamics*, 8(5):620–627, 1985.
25. L Kari. Dynamic transfer stiffness measurements of vibration isolators in the audible frequency range. *Noise Control Engineering Journal*, 49(2):88–102, 2001.
26. L Kari. On the waveguide modelling of dynamic stiffness of cylindrical vibration isolators. part I: The model, solution and experimental comparison. *Journal of Sound and Vibration*, 244(2):211–233, 2001.
27. KS Kwon and RM Lin. Frequency selection method for FRF-based model updating. *Journal of Sound and Vibration*, 278(1-2):285–306, 2004.
28. YE Lage, MM Neves, NMM Maia, and D Tcherniak. Force transmissibility versus displacement transmissibility. *Journal of Sound and Vibration*, 333(22):5708–5722, 2014.
29. RM Lin and DJ Ewins. Analytical model improvement using frequency response functions. *Mechanical Systems and Signal Processing*, 8(4):437–458, 1994.
30. FX Magrans, A Aragonès, A Rodríguez-Ferran, and O Guasch. The solution of vibroacoustic linear systems as a finite sum of transmission paths. *Mechanical Systems and Signal Processing*, 153:107464, 2021.
31. NMM Maia and JMM e Silva. *Theoretical and experimental modal analysis*. Research Studies Press Wiley, 1997.
32. T Marwala. *Finite element model updating using computational intelligence techniques: applications to structural dynamics*. Springer Science & Business Media, 2010.
33. JWR Meggitt, AS Elliott, AT Moorhouse, and HK Lai. In situ determination of dynamic stiffness for resilient elements. *Proceedings of the institution of mechanical engineers, Part C: Journal of mechanical engineering science*, 230(6):986–993, 2016.
34. JWR Meggitt. *On in-situ methodologies for the characterisation and simulation of vibro-acoustic assemblies*. PhD thesis, University of Salford (United Kingdom), 2017.
35. JWR Meggitt and AT Moorhouse. Finite element model updating using in-situ experimental data. *Journal of Sound and Vibration*, 489:115675, 2020.

36. V Meruane. Model updating using antiresonant frequencies identified from transmissibility functions. *Journal of Sound and Vibration*, 332(4):807–820, 2013.
37. AT Moorhouse, AS Elliot, and Y Heo. Intrinsic characterisation of structure-borne sound sources and isolators from in-situ measurements. In *Proceedings of Meetings on Acoustics*, volume 19. AIP Publishing, 2013.
38. AT Moorhouse and AS Elliott. The "round trip" theory for reconstruction of green's functions at passive locations. *The Journal of the Acoustical Society of America*, 134(5):3605–3612, 2013.
39. AT Moorhouse, TA Evans, and AS Elliott. Some relationships for coupled structures and their application to measurement of structural dynamic properties in situ. *Mechanical Systems and Signal Processing*, 25(5):1574–1584, 2011.
40. JE Mottershead and MI Friswell. Model updating in structural dynamics: a survey. *Journal of Sound and Vibration*, 167(2):347–375, 1993.
41. GJ O'hara. Mechanical impedance and mobility concepts. *The Journal of the Acoustical Society of America*, 41(5):1180–1184, 1967.
42. G Pavić and AS Elliott. Structure-borne sound characterization of coupled structures—part i: simple demonstrator model. *Journal of Vibration and Acoustics*, 2010.
43. B Peeters, P Guillaume, H Van der Auweraer, B Cauberghe, P Verboven, and J Leuridan. Automotive and aerospace applications of the polymax modal parameter estimation method. In *Proceedings of IMAC*, volume 22. Citeseer, 2004.
44. B Peeters, H Van der Auweraer, P Guillaume, and J Leuridan. The polymax frequency-domain method: a new standard for modal parameter estimation? *Shock and Vibration*, 11(3-4):395–409, 2004.
45. M Petyt. *Introduction to finite element vibration analysis*. Cambridge University Press, 2010.
46. JWS Rayleigh. *Some general theorems relating to vibrations*. London Mathematical Society, 1873.
47. AMR Ribeiro, JMM Silva, and NMM Maia. On the generalisation of the transmissibility concept. *Mechanical Systems and Signal Processing*, 14(1):29–35, 2000.
48. S Sehgal and H Kumar. Structural dynamic model updating techniques: A state of the art review. *Archives of Computational Methods in Engineering*, 23:515–533, 2016.
49. JC Snowdon. Vibration isolation: use and characterization. *The Journal of the Acoustical Society of America*, 66(5):1245–1274, 1979.
50. G Steenackers and P Guillaume. Finite element model updating taking into account the uncertainty on the modal parameters estimates. *Journal of Sound and Vibration*, 296(4-5):919–934, 2006.
51. G Steenackers, C Devriendt, and P Guillaume. On the use of transmissibility measurements for finite element model updating. *Journal of Sound and Vibration*, 303(3-5):707–722, 2007.
52. JWS Rayleigh. On the application of the principle of reciprocity to acoustics. *Proceedings of the Royal Society of London*, 25(171-178):118–122, 1877.
53. T Ten Wolde. On the validity and application of reciprocity in acoustical, mechano-acoustical and other dynamical systems. *Acta Acustica United with Acustica*, 28(1):23–32, 1973.
54. T Ten Wolde. Reciprocity measurements in acoustical and mechano-acoustical systems. review of theory and applications. *Acta Acustica United with Acustica*, 96(1):1–13, 2010.
55. DJ Thompson, WJ Van Vliet, and JW Verheij. Developments of the indirect method for measuring the high frequency dynamic stiffness of resilient elements. *Journal of Sound and Vibration*, 213(1):169–188, 1998.
56. S Tranter, R Glachan, R Potter, and R McGee. Using passive frequency response functions generated using the operational round trip method to make vibration predictions at remote points. In *INTER-NOISE and NOISE-CON Congress and Conference Proceedings*, volume 268, pages 4547–4558. Institute of Noise Control Engineering, 2023.
57. JW Verheij. *Multi-path sound transfer from resiliently mounted shipboard machinery: Experimental methods for analyzing and improving noise control*. PhD thesis, TU Delft, Delft University of Technology, 1982.
58. J Wang, X Hong, Y Qian, Z Wang, and L Lu. Inverse sub-structuring method for multi-coordinate coupled product transport system. *Packaging Technology and Science*, 27(5):385–408, 2014.
59. K Wienen, M Sturm, AT Moorhouse, and JWR Meggitt. Generalised round-trip identity—for the determination of structural dynamic properties at locations inaccessible or too distant for direct measurement. *Journal of Sound and Vibration*, 511:116325, 2021.

60. L Cremer and M Heckl. Structure-borne sound: structural vibrations and sound radiation at audio frequencies. Springer Science & Business Media, 2013.
61. NMM Maia, APV Urgueira and RAB Almeida. Whys and wherefores of transmissibility. *Vibration Analysis and Control: New Trends and Developments*, 197, BoD–Books on Demand, 2011.
62. M Friswell and JE Mottershead. Finite element model updating in structural dynamics. Springer, 1995.

3 Sources

In the previous chapter, components were presented as passive structures to be connected at certain points which we called interfaces. However, if all components were like this then no machine would make any sound (or do any useful work). Hence, in this chapter we introduce *source* components which contain the active 'prime mover' mechanisms that make the machine work and, in doing so, generate the dynamic forces that are the initial cause of its characteristic sound.

Active mechanisms include fluctuating aerodynamic and magnetic forces, impact, gear mesh forces, stick-slip friction, combustion and out-of-balance forces. However, the detail of the mechanisms themselves is not the main focus here, rather, we are interested in ways to characterise their net effect. Recall that the ultimate aim is no less than to replace real source components, including the above complex mechanisms, entirely by digital representations in a virtual prototype. In other words we are looking for methods that will allow us to 'switch on' the VAVP and hear it running, just like the real machine.

The challenge is therefore to characterise the 'activity' of sources realistically and in a manner consistent with the component-based approach. This implies that the quantities used are properties of the source alone, independent of the rest of the assembly and transferable to new assemblies. Furthermore, since sources, along with all other components, are to be represented as black boxes, their active properties will be defined solely at their interfaces. Their passive properties having been defined in Chapter 2, we now aim to characterise the 'source strength' of the black box.

We start in 3.1 by reviewing the suitability of various quantities for source characterisation: contact forces, free velocity, blocked forces and pseudo forces. The requirements for independence are discussed in the light of an electrical analogy. We move on to the topic of central importance to the chapter, and indeed to the whole component-based approach, namely the *in situ* blocked force method in 3.2.

Some 'engineering' approaches to source characterisation are then reviewed where measurements are significantly simplified at the expense of reduction in detail. Power-based methods (3.3) and the reception plate method (3.3.7) come into this category.

The main focus of the chapter is on structural sources since source characterisation methods have moved on considerably in the last forty years and there is now broad agreement about the leading approaches. Airborne sources are given a briefer treatment in 3.4 since, apart from sound power which has been around a long time and is well documented elsewhere, there is as yet no clear consensus on the leading methods. We provide a selective outline of some possible approaches with a clear parallel to the structural case, in particular *equivalent monopoles* and *blocked pressures*. A number of other promising approaches, holography and full field optical methods for example, are outside the scope of the book.

DOI: 10.1201/9781003399131-3

3.1 REPRESENTATIONS OF SOURCE ACTIVITY

We can think of a source as a passive component but with the addition of some embedded active mechanisms as illustrated in Figure 3.1. Normally, the point of generation is not at an interface but is buried somewhere inside the source component, hence we often talk of 'internal forces'. In a working machine, these internal forces (say $\mathbf{f}_i$) are transmitted through the interface labelled c (which might consist of multiple contact points each permitting motion in several DoFs) to a receiver structure R (which itself might be composed of multiple subcomponents) where they are observed as a structural response v_r or perhaps perceived as a sound pressure p_r. In equation form, we have that,

$$(v_r, p_r) = G(\mathbf{f}_i) \tag{3.1}$$

where $G(\square)$ is a function representing the vibro-acoustic propagation through the assembled structure.

If we assume linearity and time invariance, this problem can be expressed more conveniently in the frequency domain in terms of FRFs,

$$\begin{pmatrix} v_r(\omega) \\ p_r(\omega) \end{pmatrix} = \begin{bmatrix} \mathbf{Y}_{Cri}(\omega) \\ \mathbf{H}_{Cri}(\omega) \end{bmatrix} \mathbf{f}_i(\omega) \tag{3.2}$$

where $\mathbf{Y}_{Cri}$ is a vector of mobilities, $\mathbf{H}_{Cri}$ is a vector of vibro-acoustic transfer functions, and ω is the radian frequency (omitted hereafter for brevity). As before, the capitalised subscript C is used to denote the *coupled* assembly (i.e. with the vibration source attached to the receiver).

However, although rather general this formulation is not helpful practically since the point of generation, i, is not generally accessible or even known. Thus, neither the internal forces $\mathbf{f}_i$ nor the FRFs $\mathbf{Y}_{Cri}$ and $\mathbf{H}_{Cri}$ can be measured. Instead, using the black-box approach, the internal forces, or rather their net effect, must be characterised at the *interfaces*.

Thus, we seek an alternative representation of the source activity that:

a) is equivalent to the internal forces in terms of inducing the same response field,
b) is defined at the interfaces,
c) can be determined practically,
d) is independent and transferable.

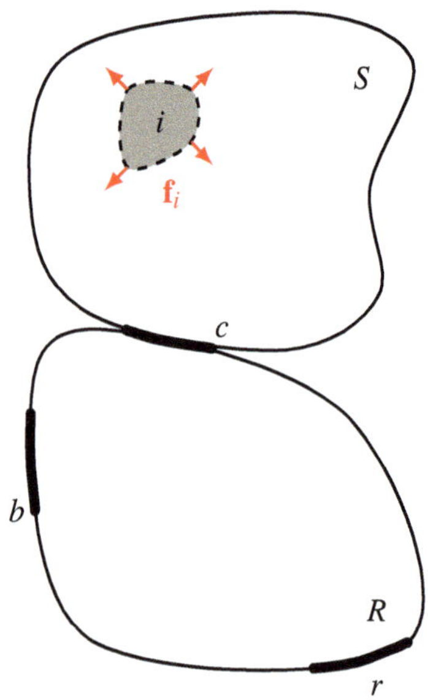

Figure 3.1 Source-receiver model for structure-borne problems. $\mathbf{f}_i$ are the internal forces generated by the operating source, c are the interface DoFs separating source and receiver, b are a set of remote DoFs on the receiver (used later for indirect force estimation) and r are the target DoFs where response predictions are sought.

The search for such a characterisation is the main topic of this chapter. Hence, in the following we consider various options, including contact forces, free velocity, blocked forces and pseudo forces. Later in the chapter (3.3) we also consider power-based representations.

At this point it should be mentioned that underlying any independent source characterisation is the assumption that the internal forces $\mathbf{f}_i$ are unaffected by changes to the source mounting. In practical terms this means that the same internal forces are assumed to occur irrespective of whether the source is 'floating', at one extreme, or coupled to a rigid receiver structure at the other, or in any intermediate state. This assumption has never really been subject to a rigorous test, not least because it would be very difficult to devise such a test. At the same time, it is possible to conceive of scenarios that might invalidate the assumption; for example, if the internal forces are due to out-of-balance forces then one could imagine that they might depend on whether the machine housing is free to move or is rigidly constrained. However, despite these provisos, the assumption has been made almost universally for the last few decades and there is an increasing body of results where the same source has been operated in very different boundary conditions with no clear sign of variance in the source mechanisms. Therefore, despite the lack of a rigorous proof, we will adopt the same assumption throughout the rest of the book.

Before we move on to discuss the various options available, we take a brief aside into the origins of source characterisation.

3.1.1 HISTORICAL PERSPECTIVE

Going back to the 1980s, much work had already been done on the modelling of passive components, particularly with FEM and SEA. Methods of coupling FEM substructures together, now known as substructuring, were also in use [17] and had even been developed for coupling measured FRFs [49, 50]. Thus, at that time, engineers could model passive components, couple them together and predict, with some confidence, their response to a *unit input force*. In other words they could predict a *relative* but not *absolute* response; in effect, they could predict FRFs of both single components and assemblies.

The missing link at that time was for methods to determine the *actual* input force. Yet, in comparison with the extensive development for modelling for passive structures, very little attention had been directed towards *sources* [34] and source mechanisms [42] at that time. Relative descriptions, such as FRFs, provide a full description of the dynamic response as well as allowing coupling into larger assemblies, but without knowing anything about the true input forces, prediction of absolute response was not possible. In short, without source characterisation the grand vision of prototyping entirely in the virtual domain would never be realised.

The problem lay not with airborne sound sources since these could already be reliably characterised with standardised sound power methods in which the source is conveniently described by a single frequency-dependent quantity, the sound power. Sound power is, to a large extent, independent of the surroundings (see 3.3) and therefore in keeping with the requirements of the component-based approach. Rather, the urgent need was for comparable methods for structural vibration sources which were almost completely lacking.

Although the importance of virtual prototyping was yet to become apparent, the lack of reliable and practical methods for structural source characterisation was picked up by ISO working group TC43. In 1986 they published a paper seeking comments from the acoustics community on a number of outline proposals for how to tackle the problem of '*structure-borne noise emission*' [39].

Describing the problem as '*dauntingly complex*', Ten Wolde and Gadefelt put forward outline proposals for methods based on directly measured velocity on the support structure, machine housing or on a reception plate, as well as indirectly measured forces. Interestingly, they did not propose direct measurement of force, probably recognising the difficulties to be described later in this chapter.

Their proposals can be grouped as those involving detailed measurements at every contact point and others using equivalent quantities which, like sound power, aimed to boil down the activity of the whole machine into a single frequency-dependent quantity. These two approaches align with what have since become known as respectively a) precision and b) engineering approaches (although see earlier comments about the use of the term 'precision' in 1.5.6).

As well as having to provide independent data, the authors stated that a suitable source characterisation should allow [39]:

1) Comparison of one source with another.
2) Comparison of sources with set limits.
3) Prediction of sound levels when installed.
4) Quantification of improvement in new low noise designs.

This list appears to be based on expectations from the use of (airborne) sound power where all these requirements can typically be met. However, as will be discussed in more detail in 3.3, the first two items are not achievable for structure-borne sound sources (except in some specific situations). Indeed, the requirements in the above list only really make sense when considering the source in an installation rather than in isolation. For example when comparing sources with one another, it is really the output of the operating machine that is of interest and it will be shown in 3.3 that this depends significantly not just on the source, but on what it is connected to. Thus, comparison with other sources is only meaningful within the context of a specific installation; the same can be said for comparison with set limits.

Item 4) in the list, use of the source characterisation for design improvements, again only really makes sense for an installed source. Thus, from the list above, only item 3), prediction of sound levels when installed, is really achievable. This further emphasises that sources are most usefully characterised in a way that facilitates prediction, in other words consistently with the component-based approach. With this in mind, in the following we discuss some potential candidates starting with contact forces.

3.1.2 CONTACT FORCE

At first sight, the effect of a source is most naturally represented by the force that it imparts on to the receiver at the connecting interface, the so-called *contact force.*

As shown in the free body diagram of the receiver structure in Figure 3.2, if the activity of the source is represented by the contact force $\mathbf{g}_c$, then the vibro-acoustic response is given by,

$$\begin{pmatrix} v_r \\ p_r \end{pmatrix} = \begin{bmatrix} \mathbf{Y}_{Rrc} \\ \mathbf{H}_{Rrc} \end{bmatrix} \mathbf{g}_c \tag{3.3}$$

where v_r and p_r are the velocity and pressure responses at the target DoFs r. Note that $\mathbf{Y}_{Rrc}$ and $\mathbf{H}_{Rrc}$ are the mobility and vibro-acoustic FRFs of the *receiver* structure alone (not of the assembly).

The receiver response in Equation 3.3 is identical to that in Equation 3.2 where excitation was by the internal force $\mathbf{f}_i$. Thus, in terms of the receiver response, this contact force representation is entirely equivalent to the coupled system excited by the internal force $\mathbf{f}_i$. The contact force therefore achieves the first two requirements in the list on page 102 in that it is defined at the interfaces and is equivalent to the internal forces in terms of receiver responses.

From item c) on the list on page 102 the question arises as to whether contact force may be determined practically. It turns out that, although not straightforward, there are at least three ways to measure contact force (see for example [5]). These methods, *direct measurement*, *mount stiffness method*, and *inverse force synthesis* are briefly described below.

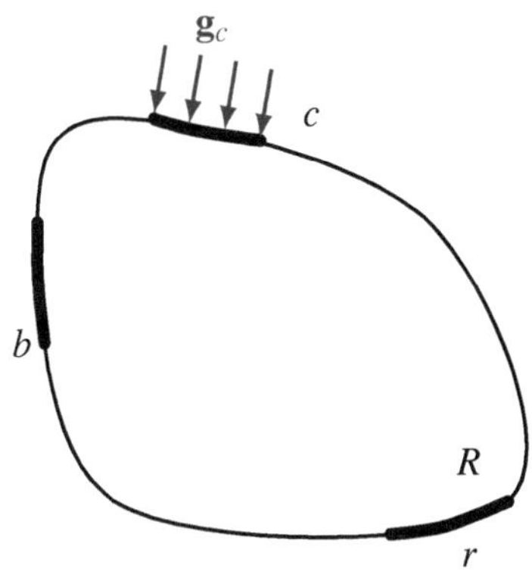

Figure 3.2 Free body diagram of a receiver substructure subject to contact force $\mathbf{g}_c$ at the source-receiver interface c. b is an additional set of DoFs used as part of inverse force synthesis.

Direct measurement

The most direct way to measure contact force is to insert force transducers between the source and receiver substructures. However, while conceptually simple, direct measurement is far from straightforward in practice. Among the difficulties are the need to measure forces in up to six DoFs (ideally simultaneously), the need to ensure that the transducer does not alter the mass and stiffness of the interface and the difficulty of registering the force exactly at the point of application. Careful design of the transducer mounting is required and even then, good measurements are not necessarily guaranteed.

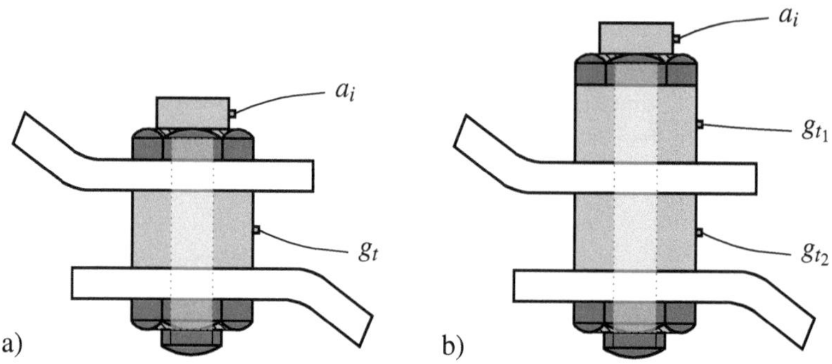

Figure 3.3 Placement of force transducers between source (top) and receiver (bottom) according to ISO 18312-1:2012. a) single, b): double force transducer.

One possible arrangement is shown in Figure 3.3 taken from ISO 18312 [3] and further examples are available in ISO 21955 [5]. Firstly, one needs to ensure that the transducer registers the actual force. As in Figure 3.3, bolted connections will often be required to achieve sufficient strength in which case a portion of the force exerted by the source on the receiver will pass through the bolt and will not be registered by the transducer. For stiff bolts, a correction to the measured (transduced) force g_t is necessary to provide an estimate of the actual force g_a,

$$g_a = g_t(1 + k_b/k_t) \tag{3.4}$$

where k_t is the stiffness of the transducer and $k_b = E_b S_b/l_b$ is the stiffness of the bolt in which E_b is its Young's modulus, S_b its cross sectional area and l_b its length. For lateral forces, the bolt stiffness is calculated using shear modulus G_b rather than Young's modulus E_b.

Figure 3.3(b) shows an alternative arrangement with a second transducer located between the bolt and the source so as to register the (reverse of the) bolt force g_{t_2}. The transduced force g_{t_2} is then corrected to give the true force according to,

$$g = g_{t_1} - g_{t_2}. \tag{3.5}$$

Next, there is the importance of capturing *all* important DoFs at the interface, a requirement that is emphasised throughout the book. ISO 18312 [3] recommends tri-axial force transducers to simultaneously capture x, y and z DoFs but makes no provision for measuring moments perhaps because commercial transducers are not available. In principle, moments can be measured with transducer pairs although this further increases the complexity of the mounting design [15].

Perhaps the greatest challenge for direct measurement though, is the need to avoid the influence of the transducers on the dynamics of the interface. It is evident from Figure 3.3 that the inserted transducers in between the source and receiver will add mass and stiffness around the interface and thereby risk altering the very phenomena one is trying to capture. For VAVP applications one is looking for transferable properties over a wide frequency range and it is often very difficult or impossible to avoid influence of the additional hardware over the whole of this range.

The physical size of the transducer also needs to be considered so that the forces are measured as close to the interface as possible. Otherwise there is a risk of a mismatch between the source and its representation in terms of measured forces.

In view of the above difficulties, while potentially useful in some situations, the direct measurement of contact force appears to be losing popularity in favour of indirect methods which are typically less invasive.

Mount stiffness method

Part 2 of ISO 18312-2:2012 describes an indirect method applicable to resiliently mounted machines employing known properties of the isolators. Essentially, one measures velocity (or acceleration) at the top (source side, c_1) and bottom (receiver side, c_2) of the isolator, their difference yielding the rate of change of compression or extension in the isolator. The force is then obtained by multiplying the velocity difference by the mount transfer impedance $\mathbf{Z}_{Ic_1c_2}$,

$$\mathbf{g}_{c_2} = -\mathbf{Z}_{Ic_1c_2}\left[\mathbf{v}_{c_1} - \mathbf{v}_{c_2}\right]. \tag{3.6}$$

This method is able to account for forces and moments although the need to know isolator properties is, depending on the frequency range required, something of a drawback. The approach is, in effect, the same as the mount stiffness method used in TPA and will be described in more detail in 5.3.2.

Inverse force synthesis

An alternative to direct measurement is known as *inverse force synthesis* or sometimes *inverse force identification*. Here, as the name suggests, the forces are obtained by *solving the inverse problem*. Such inverse approaches are of central importance in source characterisation and the main ideas will be needed later in this chapter so it is worth briefly outlining them here.

Inverse force synthesis is based around a variant of Equation 3.3. To set up the inverse problem we define a set of so-called 'indicator' DoFs consisting of points on the interface c and/or on the receiver b (see Figure 3.2). The response at these indicator DoFs is related to the contact force by,

$$\begin{pmatrix} \mathbf{v}_c \\ \mathbf{v}_b \end{pmatrix} = \begin{bmatrix} \mathbf{Y}_{Rcc} \\ \mathbf{Y}_{Rbc} \end{bmatrix} \mathbf{g}_c. \tag{3.7}$$

If the responses $\mathbf{v}_c$ and/or $\mathbf{v}_b$ are measured in an operational test, and the corresponding FRFs $\mathbf{Y}_{Rcc}$ and/or $\mathbf{Y}_{Rbc}$ are measured in a separate passive test, then the only remaining unknown in Equation 3.7 is the contact force $\mathbf{g}_c$. Provided there are at least as many equations as unknowns, which will generally be the case if there are more indicator DoFs than interface DoFs, then Equation 3.7 can in principle be solved to obtain the unknown contact forces $\mathbf{g}_c$. In words, this process addresses the question: *'what contact forces must have been present at the interface to have caused the observed responses?'*.

This inverse approach requires more sophistication than the direct measurement but there are several advantages. The main benefit is that no modification of the interface is required (apart from the need to add accelerometers which would also be required for any other method). Separation of the source and receiver substructures *is* required for measurement of the receiver FRFs, but this would also be required to insert force transducers for direct measurement. An additional advantage is that all DoFs can be included with the inverse method, including moments. The need to include particular DoFs can also be assessed using the completeness theory presented in 4.5.

How the inverse problem is solved in practice is a detailed and complex question that will be revisited at various points through the book, most notably in 7.4. For now, it is sufficient to note that, in principle, the contact forces can be estimated in this indirect way.

Independence

Having determined that contact force can be obtained practically, the last requirement from page 102 is that the source descriptor is an independent property of the source. A relationship between the internal and contact forces can be seen by equating the top lines of Equations 3.2 and 3.3, giving,

$$\mathbf{Y}_{Cci}\mathbf{f}_i = \mathbf{Y}_{Rcc}\mathbf{g}_c \tag{3.8}$$

from which (assuming the inverse $\mathbf{Y}_{Rcc}^{-1}$ exists),

$$\mathbf{g}_c = \mathbf{Y}_{Rcc}^{-1}\mathbf{Y}_{Cci}\mathbf{f}_i. \tag{3.9}$$

Unfortunately, this shows that the contact force depends on the *receiver* and is therefore not a property solely of the source as required for independence. Application of the contact force, or Equation 3.3, in virtual prototyping is therefore limited since a new contact force would need to be obtained for any modification or redesign of the assembly.

Prediction of the new contact forces is possible according to Equation 3.10 below, but lacks convenience. For example a common requirement is to predict the response in a new installation based on contact forces measured on a test bench. In this case, the contact forces in a new receiver are given by,

$$\mathbf{g}_c = [\mathbf{Y}_{Rcc} + \mathbf{Y}_{Scc}]^{-1}[\mathbf{Y}_{Tcc} + \mathbf{Y}_{Scc}]\mathbf{g}_c^{(T)} \tag{3.10}$$

where $\mathbf{g}_c^{(T)}$ are the test bench forces and $\mathbf{Y}_{Tcc}$, $\mathbf{Y}_{Scc}$ and $\mathbf{Y}_{Rcc}$ are respectively the FRFs of the test bench, source and the new receiver [5]. Thus, the new contact forces are calculable, indeed the process is outlined in ISO 21955 [5], however, it requires the test bench forces to be accompanied by the full FRF matrices of both the source and the test bench on which they were measured. Thus, lack of independence makes contact forces somewhat inconvenient as a descriptor of source activity.

Despite not being ideal for source characterisation, the contact force is encountered at various points in the book: in relation to structural power in 3.3 and the diagnostic method known as classical TPA in 5.3.1. However, for now, the search for a more suitable source characterisation continues.

3.1.3 FREE VELOCITY AND BLOCKED FORCE

In the preceding section we saw that the conceptual simplicity of contact force for VAVP applications is let down by its lack of independence. Therefore, we now continue the search for a source characterisation that achieves the four criteria on page 102, including that of independence.

In Chapter 2, we saw that in order to characterise passive components independently of their surroundings, they had to be mounted with either *free-* or *fixed-*interfaces. The same turns out to be true for source characterisation. Thus, two alternative quantities can be defined at the interfaces of a source to express its activity independently of any connected components: *free velocity* and *blocked force*. Not surprisingly, the former corresponds to free, and the latter to fixed-interface conditions.

The *free velocity* describes the velocity at the interface c of a vibration source whilst operating under a free suspension, i.e. in the absence of any connected components, as illustrated in Figure 3.4a. Mathematically, it can be defined as,

$$\tilde{\mathbf{v}}_c = \mathbf{v}_c\Big|_{\mathbf{g}_c=\mathbf{0}} \tag{3.11}$$

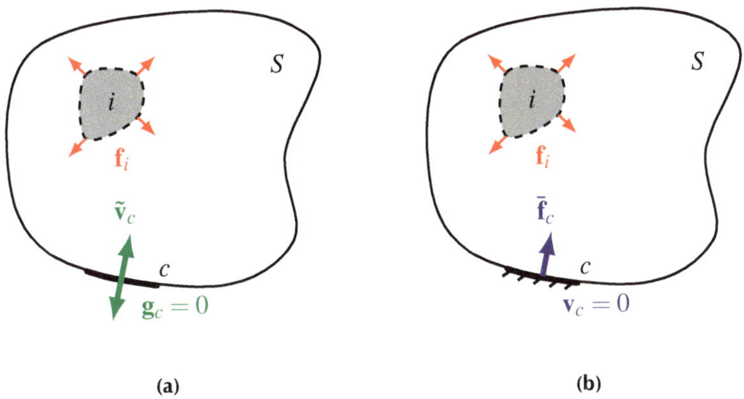

Figure 3.4 Independent active source descriptors. (a) Free velocity of a vibration source; the interface DoFs c are unconstrained and free to vibrate with their free velocity $\tilde{\mathbf{v}}_c$. (b) Blocked force of a vibration source; the interface DoFs c are rigidly constrained by the blocked force $\bar{\mathbf{f}}_c$.

where $\mathbf{v}_c$ is the velocity of the interface and $\mathbf{g}_c = \mathbf{0}$ is the constraint that no coupling/interface forces are present. In the absence of any connected components, the free velocity depends only on the vibration source: it is an independent source property.

The *blocked force* describes the reaction force at the interface c of a vibration source whilst operating under a rigid constraint, as illustrated in Figure 3.4b. Equivalently, it can be interpreted as the force which when applied externally to the interface, counteracts the motion of the interface and thus constrains its motion to zero.

Mathematically, we can define the blocked force as,

$$\bar{\mathbf{f}}_c = \mathbf{g}_c \Big|_{\mathbf{v}_c = \mathbf{0}} \tag{3.12}$$

where $\bar{\mathbf{f}}_c$ is seen as the contact force at the interface c under the rigid constraint $\mathbf{v}_c = \mathbf{0}$.

It is clear that by providing an idealised rigid constraint at the boundary, the blocked force provides an independent source description free from the influence of any specific receiver. The blocked force is therefore also an independent source property.

The free velocity and blocked force thus provide two alternative ways to describe the operational activity of a vibration source independently of its installation. The two quantities are also interrelated by the free-interface FRF of the source according to,

$$\tilde{\mathbf{v}}_c = \mathbf{Y}_{Scc} \bar{\mathbf{f}}_c. \tag{3.13}$$

The free velocity and blocked force can be further related to the internal generating force $\mathbf{f}_i$ by,

$$\tilde{\mathbf{v}}_c = \mathbf{Y}_{Sci} \mathbf{f}_i, \quad \text{and} \quad \bar{\mathbf{f}}_c = \mathbf{Y}_{Scc}^{-1} \mathbf{Y}_{Sci} \mathbf{f}_i. \tag{3.14}$$

They can also be related to the contact force $\mathbf{g}_c$ described in 3.1.2. For the free velocity we follow,

$$\mathbf{g}_c = \mathbf{Y}_{Rcc}^{-1} \mathbf{v}_c \tag{3.15}$$

$$= \mathbf{Y}_{Rcc}^{-1} \mathbf{Y}_{Cci} \mathbf{f}_i \tag{3.16}$$

$$= [\mathbf{Y}_{Rcc} + \mathbf{Y}_{Scc}]^{-1} \tilde{\mathbf{v}}_c \tag{3.17}$$

where we have used $\mathbf{Y}_{Cci} = \mathbf{Y}_{Rcc} [\mathbf{Y}_{Rcc} + \mathbf{Y}_{Scc}]^{-1} \mathbf{Y}_{Sci}$ from 5.2.1 (Equation 5.26 transposed with source DoF i in place of the receiver DoF r) along with Equation 3.14. The blocked force relation then follows straightforwardly as,

$$\mathbf{g}_c = [\mathbf{Y}_{Rcc} + \mathbf{Y}_{Scc}]^{-1} \mathbf{Y}_{Scc} \bar{\mathbf{f}}_c. \tag{3.18}$$

The above states that the independent free velocity and blocked force are related to the contact force generated at the coupling interface through the inverse sum of the source and receiver interface FRFs. Equations 3.17 and 3.18 also further illustrate the dependence of the contact force on the receiver through the FRF term $\mathbf{Y}_{Rcc}$.

As independent source quantities, the free velocity and blocked force are of fundamental theoretical importance in the field of vibro-acoustic simulation. But before going on to discuss how they are measured, we take a brief aside into electrical sources which are closely analogous and may bring some useful insight.

Electrical analogy

When analysing mechanical systems it is often convenient to consider the problem as an equivalent electrical circuit. The same can be said about the representation of vibration sources.

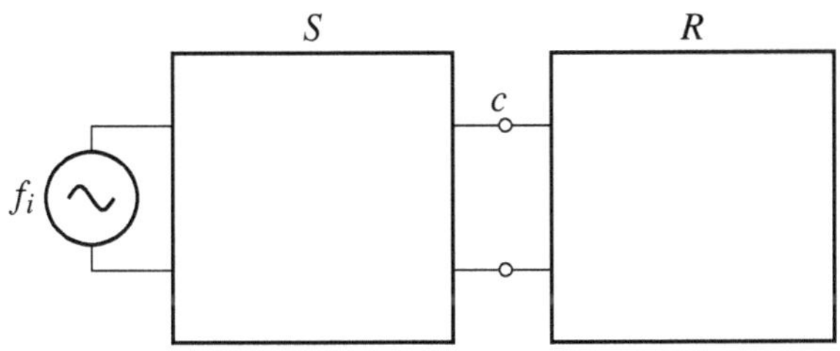

Figure 3.5 Equivalent circuit representing SR assembly. Internal forces are represented by an ideal voltage source, whilst source and receiver components are each represented by a complex network of elements (not shown) within the terminal blocks.

In the *impedance analogy*, force is taken as equivalent to voltage and velocity to current [11, 20]:

$$f \leftrightarrow V, \quad v \leftrightarrow i \quad (3.19)$$

The internal force f_i (assumed a single force for simplicity) can then be taken as an ideal voltage source driving a complex impedance network representing the dynamics of the coupled *SR* (source-receiver) assembly. An example equivalent circuit is shown in Figure 3.5.

The notion of an independent source characterisation, as provided by the blocked force or free velocity, can be understood by considering the application of Thevenin and Norton's theorem to this equivalent circuit.

Thevenin's theorem states that *any linear electrical network that contains only voltage sources, current sources and impedances can be replaced by an equivalent combination of an ideal voltage source V_{th} in series with an appropriate impedance Z_{th}* [26].

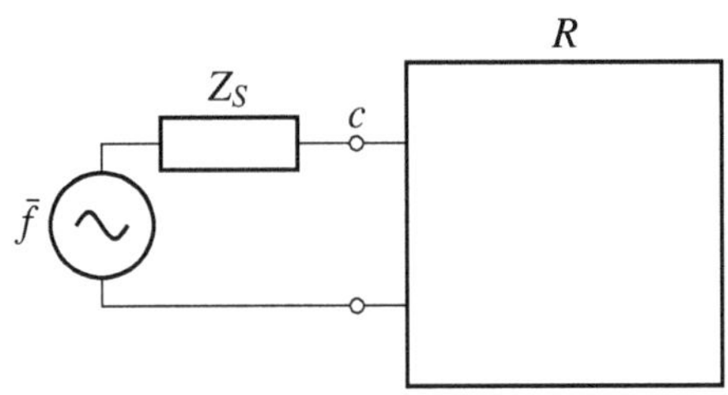

Figure 3.6 Thevenin equivalent source representation of Figure 3.5. The ideal (blocked force) voltage source in series with the impedance Z_{th} is equivalent to the complex network S and its ideal (internal force) voltage source.

The equivalent Thevenin representation of Figure 3.5 is shown in Figure 3.6. The Thevenin voltage is defined as the voltage obtained at the terminals c of the source network S when the load is removed and an open circuit condition is present. The open circuit condition represents an infinite impedance, and so no current flows. Recalling the impedance analogy ($f \leftrightarrow V$, $v \leftrightarrow i$) it is clear that this is equivalent to a rigid velocity constraint in a mechanical system. Hence, the Thevenin voltage V_{th} is equivalent to the blocked force $\bar{f}$. The impedance element Z_{th} similarly represents the mechanical impedance of the source at its interface Z_S.

The free velocity can be represented in a similar way by considering Norton's theorem, which states that *any linear electrical network that contains only voltage sources, current sources and impedances can be replaced by an ideal current source* i_{no} *in parallel with an appropriate impedance* Z_{no} [25]. The Norton current is defined as the current flowing across the terminals c of the source network S when the load is removed and a short circuit condition is present. The closed circuit condition represents a zero impedance; there is no voltage drop across it. This is clearly equivalent to the free-interface/zero force constraint in a mechanical system.

The equivalent Norton circuit is shown in Figure 3.7, where the ideal Norton current source i_{no} is equivalent to the mechanical free velocity $\tilde{v}$ and the impedance similarly represents the mechanical source at its interface $Z_{no} = Z_{th} = Z_S$. From the perspective of the receiver R, the circuits presented in Figures 3.6 and 3.7 are entirely equivalent to that of Figure 3.5. That is to say, the equivalent Thevenin voltage/blocked force produces the same electrical/mechanical response in R as the initial voltage/internal force f_i, and similarly for the equivalent Norton current.

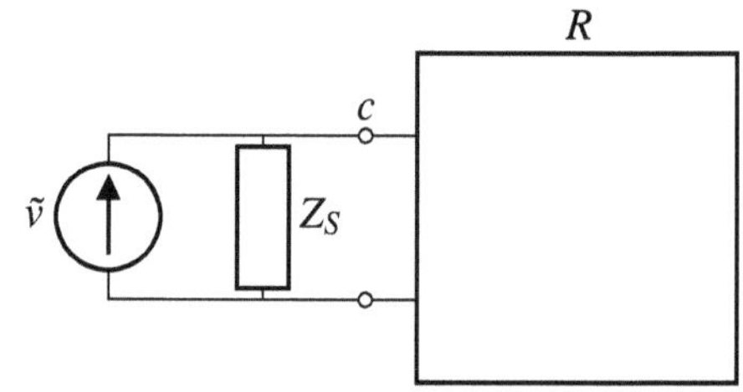

Figure 3.7 Norton equivalent source representation of Figure 3.5. The ideal (free velocity) current source in parallel with the impedance Z_{no} is equivalent to the complex network S and its ideal (internal force) voltage source.

It should be appreciated that while the electrical analogy clarifies the importance of internal source impedance, in other ways it over-simplifies the problem in that the excitation by a structural vibration source can almost never be described by a single force. Instead, the problem is inherently multi-dimensional, requiring multiple forces at different points of the connection and in different DoFs. This multi-dimensionality is however difficult to express on an equivalent circuit and is more conveniently handled with matrix-vector mathematics.

It should also be mentioned that in some publications a *mobility analogy* appears in which the equivalence of the force and velocity is reversed compared with the impedance analogy [11, 20]. However, since it will not bring any further insights we will not expand on it here. Instead, we move on to discuss the measurement of the free velocity and blocked force.

Direct measurement of free velocity

Recall that the free velocity is defined by the free-interface boundary condition where external forces are zero: $\mathbf{g}_c = 0$. It is important to recognise that this boundary condition is *idealised* and hence can never be achieved exactly in practice. If however a reasonable approximation can be obtained, then in principle the direct measurement of free velocity is possible. In practice, the free condition is approximated by resiliently supporting the source in some way. Once supported, the free velocity is then measured directly at the interface DoFs whilst the source is operating. This approach is not viable for sources that require rigid support to operate normally, for example steering systems, but is potentially useful for equipment that can be resiliently mounted, for example diesel generators, compressors, fans etc.

Guidelines for measurement of free velocity are given in ISO 9611:1996 [8]. It recommends measuring velocity in six DoFs in third octave bands at every mounting point. Accelerometer pairs are employed to give translational and rotational velocity by sum and finite difference of signals, respectively (see 4.2). As written, the procedure is not suitable for VAVPs for reasons described later, but some of its guidelines can be adapted for the component-based approach.

The free boundary condition is naturally approximated only over a specific frequency range as was discussed in the context of a passive component characterisation in 2.3.1. The low frequency limit is related to the question 'how soft is soft?' with reference to the mount stiffness. The advice of the standard [8] is that the results are valid above a lower limiting frequency, f_1 which is three times *'the highest of the six natural frequencies of the system'*, f_0. The mounted natural frequencies are calculated from a simple mass-spring model where the mass is the machine mass and the spring the soft mount (see 2.4.1). Alternatively, they may be found by performing separate FRF tests, applying excitation in different directions and looking for a peak in the magnitude. Above this lower limiting frequency it can be assumed that the reaction forces from the isolators do not change the velocity of the machine supports, in other words the mounts are sufficiently soft that the free condition is well approximated.

Another test of the free-interface condition is to run separate tests with and without the isolators in place (if possible). A difference of less than 2 dB in velocity with and without the isolator is said to indicate that the influence of the mount is negligible. This is more likely to be a practicable test using artificial excitation of the source structure with a hammer or shaker than with the real excitation.

One way to avoid altering the dynamics locally at the interface, as for the measurement of a free-interface FRF, is to suspend the source at connections away from the interface, as illustrated in Figure 3.8 (though this is not mentioned in [8]).

The upper limiting frequency, f_2 is the highest frequency for which the point-like assumption is valid. Above this frequency the connections become 'flexible' and deform so that single accelerometer pair may not capture all the significant motion. This condition is discussed in relation to interfaces in Chapter 4.

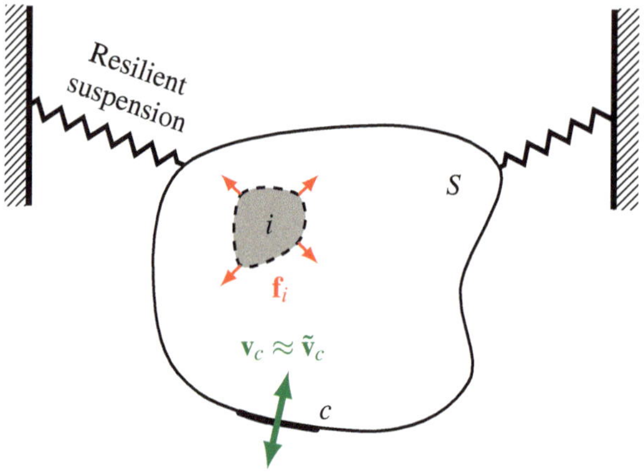

Figure 3.8 Direct measurement of free velocity.

An important practical point is to ensure that the measured free velocity is not influenced by 'flanking transmission'. This could occur due to structural connections other than the isolators, such as pipes, cables, ducting etc. The standard recommends that any such connections are fitted with flexible sections for the test to minimise flanking contributions. Flanking by airborne sound is also possible, especially for lightweight sources. The standard recommends use of sound absorbing material and potentially perforating any large flat areas of the structure to reduce their susceptibility to airborne excitation.

Having established suitable conditions, the standard recommends measurement of free velocity in third octave bands. However, source data in this form does not allow transformation to the time domain for auralisation (see 7.6.2) and hence our procedures must now depart from the standard. For VAVP purposes, free velocity must be measured in the time domain and a common phase reference retained between all DoFs, either by measuring all simultaneously or, if not possible, by leaving a reference channel in place throughout the measurements. In this way all elements in the free velocity vector can be described in terms of their magnitude and phase. Further discussion of working with operational data and conversion from third octave bands is given in 7.5 and 7.6.2.

Although theoretically valid and standardised as long ago as 1996, there are remarkably few results in the literature showing successful prediction of installed responses based on free velocity data. Apparently, transfer of resiliently mounted 'test bench' data to rigidly mounted installations is not as straightforward as it might appear. One reason is undoubtedly the impracticality for many sources of providing realistic operation when resiliently mounted. Another likely reason is that engineers could not make the method work well enough to yield results of publishable quality. It seems likely with hindsight that problems would have occurred at the substructuring step rather than because of some inherent problem with the free velocity itself. Perhaps more convincing predictions would now be possible with modern equipment and a greater understanding of the need for detailed interface descriptions in substructuring calculations. To date however, there are no signs of increased interest in the free velocity, perhaps because more attractive alternatives are now available as will be described later.

Direct measurement of blocked force

The appropriate boundary condition for blocked forces is that the interface is fully constrained, $\mathbf{v}_c = \mathbf{0}$. Again, this is an idealised condition that can only be approximated in practice, typically by attaching the source to a rigid support such as a rigid floor or a large blocking mass. If the fixed condition can be adequately approximated then the blocked forces can in principle be measured directly by inserting force transducers between the source and the support. However, in practice this is far from straightforward.

The challenges in directly measuring forces have already been described in relation to contact forces in 3.1.2. These challenges apply equally to a blocked force measurement which is, after all, simply the contact force obtained when the source is attached to a rigid receiver. The reader is therefore referred to 3.1.2 for discussion of transducer mounting.

The additional challenge for blocked forces is in approximating an infinitely rigid support over as wide a frequency range as possible. It can be assumed that a blocked condition is met if the mobility of the test bench (TB) receiver is significantly lower than that of the source, commonly taken as a factor of 10,

$$|Y_{TB}| \leq |Y_S|/10. \quad (3.20)$$

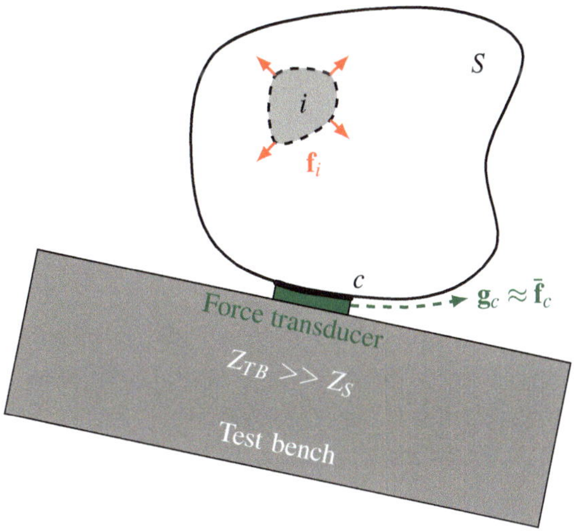

Figure 3.9 Direct measurement of blocked force.

It should be understood that this condition must apply simultaneously to all significant DoFs at every interface connection. Typically this will require special fixings to be constructed. Even if suitable mountings can be designed though, even the most rigid test bench will eventually enter a modal/flexible regime. Likewise, most sources will display FRF anti-resonances, and will hence appear rigid, at some frequencies. Thus, it is normally not possible to ensure the condition is met over a wide frequency range.

For example, in reference [40], directly measured blocked forces of a steering system were compared with *in situ* measurements (described later in 3.2). In the range up to 6 kHz, three frequency ranges were identified where the blocked condition was not met and the blocked forces were therefore underestimated. At other points, the *in situ* results were shown to be in error, which illustrates the difficulty of measurements of this type. This is one of few examples of validation of directly measured blocked forces in the literature, although there are undoubtedly other unpublished cases.

In theory, the blocked condition is most easily achieved for 'soft' sources, such as resiliently mounted machines where the mounts form part of the machine so that the source has a relatively high mobility over a wide frequency range. However, this is often not practical if the machine has to be run under load (the above mentioned steering systems for example).

To summarise so far, although conceptually simple and satisfying all the theoretical requirements, direct measurement of both free velocity and blocked force is far from straightforward in practical terms.

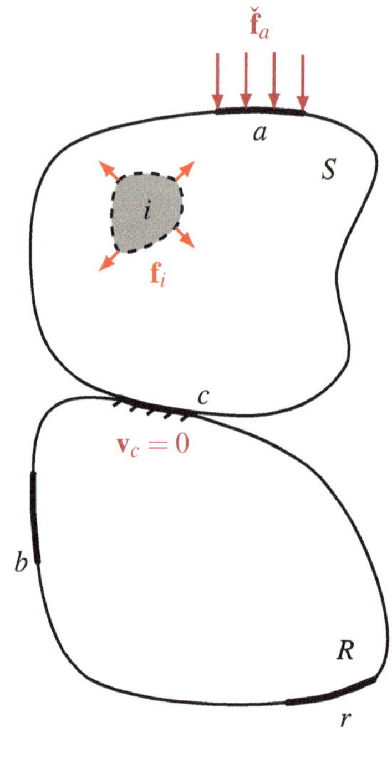

Figure 3.10 Pseudo-forces applied to remote source DoFs a combine with internal forces to cancel interface motions and nullify receiver response field.

3.1.4 PSEUDO-FORCES

The contact force, free velocity and blocked force share a common trait in that they represent source activity by physical motions or forces at the interface, i.e. ones that actually occur when the source operates under the appropriate boundary conditions. Hence, provided the boundary conditions can be adequately approximated, these quantities can be measured directly. However, there is no requirement that the quantity used to represent source activity is a physically present quantity. This is the idea behind the so-called *pseudo-force*, illustrated in Figure 3.10.

In short, the pseudo-force describes a set of hypothetical forces which, when applied on the exterior of the source, induce the same response field in the attached receiver structure as the internal forces,

$$\begin{pmatrix} v_r \\ p_r \end{pmatrix} = \begin{bmatrix} \mathbf{Y}_{Cra} \\ \mathbf{H}_{Cra} \end{bmatrix} \check{\mathbf{f}}_a. \tag{3.21}$$

These pseudo-forces $\check{\mathbf{f}}_a$ are not physical forces; they do not naturally occur whilst the source operates. Rather, they are artificial, or external, forces which equivalently represent the operational source.

In order to reproduce the exact response field, the pseudo-forces are defined such that when applied in the opposite direction whilst the source is operating, they have the effect of cancelling (i.e. constraining) the interface response,

$$\mathbf{v}_c = \mathbf{Y}_{Cci}\mathbf{f}_i - \mathbf{Y}_{Cca}\check{\mathbf{f}}_a = \mathbf{0}. \tag{3.22}$$

Note that if the interface response is constrained, the response field within the receiver is also nullified. Hence, for the same $\check{\mathbf{f}}_a$ we can also write for the remote b DoFs,

$$\mathbf{v}_b = \mathbf{Y}_{Cbi}\mathbf{f}_i - \mathbf{Y}_{Cba}\check{\mathbf{f}}_a = \mathbf{0}. \tag{3.23}$$

With the receiver response field nullified, the dynamic influence of the receiver on the source is removed, and so the pseudo-force is in fact an invariant source property.

Equation 3.22 gives rise to the pseudo/internal-force relation,

$$\check{\mathbf{f}}_a = \mathbf{Y}_{Cca}^{-1}\mathbf{Y}_{Cci}\mathbf{f}_i \tag{3.24}$$

where, for now, it is assumed that $\mathbf{Y}_{Cca} \in \mathbb{C}^{N_c \times N_a}$ with $N_c = N_a$, and its inverse exists (note that this can be generalised for cases where $N_a > N_c$ by using the matrix pseudo inverse).

When applied in the absence of the internal force, the pseudo-force gives,

$$\mathbf{v}_c = \mathbf{Y}_{Cca}\check{\mathbf{f}}_a = \mathbf{Y}_{Cca}\mathbf{Y}_{Cca}^{-1}\mathbf{Y}_{Cci}\mathbf{f}_i \tag{3.25}$$

which reduces to the original internal force response relation,

$$\mathbf{v}_c = \mathbf{Y}_{Cci}\mathbf{f}_i. \tag{3.26}$$

A similar result is obtained from Equation 3.23 for $\mathbf{v}_b$. Hence, the pseudo-force reproduces the exact response along the interface and across the entire receiver. We can thus write,

$$\begin{pmatrix} \mathbf{v}_c \\ \mathbf{v}_b \end{pmatrix} = \begin{bmatrix} \mathbf{Y}_{Cca} \\ \mathbf{Y}_{Cba} \end{bmatrix} \check{\mathbf{f}}_a. \tag{3.27}$$

which forms the basis of an indirect estimation of the pseudo-force by solving an inverse problem, similar to that described in relation to contact forces in 3.1.2.

As with the inverse contact force estimation, Equation 3.27 requires a two stage measurement procedure with an operational phase and a separate passive test.

Equation 3.24, including as it does coupled assembly FRFs, may at first appear to imply that the pseudo forces are a property of the installation rather than of the source. However, with reference to 2.4.2, the product of FRF matrices is seen to be a generalised transmissibility so that $\mathbf{Y}_{Cca}^{-1}\mathbf{Y}_{Cci} = \mathbf{Y}_{Sca}^{-1}\mathbf{Y}_{Sci}$. The right-hand side, and hence also the pseudo forces, are therefore seen to depend only on the source.

In addition to the requirement that the number of indicator DoFs is sufficient to determine the forces (i.e. N_c and N_b together should equal or exceed the number of forces to be determined), it is also necessary that the number of pseudo-force DoFs N_a is sufficient to provide complete control over the response at the interface, i.e. there must be at least as many pseudo-forces as interface DoFs, or $N_a \geq N_c$.

The inverse estimation of pseudo forces requires more sophisticated hardware and processing than direct measurement of blocked force or free velocity. However, there is a significant advantage in that the FRF required is that of the *assembly* rather than of the receiver which was the case for contact forces. This means that the pseudo-forces can be obtained entirely *in situ*, i.e. *without removing the source from its installation* and without disturbing the interface.

The pseudo-force approach thus ticks all the boxes for a source characterisation and, as an *in situ* method, has a promising additional advantage. Its one drawback is that the forces are defined at arbitrary points on the source (a) so they have no clear physical meaning. It also means that different pseudo-force positions will yield different pseudo-forces which makes direct comparison between different sources, impossible. Their abstract nature is perhaps why, despite other advantages, the pseudo-forces have not been widely taken up within industry.

3.2 IN SITU BLOCKED FORCE METHOD

Despite its obvious appeal, the direct measurement of both free velocity and blocked force has been seen to be fraught with difficulties. By the mid-2000s neither method had made much headway and there was still a long way to go before digital representations could convincingly replace real source components in a virtual prototype. At that time, although the theory was well known, the problem of transferring source data from a test bench to a real installation could still not be reliably solved. It was thought that the idealised, 'extreme' boundary conditions required for blocked force and free velocity measurement were making things more difficult by being far removed from any real installation. Engineers had therefore started to look for *in situ* methods where the characterisation could take place in a test rig that was dynamically closer to the eventual installation. The pseudo forces approach showed that *in situ* measurements were indeed possible. What was now needed was a less abstract representation. A breakthrough came around 2007 with the introduction of the *in situ* blocked force method [32].

To understand the *in situ* blocked force method we consider Figure 3.11. We have already discussed that the response field in the *free body receiver* excited by the

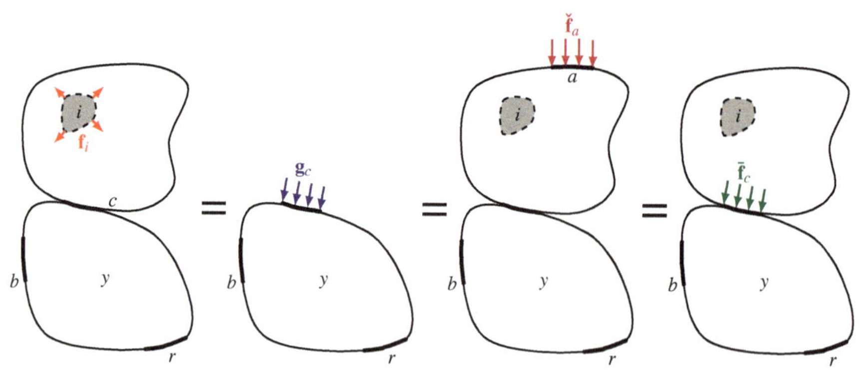

Figure 3.11 Identical response fields y are produced in the receiver with (left) the real source in operation ($\mathbf{f}_i$); (2nd left) the contact forces ($\mathbf{g}_c$) applied to the free body receiver; (3rd left) pseudo-forces ($\check{\mathbf{f}}_a$) applied to the source; (right) the *blocked forces* ($\bar{\mathbf{f}}_c$) applied to the *assembly* at the interface.

contact forces (2nd left of Figure 3.11) is identical to that with the real source in operation (1st left of Figure 3.11). The pseudo-force representation, introduced in the previous section (3rd left), also reproduces this field identically through externally applied forces on the body of the source component.

The situation on the right of Figure 3.11 is less well known but turns out to be highly significant: it says that the identical response field is again generated in the receiver when the interface of the *assembly* is excited by the *blocked forces* of the source. Mathematically, this is expressed as,

$$\begin{pmatrix} \mathbf{v}_c \\ \mathbf{v}_b \end{pmatrix} = \begin{bmatrix} \mathbf{Y}_{Ccc} \\ \mathbf{Y}_{Cbc} \end{bmatrix} \bar{\mathbf{f}}_c \tag{3.28}$$

where c and b are indicator DoFs located respectively on the interface and elsewhere on the receiver and $\bar{\mathbf{f}}_c$ is the blocked force vector.

Since the FRFs ($\mathbf{Y}_{Ccc}, \mathbf{Y}_{Cbc}$) and the operational receiver responses ($\mathbf{v}_c$, $\mathbf{v}_b$) in Equation 3.28 can be obtained by measurement, the only remaining unknown is the blocked force vector. Therefore, just as we are able to solve Equation 3.7 by inversion to obtain the contact force, we are now able to use exactly the same approach to solve Equation 3.28 for the blocked forces. The only difference is that instead of separating the source and receiver substructures for the FRF test, they are left in the assembled state. This inverse procedure forms the basis of the *in situ* blocked force method that has proved highly significant in the context of component-based simulation.

To apply the inverse procedure one first selects a number of indicator locations, either on the interface (c), elsewhere in the receiver (b), or both. A two stage measurement is then employed:

1. In the *passive* phase, the source is switched off and the FRFs are measured between all indicator and interface DoFs,
2. In the *operational* phase, the machine is operated in a desired loading regime and the response measured at all indicator DoFs,

Throughout both phases the source remains attached to the receiver so that the testing is conducted entirely *in situ*.

The final stage is the actual solution of the inverse problem, for example premultiplying both sides of Equation 3.28 by the pseudo-inverse of the FRF matrix gives,

$$\bar{\mathbf{f}}_c = \begin{bmatrix} \mathbf{Y}_{Ccc} \\ \mathbf{Y}_{Cbc} \end{bmatrix}^+ \begin{pmatrix} \mathbf{v}_c \\ \mathbf{v}_b \end{pmatrix}. \tag{3.29}$$

In practice, getting good results from the inversion is a delicate operation as will be discussed further in 7.4. For now though, it is sufficient to know that a viable route to obtaining blocked forces from *in situ* measurements does exist.

Referring back to 3.1.4, the *in situ* blocked force relation can be seen as a special case of pseudo-forces in which the latter are defined specifically at the interface rather than at some arbitrary points on the source. Unlike the more general concept of pseudo-forces though, this special case has a clear physical interpretation and is unique to a given source.

Several important advantages arise as a result of being able to test *in situ*:

- it is usually faster and more convenient not to have to separate the source and receiver substructures for the passive FRF test;
- apart from fixing accelerometers, the interface is left completely undisturbed throughout both the operational and passive phases of the test which avoids intractable difficulties of inserting force transducers without altering the forces;
- the blocked forces are obtained without having to attempt the impossible task of physically blocking the boundary; in effect, an ideal blocked condition is applied mathematically as part of the matrix inversion process;
- the ability to independently characterise the source without detaching it from the installation opens up application for a wider range of sources including embedded sources (3.2.2) and moving sources (10.4).

Example 3.1: Experimental examples of *in situ* blocked force

To illustrate the in situ blocked force relation (Equation 3.29), and to verify the invariance of the blocked force compared with the contact force, we consider two simple laboratory examples. The first uses a single interface connection between two beam-like components. By using two different length receiver beams, we demonstrate the contact force's dependence on, and the blocked force's invariance to, the receiver dynamics. In the second example a more complex, three footed, source is considered. We show that the same blocked force is obtained whether the source is rigidly or resiliently connected to a frame-like receiver, again illustrating the invariance of the blocked force.

Shown in Figure 3.12 is the experimental set-up for the first example. A steel source beam was rigidly connected to two different length receiver beams in turn. Internal forces in the 'source' beam were simulated with a hammer hit at location away from the interface. The sensor arrangement shown in Figure 3.12 (right) was designed to capture six Dof: three translational DoFs (x, y, z) and their corresponding rotations by finite difference approximation (see 4.2.1). The blocked force and contact force were both obtained by inverse methods. For the contact force, the *uncoupled* interface mobility $\mathbf{Y}_{Rcc} \in \mathbb{C}^{6\times 6}$ was used and for the blocked force the *coupled* interface mobility $\mathbf{Y}_{Ccc} \in \mathbb{C}^{6\times 6}$. The same operational velocity $\mathbf{v}_c \in \mathbb{C}^6$ was used in both cases.

Figure 3.12 Experimental case study for single interface beams (left) and interface instrumentation (right).

Shown in Figure 3.13 are the contact forces in the vertical z direction obtained from the two assemblies. It is clear that the dynamics of the receiver are present in the contact force, which differ considerably from one another. In comparison, Figure 3.14 shows the blocked forces in the vertical z direction obtained from the two assemblies. It is clear that the dynamics of the receiver are not present in the blocked forces which, within experimental error, are in agreement with one another. Figures 3.13 and 3.14 clearly illustrate the invariance of the blocked force, compared with the contact force.

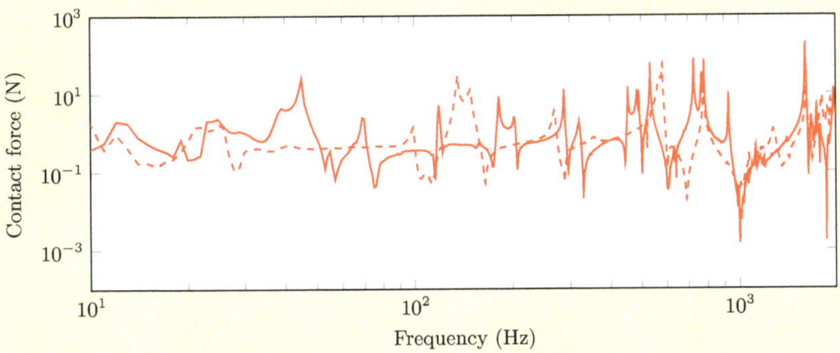

Figure 3.13 Contact forces obtained with source installed on two different receivers.

Let us now consider our second example. A three footed vibration source (a servo-motor bolted to a steel plate), was connected both rigidly and resiliently to a beam frame structure (Figure 3.15) using the same block-like interface as in Figure 3.12, yielding a total of 21 measurement DoFs. For each case the blocked force was obtained in situ using the coupled interface mobility $\mathbf{Y}_{Ccc} \in \mathbb{C}^{21\times21}$ and the operational velocity $\mathbf{v}_c \in \mathbb{C}^{21}$. Again, operational forces were simulated with a hammer hit on the source plate (which excites a wide range of frequencies to better illustrate the behaviour).

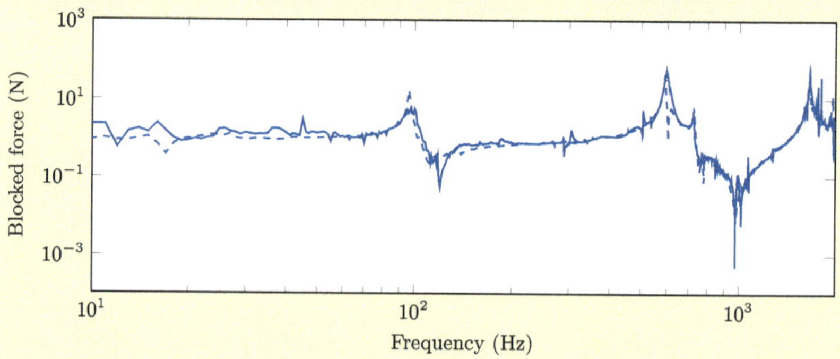

Figure 3.14 Blocked forces obtained with source installed on two different receivers.

Figure 3.15 Experimental case study for 3 footed source.

The same sensor arrangement was used as in the previous example (Figure 3.13) along with a finite difference approximation to obtain the translations and rotations. Although this may appear to involve a lot of effort, it is essential to capture *all* important DoFs at the interface as will be discussed in relation to the concept of completeness in Chapter 4.

Figure 3.16 shows the operational velocities measured at a single foot of the coupled vibration source, in the vertical z direction, for both rigid (red) and resiliently (blue) coupled cases. The effect of contrasting boundary conditions (i.e. rigid vs. resilient) is evident here with the two velocities displaying very different characteristics over the entire frequency range. Clearly, the coupled velocity is not invariant.

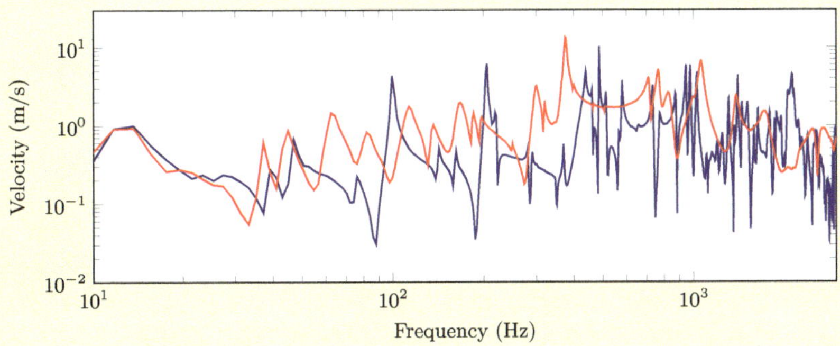

Figure 3.16 Example coupled velocities (magnitude only) in the vertical z direction at a single foot for rigid (blue) and resiliently (red) coupled assemblies.

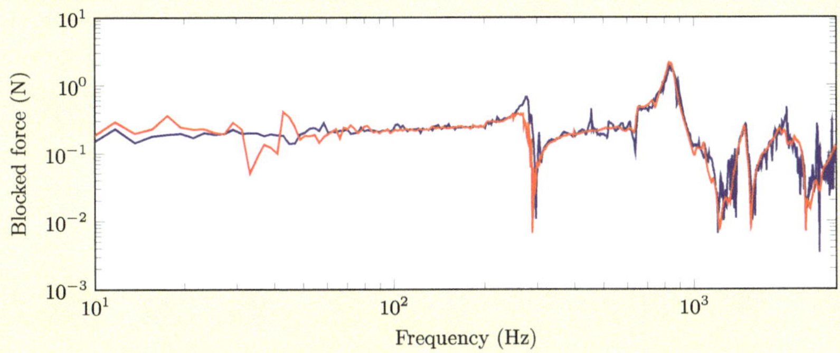

Figure 3.17 Example blocked forces (magnitude only) in the vertical z direction at a single foot obtained from rigid (blue) and resiliently (red) coupled assemblies.

Shown in Figure 3.17 are the in situ obtained blocked forces (obtained as per Equation 3.29) at the same foot, in the vertical z direction, for both rigid (red) and resiliently (blue) coupled cases. It is clear from the agreement between the two cases that the acquired blocked force is invariant to the source boundary conditions i.e. it is an independent source property. Note that similar agreement is obtained for the blocked force of the remaining DoFs, though for brevity these are not shown here [29].

Although these structures are in some ways simple, they are strongly resonant which makes them highly challenging, and of a complexity at least comparable to many industrial applications.

3.2.1 PROOF OF THE IN SITU BLOCKED FORCE RELATION

In what follows, we present first, a simple 'thought experiment' proof followed by an impedance-based derivation of the in situ blocked force relation. Various alternative derivations are also available in the literature [18, 32].

First, the somewhat counter-intuitive result in Equation 3.28 can be derived in a thought experiment by invoking the principle of linear superposition as illustrated in Figure 3.18. Recall that the blocked forces can be considered as the forces that, when applied 'in reverse' to the interface with the machine in operation, will completely suppress the response field at that interface (and therefore at all points in the receiver too). We can think of these applied blocking forces as generating a secondary wave field in the receiver that is equal in magnitude, but opposite in phase to the original wave field such that their sum is zero. If these forces are applied at the interface so as to nullify the interface response, and the machine is then switched off, as in Figure 3.18, we are left with a response field that must be simply the reverse of the original field. Reversing the sign then yields the blocked force relationship shown in Figure 3.11 and Equation 3.28.

We now consider a more extensive proof, beginning with the general system of equations that describe the behaviour of the rigidly coupled source-receiver (C) assembly, shown in Figure 3.1. With the source being an active substructure, there exist some internal vibration mechanism, represented by the set of unknown excitation forces $\mathbf{f}_i$, acting at the internal DoFs denoted i. If we consider also the two sets of DoFs representing the separating interface c and some remote receiver b positions, the equation of motion of the coupled system is given by,

$$\begin{bmatrix} \mathbf{f}_i \\ \mathbf{0} \\ \mathbf{0} \end{bmatrix} = \begin{bmatrix} \mathbf{Z}_{Sii} & \mathbf{Z}_{Sic} & \mathbf{0} \\ \mathbf{Z}_{Sci} & \mathbf{Z}_{Ccc} & \mathbf{Z}_{Rcb} \\ \mathbf{0} & \mathbf{Z}_{Rbc} & \mathbf{Z}_{Rbb} \end{bmatrix} \begin{bmatrix} \mathbf{v}_i \\ \mathbf{v}_c \\ \mathbf{v}_b \end{bmatrix}. \tag{3.30}$$

Note that the block tri-diagonal nature of the assembly's impedance matrix, i.e. $\mathbf{Z}_{Cib} = \mathbf{Z}_{Cbi} = 0$, is a consequence of the constrained interface DoFs, c, that separate i and b, as per the definition of impedance (see 2.2.3). This tri-diagonal structure is what ultimately gives rise to the in situ blocked force relation. Importantly, it assumes that the interface representation is complete, i.e. the DoF set c contains all DoFs at the interface. Any missing DoFs (i.e. an incomplete interface) will give rise to a non-tri-diagonal impedance matrix, from which no in situ relation can be derived.

Equation 3.30 may be written out explicitly and the first two rows rearranged so as to eliminate the internal response $\mathbf{v}_i$. The remaining two equations may be subsequently recast into matrix form. This reduction process will be referred to hereafter as the 'condensation of rows 1 and 2'. The reduced matrix equation is,

$$\begin{bmatrix} -\mathbf{Z}_{Sci}\mathbf{Z}_{Sii}^{-1}\mathbf{f}_i \\ \mathbf{0} \end{bmatrix} = \begin{bmatrix} -\mathbf{Z}_{Sci}\mathbf{Z}_{Sii}^{-1}\mathbf{Z}_{Sic} + \mathbf{Z}_{Ccc} & \mathbf{Z}_{Rcb} \\ \mathbf{Z}_{Rbc} & \mathbf{Z}_{Rbb} \end{bmatrix} \begin{bmatrix} \mathbf{v}_c \\ \mathbf{v}_b \end{bmatrix}. \tag{3.31}$$

Whilst Equations 3.30 and 3.31 represent the same assembly, in Equation 3.31 the excitation is represented by alternate forcing term, $-\mathbf{Z}_{Sci}\mathbf{Z}_{Sii}^{-1}\mathbf{f}_i$, which is applied to

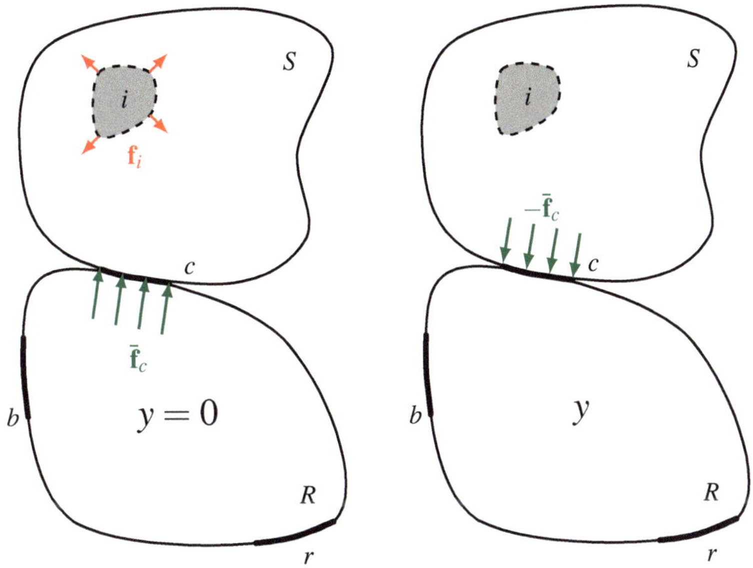

(a) Blocked forces applied in reverse to the interface of the operating machine suppress the velocity field at the interface and in the receiver

(b) Blocked forces alone reproduce the original response field y

Figure 3.18 Thought experiment to derive blocked forces relationship.

the interface DoFs c. This 'equivalent force' [18] is an independent property of the source substructure that, when acting on the interface c, results in the same dynamic behaviour as the internal force $\mathbf{f}_i$. It will now be shown that this equivalent force term is in fact the blocked force that results from a constrained coupling interface.

Consider the system of equations describing a source substructure whose interface DoFs are constrained such that their response is $\mathbf{0}$,

$$\begin{bmatrix} \mathbf{f}_i \\ \bar{\mathbf{f}}_c \end{bmatrix} = \begin{bmatrix} \mathbf{Z}_{Sii} & \mathbf{Z}_{Sic} \\ \mathbf{Z}_{Sci} & \mathbf{Z}_{Scc} \end{bmatrix} \begin{bmatrix} \mathbf{v}_i \\ \mathbf{0} \end{bmatrix}. \tag{3.32}$$

For such an assembly the force acting on the interface c is, by definition, the blocked force, $\bar{\mathbf{f}}_c$. Condensation of rows 1 and 2 whilst solving for $\bar{\mathbf{f}}_c$ yields,

$$\bar{\mathbf{f}}_c = \mathbf{Z}_{Sci}\mathbf{Z}_{Sii}^{-1}\mathbf{f}_i. \tag{3.33}$$

Referring back to Equation 3.31 it can be seen that the equivalent force acting at the interface c is in fact the negative blocked force, $-\bar{\mathbf{f}}_c$.

Lastly, it can be shown that the application of the negative blocked force at the coupling interface whilst the internal excitation, $\mathbf{f}_i$, is inactive, results in the same system of equations as that of the internally operating force, as in Equations 3.31-3.32,

$$\begin{bmatrix} \mathbf{0} \\ -\bar{\mathbf{f}}_c \\ \mathbf{0} \end{bmatrix} = \begin{bmatrix} \mathbf{Z}_{Sii} & \mathbf{Z}_{Sic} & \mathbf{0} \\ \mathbf{Z}_{Sci} & \mathbf{Z}_{Ccc} & \mathbf{Z}_{Rcb} \\ \mathbf{0} & \mathbf{Z}_{Rbc} & \mathbf{Z}_{Rbb} \end{bmatrix} \begin{bmatrix} \mathbf{v}_i \\ \mathbf{v}_c \\ \mathbf{v}_b \end{bmatrix}. \tag{3.34}$$

Let us once again condense rows 1 and 2 of Equation 3.34. The resulting matrix equation,

$$\begin{bmatrix} -\bar{\mathbf{f}}_c \\ \mathbf{0} \end{bmatrix} = \begin{bmatrix} -\mathbf{Z}_{Sci}\mathbf{Z}_{Sii}^{-1}\mathbf{Z}_{Sic} + \mathbf{Z}_{Ccc} & \mathbf{Z}_{Rcb} \\ \mathbf{Z}_{Rbc} & \mathbf{Z}_{Rbb} \end{bmatrix} \begin{bmatrix} \mathbf{v}_c \\ \mathbf{v}_b \end{bmatrix} \tag{3.35}$$

can be seen to be in complete agreement with Equation 3.31. This confirms that the application of the negative blocked force at the coupling interface, whilst the internally operating forces are inactive, yields the same dynamic response as the internally operating forces alone.

Let us now consider the inverse of Equation 3.34 where the coupled *mobility* matrix, $\mathbf{Y}_C$, is introduced,

$$\begin{bmatrix} \mathbf{v}_i \\ \mathbf{v}_c \\ \mathbf{v}_b \end{bmatrix} = \begin{bmatrix} \mathbf{Y}_{Cii} & \mathbf{Y}_{Cic} & \mathbf{Y}_{Cib} \\ \mathbf{Y}_{Cci} & \mathbf{Y}_{Ccc} & \mathbf{Y}_{Ccb} \\ \mathbf{Y}_{Cbi} & \mathbf{Y}_{Cbc} & \mathbf{Y}_{Cbb} \end{bmatrix} \begin{bmatrix} \mathbf{0} \\ -\bar{\mathbf{f}}_c \\ \mathbf{0} \end{bmatrix}. \tag{3.36}$$

By considering only the responses at the external DoFs, c and b, the above mobility formulation may be reduced to,

$$\begin{bmatrix} \mathbf{v}_c \\ \mathbf{v}_b \end{bmatrix} = -\begin{bmatrix} \mathbf{Y}_{Ccc} \\ \mathbf{Y}_{Cbc} \end{bmatrix} \bar{\mathbf{f}}_c . \tag{3.37}$$

Equation 3.37 states that the velocity of a coupled source-receiver assembly at the coupling interface and remote receiver DoFs, is given by the product of the coupled point and transfer mobility matrices, with the negative blocked force of the source substructure.

It is worth noting that the negative sign that accompanies the blocked force arises due to *our definition* of the blocked force in Equation 3.32 being positive. With the blocking force being a constraint that opposes motion, one could just as well use $-\bar{\mathbf{f}}_c$ to represent this reaction force in Equation 3.32. In this case we would lose the negative sign in Equation 3.37, which is the form most commonly presented in the literature. In practice the negative sign makes no difference, as we are typically interested in the magnitude of the response, and so it can be ignored with no consequence.

3.2.2 EMBEDDED SOURCES

In practical cases we often encounter the situation of a *source within a source*. Take a steering system for example, which may be seen as an individual source component within a larger assembly (a vehicle), but equally it may be seen as an assembly itself in which several smaller source components are embedded, an electric motor and a belt drive for example. It is of clear practical interest to relate the characterisation of the larger source to those of the embedded sources and this is the subject of this section.

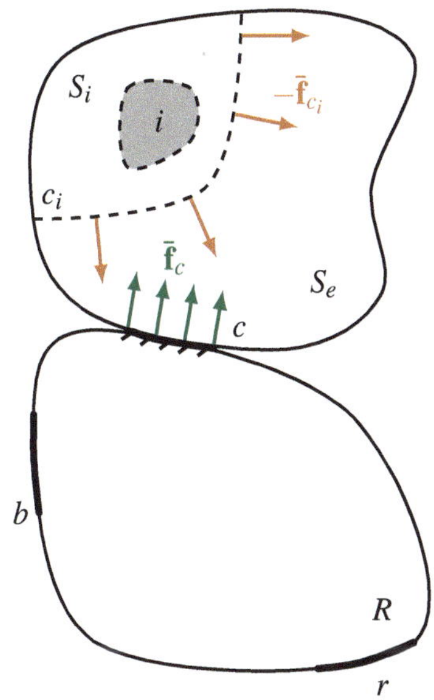

Figure 3.19 Coupled source-receiver structure with an *internal* blocked force $\bar{\mathbf{f}}_{c_i}$ propagated to the external interface c. The dotted line describes the interface of a source within a source.

Characterisation of blocked forces begins by first defining a source-receiver interface which implicitly distinguishes the source of vibration from the remainder of the assembly. However, the interface location is in principle arbitrary, as long as it completely encloses all operating mechanisms. Added to this is the fact that the *in situ* blocked force method allows characterisation of a source without physically separating the substructures either side of the interface. Together these two facts allow great flexibility in the choice of the interface even allowing non-separable portions of a structure to be defined as sources [45–47]. Thus, an embedded source, or a *source within a source* can be said to occur where there exists an internal interface within the defined source that more closely encapsulates some part of the generating mechanisms.

We are interested in representing the same operating forces of a source as blocked forces on either the external or internal interfaces, as illustrated in Figure 3.19. For example, if the source is a steering system, the external interface would comprise the connection points to the vehicle and the internal interface could be a surface somewhere inside the housing that more closely encloses one of the active components, like the motor. We are also interested in propagating the blocked forces between these two interfaces.

The source component S is excited internally by some set of operational forces at the DoFs i. From the point of view of the receiver, this excitation can be represented by an equivalent excitation at the internal source interface c_i, corresponding to a negative blocked force. The equation of motion describing the system is then,

$$\begin{pmatrix} -\bar{\mathbf{f}}_{c_i} \\ \mathbf{0} \\ \mathbf{0} \end{pmatrix} = \begin{bmatrix} \mathbf{Z}_{Sc_ic_i} & \mathbf{Z}_{S_ec_ic} & \mathbf{0} \\ \mathbf{Z}_{S_ecc_i} & \mathbf{Z}_{S_eRcc} & \mathbf{Z}_{Rcr} \\ \mathbf{0} & \mathbf{Z}_{Rrc} & \mathbf{Z}_{Rrr} \end{bmatrix} \begin{pmatrix} \mathbf{v}_{c_i} \\ \mathbf{v}_c \\ \mathbf{v}_r \end{pmatrix} \tag{3.38}$$

where $S = S_e \cup S_i$ denotes the combined internal and external portions of the source, and $\mathbf{Z}_{Sc_ic_i}$ is the point impedance matrix at the internal c_i interface. Similarly, $\mathbf{Z}_{S_eRcc}$ is the point impedance matrix at the c interface.

Choosing c_i as the source-receiver interface, Equation 3.38 considers the source definition to include only the subcomponent S_i. Choosing the source receiver interface to be c would instead define the source to include the entire source component

S. In this case the equation of motion can be written as,

$$\begin{pmatrix} \mathbf{0} \\ -\bar{\mathbf{f}}_c \\ \mathbf{0} \end{pmatrix} = \begin{bmatrix} \mathbf{Z}_{Sc_ic_i} & \mathbf{Z}_{S_ec_ic} & \mathbf{0} \\ \mathbf{Z}_{S_ecc_i} & \mathbf{Z}_{S_eRcc} & \mathbf{Z}_{Rcr} \\ \mathbf{0} & \mathbf{Z}_{Rrc} & \mathbf{Z}_{Rrr} \end{bmatrix} \begin{pmatrix} \mathbf{v}_{c_i} \\ \mathbf{v}_c \\ \mathbf{v}_r \end{pmatrix} \tag{3.39}$$

where again, the blocked force $\bar{\mathbf{f}}_c$ is used to represent the internal operating forces. Note that $\bar{\mathbf{f}}_c$ is a property of the complete source S, whilst $\bar{\mathbf{f}}_{c_i}$ is a property of the internal source S_i alone. Equation 3.38 and 3.39 both reproduce the exact response field in the attached receiver, differing only in how they define, or represent, the source.

We now look for a relation between these two representations that will allow us to propagate blocked forces from internal interfaces (c_i) to external interfaces (c) (and *vice versa*). To do so we consider Equation 3.38, where the blocked force $\bar{\mathbf{f}}_{c_i}$ is applied as an external force to the internal interface c_i. By applying an appropriate blocking force at the external interface c, we are able to enforce the constraint $\mathbf{v}_c = 0$ (and consequently $\mathbf{v}_r = 0$),

$$\begin{pmatrix} -\bar{\mathbf{f}}_{c_i} \\ \bar{\mathbf{f}}_c \\ \mathbf{0} \end{pmatrix} = \begin{bmatrix} \mathbf{Z}_{Sc_ic_i} & \mathbf{Z}_{S_ec_ic} & \mathbf{0} \\ \mathbf{Z}_{S_ecc_i} & \mathbf{Z}_{S_eRcc} & \mathbf{Z}_{Rcr} \\ \mathbf{0} & \mathbf{Z}_{Rrc} & \mathbf{Z}_{Rrr} \end{bmatrix} \begin{pmatrix} \mathbf{v}_{c_i} \\ \mathbf{0} \\ \mathbf{0} \end{pmatrix}. \tag{3.40}$$

From the top row of Equation 3.40 we obtain,

$$\mathbf{v}_{c_i} = -\left(\mathbf{Z}_{Sc_ic_i}\right)^{-1} \bar{\mathbf{f}}_{c_i} \tag{3.41}$$

which upon substitution into the second row yields,

$$\bar{\mathbf{f}}_c = -\mathbf{Z}_{S_ecc_i} \left(\mathbf{Z}_{Sc_ic_i}\right)^{-1} \bar{\mathbf{f}}_{c_i}. \tag{3.42}$$

Equation 3.42 relates the internal and external interface blocked forces (each representing the internal excitation at i), through what is essentially a generalised force transmissibility (see 2.4),

$$\bar{\mathbf{f}}_c = -\mathbf{T}^f_{Scc_i} \bar{\mathbf{f}}_{c_i} \tag{3.43}$$

with,

$$\mathbf{T}^f_{Scc_i} = \mathbf{Z}_{S_ecc_i} \left(\mathbf{Z}_{Sc_ic_i}\right)^{-1}. \tag{3.44}$$

As shown in 2.4, this transmissibility can be represented in terms of mobility FRFs instead as,

$$\mathbf{T}^f_{Scc_i} = -\mathbf{Y}^{-1}_{Ccc} \mathbf{Y}_{Ccc_i}. \tag{3.45}$$

Given an internal blocked force at c_i, Equation 3.43 propagates the blocked force downstream to the external interface c. The inverse relation is also readily available by pre-multiplying both sides of Equation 3.43 by the inverse transmissibility matrix,

$$\bar{\mathbf{f}}_{c_i} = \left(\mathbf{T}^f_{Scc_i}\right)^{-1} \bar{\mathbf{f}}_c. \tag{3.46}$$

Equation 3.46 provides an upstream propagation of the blocked force, from the external interface c to the internal interface c_i.

The above theory can be generalised to include multiple internal sources by including additional blocked forces within $\bar{\mathbf{f}}_{c_i}$ and extending the corresponding transmissibility matrices. A more complete treatment of this problem can be found in [45, 47].

The propagation of blocked forces, as described above, can also be used to investigate the influence of *structural modification* when made on subcomponents within a defined source [28]. Termed *Component Replacement TPA*, this methodology can be used, for example, to investigate how modifying an isolator effects the blocked force obtained at its base.

3.2.3 EQUIVALENT FIELD THEOREM

The *in situ* blocked force relation can be seen as a special case of a more general theorem on the representation of the fields of forced vibrations of a composite elastic system proposed by Bobrovnitskii in 2001 [13]. In it, he shows that the general problem of two subsystems (a special case being the *SR* assembly considered in this chapter) coupled along an interface, each excited by a set of internal and/or external forces, can be treated instead as the sum of two simpler problems, termed 'auxiliary problems'. This auxiliary approach was applied to two different representations which will be referred to here as the 'fixed-interface' and 'free-interface' representations as they relate to the concepts of the blocked force and free velocity respectively.

In the *fixed-interface* representation (see Figure 3.20), the initial auxiliary problem (Aux. 1) considers an assembly whose interface is rigidly constrained and the force

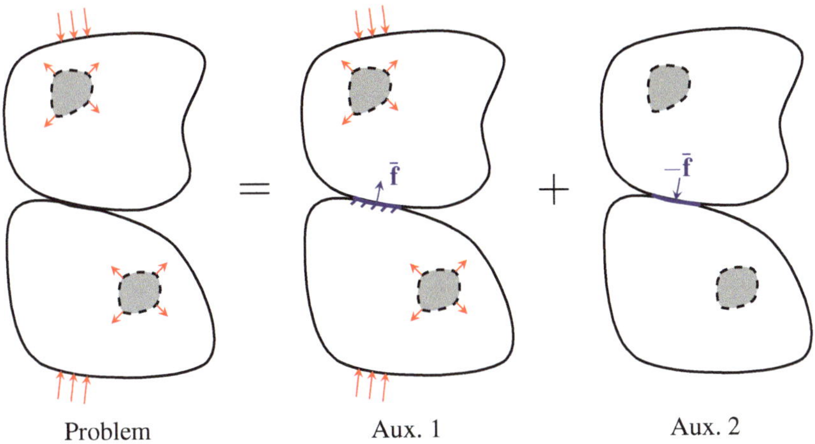

Problem Aux. 1 Aux. 2

Figure 3.20 Fixed interface representation, adapted from [13]. Internal and external forces represented by red arrows. Interface blocking force represented by blue arrow.

densities that arise along it due to the internal and external forcing in each substructure. Given the fixed-interface, each substructure's forcing can influence only its own field of vibration. In the second auxiliary problem (Aux. 2) the internal and external forcing is removed (switched off), and the (negative) blocking force is reapplied to the interface, causing a vibration field in both substructures. In [13] Bobrovnistkii shows that by adding together the two auxiliary fields, the original field is reproduced exactly in both substructures.

In the context of an *SR* assembly, where only one substructure presents internal or external forces and the other is passive, the general theory above reduces to the *in situ* blocked force method described in Section 3.2.

In the *free-interface* representation (see Figure 3.21), in the initial auxiliary problem (Aux. 1) the two substructures are separated, such that their interfaces are free. Given their separation, each substructure's forcing can again influence only its own field of vibration. In Aux. 2, the internal and external forcing is removed, and the relative velocity between the two free substructures (from Aux. 1) is applied to the interface, resulting in fields of vibration in both substructures. By adding together the two auxiliary fields, the original field is reproduced exactly. In the context of a *SR* assembly, where only one substructure presents internal or external forces, the relative velocity reduces simply to the free velocity, as introduced in 3.1.3.

The equivalent field theorem (EFT) presented above provides a connection between the invariant source representations that are the blocked force and free velocity, and the response field observed in the coupled assembly. It thus provides a theoretical basis for the in situ characterisation of these quantities. Furthermore, the

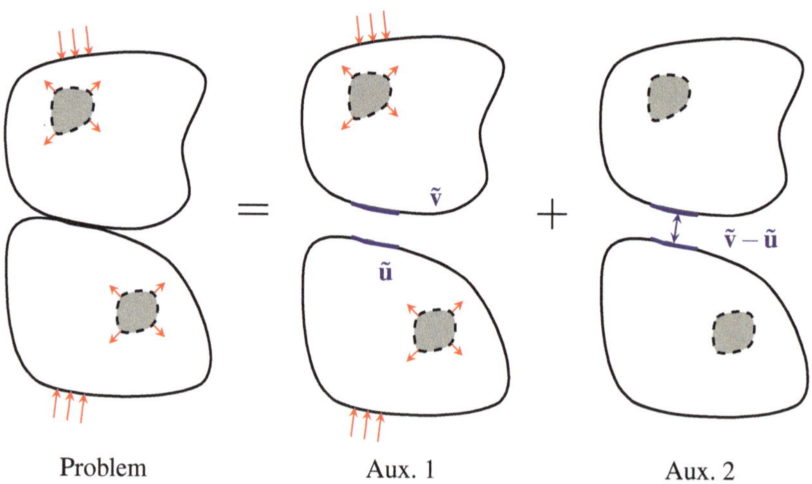

Figure 3.21 Free interface representation, adapted from [13]. Internal and external forces represented by red arrows. Free interface relative velocity represented by blue arrow.

generality of the EFT extends the above to include acoustic substructures, where the blocked force and free velocity are simply replaced by blocked pressure and free volume velocity, respectively.

With regards to its practical use, it turns out that the fixed-interface representation (blocked force) is straightforward to implement in the structural case, requiring only external forces to be applied to the interface and yielding the in situ blocked force method (see 3.2). The free-interface representation (free velocity), though straightforward in theory, is not so practical experimentally. Its implementation would require the application of a relative velocity excitation at the interface. Experimentally this is quite challenging. If we consider the acoustic case however, we have quite the opposite; the free-interface representation has a straightforward implementation, requiring a volume velocity excitation at the interface, whilst the fixed-interface representation is more challenging, requiring a direct pressure excitation.

In summary, the EFT provides a general theoretical basis for the in situ characterisation of both structural and acoustic sources by means of their blocked force and free volume velocities, respectively.

3.3 POWER-BASED METHODS

In the idealised free or fixed states considered above, no power is delivered to the receiver. Yet, intuitively, the power injected by a source must be in some way proportional to the target response at remote locations; for example the sound pressure in a ship's cabin due to an installed pump must relate to the power injected by that pump into the ship structure. Therefore, the question arises whether a source descriptor based on power may be more representative of the true importance of the source.

At first sight the power, a scalar frequency-dependent quantity, seems far simpler than vectors of interface forces or velocities. Airborne acoustic sources are most commonly characterised by their 'sound power', a single frequency-dependent quantity. The sound power is transferable to new installations, due to the impedance of air being approximately constant, and therefore allows direct comparison of sources as well as prediction of sound pressure when installed. In this section, we ask whether the same approach could be applied to structure-borne sound sources.

The sound power approach had already been standardised for several decades by the 1980s. Early attempts at characterising structure-borne sound sources attempted to harness its convenience [30, 39], but it gradually became clear that sound power is only transferable because two conditions are usually met. First, acoustic sources are typically 'hard' i.e. the impedance of the radiating surfaces is much greater than that of the receiving medium, in other words the surrounding air does not influence the velocity of radiating surfaces. Secondly, the receiving medium is always the same, air, whose impedance can be considered constant irrespective of where the source is located. This assumption breaks down inside an enclosed space and within a fraction of a wavelength of a reflective surface. Though, surfaces that are always present, such as hard ground, can be included as part of the test setup, so in practice this does not present a serious limitation. The power transmitted into the receiving air from many acoustic sources can therefore be taken as invariant, consistent with the component-based approach.

Structural vibration sources on the other hand are typically attached to solid structures with impedance that may vary strongly with frequency and from one installation to the next. Furthermore, the impedance of the source and receiver is often 'matched' that is, of similar magnitude (see 3.3.5). So neither of the above assumptions can be relied on except perhaps over a limited frequency range. The power delivered from the same vibration source may vary significantly depending on the receiver. Thus, the delivered power is a property of the *installation* rather than of the source alone and is not suitable for source characterisation. Nevertheless, power is an important fundamental quantity for installations and a detailed description is warranted.

3.3.1 POWER THROUGH A SINGLE POINT

If a single force is applied to a passive structure at a point, the power flowing into the structure is defined as the time average of the product of the contact force g and velocity v,

$$P = \langle \mathrm{Re}(g)\mathrm{Re}(v) \rangle_t = \frac{1}{2} Re(g^* v) \tag{3.47}$$

where $g = Ge^{i(\omega t + \phi_g)}$ and $v = Ve^{i(\omega t + \phi_v)}$ are the measured force and velocity at radian frequency ω, G and V are the amplitudes, ϕ_g and ϕ_v are phase angles and $\langle \square \rangle_t$ denotes average over a complete number of cycles. Physically we know that the power must flow from the active element into the passive element, so the sign convention for the force is generally taken so as to make the power positive into the receiver structure.

Note, from the right-hand form of Equation 3.47 that one must take the complex conjugate of the force vector before multiplication: $P = \frac{1}{2}\mathrm{Re}(g^* v) \neq \frac{1}{2}\mathrm{Re}(gv)$. This can be derived as follows,

$$\begin{aligned} \mathrm{Re}(g)\mathrm{Re}(v) &= \tfrac{1}{4}[g + g^*][v + v^*] = \tfrac{1}{4}[gv + g^* v^* + gv^* + g^* v] \\ &= \tfrac{1}{2}[\mathrm{Re}(gv) + \mathrm{Re}(g^* v)] \end{aligned} \tag{3.48}$$

which after after substituting for g and v yields,

$$P = \frac{1}{2} GV \langle \cos(2\omega t + \phi_g + \phi_v) + \cos(\phi_v - \phi_g) \rangle_t. \tag{3.49}$$

Figure 3.22 shows the instantaneous power which is not smooth but pulsed through the interface at twice the radian frequency. The pulsing corresponds to the first term in Equation 3.49 which however averages to zero over a complete number of cycles. It is the second, constant term that represents the time-averaged 'active' or transmitted power with $\cos(\phi_v - \phi_g)$ the important 'power factor'. Note that the instantaneous power is negative over part of the cycle in the example in Figure 3.22, a point which will be picked up in 3.3.3. From now on we will assume averaging over a complete number of cycles.

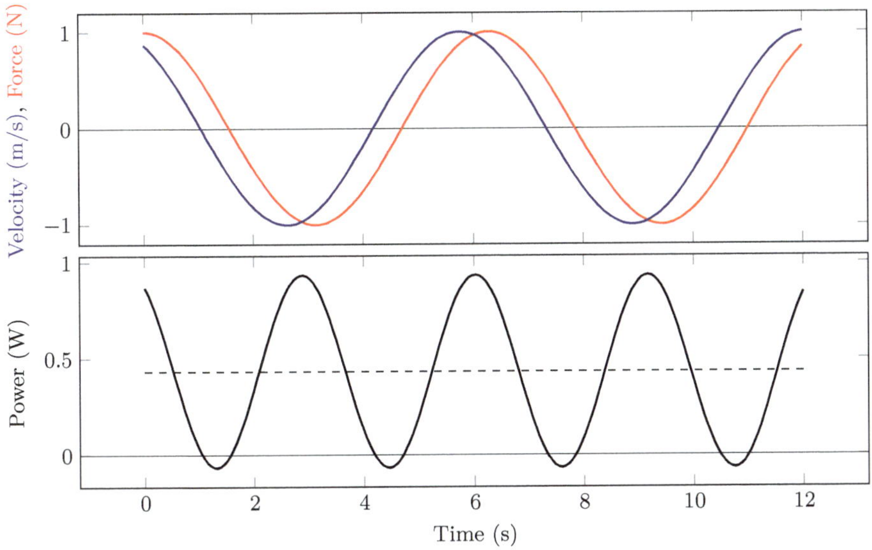

Figure 3.22 Upper: force and velocity. Lower: Instantaneous power. The instantaneous power is negative over part of the cycle in this example indicating some circulation of power, or reactive power. The time-averaged power (dashed plot) must be non-negative.

3.3.2 POWER THROUGH MULTI-DEGREE OF FREEDOM INTERFACES

The power transmitted through a multi-point interface is simply the sum of the power flowing through each point. This is conveniently expressed as,

$$P = \frac{1}{2}[g_{c_1}^* v_{c_1} + g_{c_2}^* v_{c_2} + \cdots + g_{c_n}^* v_{c_n}] = \frac{1}{2}\mathrm{Re}(\mathbf{g}_c^{\mathrm{H}} \mathbf{v}_c) \tag{3.50}$$

which is a generalisation of the right-hand side of Equation 3.47, where $\square^{\mathrm{H}}$ is the conjugate (Hermitian) transpose. The form on the right-hand side of Equation 3.50 is recognised as the scalar product of the force and velocity vectors and yields a scalar – a single frequency-dependent quantity. Note that the power through some points can be (and often is) negative, provided that the net transfer energy into the passive structure is positive.

Equation 3.50 is also valid when multiple forces occur in different directions provided that the force and velocity are correctly ordered so as to form a series of 'power pairs'. For example, a force is multiplied by a velocity in the same axis, and a moment by an angular velocity about the same axis so as to yield power in both cases. The pairing of force and response pairs is relevant to reciprocity as discussed in 2.2.7.

3.3.3 ACTIVE AND REACTIVE POWER

In Figure 3.22 it was seen that the instantaneous power may be negative over part of the cycle. This indicates some *circulation* of power backwards and forwards known as *reactive* power, in contrast to the *active* power described above which is permanently transmitted.

The active and reactive power can be interpreted from Equation 3.49: when force and velocity are in phase then the power factor from Equation 3.49 $\cos(\phi_v - \phi_g) = 1$. In this case, the transmitted power is a maximum and the transmission is purely active. When force and velocity are in quadrature however, $(\phi_v - \phi_g) = \pm\pi/2$, the power factor is zero and no power is permanently transmitted. However, the fluctuating term in Equation 3.49 still applies so that power is actually pumped backwards and forwards through the interface from source to receiver and back with no net transfer. In this situation the power transmission is purely reactive. Reactive power is well known in the field of loudspeaker design [11] since locally circulating power is not useful for listeners in the far field. In the structural domain, although no net power is delivered, the reactive power may cause fluctuation in the response locally to the source which may be perceived if the listener is in the nearfield.

Active and reactive power can be represented as the real and imaginary parts of a complex power,

$$Q = \frac{1}{2}\mathbf{g}_c^{\mathrm{H}}\mathbf{v}_c \tag{3.51}$$

The real part is the *active* power as defined above. Whenever the imaginary part is non-zero there will be negative power transmission over part of the cycle from the fluctuating term in Equation 3.49 (see Figure 3.22). Thus, the imaginary part corresponds to the *reactive* power.

The complex power captures, in a single frequency-dependent number, the overall effect of the source on a particular receiver. The contribution of each DoF can be seen by expanding the terms in the sum (see Equation 3.50).

It is helpful to write the power in terms of the receiver mobility. We start by writing the velocity in the receiver as,

$$\mathbf{v}_c = \mathbf{Y}_{Rcc}\mathbf{g}_c. \tag{3.52}$$

Substituting into Equation 3.51 we obtain,

$$Q = \frac{1}{2}\mathbf{g}_c^{\mathrm{H}}\mathbf{Y}_{Rcc}\mathbf{g}_c. \tag{3.53}$$

The result, Equation 3.53, is a quadratic form in which matrix-vector multiplications yield a scalar quantity. Rewriting the real part of Equation 3.53 we get [24],

$$P = \mathrm{Re}(Q) = \frac{1}{2}[Q + Q^*] = \frac{1}{4}\mathbf{g}_c^{\mathrm{H}}[\mathbf{Y}_{Rcc} + \mathbf{Y}_{Rcc}^*]\mathbf{g}_c \tag{3.54}$$

and hence,

$$P = \mathrm{Re}(Q) = \frac{1}{2}\mathbf{g}_c^{\mathrm{H}}\mathrm{Re}(\mathbf{Y}_{Rcc})\mathbf{g}_c. \tag{3.55}$$

The perhaps surprising result is that the real part of the power, the transmitted power, is entirely governed by the real part of the mobility matrix.

The imaginary parts too must be similarly related so we can write,

$$Q = \frac{1}{2}\mathbf{g}_c^{\mathrm{H}}[\mathrm{Re}(\mathbf{Y}_{Rcc}) + i\,\mathrm{Im}(\mathbf{Y}_{Rcc})]\mathbf{g}_c. \tag{3.56}$$

These results are helpful for interpretation of FRF data: they show that a structure's ability to absorb power is quantified by the real part of its mobility matrix. Plate-like and highly damped structures have a predominantly real mobility and tend to absorb power well especially when they are large compared with a structural wavelength. Note that the real part of mobility must always be positive since a passive substructure can only absorb, not create, power. This property can be used as a cross check on the physicality of mobility data, a technique which will be picked up in 6.3.2. Similarly, a structure's ability to store and release power, reactive power, relates to the imaginary part of its mobility. Spring or mass-like structures tend to be predominantly reactive – able to store but not absorb power. Unlike the real part, the phase of the imaginary part may be positive (corresponding to spring-like behaviour) or negative (mass-like).

A similar result to Equation 3.58 can be obtained instead in terms of velocity as,

$$Q = \frac{1}{2}\mathbf{v}_c^{\mathrm{H}}\mathbf{Z}_{Rcc}^{\mathrm{H}}\mathbf{v}_c = \frac{1}{2}\mathbf{v}_c^{\mathrm{H}}[\mathrm{Re}(\mathbf{Z}_{Rcc}) - i\,\mathrm{Im}(\mathbf{Z}_{Rcc})]\mathbf{v}_c. \tag{3.57}$$

Comparing Equations 3.56 and 3.57 we see that the active power is governed by the real parts of both the mobility and impedance, and the reactive power by their imaginary parts. This may appear counter-intuitive at first since we know that the real and imaginary parts are not simply related as inverses, i.e. $\mathrm{Re}(\mathbf{Z}) \neq [\mathrm{Re}(\mathbf{Y})]^{-1}$. The real (imaginary) parts of the mobility and impedance are in fact related as follows,

$$\mathrm{Re}(\mathbf{Z}) = (\mathbf{Y}^{\mathrm{H}}\mathbf{Y})^{-1}\mathrm{Re}(\mathbf{Y}) \quad \mathrm{Im}(\mathbf{Z}) = -(\mathbf{Y}^{\mathrm{H}}\mathbf{Y})^{-1}\mathrm{Im}(\mathbf{Y}) \tag{3.58}$$

and when we consider that the square matrix $[\mathbf{Y}^{\mathrm{H}}\mathbf{Y}]$ has purely real eigenvalues we can see that the real part of the mobility is a scaled version of the real part of impedance and similarly for the imaginary parts. This helps to explain their relationship with active and reactive power.

Note that if accelerance is used rather than mobility then one must be careful to correctly interpret the real and imaginary parts. Using Table 2.1,

$$\mathrm{Re}(\mathbf{Y}) = \mathrm{Re}\left(\frac{\mathbf{A}}{i\omega}\right) = \frac{\mathrm{Im}(\mathbf{A})}{\omega} \tag{3.59}$$

and

$$\mathrm{Im}(\mathbf{Y}) = \frac{-\mathrm{Re}(\mathbf{A})}{\omega}. \tag{3.60}$$

Thus, it is the *imaginary* part of the accelerance that governs the transmitted power and it follows that it must always be positive. The relationships for other FRF types can be derived from Table 2.1 and are summarised in Table 3.1.

Table 3.1
Frequency response functions as they relate to power

FRF	Active power	Reactive power
Receptance	$-\mathrm{Im}(R) \times \omega$	$\mathrm{Re}(R) \times \omega$
Mobility	$\mathrm{Re}(Y)$	$\mathrm{Im}(Y)$
Accelerance	$\mathrm{Im}(A)/\omega$	$-\mathrm{Re}(A)/\omega$
Dynamic stiffness	$-\mathrm{Im}(D) \times \omega$	$-\mathrm{Re}(D) \times \omega$
Impedance	$\mathrm{Re}(Z)$	$-\mathrm{Im}(Z)$
Apparent mass	$\mathrm{Im}(\hat{M})/\omega$	$\mathrm{Re}(\hat{M})/\omega$

Having laid out the basic concepts we now go on to consider how power is obtained from measurements.

3.3.4 MEASUREMENT OF POWER

Power cannot be measured directly but can be calculated as the product of measured contact forces with the corresponding velocities. Measurement is normally relatively straightforward for velocity but far more challenging for contact forces. To a large extent then, measurement of power comes down to measurement of contact forces, a subject that was addressed earlier in this chapter. In 3.1.2, the three most common methods were outlined: direct measurement, mount stiffness method and inverse force synthesis. A fourth method, calculating forces and power from component properties is particularly relevant for VAVPs and is described in the following (3.3.5).

Having obtained contact forces, by any method, one then simply combines them with the velocities to calculate power by applying Equation 3.50.

For the interested reader, ISO 18312 [3] specifically addresses the measurement of power, with part 1 considering direct measurement of forces [3] and part 2 the mount stiffness method [4].

3.3.5 COMPONENT-BASED POWER AND IMPEDANCE MATCHING

We now consider the case when the power is transmitted from an active substructure to a passive receiver substructure and how it depends on the properties of the individual components.

The derivation leads on from Equation 3.53, but first the contact forces must be expressed in terms of the independent blocked forces of the source. To do this, note that the velocity of the interface can be expressed in terms of either contact forces $\mathbf{g}_c$ or blocked forces $\bar{\mathbf{f}}_c$,

$$\mathbf{v}_c = \mathbf{Y}_{Rcc}\mathbf{g}_c = \mathbf{Y}_{Ccc}\bar{\mathbf{f}}_c. \tag{3.61}$$

Rearranging gives,

$$\mathbf{g}_c = \mathbf{Y}_{Rcc}^{-1}\mathbf{Y}_{Ccc}\bar{\mathbf{f}}_c \tag{3.62}$$

Table 3.2
Useful relationships between power, force and velocity.

Contact force in terms of free velocity	$\mathbf{g}_c = [\mathbf{Y}_{Rcc} + \mathbf{Y}_{Scc}]^{-1}\tilde{\mathbf{v}}_c$
Contact force in terms of blocked force	$\mathbf{g}_c = [\mathbf{Y}_{Rcc} + \mathbf{Y}_{Scc}]^{-1}\mathbf{Y}_{Scc}\bar{\mathbf{f}}_c$
Velocity in terms of free velocity	$\mathbf{v}_c = \mathbf{Y}_{Rcc}[\mathbf{Y}_{Rcc} + \mathbf{Y}_{Scc}]^{-1}\tilde{\mathbf{v}}_c$
Velocity in terms of blocked force	$\mathbf{v}_c = \mathbf{Y}_{Rcc}[\mathbf{Y}_{Rcc} + \mathbf{Y}_{Scc}]^{-1}\mathbf{Y}_{Scc}\bar{\mathbf{f}}_c$
Free velocity in terms of blocked force	$\tilde{\mathbf{v}}_c = \mathbf{Y}_{Scc}\bar{\mathbf{f}}_c$
Complex power in terms of velocity	$Q = \frac{1}{2}\mathbf{v}_c^{\mathrm{H}}\mathbf{Y}_{Rcc}^{-\mathrm{H}}\mathbf{v}_c$
Complex power in terms of force	$Q = \frac{1}{2}\mathbf{g}_c^{\mathrm{H}}\mathbf{Y}_{Rcc}\mathbf{g}_c$

We also need to express the coupled structure mobility matrix $\mathbf{Y}_{Ccc}$ in terms of the component source and receiver mobilities $\mathbf{Y}_{Scc}$ and $\mathbf{Y}_{Rcc}$,

$$\mathbf{Y}_{Ccc} = [\mathbf{Y}_{Rcc}^{-1} + \mathbf{Y}_{Scc}^{-1}]^{-1} = \mathbf{Y}_{Rcc}[\mathbf{Y}_{Rcc} + \mathbf{Y}_{Scc}]^{-1}\mathbf{Y}_{Scc}. \tag{3.63}$$

Substituting Equation 3.63 into 3.62 gives a relationship between the contact forces and blocked forces,

$$\mathbf{g}_c = [\mathbf{Y}_{Rcc} + \mathbf{Y}_{Scc}]^{-1}\mathbf{Y}_{Scc}\bar{\mathbf{f}}_c. \tag{3.64}$$

To do this we can recall Equation 3.18, which can be substituted into Equation 3.53 to give the complex power in terms of independent component properties,

$$Q = \frac{1}{2}(\bar{\mathbf{f}}_c^{\mathrm{H}}\mathbf{Y}_{Scc}^{\mathrm{H}}[\mathbf{Y}_{Rcc} + \mathbf{Y}_{Scc}]^{-\mathrm{H}})(\mathbf{Y}_{Rcc}[\mathbf{Y}_{Rcc} + \mathbf{Y}_{Scc}]^{-1}\mathbf{Y}_{Scc}\bar{\mathbf{f}}_c). \tag{3.65}$$

Equivalently, using Equation 3.17, in terms of free velocity,

$$Q = \frac{1}{2}(\tilde{\mathbf{v}}_c^{\mathrm{H}}[\mathbf{Y}_{Rcc} + \mathbf{Y}_{Scc}]^{-\mathrm{H}})(\mathbf{Y}_{Rcc}[\mathbf{Y}_{Rcc} + \mathbf{Y}_{Scc}]^{-1}\tilde{\mathbf{v}}_c). \tag{3.66}$$

In both the above equations the left pair of round brackets encloses the contact forces and the right pair the interface velocities for easier comparison with Equation 3.51.

Equations 3.65 and 3.66 can be interpreted in several ways for which the relationships in Table 3.2 may be helpful.

Equations 3.65 and 3.66 demonstrate how the power transmitted through an interface depends on the mobilities of the source and receiver structures. The trends are most easily observed by considering the power through a single point in which case Equation 3.66 reduces to,

$$Q = \frac{1}{2}|\tilde{v}_c|^2 \frac{Y_{Rcc}}{|Y_{Rcc} + Y_{Scc}|^2} \tag{3.67}$$

as plotted in Figure 3.23.

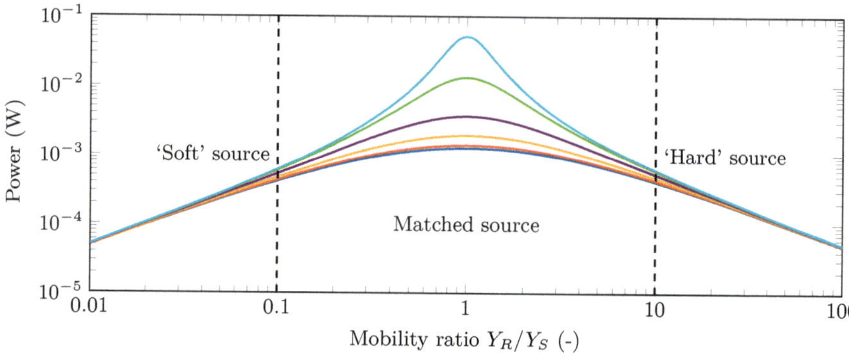

Figure 3.23 Transmitted power as a function of source and receiver mobility matching and relative phase $e^{i\theta}$ (from lowest to highest: $\theta = 0, 0.2\pi, 0.4\pi, 0.6\pi, 0.8\pi, 0.99\pi$). Note the dependence on the relative phase of the mobilities when their magnitudes are closely matched.

Figure 3.23 can be referred to as the *impedance matching* plot in which three regions can be identified: sources can be 'soft', 'hard' or 'matched' depending on the ratio of source to receiver mobility.

An example of a *soft* source is a flexible machine on a rigid receiver structure where the force imparted to the receiver approximates the *blocked force*. Here, we assume that the mobility of the receiver is negligible compared with that of the source so that Equation 3.65 reduces to,

$$Q \approx \frac{1}{2}\bar{\mathbf{f}}_c^{\mathrm{H}}\mathbf{Y}_{Rcc}\bar{\mathbf{f}}_c. \tag{3.68}$$

An example of a *hard* source is a heavy machine mounted on a flexible support structure such as a resilient mat (where the mat forms part of the receiver). In this case the source is much less mobile than the receiver and the velocity of the machine at the interface then approximates the *free velocity*. Equation 3.66 then reduces to,

$$Q \approx \frac{1}{2}\tilde{\mathbf{v}}_c^{\mathrm{H}}\mathbf{Y}_{Rcc}^{-\mathrm{H}}\tilde{\mathbf{v}}_c = \frac{1}{2}\tilde{\mathbf{v}}_c^{\mathrm{H}}\mathbf{Z}_{Rcc}^{\mathrm{H}}\tilde{\mathbf{v}}_c. \tag{3.69}$$

Soft and hard vibration sources are also known respectively as 'constant force' and 'constant velocity' sources, or simply 'force' and 'velocity' sources. There is a clear analogy with constant current and constant voltage sources in electrical engineering (see 3.1.3). Within both regimes the calculation of power is significantly simplified since, as seen in Equations 3.68 and 3.69, the source mobility is not required. Thus, hard and soft sources can be characterised solely by their active properties, free velocity and blocked force respectively, although the power does not characterise the source in either case being dependent also on the receiver.

The above gives some insight into why the (airborne) sound power is able to be used as an (approximately) independent source characterisation. First, we note that acoustic sources usually operate in the hard regime, for example, the radiating surfaces of an engine block closely approximate a velocity source since fluid loading by the surrounding air is negligible. Assuming the general form of Equation 3.69 can be applied to continuous interfaces (see Chapter 4), if the impedance of the surrounding air $\mathbf{Z}_{Rcc}$ does not vary with the location of the engine block, then everything on the right-hand side of Equation 3.69 is invariant. Hence, for many airborne sources the radiated power is invariant and can be employed as an independent and transferable source property. The same can never be true for vibration sources because the receiver mobility for structures $\mathbf{Y}_{Rcc}$ will in general be variable, often highly so.

Matched regime

The matched regime occurs when the magnitudes of the source and receiver mobilities are of the same order (they are matched). The behaviour in this regime is significantly more complicated but is important in the context of VAVPs since most structural sources are matched with their receivers at least over part of their frequency range. Power transmission is relatively efficient in the matched region (Figure 3.23) and reaches a maximum when the source and receiver mobilities are equal in magnitude, the sharpness of the peak depending on the relative phase of the source and receiver mobilities. This impedance matching phenomenon is well-known in electro-acoustics where an amplifier and its load must be impedance matched to maximise the transmitted power (usually the desired effect, unlike in vibro-acoustics where the aim is usually to minimise power).

Referring to Equation 3.67, the denominator $|Y_{Rcc} + Y_{Scc}|$ is dominated by Y_{Rcc} in the hard source regime and by Y_{Scc} for soft sources. When the magnitudes of source and receiver mobilities are equal however, $|Y_{Rcc}| = |Y_{Scc}|$, Equation 3.67 becomes,

$$Q = \frac{1}{2}|\tilde{v}_c|^2 \frac{Y_{Rcc}}{|Y_{Rcc}|^2 |e^{i\phi_R} + e^{i\phi_S}|^2} = \frac{1}{2}|\tilde{v}_c|^2 \frac{Z_{Rcc}}{|e^{i\phi_R} + e^{i\phi_S}|^2} \tag{3.70}$$

where ϕ_R and ϕ_S are the phases of receiver and source mobility. The imaginary parts in the denominator can be positive or negative and may cancel each other out entirely. Complete cancellation of the real part never occurs however since there is always some damping such that both mobilities always have a positive real part (see 3.3.3).

The most extreme case occurs where either source or receiver is mass-like and the other is spring-like in which case there is complete cancellation of the phase terms in the denominator and a sharp peak in power. Only the damping term remains in the denominator and the smaller this term, the sharper the peak. In effect this is a resonance of the source-receiver assembly.

This behaviour is illustrated in Figure 3.24; note that wherever the component mobility magnitudes cross there is a sharp peak in the coupled mobility. A negative slope indicates mass-like behaviour and a positive slope stiffness-like behaviour, so there is complete cancellation of the imaginary parts at these crossing points. Such behaviour is often highly important in practice and hence frequency regions where source and receiver mobilities are of similar magnitude require particularly close

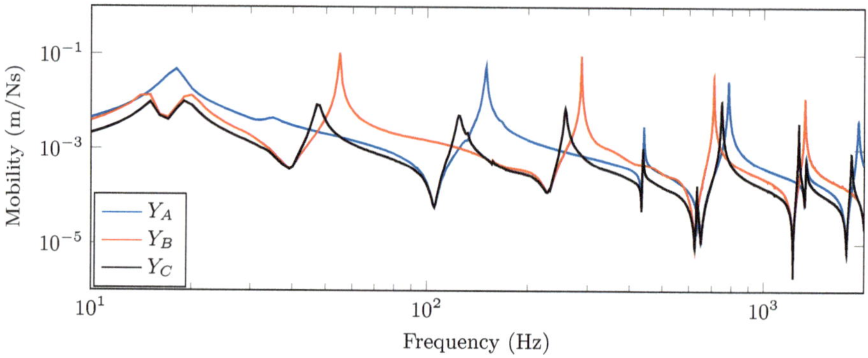

Figure 3.24 Uncoupled point mobility of two beams (Y_A, Y_B) and their coupled point mobility (Y_C) showing sharp peaks at the crossing points of the component mobilities.

attention. Also note that, away from crossing points, the coupled mobility tends to track the lower of the two component mobilities.

Among the implications of the above we see that:

- the power *emission* of a structural source depends on what it is connected to;
- structural sources cannot therefore be directly compared out of the context of their installation as is possible with (airborne) sound power;
- to compare sources, one must calculate the emission in the installed condition.

Thus for vibration sources, power does not enable comparison of sources with other sources or with set limits and therefore does not meet the requirements of the list on page 104. Therefore, if power is required, it is calculated from the free velocity or blocked force combined with the appropriate component FRFs which *do* meet the conditions of the component-based paradigm.

3.3.6 POWER-BASED SOURCE CHARACTERISATION

Although *transmitted* power is a property of the installation, power-based descriptors do exist which are solely properties of the source. The first such descriptor was introduced in [30] where Mondot and Petersson recast the single point power Equation 3.67 as the product of a 'source descriptor' and a 'coupling function',

$$Q = \frac{1}{2}|\tilde{v}_c|^2 \frac{Y_{Rcc}}{|Y_{Rcc} + Y_{Scc}|^2} = \frac{1}{2} S C_f \tag{3.71}$$

in which,

$$S = \frac{1}{2}\frac{|\tilde{v}_c|^2}{Y^*_{Scc}} \tag{3.72}$$

is the *source descriptor*, with units of power and,

$$C_f = \frac{Y_{Scc}^* Y_{Rcc}}{|Y_{Rcc} + Y_{Scc}|^2} = \frac{\alpha e^{i\Delta\phi}}{[\alpha^2 + 2\alpha\cos\Delta\phi + 1]} \tag{3.73}$$

the dimensionless *coupling function*, where $\alpha = \frac{|Y_{Rcc}|}{|Y_{Scc}|}$ and $\Delta\phi = \phi_R - \phi_S$ is the relative phase of the receiver and source mobility. Note that reference [30] uses rms velocity so the factor of $\frac{1}{2}$ is omitted.

The source descriptor has units of power and incorporates both the activity, in the form of free velocity, and the source mobility. Furthermore, since it is independent of the receiver, it is transferable and therefore compatible with the component-based approach. The source descriptor concept was therefore a promising development, although for use with practical sources it would need extending to the multi-DoF case. A suitable extension was proposed in [31] where the concepts of 'mirror power' and 'maximum available power' were also introduced.

Mirror power is based on the idea of mounting the source on a 'mirror' structure, i.e. a receiver with mobilities identical to that of the source, $\mathbf{Y}_{Rcc} = \mathbf{Y}_{Scc}$. Substituting this condition into Equations 3.65 and 3.66 gives,

$$Q_{mirror} = \frac{1}{8}\bar{\mathbf{f}}_c^{\mathrm{H}}\mathbf{Y}_{Scc}\bar{\mathbf{f}}_c = \frac{1}{8}\tilde{\mathbf{v}}_c^{\mathrm{H}}\mathbf{Y}_{Scc}^{-\mathrm{H}}\tilde{\mathbf{v}}_c \tag{3.74}$$

in other words a (complex) power quantity that is wholly a property of the source and therefore an independent source characterisation. Physically, this represents the power transmitted to a mirror receiver.

In [31] the *maximum available power* is, as the name suggests, the maximum power that can be delivered by a given source. It is stated to occur for a receiver whose mobilities are the complex conjugate of those of the source, i.e. $\mathbf{Y}_{Rcc} = \mathbf{Y}_{Scc}^*$. Substituting this condition into Equations 3.65 and 3.66 yields,

$$Q_{MAP} = \frac{1}{8}\tilde{\mathbf{v}}_c^{\mathrm{H}}\mathrm{Re}(\mathbf{Y}_{Scc})^{-1}\mathbf{Y}_{Scc}^*\mathrm{Re}(\mathbf{Y}_{Scc})^{-1}\tilde{\mathbf{v}}_c \tag{3.75}$$

the real part of which simplifies to,

$$Q_{MAP} = \frac{1}{8}\tilde{\mathbf{v}}_c^{\mathrm{H}}\mathrm{Re}(\mathbf{Y}_{Scc})^{-1}\tilde{\mathbf{v}}_c. \tag{3.76}$$

Another way to derive an independent, power-based descriptor is to take the inner product of the free velocity and blocked force vectors (both of which are properties of the source) to give,

$$Q_{CP} = \frac{1}{2}\bar{\mathbf{f}}_c^{\mathrm{H}}\tilde{\mathbf{v}}_c = \frac{1}{2}\bar{\mathbf{f}}_c^{\mathrm{H}}\mathbf{Y}_{Scc}\bar{\mathbf{f}}_c = \frac{1}{2}\tilde{\mathbf{v}}_c^{\mathrm{H}}\mathbf{Y}_{Scc}^{-\mathrm{H}}\tilde{\mathbf{v}}_c. \tag{3.77}$$

This quantity has been termed the *Characteristic Power* and is a direct extension of the source descriptor from Equation 3.72. Comparing Equations 3.74 and 3.77 shows that it is identically equal to 4 times the mirror power. The power is also shown in [31]

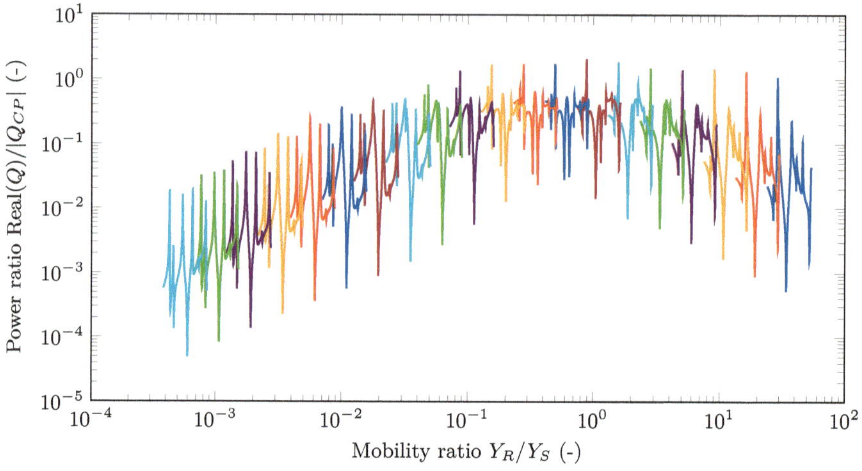

Figure 3.25 Ratio of transmitted power (real part of complex power) to absolute characteristic power for an idealised beam-like source coupled to 20 different steel plate receivers. Each colour corresponds to a plate receiver of thickness varying from 0.2 mm (rightmost curve) to 50 mm (leftmost curve) in logarithmic steps. The abscissa is the ratio of characteristic mobility of the receiver plate to that of the source beam.

to be $2i\omega$ times the kinetic energy for a mass-like source and $2i\omega$ times the strain energy for a stiffness-like source. Characteristic power does not provide a formal upper bound for the power delivered, but in [31] it is argued that it does provide an effective practical maximum for the power in most situations. It is also said to be more reliable to obtain by measurement than the maximum available power which is based on inverting the real part of a mobility matrix, itself subject to measurement errors.

Some general trends can be seen in Figure 3.25 which shows the power delivered from a source to a range of similar receivers as a function of the ratio of receiver to source mobility. This is a similar plot to Figure 3.23 except that the transmitted power is normalised to the magnitude of the characteristic power. The source is an idealised free-free beam and there are 20 different receivers comprised of infinite steel plates with thickness varying from 0.2 to 50 mm so as to generate a range of matching conditions. The sharp peaks and troughs are caused by resonances in the source beam and interference effects in the plate between the two connection points. However, the general trends follow those seen earlier in the matching plot, Figure 3.23 (a similar plot was presented in [31]). The normalised power is greatest for matched receivers and its maximum value is approximately equal to unity indicating that the characteristic power provides a useful, albeit approximate, estimate of the maximum power delivered by the source to any of the receivers.

An *advantage* of power-based descriptors is that they provide an overall, absolute measure of the strength of a source – the source activity is summarised in a single frequency-dependent number. This advantage is not exclusive to power-based descriptors however since any vector can be reduced to a single number, i.e. by taking a vector norm. Thus, free velocity and blocked force vectors can be boiled down to a single number to aid comparison (although careful thought is required if both rotational and translational DoF are present).

However, free velocity and blocked force do not provide an *absolute* measure of source strength in the same way as power. For example, a physically small, lightweight source may have a high free velocity but it will never be able to deliver much power because its mobility is also high. Without reference to the source mobility, the small size of the source would not be evident from its free velocity alone but is inherent in any of the power-based descriptors described above. Similarly, in the absence of source mobility, blocked force does not provide an absolute indicator of the importance of a source. In general, rank ordering of sources purely in terms of free velocity is only reliable within the hard source regime and for blocked forces only within the soft source regime. On the other hand, comparisons based on power are not restricted to any particular regime. Ultimately though, sources can only really be compared in the context of an installation on a given receiver. This implies that prediction is required, whether in terms of power delivered or receiver response and this brings us to one of the disadvantages of power-based descriptors.

A *disadvantage* of power-based descriptors is that information is lost in the reduction to a single overall value. Exact prediction of receiver response cannot then be achieved and one must rely on assumptions even to make an approximate prediction (see 3.3.7). This drawback is due to the reduction of a multi-DoF interface to a single overall value rather than being an inherent limitation of power as a descriptor.

To summarise, power-based descriptors are advantageous in providing an overall, absolute measure of source activity allowing meaningful comparisons, even for example comparison of airborne sound power with structure-borne source strength for a given source (see [31]). On the negative side, the single figure descriptors lack the detail required for the prediction of a receiver response. However, using blocked force or free velocity vectors one can enjoy the best of both worlds: the power-based descriptors can be calculated for purposes of comparison while retaining the detail necessary for prediction. Again, this demonstrates the flexibility of the component-based approach.

3.3.7 RECEPTION PLATE METHOD

With the reception plate method, a source is characterised by attaching it to some standardised plate receiver. While not strictly a power-based approach, it attempts to harness the convenience of power by defining a way to characterise sources based on *single equivalent quantities*, hence its inclusion in this section. Use of reception plates was proposed in [39]. A number of variations are possible (see [21] for example) but here we will focus on the version detailed in European standard

EN15657:2017 [1]. Readers without a particular interest in this method may skip this section.

With the reception plate method averaging is performed physically by attaching the source to a reverberant plate. The average response of the plate is then measured and corrected according to the plate's reverberation time in essentially the same way as the classical measurement of (airborne) sound power in a reverberant room [7]. The advantage of this approach is that the measurements and processing is made relatively simple, for example no matrix inversions are required. The disadvantage is that detail is lost in the averaging so that any subsequent prediction of the source on a different receiver is inevitably only an approximation. Hence, the reception plate method can be classified as an *engineering* approach (see Chapter 1).

The full indirect procedure defined in BS EN 15657 requires two reception plates and the following steps:

- Single equivalent *blocked force* is obtained with the source attached to a *heavy* reception plate;
- Single equivalent *free velocity* may be obtained with the source attached to a *lightweight* reception plate;
- Single equivalent *source mobility* is obtained by dividing the single equivalent free velocity by the blocked force.

All values are obtained in third octave bands at frequencies between 50 Hz and 3150 Hz. The equations in the standard are given in dB but have been transformed to linear units below for consistency with the rest of the book. The notation is however consistent with the standard [1].

Single equivalent blocked force

The blocked force is obtained with the source mounted on a *heavy* reception plate and running under whatever loading regime is required. The velocity of the plate is measured at no less than 6 locations, at least 0.5 m apart and at least 0.1 m from the equipment connection points. The power injected into the reception plate from the source is then calculated from

$$W_S = \frac{1}{2}\omega M\eta\langle v\rangle^2 \tag{3.78}$$

where ω is the frequency, M the total mass of the plate and η the loss factor. The measured reception plate velocity is averaged energetically over $N \geq 6$ points as

$$\langle v\rangle^2 = \frac{1}{N}\sum_{i=1}^{N}|v_i|^2. \tag{3.79}$$

Note that the standard uses the rms velocity but here for consistency with the rest of the book we take $\langle v\rangle$ as the Fourier spectrum magnitude so a factor of 2 appears on the right-hand side of Equation 3.78. The plate loss factor is estimated from its reverberation time, T_S, as,

$$\eta = \frac{2.2}{fT_S}. \tag{3.80}$$

Alternatively, the plate can be calibrated by a *power substitution method* in which known power from a hammer or shaker is injected, W_{cal} and the resulting mean square velocity, $\langle v_{cal}\rangle^2$, measured. The ratio $W_{cal}/\langle v_{cal}\rangle^2$ then provides a calibration factor for the plate so that replacing the hammer or shaker with the source under test gives the source power in terms of measured mean square velocity,

$$W_S = \langle v\rangle^2 \frac{W_{cal}}{\langle v_{cal}\rangle^2}. \tag{3.81}$$

Once W_S has been obtained, the single equivalent blocked force is calculated by invoking the assumption of a soft source (low mobility receiver), in effect inverting the real part of Equation 3.68, to give

$$|\bar{f}|^2 = 2\frac{W_S}{\mathrm{Re}(Y_{R,low,eq})} \tag{3.82}$$

where $|\bar{f}|$ is the single equivalent blocked force. Again, as with velocity, the standard assumes rms values for blocked forces whereas here we take $|\bar{f}|$ as the magnitude of the Fourier spectrum (hence the factor of two on the right-hand side). $Y_{R,low,eq}$ is the single equivalent real part of the mobility obtained by averaging over the plate according to

$$\mathrm{Re}(Y_{R,low,eq}) = \frac{1}{N}\sum_{i=1}^{N} \mathrm{Re}(Y_{R,low,i}). \tag{3.83}$$

The low mobility plates are made of 10±1 cm thick concrete, of minimum area 5 m^2 and a minimum low frequency loss factor of 0.08. Three such plates, isolated from each other, can be arranged at right angles to accommodate sources intended for floor or wall mounting.

Single equivalent free velocity

To obtain the single equivalent free velocity, a similar procedure is followed with the source mounted to a *lightweight* reception plate. Once attached, the source is operated, the velocities are measured and averaged as before and the injected power calculated according to Equation 3.78 (but using the properties of the light plate this time). The free velocity is then calculated from the injected power according to

$$|\tilde{v}|^2 = 2\frac{W_S|Y_{R,high,eq}|^2}{\mathrm{Re}(Y_{R,high,eq})} \tag{3.84}$$

where the factor of 2 appears as before because $|\tilde{v}|$ is the Fourier spectrum magnitude. The magnitude of the equivalent single mobility is obtained by a spatial averaging according to

$$|Y_{R,high,eq}| = \frac{1}{N}\sum_{i=1}^{N} |Y_{R,high,i}|. \tag{3.85}$$

Equation 3.84 is the single equivalent (and rearranged) version of the hard source Equation 3.69.

The high mobility plate should be a 1 mm or 1.5 mm metal plate with a mobility of at most $10^{-2}\text{ms}^{-1}\text{N}^{-1}$ mounted in a frame. To reduce acoustic feedback it should be perforated by approximately 50%.

Source single equivalent mobility

Having obtained the single equivalent blocked force and free velocity, the single equivalent mobility is obtained as follows:

$$|Y_{S,eq}|^2 = \frac{|\tilde{v}|^2}{|\bar{f}|^2}. \tag{3.86}$$

Prediction of installed power

Having obtained the source properties, the installed power, i.e. the power injected by the source into the rest of the assembly may be calculated according to,

$$W_{inst} = \frac{1}{2}\frac{|\tilde{v}|^2 \text{Re}(Y_{R,eq})}{|Y_{R,eq}|^2 + |Y_{S,eq}|^2} \tag{3.87}$$

where $Y_{R,eq}$ is the mobility of the installation receiver (not the reception plate). This simplifies to,

$$W_{inst} = \frac{1}{2}|\bar{f}|^2 \text{Re}(Y_{R,eq}) \tag{3.88}$$

in cases where the receiver mobility is lower than that of the source by at least a factor of 10 ($|Y_{S,eq}| \geq 10|Y_{R,eq}|$). This condition is often met for equipment mounted on heavy concrete slabs whose mobility is typically in the range $10^{-6} \geq Y_R \geq 10^{-5}\text{ms}^{-1}\text{N}^{-1}$ [19, 33]. Note that only one source quantity is required in this case, the blocked force.

For completeness, a similar simplification can of course be made for very light receivers ($|Y_{R,eq}| \geq 10|Y_{S,eq}|$) in which case only the free velocity is required of the source,

$$W_{inst} = \frac{1}{2}\frac{|\tilde{v}|^2}{\text{Re}(Y_{R,eq})} \tag{3.89}$$

although in practice this condition is rarely achieved.

The installed power can provide a basis to predict other downstream responses. One approach, described in Chapter 5, is to use the power as the input to an SEA subsystem which could itself form part of a larger SEA or hybrid model (see 5.4.2) of the rest of the assembly. Another approach is to use the method outlined in BS EN 12354 [2] which provides a prediction framework similar to SEA designed for prediction of vibration and sound levels inside buildings based on readily available sound insulation properties of building elements.

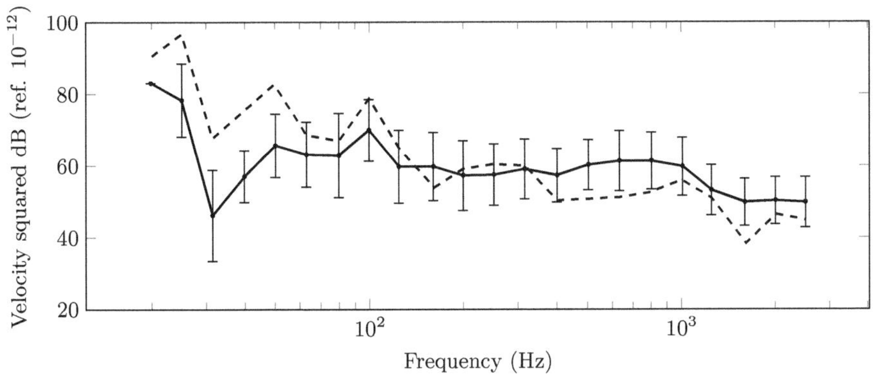

Figure 3.26 Mean square free velocity in a receiver: measured (dashed) and reception plate estimate (solid line) with standard uncertainty. Reproduced from [22].

Directly measured properties

For completeness we mention here that the standard [1] also allows for direct measurement of free velocity and mobility as an alternative to using the reception plate. The measured values are then reduced to single equivalent values by averaging prior to calculating any downstream responses. Direct measurement of forces is not encouraged due to the difficulty of inserting force transducers as described in 3.3.

Uncertainties

Uncertainties from the reception plate method arise first, due to measurement of the single equivalent quantities themselves and secondly, because of the assumptions involved in the simplification to single values. Regarding the former, there is good data from two round robin tests, reported in 2011 and 2022 [43, 44], to show that the reproducibility across a number of laboratories is around 3 dB for single equivalent blocked force and 4 dB for single equivalent free velocity. For single equivalent mobility measured directly it was typically 1 dB but this would be significantly higher for indirect measurement [22]. These values are for standard deviation accounting for 68 % of cases and should be respectively doubled and tripled in the usual way to obtain 95% and 99.7% confidence intervals.

Additional uncertainty arises when the single equivalent quantities are used for prediction because of the loss of information about phase and the interaction between feet inherent in the approach using single equivalent values. Typical trends are shown in Figure 3.26, reproduced from [22], where standard uncertainty of around 8-14dB at low frequencies and 5-8dB at high frequencies is typical, depending on the properties of the structures involved.

3.4 ACOUSTIC SOURCES

While mainly focussed on structure-borne sound sources, this chapter would not be complete without mention of acoustic sources. By the 1980s, acoustic source characterisation was already standardised [6, 7] and in widespread use for prediction, while comparable methods for structure-borne sources were almost non-existent [39]. In some respects there has been a reversal of this situation since that time. For structure-borne sound sources, we now have standardised characterisation approaches that are beginning to work well enough to achieve the ambitious overall aim of replacing real sources with digital copies. For airborne sources however, while some promising new approaches have been proposed, no clear front runner has emerged to meet the developing demands of component-based simulation. In this section, we introduce some of the approaches most closely aligned with the component-based paradigm although at current state-of-the-art, acoustic source characterisation must still be considered 'work in progress'.

3.4.1 SOUND POWER

Sound power remains the most popular method of characterising acoustic sources at the current time. In its raw form, the data is not compatible with the component-based approach although we will discuss later how it might be adapted for incorporation into a VAVP. First though, it is worth mentioning why the sound power approach, which has proved so successful for whole machines, is subject to greater limitations for characterising *components* of machines.

One of the main limitations is that the power output is not actually invariant. A reasonable assumption for 'hard' sources is that the *volume velocity* is unaffected by the surroundings (consider an engine block for example whose surface velocity is negligibly affected by the surrounding air). In many situations, the power output will also be constant, but when the source is within a fraction of a wavelength of a reflecting surface this is not the case. For whole machines this is typically not a significant problem since reflective surfaces are either distant or are always present, such as the ground, in which case they can be included as part of the test setup. *Components* however, are more likely to be installed in a tightly packed assembly with a variety of nearby surfaces and their power output is therefore more likely to vary from one installation to another.

A second issue (discussed in 7.6.2) is that sound power is based on a spatial average over the whole source so that phase and directivity information is lost. This makes it incompatible with the complex FRFs and source quantities presented so far.

Nevertheless, if sound power is all that is available, it is possible to incorporate it into a component-based simulation by treating the source as an *equivalent monopole* – a breathing sphere of infinitesimal size. The sound power for such a source in free field is given by,

$$W = \frac{\omega^2 \rho}{8\pi c} |q|^2 \tag{3.90}$$

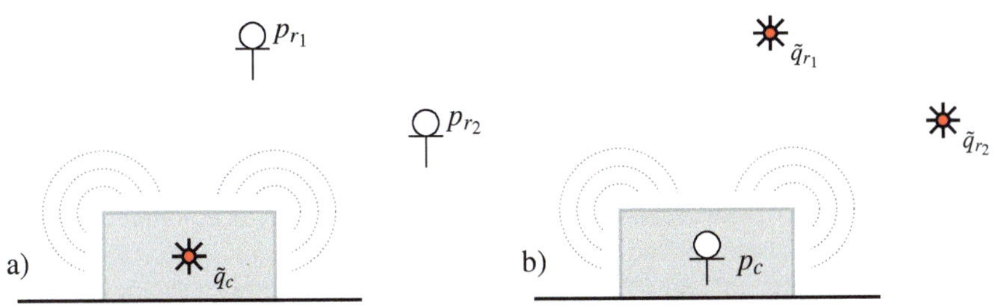

Figure 3.27 Measurement of acoustic FRFs where Q is a known volume velocity. a) direct, b) reciprocal measurement.

where W is the sound power in W, ω the radian frequency, ρ the density of air (1.204 kg m^{-3} at 20$^{\circ C}$), c the speed of sound in air (343 m s^{-1} at 20$^{\circ C}$), and $|q|$ is the magnitude of the *volume velocity*, i.e. the volume displaced by the breathing sphere per cycle with units of m^3 s^{-1}. By rearranging we can calculate the volume velocity of a monopole that is *equivalent* to the real source in terms of its free field sound power,

$$|q| = \left(\frac{8\pi c}{\omega^2 \rho} W\right)^{\frac{1}{2}}. \tag{3.91}$$

The magnitude of the sound pressure resulting from the source can now be expressed in vector form as,

$$|\mathbf{p}_r| = |\mathbf{H}_{Crc}||q_c| \tag{3.92}$$

where $\mathbf{H}_{Crc}$ is a column vector of acoustic FRFs relating the single source point c and the receiver points, r.

As yet, Equation 3.92 is lacking phase information and so does not allow transformation to the time domain for auralisation in a VAVP. The question of how to 'put back' a phase spectrum will be addressed in 7.6.2 but for now we will assume this has been done, i.e. $|q_c| \to |q_c|e^{i\phi} = q_c$, with ϕ the phase. Equation 3.92 can then be written in complex form compatible with earlier formulations,

$$\mathbf{p}_r = \mathbf{H}_{Crc} q_c. \tag{3.93}$$

The acoustic FRFs $\mathbf{H}_{Crc}$ are measured either directly or reciprocally as shown in Figure 3.27. In direct measurement, a known volume velocity source (see Chapter 6) is placed at the centre of the source component (usually with the component removed) and the sound pressure at the remote receiver points r recorded with a microphone (Figure 3.27 left). Alternatively, using reciprocity, the locations are reversed so the microphone is placed at the source centre and the volume velocity source at the remote r positions (Figure 3.27 right). Reciprocal measurement is often advantageous in practice because a microphone is generally easier than a volume velocity source to place in a confined space such as a machine enclosure.

There are two main reasons why this single monopole approach might not prove adequate. First, being a single frequency-dependent quantity, the sound power contains no information on source directivity which might be important for some components. Secondly, whereas a real source occupies some volume, this is not taken into account in the equivalent monopole representation since the component is generally removed for measurement of FRFs. In effect, the interface is then defined as the infinitesimal surface of the monopole. For physically small components, a 'volumeless' source like this may be a reasonable representation, but many source components occupy a significant volume and the assumption may be in error (consider a large engine inside a small engine bay for example).

3.4.2 CORRELATED MONOPOLES

A source description that includes directivity can be provided by representing the source by a number of monopoles. Typically, these are located on the surface of the component as in Figure 3.28 which has the additional advantage of taking the source volume into account. The radiating surface of the source is divided into patches, typically less than an acoustic wavelength in size, whose volume velocity is approximated by a point source in the centre of each.

Equation 3.93 can then be extended to accommodate multiple monopoles $q_c \to \mathbf{q}_c$, giving the sound pressure at receiver locations r as,

$$\mathbf{p}_r = \mathbf{H}_{Crc}\mathbf{q}_c \tag{3.94}$$

where $\mathbf{H}_{Crc}$ is now a matrix of acoustic FRFs.

As before, the FRFs can be measured either directly or reciprocally as shown in Figure 3.28. In the direct method, a calibrated volume velocity source is held against the source surface and the pressure at each receiver point r is recorded with

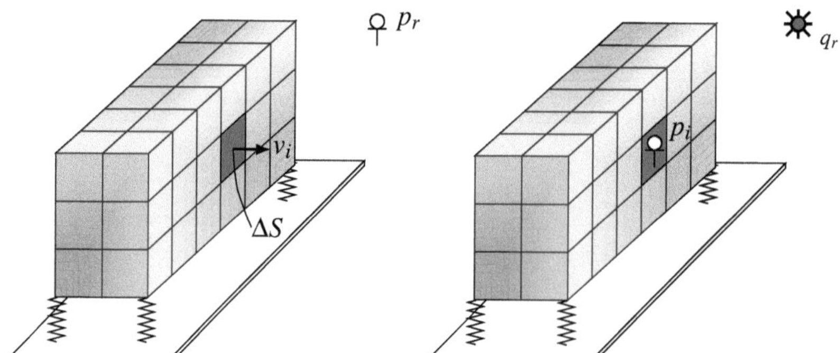

Figure 3.28 Uncorrelated monopoles located on the surface of an acoustic source, e.g. an engine block. Left: direct measurement of vibro-acoustic FRFs. Right: reciprocal measurement of FRFs.

microphones. In reciprocal measurement, the locations are reversed. As before, the small size of microphones makes this advantageous for many sources.

The *source strengths* $\mathbf{q}_c$ in Figure 3.28 may be obtained from accelerometer measurements in the centre of each patch. The equivalent volume velocity of each point source is then $q_i = v_i S_i$, where S_i is the surface area of the patch and v_i the velocity at its centre [41]. Note that, as with other multi-point velocity measurements a common phase reference must be retained, for example by measuring simultaneously.

Alternatively, source strengths may be obtained by inverse methods: it may be recognised that Equation 3.96 is directly analogous to the *in situ* blocked force Equation (3.28), so the source strengths can in principle be obtained by solving the inverse problem,

$$\mathbf{q}_c = \mathbf{H}_{Crc}^{+}\mathbf{p}_r. \tag{3.95}$$

An advantage of the inverse determination of source strength is that the volume velocities obtained are independent properties of the source. Although perhaps not obvious, there is a direct parallel between the *in situ* measured acoustic FRFs and the assembly FRFs used in the *in situ* blocked force method so that, by analogy, the source strengths obtained are independent source properties in the same way as blocked forces are. Directly measured velocity on the other hand is analogous to the direct measurement of free velocity and is only strictly transferable when the receiver mobility is high compared with that of the source (hard source). Independence of the source strengths $\mathbf{q}_c$ obtained as per Equation 3.96 will be shown explicitly with more rigour in 3.4.4 by considering the equivalent field theorem (3.2.3) applied to an acoustic source.

As for blocked forces, the inverse approach requires an operational and a passive test phase. Also as for blocked forces, obtaining a reliable solution is a delicate process (see 7.4), in fact, there are several reasons why it is usually more so for airborne than for structure-borne sources. First, vibro-acoustic interfaces are continuous and require discretisation which introduces an element of approximation. Secondly, phase tends to vary more rapidly with position in the acoustic case, particularly as frequency increases; in part, this is because measurement points are located in 3-dimensional space as opposed to on a solid surface. Thirdly, whereas in the structural case one can monitor responses directly at the interface, in the acoustic case response positions cannot be co-located with a volume velocity source, meaning that point FRFs cannot be measured. Instead, response locations must be located away from the interface, the additional distance adding to the variance of the phase.

The correlated monopoles approach is appropriate for low frequencies and/or small distances such as at low frequencies within a vehicle compartment. For high frequencies however, a different approach is required.

3.4.3 UNCORRELATED MONOPOLES

As frequency increases, phase varies more and more rapidly to the point where it can no longer be determined reliably. In this case, a more stable solution is often obtained by assuming the phase to be random, in other words by assuming the sources to be *uncorrelated.*

The method of *uncorrelated monopoles* ([41, 48]) divides the source surface into patches in the same way as before (Figure 3.28) but differs in the way the source strengths are defined and processed.

The equations can be derived from the relation for correlated monopoles, Equation 3.95. Post-multiplying each side by its own Hermitian gives,

$$\mathbf{p}_r\mathbf{p}_r^{\mathrm{H}} = \mathbf{H}_{Crc}\mathbf{q}_c\mathbf{q}_c^{\mathrm{H}}\mathbf{H}_{Crc}^{\mathrm{H}} \tag{3.96}$$

or after time averaging,

$$\mathbf{S_{pp}} = \mathbf{H}_{Crc}\mathbf{S_{qq}}\mathbf{H}_{Crc}^{\mathrm{H}} \tag{3.97}$$

where $\mathbf{S_{pp}}$ and $\mathbf{S_{qq}}$ are the cross spectral matrices of the resultant sound pressure and the source volume velocity respectively (see 7.2.3 for more on cross spectra).

Because the sources are assumed uncorrelated, the cross terms average out to zero. Taking the diagonals of each side gives the sound pressure squared at each location as,

$$|p_i|^2 = |H_{i1}|^2|q_1|^2 + |H_{i2}|^2|q_2|^2 + \cdots |H_{iN}|^2|q_N|^2. \tag{3.98}$$

In words, this says that the sound pressure energy at a point is the energetic sum of the contributions from all sources. It is evident that only the magnitudes of the source strength and FRF are required.

The acoustic FRFs may be obtained in the same was as for correlated monopoles, although Equation 3.98 shows that the phase is not required for subsequent calculations.

To obtain the source strengths, in principle one could use directly measured velocity in the same as for correlated monopoles. Also, since phase is not required, the velocity can be energetically averaged over each patch to improve reliability. However, it is more usual (and probably more reliable) to use sound intensity measurements over each patch, obtaining an average either by scanning or with repeated point measurements distributed over the patch [9, 10]. The measured sound intensity for each patch, I_i is multiplied by the patch area S_i to give a partial sound power for each patch,

$$W_i = I_i S_i \tag{3.99}$$

which can be converted to an equivalent volume velocity in a similar way as for sound power,

$$|q| = \left(\frac{4\pi c}{\omega^2 \rho} W\right)^{\frac{1}{2}}. \tag{3.100}$$

Note that the term inside the brackets is half the value of that in Equation 3.91 which accounts for the presence of the hard surface. (Also note that there is also a factor of 2 compared with the equation given by Verheij in [41], presumably the difference between rms and spectral magnitudes as encountered in 3.3.7.)

The uncorrelated monopoles method has yielded good results in practice for sources such as engines [41, 48]. Reciprocity makes measurements possible even in tightly enclosed spaces and the requirements are simplified by not needing phase. However, in its raw form the power-based data obtained from uncorrelated monopoles, as with sound power, is not compatible with the prediction schemes presented so far. The question of how to incorporate power-based data into the component-based framework, will be addressed in Chapter 7.

In the above we have focused on methods as described by Verheij [41] but a number of variations have been explored since that time, for example for tyre and pass-by noise [12, 36], small motors [35], fans [23, 27], marine and general sources [37, 38]. There is increasing use of partially correlated sources characterised by a cross spectral matrix which effectively combines the correlated (low frequency) and uncorrelated (high frequency) approaches (see 7.2.3 for more on cross spectra). Another possible approach is to use the overlap-add method, to be described in 7.5, which is equally valid for correlated and uncorrelated sources because every measurement window is processed without first obtaining an average.

3.4.4 EQUIVALENT FIELD APPROACHES

As shown above, the correlated monopole representation when treated as an inverse problem can be used to solve for the source strengths $\mathbf{q}_c$, much like the in situ blocked force relation. What is less obvious is that these source strengths are in fact also independent source properties when obtained from in situ measurements. This is made concrete by considering the Equivalent Field Theorem (3.2.3) when applied to a pair of coupled acoustic domains.

Free interface (equivalent monopoles)

As discussed in 3.2.3, the Equivalent Field Theorem (EFT) provides a theoretical basis for the in situ characterisation of both structural and acoustics sources. For a structural source we rely on its first representation in terms of the blocked force. For an acoustics source, we rely on its second representation, which states that the pressure response field in a receiver domain can be reproduced identically by applying an equivalent differential velocity across the source-receiver interface,

$$p(\mathbf{r}) = \int_S H(\mathbf{r}, \mathbf{r}_0)\tilde{u}(\mathbf{r}_0)dS \tag{3.101}$$

where $\mathbf{r}$ and $\mathbf{r}_0$ are the response and source positions, respectively, and $H(\square)$ denotes an acoustics Green's function (which can be interpreted for our purposes as an acoustic FRF) of the *coupled* system. As in the structural case, the location of the interface S is completely arbitrary, so long as it encloses all the operational mechanisms. In the previous correlated monopole discussion, this interface was taken to be the structural-acoustic interface, though it could just as well be a virtual interface surrounding the source, as illustrated in Figure 3.29.

This differential velocity $\tilde{u}(\mathbf{r}_0)$ should be equal to the relative velocity between the two subsystems when uncoupled/freely suspended. When only one subsystem is active, it is simply that subsystem's free velocity. Admittedly, the notion of a 'freely

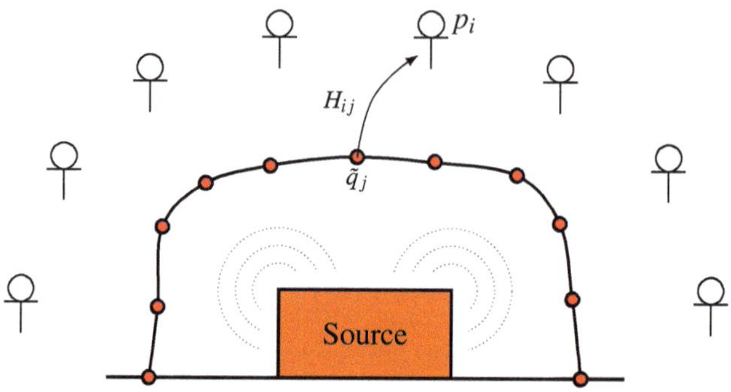

Figure 3.29 Free volume velocity representation of an acoustic source.

suspended' acoustic source and its 'free velocity' are less intuitive than their structural counterparts, nevertheless the theory remains valid.

Perhaps the main difference between structural and acoustic applications of the EFT in practical terms is that in the acoustic case, the differential velocity is applied over an interface that is a continuous surface. As will be discussed in the next chapter (Interfaces), some form of discretisation is required in order to allow Equation 3.101 to be solved experimentally. Hence,

$$p(\mathbf{r}) = \int_S H(\mathbf{r},\mathbf{r}_0)\tilde{u}(\mathbf{r}_0)dS \approx \sum_i H(\mathbf{r},\mathbf{r}_i)S_i\tilde{u}(r_i) = \mathbf{H}_{Crc}\tilde{\mathbf{q}}_c \tag{3.102}$$

where the subscript c is as usual taken to denote the interface. The source is characterised by the vector of *free volume velocities* $\tilde{q}_i = S_i\tilde{u}(r_i)$. Indeed, writing Equation 3.102 for a series of response positions, we obtain an equation identical in form to Equation 3.28,

$$\begin{pmatrix} \mathbf{p}_c \\ \mathbf{p}_b \end{pmatrix} \approx \begin{bmatrix} \mathbf{H}_{Ccc} \\ \mathbf{H}_{Cbc} \end{bmatrix} \tilde{\mathbf{q}}_c \tag{3.103}$$

where $\mathbf{p}_c \in \mathbb{C}^{N_c}$ and $\mathbf{p}_b \in \mathbb{C}^{N_b}$ are the sound pressures measured across the discretised interface c and at the remote points b, $\mathbf{H}_{Ccc} \in \mathbb{C}^{N_c \times N_c}$ and $\mathbf{H}_{Cbc} \in \mathbb{C}^{N_b \times N_c}$ are the acoustic transfer functions and $\tilde{\mathbf{q}}_c \in \mathbb{C}^{N_c}$ is the vector of free volume velocities defined across the discretised interface c. Here we have explicitly introduced the $\tilde{\square}$ accent onto $\mathbf{q}_c$ to indicate its 'free-interface' nature. The accuracy of this approximation will depend on the usual points per wavelength arguments.

Unlike the structural case, it is impractical to measure the acoustic pressure response at the same position that a volume velocity is being applied, and so we reduce Equation 3.103 to,

$$\mathbf{p}_b \approx \mathbf{H}_{Cbc}\tilde{\mathbf{q}}_c. \tag{3.104}$$

As described previously, Equation 3.104 can be solved for $\tilde{\mathbf{q}}_c$ by the (generalised) inverse of $\mathbf{H}_{Cbc}$,

$$\tilde{\mathbf{q}}_c \approx \mathbf{H}_{Cbc}^{+}\mathbf{p}_b. \tag{3.105}$$

The obtained free volume velocity $\tilde{\mathbf{q}}_c$ should be invariant (up to the accuracy of the discretisation) of the receiver subdomain, and thus independently characterise the acoustic source. Examples of this approach towards air-borne source characterisation can be found in [16, 27].

Note that whilst this approach to air-borne source characterisation shares a similar underlying principle to that of the in situ blocked force method, it is less well developed owing to the popularity and general applicability of standard sound power-based approaches.

Fixed interface (blocked pressure)

It should not be a surprise that an alternative source characterisation can also be achieved by considering the first (fixed-interface) representation of the EFT.

According to this first representation, the acoustic response surrounding a source can be represented equivalently by the *blocked pressure* field over the enclosing interface S,

$$p(\mathbf{r}) = \int_S G(\mathbf{r}, \mathbf{r}_0)\bar{p}(\mathbf{r}_0)dS. \tag{3.106}$$

Equation 3.106 could similarly be discretised and used to establish a set of linear equations relating the blocked interface and remote pressure. However, unlike $H(\square)$ the relating transfer function $G(\square)$ is not so straightforward to measure. Rather, the fixed-interface representation is better suited to a direct measurement approach. Such an approach was investigated by Bobrovnitskii and Pavic in [14], where a rigid spherical enclosure was designed with embedded microphones, thus measuring directly the blocked pressure.

Example 3.2: Acoustic source characterisation

In this example we demonstrate the in situ characterisation of an air-borne sound source using the free volume velocity and verify its independence with respect to the receiver domain. A high speed compressor unit is mounted within a short section of duct (see Figure 3.30) – this section of duct, including the compressor, is considered as the acoustic source. To either end of the source additional lengths of duct are attached to alter the acoustic impedance of the receiver domain. Three different combinations of duct lengths are considered (see [27] for further details).

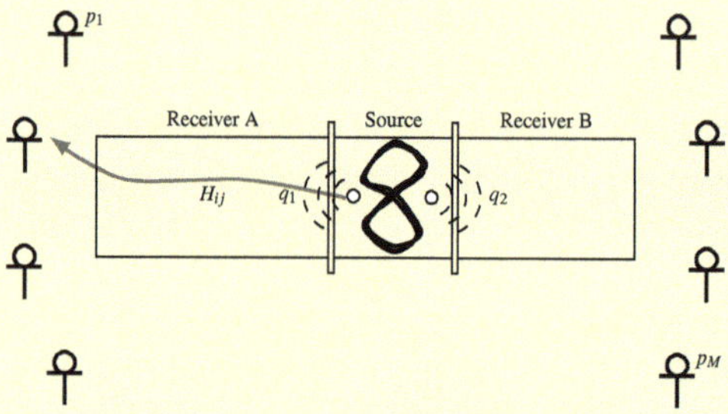

Figure 3.30 Experimental set-up (left) and photo of apparatus (right).

For each configuration the operational sound pressure level is measured at 8 microphones surrounding the assembly, such that $\mathbf{p}_r \in \mathbb{C}^8$; some example measurements are shown in Figure 3.31. The acoustic transfer function is measured between a pair of inlet and outlet points, such that $\mathbf{q}_c \in \mathbb{C}^2$, and the surrounding microphone array. This low order discretisation is valid in the planar regime of the duct, i.e. below the cut-on of complex higher order modes.

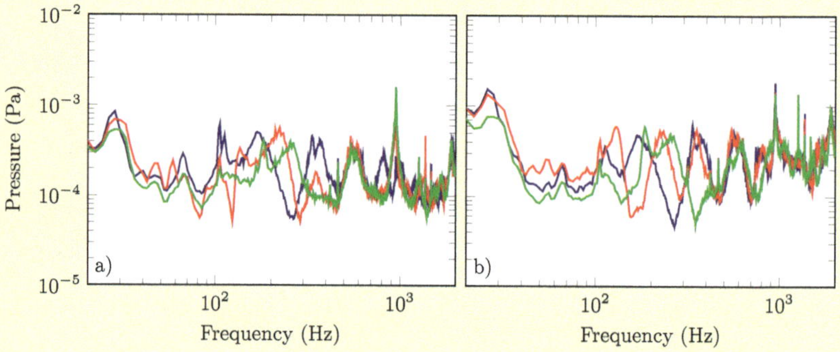

Figure 3.31 Example sound pressure measured around test rig – a) inlet, b) outlet. Red, blue and green represent different combinations of duct lengths.

Solving the over-determined problem by inverting the acoustic transfer function matrix $\mathbf{H}_{Crc} \in \mathbb{C}^{8\times 2}$, as per Equation 3.105, yields a pair of free volume velocities, as shown in Figure 3.32. It is evident from the agreement between the different duct length configurations, that the free volume velocities are indeed invariant to the receiver domain and thus independently characterise the acoustic source.

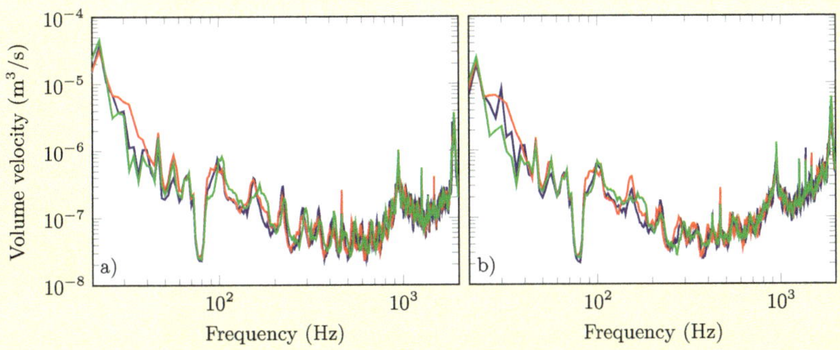

Figure 3.32 Example volume velocities obtained using in situ characterisation – a) inlet, b) outlet. Red, blue and green represent different combinations of duct lengths.

3.5 SUMMARY

Having defined their passive properties in Chapter 2, this chapter has focused on characterisation of the *active* properties of sources. For compatibility with the component-based approach, source descriptors must be: equivalent to the real source in terms of the response induced, defined at interfaces, measurable and transferable. A brief historical review revealed that methods for structure-borne source characterisation were virtually nonexistent in the 1980s while airborne source characterisation was already standardised with sound power.

We moved on to discuss possible candidates for source descriptors. *Contact force*, appears to be an obvious choice but turns out to be a property of the assembly rather than the source itself and is therefore not transferable. *Free velocity* and *blocked force are* transferable source properties, however, approximating idealised free or fixed boundary conditions in practice makes direct measurement difficult. For force measurement, there is the additional problem that inserting force transducers can alter the dynamics of the interface. A final candidate is the so-called *pseudo force* which is obtained by inverse methods from the measured response. The pseudo forces achieve all the requirements for the component-based approach but are abstract in nature and not unique for a given source.

The central concept in the chapter is the *in situ* blocked force method which provides an independent source characterisation from measurements made entirely *in situ*. The method requires solution of an inverse problem which is known to be a delicate process, as discussed in more detail in Chapter 7. However, as an *in situ* method it has several advantages: source and receiver do not need to be separated so the interface is not disturbed as with direct force measurement; sources can be characterised on a test bench or in a realistic installation even including moving sources.

We discuss how the *in situ* blocked force method can be seen as a special case of a general equivalent field theorem which can also be applied to acoustic sources.

We go on to discuss power which is intuitively related to any downstream effect of the source. Power may be *active*, i.e. permanently transmitted to the receiver or *reactive* i.e. circulated backwards and forwards once per cycle. Well-damped structures are good absorbers of (active) power but spring-like and mass-like structures are reactive in nature. For a given source, transmitted power depends on the ratio of source and receiver mobility and three regimes can be identified: soft source, matched and hard source. Soft sources exert approximately the same force irrespective of the receiver (blocked force) whereas for hard sources the velocity does not change (free velocity). In the matched regime, the power is dependent on both source and receiver. Many acoustic sources are 'hard' and are amenable to simplifying assumptions but structural sources will typically be matched, at least over part of the range and no such simplifying assumptions are possible.

Finally, a brief description of some acoustic source characterisation methods is provided, starting with sound power and going on to correlated and uncorrelated equivalent monopoles. Source strengths may be measured directly or can be solved by inverse methods in a similar way to structural sources. However with uncorrelated methods and sound power, the phase is not retained so some way of adding phase will be required before these results can be incorporated into a VAVP.

Throughout this chapter the importance of a good description of the interface is emphasised and this is considered further in the following chapter.

Bibliography

1. BS EN 15657:2017 Acoustic properties of building elements and of buildings. laboratory measurement of structure-borne sound from building service equipment for all installation conditions.
2. EN 12354-5:2023 Building acoustics – Estimation of acoustic performance of buildings from the performance of elements - part 5: Sounds levels due to the service equipment.
3. ISO 18312-1:2012 Mechanical vibration and shock – measurement of vibration power flow from machines into connected support structures part 1: Direct method.
4. ISO 18312-2:2012 Mechanical vibration and shock – measurement of vibration power flow from machines into connected support structures part 2: Indirect method.
5. ISO 21955:2021 Acoustics — experimental method for transposition of dynamic forces generated by an active component from a test bench to a receiving structure.
6. ISO 3740:1980 Acoustics — determination of sound power levels of noise sources — guidelines for the use of basic standards and for the preparation of noise test codes.
7. ISO 3741:1988 Acoustics — determination of sound power levels of noise sources — precision methods for broad-band sources in reverberation rooms.
8. ISO 9611:1996 Acoustics — characterization of sources of structure-borne sound with respect to sound radiation from connected structures. measurement of velocity at the contact points of machinery when resiliently mounted.
9. ISO 9614-2:1997 Acoustics — determination of sound power levels of noise sources using sound intensity - measurement by scanning.
10. ISO 9614-3:2009 Acoustics — determination of sound power levels of noise sources using sound intensity - precision method for measurement by scanning.
11. L Beranek and T Mellow. *Acoustics: sound fields and transducers*. Academic Press, 2012.
12. D Berckmans, P Kindt, P Sas, and W Desmet. Evaluation of substitution monopole models for tire noise sound synthesis. *Mechanical Systems and Signal Processing*, 24(1):240–255, 2010.

13. YI Bobrovnitskii. A theorem on the representation of the field of forced vibrations of a composite elastic system. *Acoustical Physics*, 47(5):507–510, 2001.
14. YI Bobrovnitskii and G Pavic. Modelling and characterization of airborne noise sources. *Journal of Sound and Vibration*, 261(3):527–555, 2003.
15. M Burkhardt, E Hensel, and WG Drossel. Test bench development and validation for blocked force measurements in six degree of freedom. In *INTER-NOISE and NOISE-CON Congress and Conference Proceedings*, volume 263, pages 3933–3943. Institute of Noise Control Engineering, 2021.
16. K Chen and DW Herrin. Determination of acoustic free sources in large cross-section ducts. *Applied Acoustics*, 181:108131, 2021.
17. RR Craig Jr and MCC Bampton. Coupling of substructures for dynamic analyses. *AIAA Journal*, 6(7):1313–1319, 1968.
18. D De Klerk. Dynamic response characterization of complex systems through operational identification and dynamic substructuring. *Delft University of Technology, The Netherlands*, 2009.
19. FJ Fahy and ME Westcott. Measurement of floor mobility at low frequencies in some buildings with long floor spans. *Journal of Sound and Vibration*, 57(1):101–129, 1978.
20. P Gardonio and MJ Brennan. Mobility and impedance methods in structural dynamics: an historical review. 2000.
21. BM Gibbs, N Qi, and AT Moorhouse. A practical characterisation for vibro-acoustic sources in buildings. *Acta Acustica United with Acustica*, 93(1):84–93, 2007.
22. BM Gibbs, G Seiffert, and KH Lai. Uncertainties in the two-stage reception plate method for source characterisation and prediction of structure-borne sound power. *Acta Acustica United with Acustica*, 102(3):441–451, 2016.
23. KR Holland and PA Nelson. The application of inverse methods to spatially-distributed acoustic sources. *Journal of Sound and Vibration*, 332(22):5727–5747, 2013.
24. S Jianxin, AT Moorhouse, and BM Gibbs. Towards a practical characterization for structure-borne sound sources based on mobility techniques. *Journal of Sound and Vibration*, 4(185):737–741, 1995.
25. DH Johnson. Origins of the equivalent circuit concept: the current-source equivalent. *Proceedings of the IEEE*, 91(5):817–821, 2003.
26. DH Johnson. Origins of the equivalent circuit concept: the voltage-source equivalent. *Proceedings of the IEEE*, 91(4):636–640, 2003.
27. JWR Meggitt, AS Elliott, AT Moorhouse, G Banwell, H Hopper, and J Lamb. Broadband characterisation of in-duct acoustic sources using an equivalent source approach. *Journal of Sound and Vibration*, 442:800–816, 2019.
28. JWR Meggitt, AS Elliott, AT Moorhouse, A Jalibert, and G Franks. Component replacement tpa: A transmissibility-based structural modification method for in-situ Transfer Path Analysis. *Journal of Sound and Vibration*, 499:115991, 2021.
29. JWR Meggitt and AT Moorhouse. On the completeness of interface descriptions and the consistency of blocked forces obtained in situ. *Mechanical Systems and Signal Processing*, 145:106850, 2020.
30. JM Mondot and BAT Petersson. Characterization of structure-borne sound sources: the source descriptor and the coupling function. *Journal of Sound and Vibration*, 114(3):507–518, 1987.
31. AT Moorhouse. On the characteristic power of structure-borne sound sources. *Journal of Sound and Vibration*, 248(3):441–459, 2001.
32. AT Moorhouse, AS Elliott, and TA Evans. In situ measurement of the blocked force of structure-borne sound sources. *Journal of Sound and Vibration*, 325(4-5):679–685, 2009.
33. AT Moorhouse and BM Gibbs. Calculation of the mean and maximum mobility for concrete floors. *Applied Acoustics*, 45(3):227–245, 1995.
34. AT. Moorhouse, JWR. Meggitt, and AS. Elliott. Virtual acoustic prototyping - a story of four decades.
35. AT Moorhouse and G Seiffert. Characterisation of an airborne sound source for use in a virtual acoustic prototype. *Journal of Sound and Vibration*, 296(1-2):334–352, 2006.
36. A Papaioannou, SJ Elliott, and J Cheer. Application of lp norm regularisation techniques in the synthesis of indoor tyre pass-by noise with the inverse method. *Journal of Sound and Vibration*, 473:115240, 2020.
37. G Pavić and A Lindberg. Airborne sound characterisation by a dummy source approach. *Journal of Sound and Vibration*, 392:91–112, 2017.

38. J Tanttari and A Hynninen. Acoustic source characterization of marine propulsors. *Journal of Marine Science and Engineering*, 10(9):1273, 2022.
39. T Ten Wolde and G Gadefelt. Development of standard measurement methods for structureborne sound emission. *Noise Control Engineering Journal*, 28(1):5–14, 1987.
40. M Van der Seijs. Experimental dynamic substructuring: Analysis and design strategies for vehicle development. 2016.
41. JW Verheij. Inverse and reciprocity methods for machinery noise source characterization and sound path quantification. part 1: Sources. *International Journal of Acoustics and Vibration*, 2(1):11–20, 1997.
42. AP Dowling and JE Ffowcs Williams. *Sound and sources of sound*. Ellis Horwood, 1983.
43. V Wittstock, J Scheck, and M Villot. Structure-borne sound sources in buildings–estimating the uncertainty of source properties and installed power from interlaboratory test results. *Acta Acustica*, 6:16, 2022.
44. V Wittstock, M Villot, and J Scheck. Results of a round robin on structure-borne sound power. In *Proceedings of Forum Acusticum, Aalborg*, volume 26, 2011.
45. DF Zabel, T Alber, M Sturm, and AT Moorhouse. Internal Transfer Path Analysis based on in-situ blocked forces and transmissibility functions.
46. DF Zabel, M Sturm, T Alber, and AT Moorhouse. Embedded mems accelerometers for the in-situ measurement of blocked forces in coupled structures. In *DAGA*, volume 1, pages 7–10, 2017.
47. DF Zabel. *Structure-borne sound transmission within electric power steering systems*. PhD thesis, University of Salford, 2018.
48. J Zheng, FJ Fahy, and D Anderton. Application of a vibro-acoustic reciprocity technique to the prediction of sound radiated by a motored ic engine. *Applied Acoustics*, 42(4):333–346, 1994.
49. B Jetmundsen, RL Bielawa, and WG Flannelly. Generalized frequency domain substructure synthesis. *Journal of the American Helicopter Society* 33(1):55–64. Vertical Flight Society, 1988.
50. D de Klerk, DJ Rixen and SN Voormeeren. General framework for dynamic substructuring: history, review and classification of techniques. AIAA Journal. 46(5): 1169–1181, 2008

4 Interfaces

4.1 INTRODUCTION

In the preceding two chapters it was described how passive and active components are characterised almost exclusively at their *interfaces*. The need for an accurate description of the interface dynamics has been emphasised but, so far, we have not discussed how this might be achieved. The subject is in fact so important that we now devote an entire chapter to it: in source characterisation and substructuring it is said that *everything goes right or wrong at the interface*!

In physical assemblies, a range of contact types occur, often comprising bolted connections through flanges, typically rectangular or circular in shape but potentially of various other forms. The size and shape of the contact may vary greatly between applications, particularly when vibro-acoustic interfaces are included in which the interface consists of the entire 'wetted' surface. In this chapter we consider how to model such contacts in a form suitable for representation by measured data.

When modelling connections, we encounter the problem that the real contact always occurs over a finite area; this 'continuous' form must be reduced to a 'discrete' form for our VAVP purposes. For a contact that is physically small compared with a structural wavelength the problem is conveniently solved by assuming it to be a discreet point. Indeed, the most commonly adopted model is one in which connections are *point-like*, and the first part of this chapter is devoted to single and multi-point interfaces (4.2 and 4.3, respectively). Interfaces or contacts that extend over more than a fraction of a wavelength, including most vibro-acoustic types, are termed 'flexible' and require some other form of discretisation which is the subject of 4.4.

Representing contact geometry is only part of the story however – the overarching aim of an interface model is to represent, or capture, *all* important dynamics occurring at the interface. In general, 6 DoFs are required for every point-like connection although this can sometimes be reduced in practice. For flexible interfaces, > 6 DoFs might be required per connection point. In 4.5 we ask the important question; how many DoFs are necessary to describe the interface? To this end, we introduce the notion of *interface completeness* and show the importance of capturing all significant DoFs through an experimental example.

Having represented the contact with an appropriate model, we must then acquire the measured data to represent it. The interface descriptions of connected components must of course be compatible, and similarly, blocked forces and FRFs at an interface must share a matching set of DoFs. Unfortunately, practical limitations, such as lack of access, mean that the required DoFs cannot usually be measured directly. Therefore, the measured data must be *transformed* from measurement coordinates to those of the interface model. An important role of this chapter then is to present *transformations* for various scenarios: finite difference approximations,

DOI: 10.1201/9781003399131-4

virtual points, flexible interfaces and equivalent multi/single point connections are all discussed before leading onto multi-point and continuous interfaces.

4.2 THE SINGLE POINT INTERFACE

The point contact is an idealisation of an interface connection point, valid when the size of the contact area is much less than the shortest (structural) wave length of interest. In this case, local flexure of the interface is minimal, and its dynamics are governed principally by its rigid body motions. Consequently, a single point-like interface is characterised by 6 DoFs; the translational x, y and z DoFs, and their respective rotations α, β, and γ. The equation governing the 6 DoF motion of a single point interface is then,

$$\begin{pmatrix} v_x \\ v_y \\ v_z \\ v_\alpha \\ v_\beta \\ v_\gamma \end{pmatrix} = \begin{bmatrix} Y_{xx} & Y_{xy} & Y_{xz} & Y_{x\alpha} & Y_{x\beta} & Y_{x\gamma} \\ Y_{yx} & Y_{yy} & Y_{yz} & Y_{y\alpha} & Y_{y\beta} & Y_{y\gamma} \\ Y_{zx} & Y_{zy} & Y_{zz} & Y_{z\alpha} & Y_{z\beta} & Y_{z\gamma} \\ Y_{\alpha x} & Y_{\alpha y} & Y_{\alpha z} & Y_{\alpha\alpha} & Y_{\alpha\beta} & Y_{\alpha\gamma} \\ Y_{\beta x} & Y_{\beta y} & Y_{\beta z} & Y_{\beta\alpha} & Y_{\beta\beta} & Y_{\beta\gamma} \\ Y_{\gamma x} & Y_{\gamma y} & Y_{\gamma z} & Y_{\gamma\alpha} & Y_{\gamma\beta} & Y_{\gamma\gamma} \end{bmatrix} \begin{pmatrix} f_x \\ f_y \\ f_z \\ f_\alpha \\ f_\beta \\ f_\gamma \end{pmatrix} \tag{4.1}$$

where for example v_α is the rotational response (say velocity) about the x axis, f_γ is the moment force about the z axis, and $Y_{\alpha\gamma}$ is the transfer FRF relating the two.

As discussed in Chapter 2, rotational DoFs are generally not available by direct measurement (though rotational response sensors are available, their use is not very common). In fact, even the direct measurement of a translational point FRF (e.g. Y_{xx} is challenging; requiring an exact co-location of the applied force and measured response. In practice, to overcome these challenges it is typical to measure a series of translational FRFs surrounding the connection point, and through an appropriate transformation, use these to estimate the translational and rotational DoFs of interest.

The matrix of measured FRF satisfies the equation,

$$\begin{pmatrix} v_1 \\ v_2 \\ \vdots \\ v_M \end{pmatrix} = \begin{bmatrix} Y_{11} & Y_{12} & \cdots & Y_{1N} \\ Y_{11} & Y_{22} & \cdots & Y_{2N} \\ \vdots & \vdots & \ddots & \vdots \\ Y_{M1} & Y_{M2} & \cdots & Y_{MN} \end{bmatrix} \begin{pmatrix} f_1 \\ f_2 \\ \vdots \\ f_N \end{pmatrix} \tag{4.2}$$

or

$$\hat{\mathbf{v}} = \hat{\mathbf{Y}}\hat{\mathbf{f}} \tag{4.3}$$

where v_m and f_n are, respectively, the response and applied force at the m/nth *measurement DoF*, and the $\hat{\square}$ accent is used to denote a *measured* quantity, as opposed to a transformed one.

Assuming a point-like interface, such that only translational and rotational motions are permitted, the set of measured translational responses $\hat{\mathbf{v}} \in \mathbb{C}^{M_v}$ ($M_v \geq 6$ denoting the number of *measured* response DoFs) can be related to the set of mixed translational/rotational responses $\mathbf{v} \in \mathbb{C}^6$ defined at some other location according to the linear transformation,

$$\mathbf{v} = \mathbf{T}_v \hat{\mathbf{v}} \tag{4.4}$$

where $\mathbf{T}_v \in \mathbb{C}^{6\times M_v}$ is the so-called response transformation matrix.

Similarly, the set of applied translational forces $\hat{\mathbf{f}} \in \mathbb{C}^{M_f}$ ($M_f \geq 6$ denoting the number of measured force DoFs) can be thought to generate a mixed set of translational and rotational forces $\mathbf{f} \in \mathbb{C}^6$ at some other location, again according to a linear transformation,

$$\mathbf{f} = \mathbf{R}_f^{\mathrm{T}} \hat{\mathbf{f}} \quad \text{or in reverse} \quad \hat{\mathbf{f}} = \mathbf{T}_f^{\mathrm{T}} \mathbf{f} \tag{4.5}$$

where with $\mathbf{T}_f = \mathbf{R}_f^{+} \in \mathbb{C}^{M_f\times 6}$ is the so-called force transformation matrix.

Substituting Equations 4.4 and 4.5 into 4.3 we obtain,

$$\mathbf{v} = \mathbf{T}_v \hat{\mathbf{Y}} \hat{\mathbf{f}} = \overbrace{\mathbf{T}_v \hat{\mathbf{Y}} \mathbf{T}_f^{\mathrm{T}}}^{\mathbf{Y}} \mathbf{f} \tag{4.6}$$

which is a relation between the transformed force and response vectors. The matrix product $\mathbf{Y} = \mathbf{T}_v \hat{\mathbf{Y}} \mathbf{T}_f^{\mathrm{T}}$ thus describes the transformed FRF.

$$\begin{bmatrix} Y_{xx} & Y_{xy} & Y_{xz} & Y_{x\alpha} & Y_{x\beta} & Y_{x\gamma} \\ Y_{yx} & Y_{yy} & Y_{yz} & Y_{y\alpha} & Y_{y\beta} & Y_{y\gamma} \\ Y_{zx} & Y_{zy} & Y_{zz} & Y_{z\alpha} & Y_{z\beta} & Y_{z\gamma} \\ Y_{\alpha x} & Y_{\alpha y} & Y_{\alpha z} & Y_{\alpha\alpha} & Y_{\alpha\beta} & Y_{\alpha\gamma} \\ Y_{\beta x} & Y_{\beta y} & Y_{\beta z} & Y_{\beta\alpha} & Y_{\beta\beta} & Y_{\beta\gamma} \\ Y_{\gamma x} & Y_{\gamma y} & Y_{\gamma z} & Y_{\gamma\alpha} & Y_{\gamma\beta} & Y_{\gamma\gamma} \end{bmatrix} = \mathbf{T}_v \begin{bmatrix} Y_{11} & Y_{12} & \cdots & Y_{1N} \\ Y_{21} & Y_{22} & \cdots & Y_{2N} \\ \vdots & \vdots & \ddots & \vdots \\ Y_{M1} & Y_{M2} & \cdots & Y_{MN} \end{bmatrix} \mathbf{T}_f^{\mathrm{T}} \tag{4.7}$$

The construction of these transformation matrices will differ depending on the chosen representation. In this chapter we will introduce two popular approaches, the Finite Difference (FD) approximation and the Virtual Point (VP) transformation.

Regardless of the approach taken, the interpretation of Equation 4.7 remains the same. Pre-multiplication by $\mathbf{T}_v$ is an operation that acts on the measured response DoFs (i.e. the rows of the FRF matrix), projecting them onto the transformed DoFs of interest. Similarly, post-multiplication by $\mathbf{T}_f$ acts to project the applied force DoFs (i.e. columns of the FRF matrix) onto the transformed DoFs. They take the general form,

$$\mathbf{T}_{v,f} = \begin{array}{c} \begin{array}{ccccc} 1 & 2 & 3 & \cdots & M \end{array} \\ \begin{bmatrix} T_{x1} & T_{x2} & T_{x3} & \cdots & T_{xM} \\ T_{y1} & T_{y2} & T_{y3} & \cdots & T_{yM} \\ T_{z1} & T_{z2} & T_{z3} & \cdots & T_{zM} \\ T_{\alpha 1} & T_{\alpha 2} & T_{\alpha 3} & \cdots & T_{\alpha M} \\ T_{\beta 1} & T_{\beta 2} & T_{\beta 3} & \cdots & T_{\beta M} \\ T_{\gamma 1} & T_{\gamma 2} & T_{\gamma 3} & \cdots & T_{\gamma M} \end{bmatrix} \end{array} \begin{array}{c} x \\ y \\ z \\ \alpha \\ \beta \\ \gamma \end{array} \tag{4.8}$$

where, for example, T_{xm} describes the projection of the mth measurement DoF (representing an applied force or measured response) to the transformed x DoF. In the sections that follow we will describe the FD and VP approaches to constructing these transformation matrices.

4.2.1 FINITE DIFFERENCE APPROXIMATION

To introduce the Finite Difference (FD) method we begin by considering the simplified scenario illustrated in Figure 4.1; a translational and rotational force is applied at some position whereabouts a corresponding translational and rotational response is observed. The finite difference method can easily be extended to more complex arrangements with multiple DoFs in x, y and z coordinates. This will be shown later.

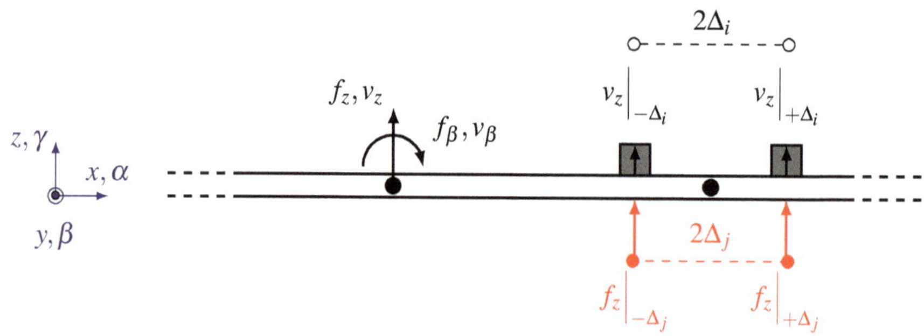

Figure 4.1 Finite difference set-up for simple configuration. Left – true force/moment and response/rotational response, right – finite difference arrangement.

The FRFs that characterise this scenario are the translational, rotational and cross FRFs,

$$Y_{zz} = \frac{v_z}{f_z}, \quad Y_{\beta\beta} = \frac{v_\beta}{f_\beta}, \quad Y_{\beta z} = \frac{v_\beta}{f_z}, \quad Y_{z\beta} = \frac{v_z}{f_\beta}. \tag{4.9}$$

To derive expressions for the rotational response FRF $Y_{\beta z}$, we begin by recalling that the instantaneous rotational displacement about a given point is equal to the spatial derivative of the displacement,

$$u_\beta = \frac{du_z}{dx} \tag{4.10}$$

and by differentiating with respect to time, we obtain the rotational velocity as,

$$v_\beta = \frac{d}{dt}\frac{du_z}{dx} = \frac{dv_z}{dx}. \tag{4.11}$$

Substituting Equation 4.11 into the definition of $Y_{\beta z}$ we get,

$$Y_{\beta z} = \frac{v_\beta}{f_z} = \frac{1}{f_z}\frac{dv_z}{dx_i} = \frac{d}{dx_i}\frac{v_z}{f_z}. \tag{4.12}$$

where we have introduced the subscript x_i to make clear that the derivative is with respect to response position, not force position x_j. Equation 4.12 states that the rotational response FRF is obtained by differentiating the translational response FRF with respect to the response position x_i,

$$Y_{\beta z} = \frac{d}{dx_i}Y_{zz}. \tag{4.13}$$

Note that for the purposes of derivation we are temporarily treating the FRF as a continuous differentiable function – our usual measured FRFs are then 'samples' of this function for given excitation and response locations.

To derive an expression for the rotational force FRF we note that a pure moment can be represented by an equivalent pair of opposing translational forces separated by the infinitesimal spacing Δ_j. The translational response due to a pure moment excitation can thus be written,

$$v_z = \lim_{\Delta_j \to 0} \left(Y_{zz}\Big|_{x_j+\Delta_j} f_z - Y_{zz}\Big|_{x_j} f_z \right) \tag{4.14}$$

where the translational FRFs are evaluated at, respectively, $x_j + \Delta_j$ and x_j. Multiplying top and bottom by Δ_j and evaluating the limit leads to,

$$v_z = \lim_{\Delta_j \to 0} \left(\frac{Y_{zz}\Big|_{x_j+\Delta_j} - Y_{zz}\Big|_{x_j}}{\Delta_j} \Delta_j f_z \right) = \frac{d}{dx_j} Y_{zf} f_\beta \tag{4.15}$$

where $f_\beta = \Delta_j f_z$ is the moment force about the y axis. From Equation 4.15 we see that the rotational force FRF is obtained by differentiating the translational force FRF with respect to the force position x_j,

$$Y_{z\beta} = \frac{d}{dx_j} Y_{zz}. \tag{4.16}$$

To obtain an expression for the rotational response and rotational force FRF, we simply combine the above steps, giving,

$$Y_{\beta\beta} = \frac{d^2}{dx_i dx_j} Y_{zz}. \tag{4.17}$$

Equations 4.11, 4.16 and 4.17 show that rotational FRFs are related to translational FRFs by differentiation with respect to the response and/or excitation position. Experimentally, these derivatives can be evaluated through a finite difference approximation. Schemes include forward, backward and central differences, of different order [7, 17]. The first order central difference scheme has seen the most use, providing a good balance between approximation error and ease of implementation [9].

According to the first order central difference scheme, the rotational FRFs derived above can be approximated as,

$$Y_{\beta z} \approx \frac{Y_{zz}\Big|_{+\Delta_i} - Y_{zz}\Big|_{-\Delta_i}}{2\Delta_i} \tag{4.18}$$

$$Y_{z\beta} \approx \frac{Y_{zz}\Big|_{+\Delta_j} - Y_{zz}\Big|_{-\Delta_j}}{2\Delta_j} \tag{4.19}$$

$$Y_{\beta\beta} \approx \frac{Y_{zz}\Big|_{+\Delta_i,+\Delta_j} - Y_{zz}\Big|_{-\Delta_i,+\Delta_j} - Y_{zz}\Big|_{+\Delta_i,-\Delta_j} + Y_{zz}\Big|_{-\Delta_i,-\Delta_j}}{4\Delta_i\Delta_j} \tag{4.20}$$

where $\square|_{\pm\Delta_{i,j}}$ indicates that the FRF's response and force are measured at a distance $\pm\Delta_{i,j}$ from the position i and j, respectively.

The finite difference approximation requires measuring the individual translational FRFs between a pair of spaced excitations (to compute moments) and responses (to compute rotational responses), as illustrated in Figure 4.1. Noting that the central response/force can be obtained by averaging the spaced measurements,

$$v_z \approx \frac{v_z\Big|_{+\Delta_i} + v_z\Big|_{-\Delta_i}}{2}, \quad f_z \approx \frac{f_z\Big|_{+\Delta_j} + f_z\Big|_{-\Delta_j}}{2} \tag{4.21}$$

the translational FRF can be approximated using the same set of spaced measurements,

$$Y_{zz} \approx \frac{Y_{zz}\Big|_{+\Delta_i} + Y_{zz}\Big|_{-\Delta_i} + Y_{zz}\Big|_{+\Delta_j} + Y_{zz}\Big|_{-\Delta_j}}{4}. \tag{4.22}$$

Note that in their current form, the finite difference approximation of the cross-FRFs in Equation 4.18 and 4.19 require, respectively, translational excitation and response measurements directly at the central position. To rely on the spaced measurements alone (i.e. avoid measuring directly at the central point), Equations 4.18 and 4.19 must be modified; for Equation 4.18 the force is replaced by the average of forces between $\pm\Delta_j$ and similarly, for Equation 4.19 the response is replaced by the average response between $\pm\Delta_i$. This yields,

$$Y_{\beta z} \approx \frac{Y_{zz}\Big|_{+\Delta_i,+\Delta_j} + Y_{zz}\Big|_{+\Delta_i,-\Delta_j} - Y_{zz}\Big|_{-\Delta_i,+\Delta_j} - Y_{zz}\Big|_{-\Delta_i,-\Delta_j}}{4\Delta_i}. \tag{4.23}$$

$$Y_{z\beta} \approx \frac{Y_{zz}\Big|_{+\Delta_i,+\Delta_j} + Y_{zz}\Big|_{-\Delta_i,+\Delta_j} - Y_{zz}\Big|_{+\Delta_i,-\Delta_j} - Y_{zz}\Big|_{-\Delta_i-\Delta_j}}{4\Delta_j} \tag{4.24}$$

Equation 4.22, 4.20, and 4.23-4.24 are, respectively, the first order central difference equations for the translational, rotational and cross-FRFs. They require a pair of spaced forces to be applied symmetrically about the central point, and the response measured by a pair of sensors, also spaced symmetrically. The set of finite difference equations can be expressed in a more concise matrix form as,

$$\begin{bmatrix} Y_{zz} & Y_{z\beta} \\ Y_{\beta z} & Y_{\beta\beta} \end{bmatrix} = \begin{bmatrix} \frac{1}{2} & \frac{1}{2} \\ \frac{1}{2\Delta_i} & \frac{-1}{2\Delta_i} \end{bmatrix} \begin{bmatrix} Y_{zz}\Big|_{+\Delta_i,+\Delta_j} & Y_{zz}\Big|_{+\Delta_i,-\Delta_j} \\ Y_{zz}\Big|_{-\Delta_i,+\Delta_j} & Y_{zz}\Big|_{-\Delta_i,-\Delta_j} \end{bmatrix} \begin{bmatrix} \frac{1}{2} & \frac{1}{2\Delta_j} \\ \frac{1}{2} & \frac{-1}{2\Delta_j} \end{bmatrix} \tag{4.25}$$

or

$$\mathbf{Y} = \mathbf{T}_v \hat{\mathbf{Y}} \mathbf{T}_f^{\mathrm{T}} \tag{4.26}$$

where $\hat{\mathbf{Y}}$ is the matrix of measured (spaced translational) FRFs, and $\mathbf{T}_{v,f}$ are the FD transformation matrices that project these onto the translational and rotational DoFs.

To extend the finite difference method to more complex arrangements it is only necessary to understand how to build the transformation matrices $\mathbf{T}_{v,f}$ using the finite difference scheme. From the simple case above, we see that to estimate a translational FRF, the transformation matrix must compute an average of all measured FRFs, whilst for a rotational FRF it must compute the difference weighted by the reciprocal of the spacing used (note that this spacing can differ for the forces and response about each axis).

To demonstrate the construction of the finite difference transformation matrices we consider two practical configurations; the first, a three DoF set-up (z, α and β) that neglects the in-plane rotation and translations, and the second, a full six DoF arrangement (x, y, z, α, β, γ). The corresponding finite difference force/sensor arrangements are shown respectively on the left and right-hand side of Figure 4.2.

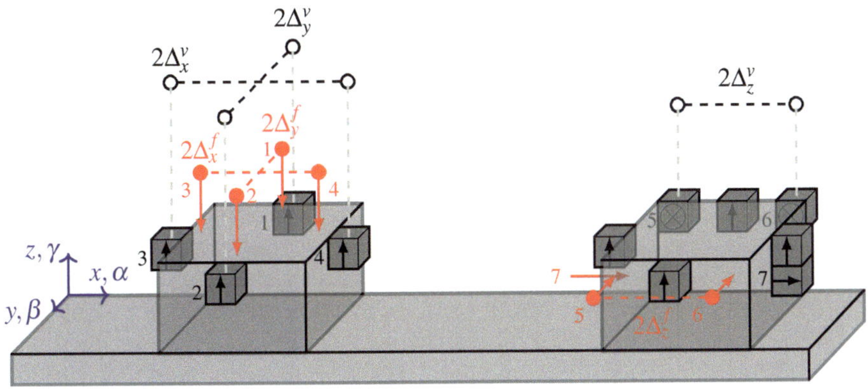

Figure 4.2 Example response/force configurations for two interface representations; left – reduced 3 DoF set-up, right – full 6 DoF arrangement.

The three DoF set-up (Figure 4.2, left) utilises 4 accelerometers and 4 applied forces to capture the DoFs of interest. Following the sensor/force ordering shown in Figure 4.2, the corresponding FD transformation matrix takes the form,

$$\mathbf{T}_{v,f} = \begin{array}{c} \\ \left[\begin{array}{cccc} 1 & 2 & 3 & 4 \\ \frac{1}{4} & \frac{1}{4} & \frac{1}{4} & \frac{1}{4} \\ \frac{1}{2\Delta_x} & -\frac{1}{2\Delta_x} & 0 & 0 \\ 0 & 0 & \frac{1}{2\Delta_y} & -\frac{1}{2\Delta_y} \end{array} \right] \end{array} \begin{array}{c} \\ z \\ \alpha \\ \beta \end{array} \tag{4.27}$$

where the force and response spacings might differ. From Equation 4.27 we see that the translational z DoF is estimated by the average of the 4 responses/forces, whilst the rotational α and β DoFs are estimated using the weighted difference between spaced response/force pairs.

The full six DoF set-up (Figure 4.2, right) utilises 7 accelerometers and applied forces to estimate/capture the DoFs of interest. Following the sensor/force ordering shown in Figure 4.2, the corresponding FD transformation matrices takes the form,

$$\mathbf{T}_{v,f} = \begin{array}{c} \begin{matrix} 1 & 2 & 3 & 4 & 5 & 6 & 7 \end{matrix} \\ \begin{bmatrix} 0 & 0 & 0 & 0 & 0 & 0 & 1 \\ 0 & 0 & 0 & 0 & \frac{1}{2} & \frac{1}{2} & 0 \\ \frac{1}{4} & \frac{1}{4} & \frac{1}{4} & \frac{1}{4} & 0 & 0 & 0 \\ \frac{1}{2\Delta_x} & -\frac{1}{2\Delta_x} & 0 & 0 & 0 & 0 & 0 \\ 0 & 0 & \frac{1}{2\Delta_y} & -\frac{1}{2\Delta_y} & 0 & 0 & 0 \\ 0 & 0 & 0 & 0 & \frac{1}{2\Delta_z} & -\frac{1}{2\Delta_z} & 0 \end{bmatrix} \end{array} \begin{matrix} x \\ y \\ z \\ \alpha \\ \beta \\ \gamma \end{matrix} . \tag{4.28}$$

From Equation 4.28 we see that the translational z and rotational α and β DoFs are estimated in the same way as before. The translational x DoF is obtained from a single force/sensor, the translational y DoF is estimated from the average of a spaced pair, and the rotational γ DoF from the weighted difference of this same spaced pair.

Whilst the examples presented above use transformation matrices of the same form for both forces and responses, this isn't a requirement. For example, the x force (number 7) of the 6 DoF set-up might be challenging to apply in practice, due to its proximity to sensor 3. Instead, a spaced pair could be applied either side. This would then require a slightly modified transformation matrix $\mathbf{T}_f$, whilst $\mathbf{T}_v$ would remain the same.

Key to the success of the FD method is an appropriate choice of sensor/force spacing. For a given frequency, too small a spacing will yield very similar response/force values between the spaced pairs. On computing their difference, so to estimate a rotational DoF, subtraction of two very similar values will have a very high sensitivity to measurement noise and error. This source of error puts a low frequency limit on the application of the FD method. In contrast, too wide a spacing at the sensor/force pairs will no longer be rigidly linked, and the first order finite difference scheme is no longer valid. This error puts a high frequency limit on the FD method. Though bounded in this sense, the finite difference approximation has seen much success. Some example applications can be found in [8, 12]. It is, however, somewhat limited from a practical perspective; it requires symmetrically spaced sensor/force pairs. Whilst for many practical interfaces this does not present a problem, for some it does. Even with a relatively simple interface, application of the FD approximation is unlikely to see the estimated translational and rotational DoFs co-located. This generally isn't a problem when conducting diagnostic procedures, like in situ TPA, though it can become so when substructuring is introduced (as for component-based TPA), since this requires the precise co-location of DoFs to correctly satisfy equilibrium and continuity. This limitation of the FD approach is overcome by the more flexible Virtual Point transformation, discussed in the next section.

4.2.2 VIRTUAL POINT TRANSFORMATION

The Virtual Point (VP) transformation is a rigid body kinematic transformation that relates a set of arbitrarily positioned (translational) response sensors and excitation forces, to the translational and rotational dynamics of a so-called *virtual point* [18].

If we assume that, locally, each interface connection behaves rigidly (the same assumption required by the finite difference approximation), the translational response of an arbitrary sensor $\hat{v}_m \in \mathbb{C}$ can be expressed in terms of the combined translational and rotational motion of a virtual point $\mathbf{v} \in \mathbb{C}^6$ as,

$$\hat{v}_m = \overbrace{\begin{bmatrix} e_x & e_y & e_z \end{bmatrix}}^{\mathbf{e}_m} \overbrace{\begin{bmatrix} 1 & 0 & 0 & 0 & r_z & -r_y \\ 0 & 1 & 0 & -r_z & 0 & r_x \\ 0 & 0 & 1 & r_y & -r_x & 0 \end{bmatrix}}^{\mathbf{r}_m} \begin{pmatrix} v_x \\ v_y \\ v_z \\ v_\alpha \\ v_\beta \\ v_\gamma \end{pmatrix} + \varepsilon = \mathbf{R}_{v_m}\mathbf{v} + \varepsilon \tag{4.29}$$

In the above, $\mathbf{R}_{v_m}$ represents a pair of linear transformations that account for the position and orientation of the sensor relative to the virtual point, or more precisely its coordinate system. The rightmost matrix $\mathbf{r}_m$ is the so-called Interface Deformation Mode (IDM) matrix that projects the motion of the VP onto the position of the sensor. The first three columns, which together form an identity matrix, simply state that an x, y, or z translation of the VP will yield identical translation of the measurement position. The remaining three columns describe how a rotational about each axis of the VP will induce a translation in the x, y and z coordinates of the measurement position. The magnitude of the resulting translation clearly depends on the relative distance (specified in meters) between the VP and measurement position, described in the x, y and z directions as r_x, r_y and r_z. The further away the VP and measurement position, the greater the translation will be from a given rotation.

The vector $\mathbf{e}_m$ then accounts for the orientation of the sensor, which may or may not be inline with the coordinate system used by the VP. Suppose for example the sensor is pointing vertically upwards in the z direction of the VP. The orientation vector would then be,

$$\mathbf{e}_z = \begin{bmatrix} 0 & 0 & 1 \end{bmatrix}. \tag{4.30}$$

More generally, the orientation vector can be determined from the sensor's inclination and azimuthal angles as in Figure 4.3,

$$\begin{bmatrix} e_x & e_y & e_z \end{bmatrix} = \begin{bmatrix} \sin\theta\cos\phi & \sin\theta\sin\phi & \cos\theta \end{bmatrix} \tag{4.31}$$

or obtained from a CAD model using the normalised cross-product between two vectors $\mathbf{a}$ and $\mathbf{b}$ that define the orientation of the surface to which the sensor is attached,

$$\mathbf{e} = \frac{\mathbf{a}\times\mathbf{b}}{||\mathbf{a}||\cdot||\mathbf{b}||} \tag{4.32}$$

as in Figure 4.3.

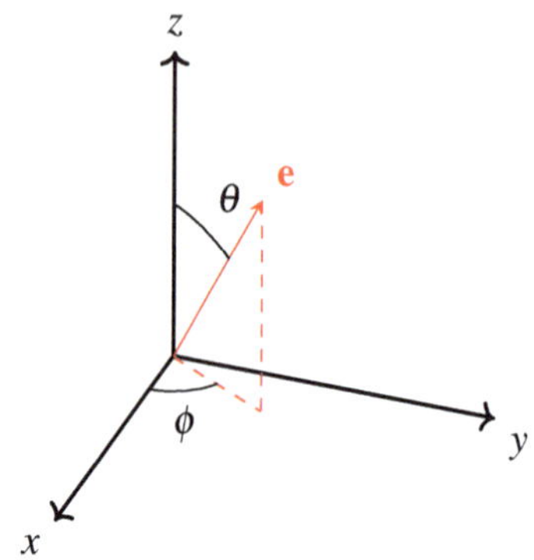

Figure 4.3 Orientation vector **e** from spherical coordinates.

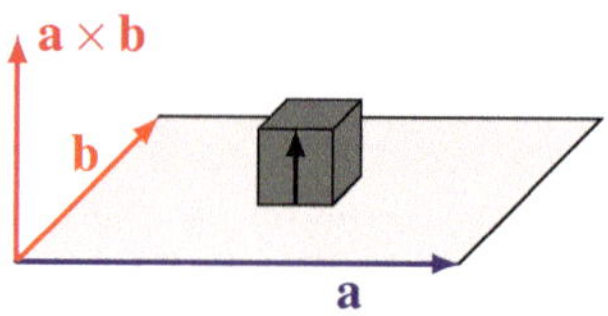

Figure 4.4 Orientation vector **e** from cross-product.

The final term in Equation 4.29, ε, represents the component of the response $\hat{v}_m$ that cannot be reproduced by the virtual point with its underlying rigid assumption. This can be considered a *residual flexibility*.

Equation 4.29 can be written for multiple sensors surrounding the connection point, leading to the system of equations,

$$\begin{pmatrix} \hat{v}_1 \\ \hat{v}_2 \\ \vdots \\ \hat{v}_{M_v} \end{pmatrix} = \begin{bmatrix} \mathbf{e}_1 & & & \\ & \mathbf{e}_2 & & \\ & & \ddots & \\ & & & \mathbf{e}_{M_v} \end{bmatrix} \begin{bmatrix} \mathbf{r}_1 \\ \mathbf{r}_2 \\ \vdots \\ \mathbf{r}_{M_v} \end{bmatrix} \begin{pmatrix} v_x \\ v_y \\ v_z \\ v_\alpha \\ v_\beta \\ v_\gamma \end{pmatrix} + \begin{pmatrix} \varepsilon_1 \\ \varepsilon_2 \\ \vdots \\ \varepsilon_{M_v} \end{pmatrix} \tag{4.33}$$

or simply

$$\hat{\mathbf{v}} = \mathbf{R}_v \mathbf{v} + \varepsilon \tag{4.34}$$

where $\hat{\mathbf{v}}$ is a vector of measured (translation) responses, and $\mathbf{v}$ is the vector of translational and rotational responses at the virtual point.

When using the VP transformation, it is particularly convenient to use tri-axial accelerometers, as in this case, the accelerometer responses recorded in x, y and z are done so at (approx.) the same location, and so they share r_x, r_y and r_z. For a single

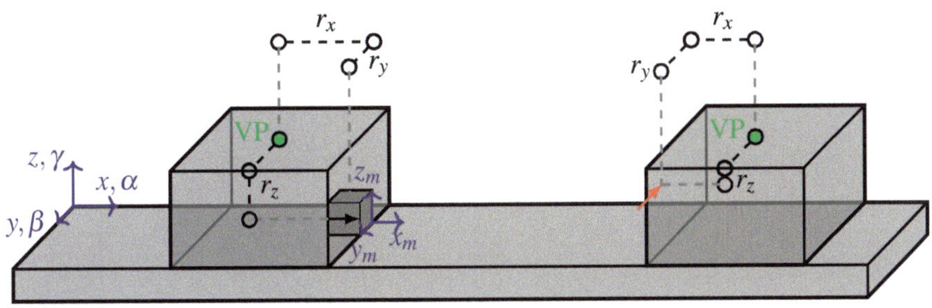

Figure 4.5 Example relative coordinates for virtual point transformations.

tri-axial sensor, Equation 4.29 then becomes,

$$\begin{pmatrix} \hat{v}_{x,m} \\ \hat{v}_{y,m} \\ \hat{v}_{z,m} \end{pmatrix} = \overbrace{\begin{bmatrix} e_{x,x} & e_{x,y} & e_{x,z} \\ e_{y,x} & e_{y,y} & e_{y,z} \\ e_{z,x} & e_{z,y} & e_{z,z} \end{bmatrix}}^{\mathbf{E}_m} \begin{bmatrix} 1 & 0 & 0 & 0 & r_z & -r_y \\ 0 & 1 & 0 & -r_z & 0 & r_x \\ 0 & 0 & 1 & r_y & -r_x & 0 \end{bmatrix} \begin{pmatrix} v_x \\ v_y \\ v_z \\ v_\alpha \\ v_\beta \\ v_\gamma \end{pmatrix} + \begin{pmatrix} \varepsilon_{x,m} \\ \varepsilon_{y,m} \\ \varepsilon_{z,m} \end{pmatrix} \tag{4.35}$$

where $\mathbf{E}_m$ is now a matrix describing the orientation of the mth tri-axial sensor's coordinate axes relative to that of the VP. For three tri-axial sensors (the minimum required to capture all 6 DoFs), Equation 4.33 becomes,

$$\begin{pmatrix} \hat{\mathbf{v}}_1 \\ \hat{\mathbf{v}}_2 \\ \hat{\mathbf{v}}_3 \end{pmatrix} = \begin{bmatrix} \mathbf{E}_1 & & \\ & \mathbf{E}_2 & \\ & & \mathbf{E}_3 \end{bmatrix} \begin{bmatrix} \mathbf{r}_1 \\ \mathbf{r}_2 \\ \mathbf{r}_3 \end{bmatrix} \begin{pmatrix} v_x \\ v_y \\ v_z \\ v_\alpha \\ v_\beta \\ v_\gamma \end{pmatrix} + \begin{pmatrix} \varepsilon_1 \\ \varepsilon_2 \\ \varepsilon_3 \end{pmatrix} \tag{4.36}$$

where $\hat{\mathbf{v}}_m \in \mathbb{C}^3$ is the vector of tri-axial responses for sensor m.

We can now apply a similar idea to the external forces. Application of an arbitrary external force f_n (at coordinates r_x, r_y, r_z relative to the VP) generates a set of translational and rotational forces at the VP. These forces are related by the equation,

$$\begin{pmatrix} f_x \\ f_y \\ f_z \\ f_\alpha \\ f_\beta \\ f_\gamma \end{pmatrix} = \overbrace{\begin{bmatrix} 1 & 0 & 0 \\ 0 & 1 & 0 \\ 0 & 0 & 1 \\ 0 & -r_z & r_y \\ r_z & 0 & -r_x \\ -r_y & r_x & 0 \end{bmatrix}}^{\mathbf{r}_n^{\mathrm{T}}} \overbrace{\begin{bmatrix} e_x \\ e_y \\ e_z \end{bmatrix}}^{\mathbf{e}_n^{\mathrm{T}}} \hat{f}_n = \mathbf{R}_{f_n}^{\mathrm{T}} \hat{f}_n. \tag{4.37}$$

where $\mathbf{R}_{f_n}^{\mathrm{T}}$ again represents a pair of linear transformations that account for the position and orientation of the applied force. An equation of this form can be written for each excitation, leading to,

$$\begin{pmatrix} f_x \\ f_y \\ f_z \\ f_\alpha \\ f_\beta \\ f_\gamma \end{pmatrix} = \begin{bmatrix} \mathbf{r}_1^{\mathrm{T}} & \mathbf{r}_2^{\mathrm{T}} & \cdots & \mathbf{r}_{M_f}^{\mathrm{T}} \end{bmatrix} \begin{bmatrix} \mathbf{e}_1^{\mathrm{T}} & & & \\ & \mathbf{e}_2^{\mathrm{T}} & & \\ & & \ddots & \\ & & & \mathbf{e}_{M_f}^{\mathrm{T}} \end{bmatrix} \begin{pmatrix} \hat{f}_1 \\ \hat{f}_2 \\ \vdots \\ \hat{f}_{M_f} \end{pmatrix} \tag{4.38}$$

or simply,

$$\mathbf{f} = \mathbf{R}_f^{\mathrm{T}} \hat{\mathbf{f}} \tag{4.39}$$

where $\hat{\mathbf{f}}$ is a vector of applied/measured translation forces, and $\mathbf{f}$ is the vector of resulting translational and moment forces at the VP. Note that whilst tri-axial accelerometers can be used to simplify the VP response measurement, no equivalent is available for excitations. These must be applied individually.

Through a least squares pseudo-inverse, the above equations yield a pair of transformation matrices $\mathbf{T}_v$ and $\mathbf{T}_f$,

$$\mathbf{v} = (\mathbf{R}_v)^+ \hat{\mathbf{v}} = \mathbf{T}_v \hat{\mathbf{v}} \tag{4.40}$$

$$\hat{\mathbf{f}} = \left(\mathbf{R}_f^{\mathrm{T}}\right)^+ \mathbf{f} = \mathbf{T}_f^{\mathrm{T}} \mathbf{f}. \tag{4.41}$$

To apply the VP transformation to a measured FRF matrix,

$$\hat{\mathbf{v}} = \hat{\mathbf{Y}} \hat{\mathbf{f}} \tag{4.42}$$

we begin by substituting for $\hat{\mathbf{f}}$ from Equation 4.41,

$$\hat{\mathbf{v}} = \hat{\mathbf{Y}} \mathbf{T}_f^{\mathrm{T}} \mathbf{f} \tag{4.43}$$

before pre-multiplying both sides of Equation 4.43 by $\mathbf{T}_v$ to obtain,

$$\mathbf{v} = \mathbf{T}_v \hat{\mathbf{Y}} \mathbf{T}_f^{\mathrm{T}} \mathbf{f}. \tag{4.44}$$

Equation 4.44 relates the VP transformed force $\mathbf{f}$ to the VP transformed response $\mathbf{v}$. As such, the matrix product must describe the VP transformed FRF matrix $\mathbf{Y}$,

$$\mathbf{Y} = \mathbf{T}_v \hat{\mathbf{Y}} \mathbf{T}_f^{\mathrm{T}}. \tag{4.45}$$

The VP transformation has several advantages over the equivalent multi-point connection (discussed below in 4.2.4) and finite difference approximation. Firstly, the translational and rotational DoFs of the VP are estimated based on the response of all sensors. This enables an over-determination of the problem, which can provide a more accurate estimate. A second advantage is that the resulting DoFs are, by definition, exactly co-located at the VP. This is particularly important when trying to substructure together FRFs of different components where all DoFs must be correctly aligned. This is also useful for combining measurements with models. The VP transformation has been extended further to include flexible interface modes (discussed below in 4.2.5) [15], and rotational sensors [5].

Whilst a powerful tool, the VP transformation has a notable disadvantage. It requires accurate information on both the position and orientation of each sensor and applied force (so to construct the transformation matrices $\mathbf{T}_v$ and $\mathbf{T}_f$). If a CAD model of the interface is available this information may be readily obtained. In many practical scenarios, however, CAD models are not available, and the VP transformation may be subject to large uncertainties or simply not applicable.

4.2.3 PERFORMANCE INDICATORS

The transformations described above are based on the assumption that the interface can be described entirely by its rigid body dynamics. This is typically a valid assumption in the low/mid frequency range, where the structural wavelength is much larger than the size of the interface. However, as frequency increases so does the deformation of the interface and at higher frequencies there may be a need to assess the validity of the rigid body assumption.

Transforming from a set of measured DoFs involves projecting an M dimensional space representing the measurement DoFs, on to an $N = 6$ dimensional space representing the transformed rigid body DoFs. In most scenarios, $M > N$ and so this projection potentially involves the loss of information.

If the interface in question does in fact behave rigidly, then the rank of its measured FRF matrix $\hat{\mathbf{Y}} \in \mathbb{C}^{M_v \times M_f}$ must be limited to 6 (note that in general the number of measured force and response DoFs, M_f and M_v, might differ). In this case, the interface transformations can be applied in reverse to reconstruct the original measured FRF matrix from the transformed matrix,

$$\hat{\mathbf{Y}}' = \mathbf{T}_v^+ \mathbf{Y} \mathbf{T}_f^{+\mathrm{T}} = \mathbf{T}_v^+ \left(\mathbf{T}_v \hat{\mathbf{Y}} \mathbf{T}_f^{\mathrm{T}}\right) \mathbf{T}_f^{+\mathrm{T}} = \hat{\mathbf{Y}} \qquad \text{if} \quad \operatorname{rank}\left(\hat{\mathbf{Y}}\right) \leq 6 \tag{4.46}$$

where we use the $\square'$ accent to denote a reconstruction of the original FRF matrix.

If however, the interface possesses some flexibility, the rank of $\hat{\mathbf{Y}}$ has the potential to be greater than 6. Consequently, the measured FRF matrix can only be reconstructed approximately,

$$\hat{\mathbf{Y}}' = \mathbf{T}_v^+ \mathbf{Y} \mathbf{T}_f^{+\mathrm{T}} = \mathbf{T}_v^+ \left(\mathbf{T}_v \hat{\mathbf{Y}} \mathbf{T}_f^{\mathrm{T}}\right) \mathbf{T}_f^{+\mathrm{T}} \approx \hat{\mathbf{Y}} \qquad \text{if} \quad \operatorname{rank}\left(\hat{\mathbf{Y}}\right) > 6 \tag{4.47}$$

Note that the reconstructed FRF matrix $\hat{\mathbf{Y}}'$ includes only the rigid body dynamics; flexible interface dynamics in $\hat{\mathbf{Y}}$ have been filtered out.

It is important to note that this action of 'filtering out flexible dynamics' implicitly requires an over-determined transformation with $M_{f,v} > N$. In the determined case the transformation matrix becomes square ($M_{f,v} = N$), and assuming its inverse exists, it is exactly reversible. In other words, there are not enough measurement DoFs to observe flexibility; the interface is 'under sampled'.

In the over-determined case, the above notion can be used to establish sensor and force consistency criteria that indicate the validity of the rigid interface assumption, as well as the quality of the sensor arrangement and applied forces [18].

Sensor consistency

To assess the consistency of a sensor arrangement we consider a unit force $f_h = 1$ applied to the component in question at the arbitrary position h (not necessarily at the interface). The directly measured response satisfies the equation,

$$\hat{\mathbf{v}} = \hat{\mathbf{Y}}_h f_h = \hat{\mathbf{Y}}_h \tag{4.48}$$

where $\hat{\mathbf{Y}}_h$ is a column vector of measured FRFs between force position h and the interface responses. Similarly, the transformed response satisfies,

$$\mathbf{v} = \mathbf{Y}_h f_h = \mathbf{Y}_h = \mathbf{T}_v \hat{\mathbf{Y}}_h. \tag{4.49}$$

Applying the inverse transformation to the transformed velocity projects the response back onto the measured DoFs,

$$\hat{\mathbf{Y}}'_h = \mathbf{T}_v^+ \mathbf{T}_v \hat{\mathbf{Y}}_h. \tag{4.50}$$

where $\square'$ is again used to denote filtering of rigid body dynamics. The matrix product $\mathbf{T}_v^+ \mathbf{T}_v$ has the effect of filtering out flexible interface modes, retaining only the rigid body dynamics.

The sensor consistency can be assessed by simply comparing the directly measured $\hat{\mathbf{v}}$ and reconstructed (rigid body only) $\hat{\mathbf{v}}'$ responses. This comparison can be done in a variety of ways. A popular approach is to use a correlation-based measure of similarity,

$$\rho_{v,h} = \text{Corr}\left(\hat{\mathbf{Y}}_h, \hat{\mathbf{Y}}'_h\right) \tag{4.51}$$

$$\text{Corr}\,(\mathbf{a}, \mathbf{b}) = \frac{\left|\mathbf{a}^{\mathrm{H}}\mathbf{b}\right|^2}{\mathbf{a}^{\mathrm{H}}\mathbf{a} \cdot \mathbf{b}^{\mathrm{H}}\mathbf{b}} \tag{4.52}$$

The sensor consistency $\rho_{v,h}$ is a frequency dependent scalar quantity that varies between 0 and 1. A value of 1 indicates rigid interface behaviour, i.e. the reconstructed (rigid body filtered) responses are identical to those measured directly. A value less than 1 indicates some degree of interface flexibility, which is lost through the applied transformation.

Note that $\rho_{v,h}$ is evaluated for the force excitation at position h, and that multiple force locations might be considered. In this case each force could be used to determine an independent $\rho_{v,h}$. Alternatively, the collection of measured and reconstructed responses can be vectorised (i.e. stacked into a single large column vector, denoted by $\underset{\rightarrow}{\square}$, see Equation 8.28) and used to determine an overall sensor consistency,

$$\rho_v = \text{Corr}\left(\underset{\rightarrow}{\hat{\mathbf{Y}}}, \underset{\rightarrow}{\hat{\mathbf{Y}}'}\right). \tag{4.53}$$

Whilst the location of the applied forces is, in theory arbitrary, in [18] it is recommended to avoid using the interface so as to ensure that a global response is generated, as opposed to a local response that might be highly sensitive to the impact position.

Force consistency

Based on the above sensor criterion, it is straightforward to develop an equivalent *impact consistency* criterion so as to assess the quality of the applied forces.

In this case consider the response of a single interface sensor to all of the forces applied around the interface,

$$\hat{v}_i = \hat{\mathbf{Y}}_i \hat{\mathbf{f}} \tag{4.54}$$

where $\hat{\mathbf{Y}}_i$ is a row vector of directly measured FRFs between the applied forces and sensor i.

Similarly, the response at sensor i due to the rigid body filtered forces is given by,

$$\hat{v}_i'' = \hat{\mathbf{Y}}_i \mathbf{T}_f^{\mathrm{T}} \mathbf{f} = \hat{\mathbf{Y}}_i \mathbf{T}_f^{\mathrm{T}} \mathbf{T}_f^{+\mathrm{T}} \hat{\mathbf{f}} = \hat{\mathbf{Y}}_i'' \hat{\mathbf{f}} \tag{4.55}$$

where $\square''$ is used to denote filtering of rigid body forces. The impact consistency is then defined as,

$$\rho_{f,i} = \mathrm{Corr}\left(\hat{\mathbf{Y}}_i, \hat{\mathbf{Y}}_i''\right). \tag{4.56}$$

Like the sensor consistency, $\rho_{f,i}$ is bound between 0 and 1, with a value of 1 indicating perfect transmission of force between the applied impacts and the VP.

To obtain a single overall force consistency metric, we can again collect and vectorise the measured and reconstructed responses across all applied forces,

$$\rho_f = \mathrm{Corr}\left(\underset{\rightarrow}{\hat{\mathbf{Y}}}, \underset{\rightarrow}{\hat{\mathbf{Y}}''}\right). \tag{4.57}$$

Combined interface consistency

Having introduced the sensor and force consistency metrics above, it is perhaps obvious that one could formulate in the same manner a single overall consistency that combines the two. Indeed, applying both force and response transformations, alongside their respective inverses, one obtains the filtered FRF matrix,

$$\hat{\mathbf{Y}}''' = \mathbf{T}_v^{+} \left(\mathbf{T}_v \hat{\mathbf{Y}} \mathbf{T}_f^{\mathrm{T}}\right) \mathbf{T}_f^{+\mathrm{T}} \tag{4.58}$$

which, by comparison against the original FRF matrix, forms an overall interface consistency metric,

$$\rho = \mathrm{Corr}\left(\underset{\rightarrow}{\hat{\mathbf{Y}}}, \underset{\rightarrow}{\hat{\mathbf{Y}}'''}\right). \tag{4.59}$$

4.2.4 EQUIVALENT MULTI-POINT INTERFACE

Often when conducting a diagnostic test, such as in situ TPA, interest does not lie with the contribution of the translational and rotational DoFs themselves, but rather with the contribution of the different connection points as a whole. In this case, there is no need to explicitly calculate the rotational DoFs, though their contribution must still be captured. This leads to the so-called *equivalent multi-point* (EMP) connection.

The equivalent multi-point connection is the most straightforward approach to characterise an interface, including the contribution of rotational DoFs. The interface is instrumented/excited by 6 or more sensors/forces, the position and orientation of which is arbitrarily provided they suitably span the x, y and z coordinates. By measuring 6+ translational DoFs, all 6 rigid body DoFs (translations and rotations) are implicitly captured. Importantly however, the averaged translational and rotational components are not calculated explicitly; their influence is encoded within the 6+ individual translations. Hence we treat the connection as 6+ equivalent translational DoFs. Take for example the interface configuration in Figure 4.2 (right). Seven translational sensors are positioned around the interface. These are accompanied by 7 translational force excitations. According to the equivalent multi-point connection the interface FRF matrix for this contact point would be of dimensions 7×7. Rotational information about the x, y and z axes are encoded by the sensors spaced about their respective axes.

Note that, the sensors/applied forces do not have to be included as spaced pairs. So long as no two sensors/forces are co-located, and those that are included sufficiently span x, y and z, all translational and rotational interface motions will be captured.

The main advantages of the equivalent multi-point connection approach are its straight forward implementation (the transformation matrices simply reduce to the identity matrix, $\mathbf{T}_{v,f} = \mathbf{I}$) and ability to capture rotational DoFs. Furthermore, in the case that >6 measurement points are used, it has the ability to capture flexible interface dynamics. Its main disadvantage is that the true translational and rotational motions are not explicitly calculated. This can cause some issues when trying to combine data sets, for example measured blocked forces with transfer functions obtained from substructuring or other experimental set-ups.

Though straightforward to implement, it is important to note that if more than 6 translations are used in the EMP representation, some regularisation may be required to avoid a rank deficient FRF matrix; if there are more sensors/forces present than DoFs to capture (i.e. more parameters than can be justified by the data) and the connection point indeed behaves rigidly, the inversion of the FRF matrix can become ill-conditioned and large errors returned. A simple solution is to perform a truncated Singular Value Decomposition to invert the matrix, retaining only the first 6 singular values.

4.2.5 FLEXIBLE INTERFACES

The notion of a point interface is typically valid in the low-mid frequency range. Fortunately, this is the range where most structural vibration problems are concerned, and so the point interface assumption has served the community well over the decades. However, new technologies (such as vehicle electrification) are ever extending the frequency of interest [11]. With the corresponding wavelength decreasing, structural connections that might once have appeared 'point-like' are becoming comparable in size to wavelength. As a consequence, the assumption of locally rigid dynamics breaks down as local deformations occur. To deal with this, more complex representations of interface dynamics are required. Below we briefly summarise two approaches; a flexible extension of the virtual point transformation, and a generalised polynomial interface model.

Virtual point extension

The VP transformation, as presented in 4.2.2, relies on the construction of the so-called Interface Deformation Mode (IDM) matrices, $\mathbf{R}_m$ that describe how the measured DoF would respond to the motion of a virtual point, assuming rigid body motion. The rows of $\mathbf{R}_m$ correspond to the x, y and z coordinates of the measurement DoF, whilst the columns relate to the permissible motions of the VP (three translations, and three rotations),

$$\mathbf{R}_m = \left[\begin{array}{ccc|ccc} \multicolumn{3}{c|}{\overbrace{\hphantom{1\;0\;0}}^{\text{Translation}}} & \multicolumn{3}{c}{\overbrace{\hphantom{0\;r_z\;-r_y}}^{\text{Rotation}}} \\ 1 & 0 & 0 & 0 & r_z & -r_y \\ 0 & 1 & 0 & -r_z & 0 & r_x \\ 0 & 0 & 1 & r_y & -r_x & 0 \end{array}\right]. \tag{4.60}$$

Importantly, any residual motion (i.e. flexibility) not captured by the 6 VP DoFs is discarded and not included in the interface representation. To accommodate for flexibility of the interface, additional modes must be added to the IDM matrix. In [15] a 'bottom-up' definition of flexible interface behaviour is suggested, which avoids the need for a priori knowledge of the interface dynamics and is therefore more generally applicable.

The following example is given of an extended VP IDM matrix that is capable of capturing an additional three extension (i.e. 'strain') and torsion deformations, alongside the standard rigid modes,

$$\mathbf{R}_m^{ext} = \left[\begin{array}{ccc|ccc|ccc|ccc} \multicolumn{3}{c|}{\text{Translation}} & \multicolumn{3}{c|}{\text{Rotation}} & \multicolumn{3}{c|}{\text{Extension}} & \multicolumn{3}{c}{\text{Torsion}} \\ 1 & 0 & 0 & 0 & r_z & -r_y & r_x & 0 & 0 & 0 & r_y r_z & -r_z r_y \\ 0 & 1 & 0 & -r_z & 0 & r_x & 0 & r_y & 0 & -r_x r_z & 0 & r_z r_x \\ 0 & 0 & 1 & r_y & -r_x & 0 & 0 & 0 & r_z & r_x r_y & -r_y r_x & 0 \end{array}\right]. \tag{4.61}$$

Based on this extended IDM matrix, a measured response is represented as,

$$\hat{v}_m = \begin{bmatrix} e_x \\ e_y \\ e_z \end{bmatrix}^{\mathrm{T}} \mathbf{R}_m^{ext} \begin{bmatrix} \mathbf{v}_{tr.} \\ \mathbf{v}_{rot.} \\ \mathbf{v}_{ext.} \\ \mathbf{v}_{tor.} \end{bmatrix} \tag{4.62}$$

where the additional VP motions $\mathbf{v}_{ext.} = [v_{ext.,x}, v_{ext.,y}, v_{ext.,z}]^{\mathrm{T}}$ and $\mathbf{v}_{tor.} = [v_{tor.,x}, v_{tor.,y}, v_{tor.,z}]^{\mathrm{T}}$ have been included in the VP response vector $\mathbf{v}$. Note that a similar procedure can be applied to the force IDM matrix.

The flexible IDMs can be defined with similar ease as rigid IDMs, requiring only the product of measurement DoF coordinates. When adding additional flexible IDMs, it is important to always ensure that the resultant matrix $\mathbf{R}$ is full-rank, i.e. additional modes are only effective if there is sufficient information available from the measured DoFs. In practice, it is necessary to 'cherry-pick' the correct set of IDM modes for the problem at hand, whilst avoiding any unnecessary IDMs in the interface representation.

Polynomial interface models

An alternative, though closely related, method based on general polynomial interface models has also been proposed [1]. An advantage of this approach is its ability to base the interface representation on the physics of the structure (i.e. rigid, beam-like, plate-like, etc.).

It is proposed that to reconstruct the response field about a region (e.g. surrounding an interface), a polynomial model can be used. Specifically, the translational x, y, and z response measured at the position $\mathbf{r} = [r_x, r_y, r_z]$ (relative to the coordinates of the polynomial model) can be represented by a polynomial of increasing degree relative to the complexity of the local flexure,

$$\begin{bmatrix} v_x \\ v_y \\ v_z \end{bmatrix} = \begin{bmatrix} \alpha_{x1} + \beta_{x1} r_x + \beta_{x2} r_y + \beta_{x3} r_z + \gamma_{x1} r_x^2 + \gamma_{x2} r_y^2 + \gamma_{x3} r_z^2 + \gamma_{x4} r_x r_y + \cdots \\ \alpha_{y1} + \beta_{y1} r_x + \beta_{y2} r_y + \beta_{y3} r_z + \gamma_{y1} r_x^2 + \gamma_{y2} r_y^2 + \gamma_{y3} r_z^2 + \gamma_{y4} r_x r_y + \cdots \\ \alpha_{z1} + \beta_{z1} r_x + \beta_{z2} r_y + \beta_{z3} r_z + \gamma_{z1} r_x^2 + \gamma_{z2} r_y^2 + \gamma_{z3} r_z^2 + \gamma_{z4} r_x r_y + \cdots \end{bmatrix} \tag{4.63}$$

where α, β and γ are the polynomial coefficients.

The response measured by a single translational sensor, $\hat{v}_m$, can be related to this expansion through its orientation, described as before by the vector $\mathbf{e} = [e_x, e_y, e_z]$,

$$\hat{v}_m = \begin{bmatrix} e_x \\ e_y \\ e_z \end{bmatrix}^{\mathrm{T}} \begin{bmatrix} \alpha_{x1} + \beta_{x1} r_x + \beta_{x2} r_y + \beta_{x3} r_z + \gamma_{x1} r_x^2 + \gamma_{x2} r_y^2 + \gamma_{x3} r_z^2 + \gamma_{x4} r_x r_y + \cdots \\ \alpha_{y1} + \beta_{y1} r_x + \beta_{y2} r_y + \beta_{y3} r_z + \gamma_{y1} r_x^2 + \gamma_{y2} r_y^2 + \gamma_{y3} r_z^2 + \gamma_{y4} r_x r_y + \cdots \\ \alpha_{z1} + \beta_{z1} r_x + \beta_{z2} r_y + \beta_{z3} r_z + \gamma_{z1} r_x^2 + \gamma_{z2} r_y^2 + \gamma_{z3} r_z^2 + \gamma_{z4} r_x r_y + \cdots \end{bmatrix} \tag{4.64}$$

or equivalently by,

$$\hat{v}_m = \begin{bmatrix} e_x \\ e_y \\ e_z \end{bmatrix}^{\mathrm{T}} \left\{ \begin{bmatrix} \alpha_{x1} \\ \alpha_{y1} \\ \alpha_{z1} \end{bmatrix} + \begin{bmatrix} \beta_{x1} & \beta_{x2} & \beta_{x3} \\ \beta_{y1} & \beta_{y2} & \beta_{y3} \\ \beta_{z1} & \beta_{z2} & \beta_{z3} \end{bmatrix} \begin{pmatrix} r_x \\ r_y \\ r_z \end{pmatrix} + \cdots \right.$$
$$\left. + \begin{bmatrix} \gamma_{x1} & \gamma_{x2} & \gamma_{x3} \\ \gamma_{y1} & \gamma_{y2} & \gamma_{y3} \\ \gamma_{z1} & \gamma_{z2} & \gamma_{z3} \end{bmatrix} \begin{pmatrix} r_x^2 \\ r_y^2 \\ r_z^2 \end{pmatrix} + \cdots \right\} \tag{4.65}$$

or more simply,

$$\hat{v}_m = \mathbf{e} \overbrace{\begin{bmatrix} \mathbf{I} & \mathbf{I}r_x & \mathbf{I}r_y & \mathbf{I}r_z & \mathbf{I}r_x^2 & \cdots \end{bmatrix}}^{\mathbf{r}_m} \overbrace{\begin{pmatrix} \alpha_1 \\ \beta_1 \\ \beta_2 \\ \beta_3 \\ \gamma_1 \\ \vdots \end{pmatrix}}^{\mu} \tag{4.66}$$

where $\mathbf{I}$ is the 3×3 identity matrix, $\alpha_n = [\alpha_{xn}, \alpha_{yn}, \alpha_{zn}]^{\mathrm{T}}$, $\beta_n = [\beta_{xn}, \beta_{yn}, \beta_{zn}]^{\mathrm{T}}$, and $\gamma_n = [\gamma_{xn}, \gamma_{yn}, \gamma_{zn}]^{\mathrm{T}}$.

Equation 4.66 takes a similar form to Equation 4.29. It relates the measured response $\hat{v}_m$ to a series of polynomial coefficients μ (in Equation 4.29 these were the VP motions) via a transformation matrix $\mathbf{r}_m$. As for the VP, the transformation matrix $\mathbf{r}_m$ contains only geometric information regarding the sensor position, whilst μ contains the information about the response field, albeit decomposed into polynomial coefficients.

To estimate the polynomial coefficients we consider multiple response measurements, and stack Equation 4.66 accordingly,

$$\begin{pmatrix} \hat{v}_1 \\ \hat{v}_2 \\ \vdots \\ \hat{v}_{M_v} \end{pmatrix} = \overbrace{\begin{bmatrix} \mathbf{e}_1 & & & \\ & \mathbf{e}_2 & & \\ & & \ddots & \\ & & & \mathbf{e}_{M_v} \end{bmatrix} \begin{bmatrix} \mathbf{r}_1 \\ \mathbf{r}_2 \\ \vdots \\ \mathbf{r}_{M_v} \end{bmatrix}}^{\mathbf{R}_v} \begin{pmatrix} \alpha_1 \\ \beta_1 \\ \beta_2 \\ \beta_3 \\ \gamma_1 \\ \vdots \end{pmatrix}. \tag{4.67}$$

Providing an equal or greater number of measurements are taken than polynomial coefficients are sought, Equation 4.67 can be solved as,

$$\mu = \mathbf{R}_v^+ \hat{\mathbf{v}} = \mathbf{T}_v \hat{\mathbf{v}} \tag{4.68}$$

Following a similar procedure an equivalent representation of the applied forces can be made, as discussed in [1].

From this general set-up, various interface representations can be developed. As an example let us consider the simple rigid interface case. For this case we should have:

- No higher order deformations: $\gamma_n = 0$
- No normal strain: $\partial u_x/\partial x = \partial u_y/\partial y = \partial u_z/\partial z = 0$, which from Equation 4.63 leads to $\beta_{x1} = \beta_{y2} = \beta_{z3} = 0$.
- No shear strain: $\partial u_x/\partial y + \partial u_y/\partial x = \partial u_x/\partial z + \partial u_z/\partial x = \partial u_y/\partial z + \partial u_z/\partial y = 0$, which from Equation 4.63 leads to, $\beta_{x2} + \beta_{y1} = \beta_{x3} + \beta_{z1} = \beta_{y3} + \beta_{z2} = 0$

Substituting these requirements into Equation 4.65 yields,

$$\hat{v}_m = \begin{bmatrix} e_x \\ e_y \\ e_z \end{bmatrix}^{\mathrm{T}} \left\{ \begin{bmatrix} \alpha_{x1} \\ \alpha_{y1} \\ \alpha_{z1} \end{bmatrix} + \begin{bmatrix} 0 & -\beta_{y1} & \beta_{x3} \\ \beta_{y1} & 0 & \beta_{z2} \\ -\beta_{x3} & \beta_{z2} & 0 \end{bmatrix} \begin{pmatrix} r_x \\ r_y \\ r_z \end{pmatrix} \right\} \tag{4.69}$$

which can be written equivalently as,

$$\hat{v}_m = \begin{bmatrix} e_x \\ e_y \\ e_z \end{bmatrix}^{\mathrm{T}} \left[\begin{array}{ccc|ccc} 1 & 0 & 0 & 0 & r_z & -r_y \\ 0 & 1 & 0 & -r_z & 0 & r_x \\ 0 & 0 & 1 & r_y & -r_x & 0 \end{array} \right] \begin{pmatrix} \alpha_{x1} \\ \alpha_{y1} \\ \alpha_{z1} \\ \beta_{z2} \\ \beta_{x3} \\ \beta_{y1} \end{pmatrix}. \tag{4.70}$$

This result is identical to the conventional VP transformation with α_{x1}, α_{y1} and α_{z1} representing the interface translations, and β_{x1}, β_{y1} and β_{z1} their corresponding rotations.

Note that by changing the physical constraints, different degrees of flexibility can be captured. As an example, allowing non-zero normal strain leads to $\beta_{x1} = \beta_{y2} = \beta_{z3} \neq 0$, and consequently,

$$\hat{v}_m = \begin{bmatrix} e_x \\ e_y \\ e_z \end{bmatrix}^{\mathrm{T}} \left[\begin{array}{ccc|ccc|ccc} 1 & 0 & 0 & 0 & r_z & -r_y & r_x & 0 & 0 \\ 0 & 1 & 0 & -r_z & 0 & r_x & 0 & r_y & 0 \\ 0 & 0 & 1 & r_y & -r_x & 0 & 0 & 0 & r_z \end{array} \right] \begin{pmatrix} \alpha_{x1} \\ \alpha_{x2} \\ \alpha_{x3} \\ \beta_{z2} \\ \beta_{x3} \\ \beta_{y1} \\ \beta_{x1} \\ \beta_{y2} \\ \beta_{z3} \end{pmatrix} \tag{4.71}$$

where β_{x1}, β_{y2} and β_{z3} describe extension in the polynomial x, y and z coordinates, as with the extended virtual point (see Equation 4.62).

Based on Euler—Bernoulli and Kirchhoff—Love theories of beam and plate elements, two alternative polynomial representations are presented in [1] that take into account local flexure.

4.2.6 EQUIVALENT SINGLE POINT CONNECTION

The idea of *single equivalent quantities* was introduced with the reception plate method in 3.3.7 and leads naturally to the idea of a *Equivalent Single Point Connection* (ESPC). Whereas with a VP each connection point on the interface is represented by 6 DoFs (or greater if its flexible IDMs are used), with an ESPC the *entire* interface is represented by a single equivalent DoF. The advantage is that it allows the important gross behaviour to be represented by scalar values which can be easily manipulated; the disadvantage is that it is only an approximation.

To see how this works, we expand the real part of Equation 3.68 for the power from a soft source to give the power as a double sum,

$$P = \frac{1}{2}\sum_{i=1}^{N}\sum_{j=1}^{N} \bar{f}_i^* \mathrm{Re}(Y_{Rij})\bar{f}_j \tag{4.72}$$

which can be grouped as the sum of N diagonal terms and the remaining $N(N-1)$ off-diagonals,

$$P = \frac{1}{2}\sum_{i=1}^{N} |\bar{f}_i|^2 \mathrm{Re}(Y_{Rii}) + \frac{1}{2}\sum_{i=1}^{N}\sum_{j=1\neq i}^{N} \bar{f}_i^* \bar{f}_j \mathrm{Re}(Y_{Rij}). \tag{4.73}$$

Assuming that the point mobilities and blocked force magnitudes do not both vary greatly from point to point we can write the first term of Equation 4.73 in terms of single equivalent quantities, giving,

$$P \approx \frac{1}{2}|\bar{\mathbf{f}}|^2 \mathrm{Re}(Y_{R,eq}) + \frac{1}{2}\sum_{i=1}^{N}\sum_{j=1\neq i}^{N} \bar{f}_i^* \bar{f}_j \mathrm{Re}(Y_{Rij}) \tag{4.74}$$

where,

$$|\bar{\mathbf{f}}|^2 = \sum_{i=1}^{N} |\bar{f}_i|^2 \quad \text{and} \quad \mathrm{Re}(Y_{R,eq}) = \frac{1}{N}\sum_{i=1}^{N} \mathrm{Re}(Y_{Rii}) \tag{4.75}$$

are the single equivalent blocked force and single equivalent mobility real part, respectively. The first term on the right-hand side of Equation 4.73 is then the power as calculated by the reception plate method (3.3.7) and is always positive.

Whereas the first term in Equation 4.74 accounts for the diagonal terms, the second term consists of the off-diagonals representing the interaction between connection points. This term is neglected in the reduction to a single equivalent connection. Noting that each off-diagonal term in the double sum of Equation 4.74 has a reciprocal partner we add a typical pair to give,

$$\mathrm{Re}\left(Y_{Rij}\right)|\bar{f}_i||\bar{f}_j|[e^{i\phi_{ij}} + e^{i\phi_{ji}}] = 2\mathrm{Re}\left(Y_{Rij}\right)|\bar{f}_i||\bar{f}_j|\cos(\phi_{\bar{f}_j} - \phi_{\bar{f}_i}) \tag{4.76}$$

where $\phi_{ij} = \phi_{\bar{f}_j} - \phi_{\bar{f}_i}$ is the relative phase of the blocked forces. The sum of these paired terms is thus always real but may be positive or negative.

From Equation 4.76, the sum of the cross terms will be small if the transfer mobilities Y_{Rij} are small, which is more likely to be the case for widely spaced points (in wavelength terms). Additionally, the cross term contribution will be small if there is cancellation of positive and negative terms, in effect, if they are uncorrelated (in a similar manner to uncorrelated monopoles described in 3.4.3). Again, this is more likely for widely spaced points since, broadly speaking, the real part of a transfer mobility will be positive for points within half a structural wavelength, negative, between a half and a whole wavelength and will then fluctuate in sign as the wavelength distance increases further. In other words, the interactions between feet tend to wash out at higher frequencies which helps explain the trend illustrated in Figure 3.26 of uncertainty reducing with frequency. Frequency averaging, for example over a third octave band will also tend to increase cancellation [22].

Strict upper and lower bounds can be expressed in terms of matrix eigenvalues [13]. Writing the quadratic form for power from a soft source, Equation 3.68, as an eigenvalue expansion leads to,

$$\frac{1}{2}|\bar{\mathbf{f}}|^2\rho_{min} \leq P \leq \frac{1}{2}|\bar{\mathbf{f}}|^2\rho_{max} \tag{4.77}$$

where $\rho_i = \rho\left[\text{Re}(Y_{Rij})\right]$ are the eigenvalues of the real part of the receiver mobility matrix. A similar form applies to the squared velocity,

$$\frac{1}{2}|\bar{\mathbf{f}}|^2\sigma_{min}^2 \leq |v|^2 \leq \frac{1}{2}|\bar{\mathbf{f}}|^2\sigma_{max}^2 \tag{4.78}$$

where $\sigma_i = \sigma\left[Y_{Rij}\right]$ are the singular values of the (complex) receiver mobility matrix.

4.3 MULTI-POINT INTERFACES

In the previous section we considered the single point interface, and introduced a number of methods for capturing the 6+ DoF motion that characterise it. In practice, it is rare that a structural component couples to another via a single connection point. Hence, it is necessary to extend the single-point treatment above, to the general multi-point case.

For the single-point interface, the full 6 DoF FRF matrix (see Equation 4.1) is obtained by,

$$\mathbf{Y} = \mathbf{T}_v\hat{\mathbf{Y}}\mathbf{T}_f^{\mathrm{T}} \tag{4.79}$$

where $\hat{\mathbf{Y}}$ is a matrix of measured translational FRFs (see Equation 4.2), and $\mathbf{T}_{v,f}$ are a pair of transformation matrices for the measured responses and forces, respectively. To extend this treatment to a multi-point interface we simply note that the transformed DoFs for each connection point should depend only on the measured DoFs surrounding it, i.e. the virtual point describing connection 1 depends only on the measured DoFs surrounding point 1. For an interface with N connection points, the transformed response and force DoFs are related to the measured response and force DoFs by,

$$\begin{pmatrix} \mathbf{v}_{C_1} \\ \vdots \\ \mathbf{v}_{C_N} \end{pmatrix} = \begin{bmatrix} \mathbf{T}_{v,C_1} & & \\ & \ddots & \\ & & \mathbf{T}_{v,C_N} \end{bmatrix} \begin{pmatrix} \hat{\mathbf{v}}_{C_1} \\ \vdots \\ \hat{\mathbf{v}}_{C_N} \end{pmatrix} \tag{4.80}$$

and

$$\begin{pmatrix} \hat{\mathbf{f}}_{c_1} \\ \vdots \\ \hat{\mathbf{f}}_{c_N} \end{pmatrix} = \begin{bmatrix} \mathbf{T}_{f,c_1}^{\mathrm{T}} & & \\ & \ddots & \\ & & \mathbf{T}_{f,c_N}^{\mathrm{T}} \end{bmatrix} \begin{pmatrix} \mathbf{f}_{c_1} \\ \vdots \\ \mathbf{f}_{c_N} \end{pmatrix} \tag{4.81}$$

where $\hat{\square}$ denotes the measured DoFs, and $\mathbf{T}_{v,c_n}$ and $\mathbf{T}_{f,c_n}$ are the response and force transformation matrices for the nth connection point. This leads to, for the full multi-point interface FRF,

$$\mathbf{Y}_{cc} = \begin{bmatrix} \mathbf{T}_{v,c_1} & & \\ & \ddots & \\ & & \mathbf{T}_{v,c_N} \end{bmatrix} \begin{bmatrix} \hat{\mathbf{Y}}_{c_1c_1} & \cdots & \hat{\mathbf{Y}}_{c_1c_N} \\ \vdots & \ddots & \vdots \\ \hat{\mathbf{Y}}_{c_Nc_1} & \cdots & \hat{\mathbf{Y}}_{c_Nc_N} \end{bmatrix} \begin{bmatrix} \mathbf{T}_{f,c_1} & & \\ & \ddots & \\ & & \mathbf{T}_{f,c_N} \end{bmatrix}^{\mathrm{T}} \tag{4.82}$$

where it is noted that each transformation matrix can be constructed using the most appropriate method for that connection point.

Finally, to include a set of additional remote DoFs r where no interface transformations are required, Equation 4.82 can be extended simply as,

$$\begin{bmatrix} \mathbf{Y}_{cc} & \mathbf{Y}_{cr} \\ \mathbf{Y}_{rc} & \mathbf{Y}_{rr} \end{bmatrix} = \begin{bmatrix} \mathbf{T}_{v,c} & \\ & \mathbf{I} \end{bmatrix} \begin{bmatrix} \hat{\mathbf{Y}}_{cc} & \hat{\mathbf{Y}}_{cr} \\ \hat{\mathbf{Y}}_{rc} & \hat{\mathbf{Y}}_{rr} \end{bmatrix} \begin{bmatrix} \mathbf{T}_{f,c} & \\ & \mathbf{I} \end{bmatrix}^{\mathrm{T}} \tag{4.83}$$

where $c = \{c_1, c_2, \cdots, c_N\}$.

4.4 CONTINUOUS INTERFACES

The point-like contact considered up to now is of course an idealisation that will start to break down for physically large contacts, and at high frequencies where the contact dimensions are larger than a fraction of a wavelength of the dominant structural waves. In such cases, the contact can be considered to be *extended* rather than *point-like*, in which case the force and velocity are described by continuously varying distributions rather than by discrete values. Continuous functions in this form are not immediately compatible with the overall approach outlined in this book but need to be discretised in some way. In what follows we consider two particular approaches: a truncated series expansion and a piece-wise constant patch. By way of examples, we consider the former as applied to a circular interface, and the latter to a vibro-acoustic interface.

4.4.1 CIRCULAR INTERFACES

Circular interfaces are often encountered in practice with circular flanges and connections of pipes and tubes. In this case we effectively have a line contact around a closed contour, typically a circle or ellipse. Small tubes and pipes will approximate to point contact but for larger components we may need to account for continuously distributed velocity and force around the contour. Some way of discretising these functions is required for our purposes. This case has been treated by Bonhoff and Petersson [2–4].

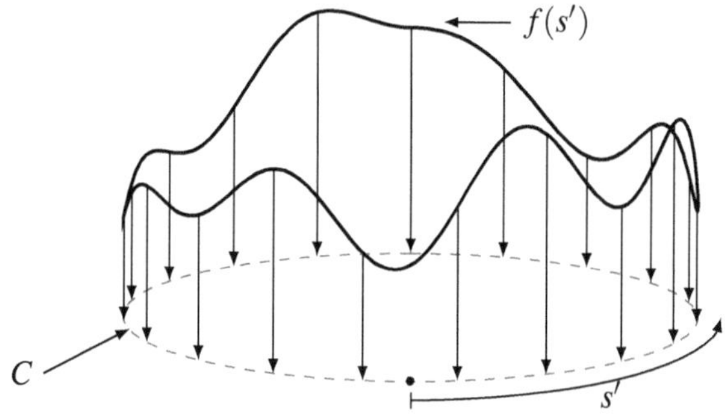

Figure 4.6 Continuous force distribution on a circular interface.

Any continuous distribution can of course be represented as a series expansion and in this case, because the distributions around the circle are naturally periodic, the situation lends itself to the use of spatial Fourier series. The response and force distributions are therefore written as an infinite sum of Fourier terms,

$$v(s) = \sum_{p=-\infty}^{\infty} \tilde{v}_p e^{ik_p s}, \quad f(s') = \sum_{q=-\infty}^{\infty} \tilde{f}_q e^{ik_q s'} \tag{4.84}$$

where $v(s)$ and $f(s')$ are the velocity and force distribution, $k_p = \frac{2\pi p}{C}$ and $k_q = \frac{2\pi q}{C}$ in which p and q represent the number of full cycles around the contour of length C, and s and s' the distance around the contour. These functions are naturally periodic (see Figure 4.6). The coefficients $\tilde{v}_p$ and $\tilde{f}_q$ quantify the strength of the various orders in the continuous distribution and are obtained in the normal way for Fourier series,

$$\tilde{v}_p = \frac{1}{C}\int_0^C v(s)e^{-ik_p s}\,ds, \quad \tilde{f}_q = \frac{1}{C}\int_0^C f(s')e^{-ik_q s'}\,ds'. \tag{4.85}$$

Since the forcing and response locations are considered as continuous variables then the FRF takes the form of a Green's function $Y(s|s')$, which is expanded as an infinite 2-dimensional Fourier series,

$$Y(s|s') = \sum_{p=-\infty}^{\infty}\sum_{q=-\infty}^{\infty} \tilde{Y}_{pq} e^{ik_p s} e^{ik_q s'} \tag{4.86}$$

where s and s' are the position on the contour of the response and excitation respectively and the coefficients are given by,

$$\tilde{Y}_{pq} = \frac{1}{C^2}\int_0^C\int_0^C Y(s|s')e^{-ik_p s}e^{-ik_q s'}\,ds\,ds' \tag{4.87}$$

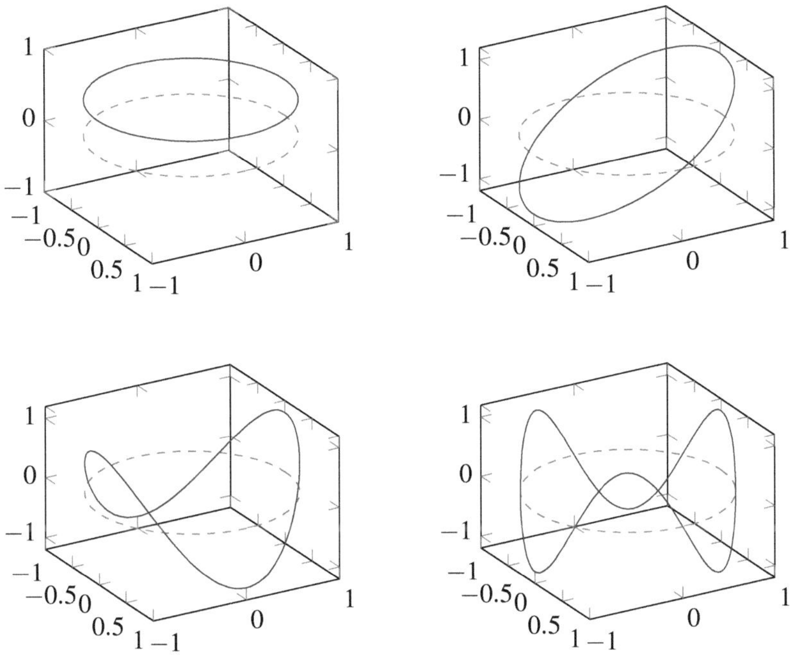

Figure 4.7 Illustration of the Fourier orders round a circular interface.

The $\tilde{Y}_{pq}$ coefficients were termed 'interface mobilities' by Bonhoff and Petersson ([2–4]).

A little interpretation of the terms in the interface mobility matrix may be helpful here (see Figure 4.7). The zeroth order term e^0 is a constant around the contour and represents an in-phase force or prescribed motion, rather like a single equivalent normal force or velocity. The first order terms $e^{\pm i2\pi s/C}$ consist of a single sinusoid around the contour and represent rotation about an axis in the plane of the contour. The phase of the first order terms allows rotation about any axis in the plane of the contour to be represented. Thus, the zero and first order terms represent rigid translation and rotation, effectively forming a virtual point (see 4.2.2) at the centre of the circular interface (although omitting in-plane forces and the out-of-plane moment). Higher orders represent wave-like distributions corresponding to flexible interfaces (4.2.5), for example the second order distribution $e^{\pm j4\pi s/C}$ has a saddle shape distribution with two positive and two negative peaks around the contour, reminiscent of a quadrupole.

For continuously distributed functions the relationship of response to force (analogous to Equation 2.7) is given by the convolution integral of the Green's function,

$$v(s) = \int_0^C Y(s|s')f(s')ds' \tag{4.88}$$

which can be transformed into the Fourier domain by substituting in Equations 4.84 and 4.86, and using Equation 4.85 to give,

$$\tilde{v}_p = C \sum_{q=-\infty}^{\infty} \tilde{Y}_{pq} \tilde{f}_q \tag{4.89}$$

To implement this experimentally, the infinite series must be truncated, say to N terms (in theory the force and response series can be different lengths although it simplifies things to keep them the same). Then by stacking the $\tilde{v}_p$ into a column vector $\tilde{\mathbf{v}}_p$, the continuous interface can be approximated in matrix form:

$$\tilde{\mathbf{v}} = \tilde{\mathbf{Y}}\tilde{\mathbf{f}} \tag{4.90}$$

where $\tilde{\mathbf{Y}} = [\tilde{Y}]_{pq}$ and $\tilde{\mathbf{f}}$ is column vector of $\tilde{f}_q$ terms.

Bonhoff and Petersson have shown [2–4] that in many cases the interface mobility matrix is approximately diagonal. Physically this means that an excitation in one order leads to a dominant response in the same order, for example a first order excitation (a moment) will tend to lead to a pure rotational response around the same axis and so on. However, this is not always the case and one can envisage situations where cross order terms (off diagonals) are significant, particularly for non-symmetrical structures. For example, consider a circular interface close to the end of a cantilever beam: a zero order excitation (effectively an out-of-plane force) will tend to cause a rotation of the interface due to bending of the cantilever. In other words, we expect a first order response to a zero order excitation and, by reciprocity, the reverse is true, so we expect a translational response to a rotational excitation. In other words, the cross order terms may be significant, particularly for non-symmetric structures.

Measurement

Up to this point we have seen that the continuous force and velocity distributions can be represented in matrix form by truncating the series expansions. We now consider how the series coefficients (the interface mobilities) might be obtained from measurements. Measurements are of course only able to *sample* the continuous distributions at discrete points around the contour which, for regularly spaced points, is exactly analogous to the familiar digital sampling of continuous time data. We omit the derivation for brevity, but it is then evident by analogy that the Fourier coefficients can be obtained with a spatial Discrete Fourier transformation (DFT),

$$\tilde{\mathbf{v}} = \mathbf{T}_v \hat{\mathbf{v}} \tag{4.91}$$

where,

$$\mathbf{T}_v = \frac{1}{M} \begin{bmatrix} e^0 & e^0 & \dots & e^0 \\ e^{-i2\pi/M} & e^{-i4\pi/M} & \dots & e^{-i2M\pi/M} \\ \vdots & \vdots & \ddots & \vdots \\ e^{-i2M\pi/M} & e^{-i4M\pi/M} & \dots & e^{-i2MM\pi/M} \end{bmatrix} \tag{4.92}$$

or

$$[T_v]_{mn} = \frac{1}{M} e^{-i2mn\pi/M} |_{m,n=0,\cdots,M-1} \tag{4.93}$$

where m is the measurement location and n the order. The vector of applied forces can similarly be related to the coefficients of its DFT by the transformation matrix $\mathbf{R}_f^{\mathrm{T}}$ (built in the same way as $\mathbf{T}_v$ in Equation 4.92),

$$\tilde{\mathbf{f}} = \mathbf{R}_f^{\mathrm{T}} \hat{\mathbf{f}} \tag{4.94}$$

Substituting $\hat{\mathbf{v}} = \hat{\mathbf{Y}}\hat{\mathbf{v}}$ and the inverse of Equation 4.94 ($\mathbf{f} = \mathbf{T}_f^{\mathrm{T}}\tilde{\mathbf{f}}$ with $\mathbf{T}_f = \mathbf{R}_f^{-1}$) into Equation 4.91 gives,

$$\tilde{\mathbf{v}} = \mathbf{T}_v \hat{\mathbf{Y}} \mathbf{T}_f^{\mathrm{T}} \tilde{\mathbf{f}}. \tag{4.95}$$

The interface mobilities can be obtained from the measured FRF matrix as,

$$\tilde{\mathbf{Y}} = \mathbf{T}_v \hat{\mathbf{Y}} \mathbf{T}_f^{\mathrm{T}} \tag{4.96}$$

which is the same form as transformations given earlier in the chapter (such as Equations 4.26, 4.44 and 4.79).

Equations 4.91, 4.94 and 4.96 show how the measured response force and FRF can all be transformed into what is effectively the spatial Fourier domain. The advantage of this format is that it may provide insight into the important behaviour of the interface. A disadvantage is the additional processing required with the possibility of making a mistake. For the purposes of building a VAVP it is a matter of personal preference whether to work with the original measured DoFs, or those of a transformed domain. If the original measured data is used then the transducer arrays effectively form an equivalent multi-point interface as described in 4.2.4; the important thing in this case is to ensure that the sensors are placed so as to capture all important types of behaviour at the interface.

Other geometries

In the above we have described an approach to characterising continuous interfaces comprising the following main steps:

- a set of orthogonal functions is selected, which is able to represent the continuous force and response distribution over the interface,
- a sensor array is designed to capture all important degrees of freedom,
- the measured force and response vectors and FRF matrix are transformed to obtain the coefficients in the series expansion,
- the expansion is truncated and cast into matrix form.

In theory, this general approach can be applied to extended interfaces with other geometries. For example, continuous contact over a rectangle could be described as the product of a series of orthogonal functions in x and y and the measured data transformed into this domain. However, there do not appear to be examples of this approach being used in the literature. As with circular interfaces, the advantage would be to gain insight into the dynamics of the interface but, provided the sensors are placed so as to capture all the necessary behaviour, an EMPC or VP interface would serve equally well for the purposes of building a VAVP.

4.4.2 VIBRO-ACOUSTIC INTERFACES

For the most part, structural components couple via point-like connections, each described entirely by 6 DoFs. In extending the VAVP framework to include vibro-acoustic components, it is not uncommon to encounter components that couple to one another over surfaces, rather than though discrete points. As an example, sound radiation from a plate involves the coupling of a structural component (plate) to an acoustic domain over a continuous surface. With a small extension to the concept of mobility, it happens that many of the methodologies presented in this book can be applied to surface coupled vibro-acoustic problems also. The necessary extension is the so-called *patch mobility* (or more generally, patch FRF), described in detail below.

When dealing with a single point-like excitation of a component, the response at position i is given by the usual equation,

$$v_i = Y_{ij} f_j. \tag{4.97}$$

For multiple point-like excitations, the above equation is simply extended to a matrix/vector form, with $\mathbf{f}$ describing the multiple point-like forces.

When moving away from point-like excitation and towards extended or distributed excitation, it is no longer appropriate to consider the force f, per se. Rather, the continuous excitation is described in terms of its force density $p(j)$, with units N/m^2. The response at position i is now obtained by integrating over S_f, the forcing area,

$$v(i) = \int_{S_f} Y(i,j)p(j)dj. \tag{4.98}$$

To deal with this integral problem we use a co-location method. The excitation surface is separated into a series of N_f non-overlapping patches. The response at position i can be obtained by integrating over each patch S_j, before summing over all patches,

$$v(i) = \sum_{j=1}^{N_f} \int_{S_j} Y(i,j)p(j)dj. \tag{4.99}$$

Assuming that over each patch the force density remains constant, that is $p(j) = \langle p \rangle_j$ for $j \in S_j$, the force density term can come out of the integral leaving,

$$v(i) = \sum_{j=1}^{N_f} \langle p \rangle_j \int_{S_j} Y(i,j)dj. \tag{4.100}$$

The remaining integral can then be replaced using mean value theorem $\int_{S_j} Y(i,j)dj = S_j\langle Y(i)\rangle_j$,

$$v(i) = S_j \sum_{j=1}^{N_f} \langle Y(i)\rangle_j \langle p\rangle_j. \tag{4.101}$$

The force density and area terms combine to give the total force over the patch, $S_j\langle p\rangle_j = \langle f\rangle_j$.

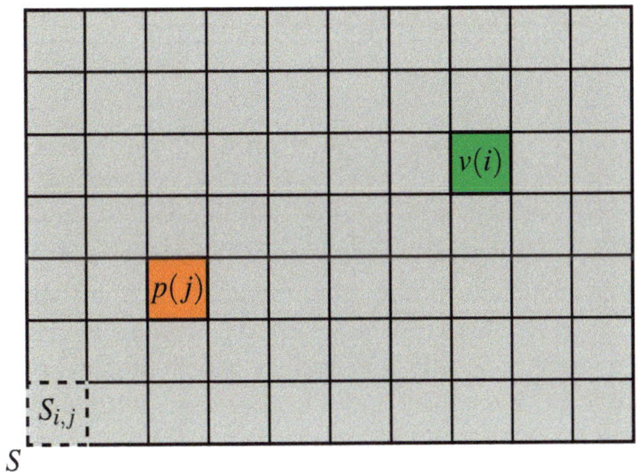

Figure 4.8 Patch discretised surface subject to piece-wise constant forcing over patch j, with mean response over patch i.

We now consider the response not at a single point, but the average over an equivalent response patch S_i,

$$\langle v\rangle_i = \frac{1}{S_i}\int_{S_i} v(i)di = \frac{1}{S_i}\langle f\rangle_j \int_{S_i} \sum_{j=1}^{N_f} \langle Y(i)\rangle_j di. \tag{4.102}$$

Noting that $\frac{1}{S_i}\int_{S_i}\langle Y(i)\rangle_j di = \langle\langle Y\rangle_j\rangle_i$, the above becomes,

$$\langle v\rangle_i = \sum_{j=1}^{N_f} \langle\langle Y\rangle_j\rangle_i \langle f\rangle_j. \tag{4.103}$$

From Equation 4.103 we can define the so-called patch FRF as,

$$\langle\langle Y\rangle_j\rangle_i = \frac{\langle v\rangle_i}{\langle f\rangle_i} \tag{4.104}$$

that is, the ratio of the spatially averaged response over patch i to the force over patch j.

The patch FRF provides a natural extension of the point-based FRF to 2D surfaces. It can be used to construct *patch FRF matrices*, and thus represent continuous areas in a piece-wise constant manner consistent with the methods discussed and described throughout this book. Clearly, the accuracy of its approximation improves as the number of patches increases. For an appropriate level of discretisation, it can be applied to both structural, acoustic and vibro-acoustic problems. For example, surface coupling of two components can be approximated by performing substructuring with patch FRFs. Similarly, patch-based substructuring could be used to couple structural and acoustic components.

Some notable applications of the patch FRF include the coupling of acoustic domains [14], the prediction of sound transmission [6, 10], and acoustic source characterisation [16].

4.5 INTERFACE COMPLETENESS

The importance of accurately representing all dominant interactions taking place at the interface has been emphasised in the above. For components with discrete point-like interfaces, these interactions are represented by the translational and rotational DoFs at each coupling point, all of which have the potential to contribute towards the dynamics of the assembled VAVP. An interface description that is missing important DoFs can be said to be *incomplete* and thus the main contribution of this section is to introduce the concept of *interface completeness*. We also present an example of what can go wrong when the interface description is incomplete. First however, we consider a way to determine the *number* of dominant DoFs at the interface.

4.5.1 INTERFACE RANK

An important relationship came up in the derivation of the Round Trip method in Chapter 2, namely Equation 4.105 (repeated here for convenience),

$$\text{rank}(\mathbf{Y}_{Cba}) \leq \text{rank}(\mathbf{Y}_{Ccc}) \tag{4.105}$$

where c are the interface DoFs that separate two components on which the DoFs a and b are located (see Figure 4.9). The rank of $\mathbf{Y}_{Ccc}$ is an expression of the number of DoFs at the interface, important information in this context since it determines the size of the blocked force vector and the FRF matrices we need for a complete characterisation. Equation 4.105 states that this rank can be determined from $\mathbf{Y}_{Cab}$ which is relatively simple to measure by exciting at points a on one side of the interface and recording the response at points b on the other side.

The physical interpretation of Equation 4.105 is that the interface acts as a 'bottleneck' since adding more points at a or b does not increase the rank of $\mathbf{Y}_{Cba}$.

These observations of rank enable us to determine how many DoFs are required for a complete characterisation of the interface, but as yet they provide no information with regards to what those DoFs are. For this, a more detailed method is required, as described in the following in relation to blocked force determination.

4.5.2 COMPLETENESS IN BLOCKED FORCE DETERMINATION

To demonstrate the importance of completeness, we will consider it from the perspective of an in situ blocked force characterisation, as detailed in 3.2. Note however, that the concept of completeness applies more generally to substructuring and substructure decoupling, etc.

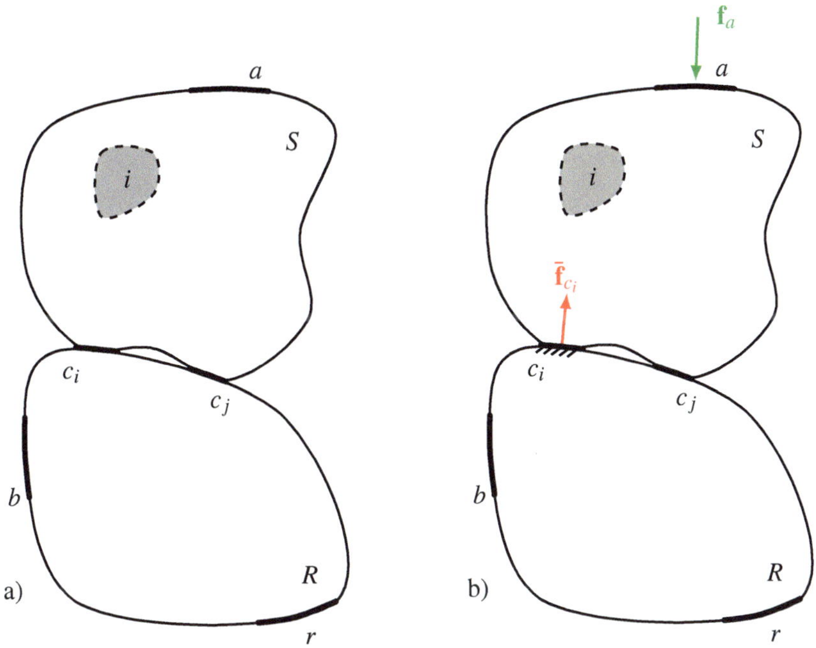

Figure 4.9 a) Source receiver assembly with separated interface DoFs. b) Incomplete interface representation where only a subset of the interface DoFs are constrained.

Recall from Chapter 3 the in situ relation for the blocked force,

$$\mathbf{v}_b = \mathbf{Y}_{Cbc}\bar{\mathbf{f}}_c \tag{4.106}$$

where $\mathbf{Y}_{Cbc}$ is the transfer FRF of the coupled assembly, $\mathbf{v}_b$ is an operational response and $\bar{\mathbf{f}}_c$ is the blocked force.

In our previous exposition of the in situ blocked force method, the source-receiver interface c was considered complete, in that it contained all DoFs that are physically present at the interface. In what follows we will consider the effect of an incomplete interface description on the characterisation of a blocked force. We do this by treating the interface c as two sets of DoFs: those that are known and measurable, c_i; and those that are unknown, or known but unmeasurable, c_j. Equation 4.106 can then be rewritten as the sum of their respective contributions,

$$\mathbf{v}_b = \mathbf{Y}_{Cbc_i}\bar{\mathbf{f}}_{c_i} + \mathbf{Y}_{Cbc_j}\bar{\mathbf{f}}_{c_j}. \tag{4.107}$$

As discussed in Chapter 3, the in situ characterisation of the blocked force requires the operational response to be pre-multiplied by the inverse of a measured FRF matrix. Supposing that the c_j DoFs were inaccessible for measurement, or perhaps unknown, only the $\mathbf{Y}_{Cbc_i}$ FRF is available. The blocked force is then given by,

$$\mathbf{Y}_{Cbc_i}^{-1}\mathbf{v}_b = \tilde{\bar{\mathbf{f}}}_{c_i} = \bar{\mathbf{f}}_{c_i} + \mathbf{Y}_{Cbc_i}^{-1}\mathbf{Y}_{Cbc_j}\bar{\mathbf{f}}_{c_j} \tag{4.108}$$

where $\tilde{\bar{\mathbf{f}}}_{c_i}$ represents the acquired blocked force and $\bar{\mathbf{f}}_{c_i}$ its true value. Notice that the true blocked force is accompanied by an additional term, $\mathbf{Y}_{Cbc_i}^{-1}\mathbf{Y}_{Cbc_j}\bar{\mathbf{f}}_{c_j}$, that has arisen due to the neglected DoFs c_j. This term constitutes a *bias error* on the true blocked force. Also notice that this additional term is dependent upon the dynamics of the coupled assembly. Consequently, the acquired blocked force is no longer an independent property of the source and is no longer transferable between assemblies. As discussed in Chapter 3, one of the principle advantages of the blocked force over other source characterisation methods is that the acquired result may be transferred from one assembly, say a test bench, to another, say a virtual prototype. In the presence of an incomplete interface representation though, the acquired blocked force becomes dependent on the assembly in which is it is characterised.

When characterising a blocked force, it is beneficial to minimise the influence of bias errors due to incompleteness; all significant DoFs should be included in the interface description and the subsequent characterisation. For well-defined interfaces with point-like connections this may be possible with the help of methods such as the Virtual Point Transformation, Equivalent Multi-Point Connections, or the Finite Difference Approximation, albeit at the cost of additional experimental effort. Often however, such efforts are not necessary. Typically only a few important DoFs contribute to the operational response of an assembly. Which DoFs, and how many of them, will vary depending on interface dynamics, in particular whether a rigid or resilient coupling is present; rigid interface coupling typically requires a greater number of DoFs to achieve a sufficient level of completeness.

Before performing any measurement campaign, it is sensible to ask, how 'complete' is the chosen interface description? The answer to this question may aid in the refining or relaxing the chosen interface description to achieve a sufficient level of completeness for an acceptable level of experimental effort. This is particularly useful for less well defined interfaces, such as continuous joints e.g. due to welding, gluing, etc.

In Chapter 8, we will introduce a method of quantifying completeness as part of our treatment of uncertainties, but for now we consider an illustrative example of what can go wrong when interface descriptions are not complete.

Example 4.1: Influence of interface completeness

To demonstrate the concept of completeness, and its importance in the context of vibro-acoustic simulation, we will consider the specific scenario that is an in situ blocked force characterisation.

To do so we will revisit and extend Example 3.1, where blocked forces obtained from rigid and resilient boundary conditions were shown to be the same (i.e. independent of the receiver) even though the responses from which they were calculated varied drastically (see blue plots in Figures 4.10 and 4.11 for the reproduced results). Importantly, in this previous example the blocked forces where characterised using a *complete* interface representation, i.e. 6 DoFs were included at each coupling point.

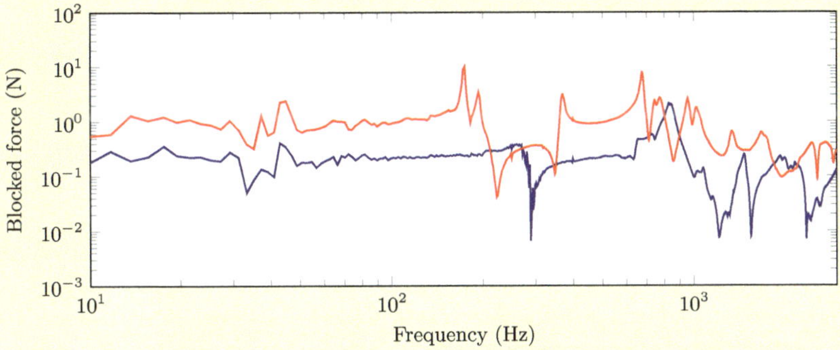

Figure 4.10 Blocked force in the vertical z direction obtained from a resiliently coupled vibration source for a complete (blue, including all DoFs) and incomplete (red, including only vertical z DoFs) interface representation.

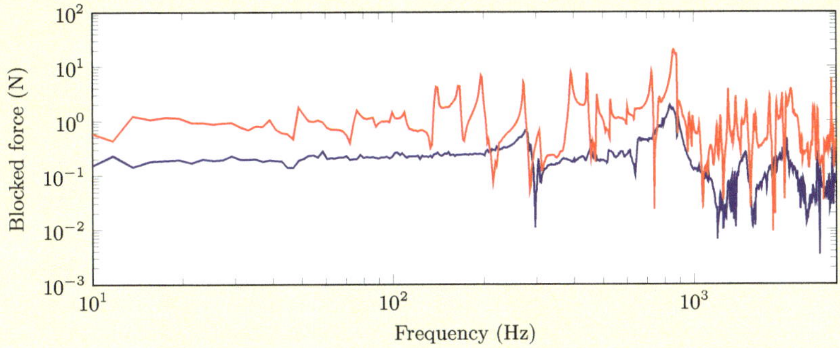

Figure 4.11 Blocked force in the vertical z direction obtained from a rigidly coupled vibration source for a complete (blue, including all DoFs) and incomplete (red, including only vertical z DoFs) interface representation.

The importance of completeness is demonstrated by recalculating the blocked forces. However, this time only the vertical z DoF at each foot are retained, i.e. using an 'incomplete' interface representation. The resulting blocked forces are presented in Figures 4.10 and 4.11 for the resilient and rigid assemblies, respectively, and are compared against those obtained using a complete interface.

It is clear that the incomplete interface results in a blocked force which, when compared against the complete case, is now contaminated by additional resonances. These resonances arise from the unconstrained DoFs that were omitted in the characterisation, i.e. the second term in Equation 4.108. It can be seen that these artifacts differ between the rigid and resilient coupling cases so the obtained blocked force is no longer invariant or transferable.

4.6 SUMMARY

In this chapter various ways of representing interfaces have been presented. Real interfaces are always continuous, but this is not compatible with our component-based approach so some form of idealisation is inevitably required. The most common is to assume the contact is small enough to be considered a *point contact* which in theory is fully described by six DoFs, including three translations and three rotations. We then have to consider that measurements can rarely be made exactly at the contact and furthermore that rotations are usually not measurable directly. Thus, a matrix transformation is required to transform the matrix of measured DoFs to those of the idealised point. The first part of the chapter is largely concerned with providing suitable transformations for different scenarios.

Single point contact has been considered first. Suitable transformations are presented, including the use of *Finite Difference* approximations to transform uni-axial accelerometer readings into rotations. The point contact idea is generalised further with the *Virtual Point Transformation* which provides a useful way to ensure co-located interface DoFs, and thus ensure correct compatibility between connected components, whether measured or modeled. A series of performance indicators were also introduced to assess the ability of a given interface transformation to suitably capture the observed interface dynamics. If unable to, then it may be necessary to consider a flexible interface representation, for example using the extended VP.

Provided enough accelerometers are positioned in appropriate directions, then rotations and other DoFs are implicitly captured even without being calculated explicitly. An *Equivalent Multi-Point Interface* formulation uses this idea and has the advantage of simplicity in that no transformation is required, although compatibility with neighbouring components might be difficult to achieve. Finally, still under the heading of single point contact, we introduced the idea of *Equivalent Single-Point Connection*, an approximate engineering approach where an entire multi-point interface is reduced to a single equivalent connection so as to simplify the calculations.

Moving on to other types, we first considered multi-point interfaces which can be derived from the single point representations fairly straightforwardly. We also considered continuous interfaces where the continuously varying force and response distributions are expressed as: a series expansion, which is then truncated to finite number of terms; or as a piece-wise constant 'patch'. In the example of a circular interface, the series expansions represents that of a discrete Fourier Transform. For 2D interfaces, such vibro-acoustic interfaces, the patch representation is also applicable. The continuous interface is discretised into a set of patches which are then treated as a set of equivalent points. The equivalent point properties are obtained by integration across the patch and can be approximated by measurements at the centre.

It has been emphasised throughout that a good interface description needs to account for all important DoFs at the interface. The final section introduces the concept of *interface completeness*. An example is given of how things can go wrong when the interface description is incomplete. A numerical criterion for completeness will be given later.

Having defined passive and active components in Chapters 2 and 3 and the interfaces between them in this chapter, we are now ready to start building assemblies.

Bibliography

1. JO Almirón, F Bianciardi, and W Desmet. Flexible interface models for force/displacement field reconstruction applications. *Journal of Sound and Vibration*, 534:117001, 2022.
2. HA Bonhoff and BAT Petersson. The influence of cross-order terms in interface mobilities for structure-borne sound source characterization: Plate-like structures. *Journal of Sound and Vibration*, 311(1-2):473–484, 2008.
3. HA Bonhoff and BAT Petersson. The influence of cross-order terms in interface mobilities for structure-borne sound source characterization: Frame-like structures. *Journal of Sound and Vibration*, 319(1-2):305–319, 2009.
4. HA Bonhoff and BAT Petersson. The influence of cross-order terms in interface mobilities for structure-borne sound source characterization. *Journal of Sound and Vibration*, 329(16):3280–3303, 2010.
5. T Bregar, N Holeček, G Čepon, DJ Rixen, and M Boltežar. Including directly measured rotations in the virtual point transformation. *Mechanical Systems and Signal Processing*, 141:106440, 2020.
6. JD Chazot and JL Guyader. Prediction of transmission loss of double panels with a patch-mobility method. *The Journal of the Acoustical Society of America*, 121(1):267–278, 2007.
7. MLM Duarte and DJ Ewins. Rotational degrees of freedom for structural coupling analysis via finite-difference technique with residual compensation. *Mechanical Systems and Signal Processing*, 14(2):205–227, 2000.
8. AS Elliott. *Characterisation of structure borne sound sources in-situ.* PhD thesis, University of Salford (United Kingdom), 2009.
9. AS Elliott, AT Moorhouse, and G Pavić. Moment excitation and the measurement of moment mobilities. *Journal of Sound and Vibration*, 331(11):2499–2519, 2012.
10. K Kesour and N Atalla. A hybrid patch transfer-green functions method to solve transmission loss problems of flat single and double walls with attached sound packages. *Journal of Sound and Vibration*, 429:1–17, 2018.
11. D Lennström, A Ågren, and A Nykänen. Sound quality evaluation of electric cars: preferences and influence of the test environment. In *Proceedings of the Aachen Acoustics Colloquium*, pages 95–100, 2011.
12. JWR Meggitt. *On in-situ methodologies for the characterisation and simulation of vibro-acoustic assemblies.* PhD thesis, University of Salford (United Kingdom), 2017.

13. AT Moorhouse and BM Gibbs. Simplified characterisation of multiple point excited structures using mobility matrix eigenvalues and eigenvectors. *Acta Acustica United with Acustica*, 84(5):843–853, 1998.
14. M Ouisse, L Maxit, C Cacciolati, and JL Guyader. Patch transfer functions as a tool to couple linear acoustic problems. *Journal of Vibration and Acoustics*. 127(5): 458–466, 2005.
15. EA Pasma, MV van der Seijs, SWB Klaassen, and MW van der Kooij. Frequency based substructuring with the virtual point transformation, flexible interface modes and a transmission simulator. In *Dynamics of Coupled Structures, Volume 4: Proceedings of the 36th IMAC*, pages 205–213. Springer, 2018.
16. G Pavić. Air-borne sound source characterization by patch impedance coupling approach. *Journal of Sound and Vibration*, 329(23):4907–4921, 2010.
17. SS Sattinger. Method for experimentally determining rotational mobilities of structures. *Shock Vib. Bull.;(United States)*, 50, 1980.
18. M van der Seijs, D van den Bosch, DJ Rixen, and D de Klerk. An improved methodology for the virtual point transformation of measured frequency response functions in dynamic substructuring. In *4th ECCOMAS thematic conference on computational methods in structural dynamics and earthquake engineering*, volume 4, 2013.

5 Assemblies

5.1 INTRODUCTION

Up to this point we have fully defined all the components of our VAVP: active properties, describing the initial excitation, in Chapter 3 and passive properties, defining transmission and filtering of the excitation on its way to the listener in Chapter 2. We also defined interfaces in Chapter 4 aiming to ensure the suitability of the representation and compatibility between components. If we were constructing a physical prototype, we would now be at the point where all the components had been delivered and the remaining step would be to assemble them into a working whole. Thus, what we now need for our VAVP is the digital equivalent of bolting components together to form an *assembly*. The main subject of this chapter is therefore the coupling of digital components. We also consider a range of measurement and analysis techniques known as Transfer Path Analysis (TPA) that are carried out on existing assemblies and are relevant to the construction of VAVPs.

We begin by laying out the main equations for coupling within the general framework known as 'frequency-based substructuring', 'dynamic substructuring' or simply 'substructuring'. This topic has been thoroughly covered elsewhere [2] so a comprehensive treatment is not required here, rather we focus on its application to virtual prototyping. We present three alternative substructuring formulations: direct, primal and dual. Each method is derived from applying equilibrium and continuity constraints between the interface DoFs of connecting components. Practical application to assemblies with resilient couplings (i.e. vibration isolators) is also discussed. Lastly, we consider the problem of extracting a component's free-interface FRF from those of an assembly by 'substructure decoupling'.

We go on to consider the range of techniques known collectively as Transfer Path Analysis (TPA). In TPA, measurements are conducted on an existing assembly and, in early variants, its application was purely diagnostic being limited to understanding transmission through the assembly. We will show how a more recent variant, based on the *in situ* blocked force method from Chapter 3, provides similar diagnostic information but also opens up far wider possibilities as a predictive method. Indeed, it will be shown that 'blocked force TPA' opens the door into the component-based approach, and thus the development of VAVPs.

In the last part of the chapter, we consider the problem of integrating measured and modelled data from different origins into a common framework for component-based simulation. While classical substructuring allows FEM and experimental data to be combined with great flexibility, higher frequency applications often demand a statistical approach which is not compatible with any of the formulations presented so far. We therefore present an extension of the substructuring approach that also allows statistical subsystems to be incorporated into a VAVP. The so-called 'hybrid Finite Element – Statistical Energy Analysis – eXperimental (FE-SEA-X) model'

DOI: 10.1201/9781003399131-5

is thus introduced and will be seen to provide a broader framework than classical substructuring.

5.2 COMPONENT ASSEMBLY – DYNAMIC SUBSTRUCTURING

Typically, the assembly process involves rigidly coupling components together by satisfying the conditions of continuity and equilibrium between all connecting interfaces.

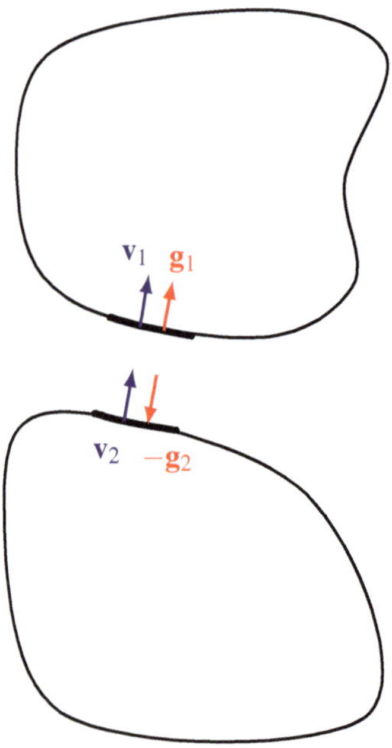

Figure 5.1 Conditions of continuity and equilibrium.

The condition of equilibrium is simply a restatement of Newton's 3rd law – when two objects are in contact they exert equal and opposite forces on one another. For two substructures in rigid contact at a single point it may be expressed in the form,

$$g_1 = -g_2 \tag{5.1}$$

where g_1 and g_2 represent the interface coupling forces at the connecting DoFs of the two substructures.

The condition of continuity (also known as compatibility) states that the substructural DoFs in rigid contact must move in unison, i.e. with equal velocity (also displacement and acceleration), and may be expressed in the form,

$$v_1 = v_2 = v \tag{5.2}$$

where v_1 and v_2 represent the contact velocities of the two substructures, and v the velocity of the coupled structure.

In the presence of several substructures and/or connecting DoFs, the above conditions can be expressed more generally in vector form as,

$$\mathbf{g}_1 = -\mathbf{g}_2 \tag{5.3}$$

and

$$\mathbf{v}_1 = \mathbf{v}_2 = \mathbf{v} \tag{5.4}$$

where Equation 5.3 and 5.4 are now generalised in the sense that they can also include moment forces and angular velocities, respectively.

Dynamic substructuring, the process of assembling substructural properties to form an assembly, amounts to the application of the above conditions to the connecting interfaces between each substructure (or equivalently, component). Depending on how these substructures are represented (i.e. in the physical domain, modal domain, frequency domain, or state-space) the substructuring formulation may differ, though the underlying physics remains the same.

In this book we are principally focused on frequency domain methods, and so we will consider *FRF-based* substructuring methods. That said, the substructuring methods presented below are not limited to experimental FRFs. FRF matrices determined by analytical or numerical means can also be used to represent individual substructures. In-fact, the ability to combine experimental and numerical substructures to form ‘hybrid assemblies’ is one of the principal advantages of the component-based approach to simulation.

An extensive review of all substructuring formulations is considered beyond the scope of this chapter. Rather, we present what we believe to be the two most practical approaches in the context of virtual prototyping; the direct formulation, and the three field formulation.

5.2.1 DIRECT FORMULATION

The direct formulation considers the special case in which just two substructures are being coupled (typically a source S and receiver R). Each substructure is characterised independently by their free-interface FRF matrices, $\mathbf{Y}_{Scc} \in \mathbb{C}^{N_c \times N_c}$ and $\mathbf{Y}_{Rcc} \in \mathbb{C}^{N_c \times N_c}$. If remote DoFs are considered on the receiver structure the transfer FRF $\mathbf{Y}_{Rrc} \in \mathbb{C}^{N_r \times N_c}$ (or the vibro-acoustic FRF $\mathbf{H}_{Rrc} \in \mathbb{C}^{N_r \times N_c}$, if the sound pressure response is of interest) is also necessary. Based on these free-interface FRF matrices, the point and transfer FRFs of the assembled structure, $\mathbf{Y}_{Ccc} \in \mathbb{C}^{N_c \times N_c}$ and $\mathbf{Y}_{Crc} \in \mathbb{C}^{N_r \times N_c}$ ($\mathbf{H}_{Crc} \in \mathbb{C}^{N_r \times N_c}$), are predicted. An important requirement of the direct formulation is the that FRF matrices for source and receiver are arranged congruently; the row and column DoFs at matched between $\mathbf{Y}_{Scc}$, $\mathbf{Y}_{Rcc}$ and $\mathbf{Y}_{Rrc}$.

Coupled interface FRF

Let us first consider the point FRF of the assembly interface, $\mathbf{Y}_{Ccc}$. The interface responses of the (uncoupled) substructures S and R due to an external interface forcing are given by,

$$\mathbf{v}_{Sc} = \mathbf{Y}_{Scc}\left(\mathbf{f}_{Sc} + \mathbf{g}_{Sc}\right), \quad \mathbf{v}_{Rc} = \mathbf{Y}_{Rcc}\left(\mathbf{f}_{Rc} + \mathbf{g}_{Rc}\right) \tag{5.5}$$

where $\mathbf{f}$ represents the externally applied force, and $\mathbf{g}$ a set of interface coupling forces (equal to $\mathbf{0}$ whilst the substructures are uncoupled).

We assume the inverse matrices $\mathbf{Y}_{Scc}^{-1}$ and $\mathbf{Y}_{Rcc}^{-1}$ exist, and rearrange the above equations to isolate the interface forces,

$$\mathbf{Y}_{Scc}^{-1}\mathbf{v}_{Sc} - \mathbf{f}_{Sc} = \mathbf{g}_{Sc}, \quad \mathbf{Y}_{Rcc}^{-1}\mathbf{v}_{Rc} - \mathbf{f}_{Rc} = \mathbf{g}_{Rc}. \tag{5.6}$$

Equilibrium and continuity are then enforced as per Equations 5.3 and 5.4,

$$\mathbf{Y}_{Scc}^{-1}\mathbf{v}_c - \mathbf{f}_{Sc} = -\mathbf{Y}_{Rcc}^{-1}\mathbf{v}_c + \mathbf{f}_{Rc} \tag{5.7}$$

where it is noted that the coupled response $\mathbf{v}_c$ has been substituted for both the source and receiver responses.

Isolating the external force terms and introducing the total external force $\mathbf{f}_c = \mathbf{f}_{Rc} + \mathbf{f}_{Sc}$ yields,

$$\left(\mathbf{Y}_{Scc}^{-1} + \mathbf{Y}_{Rcc}^{-1}\right)\mathbf{v}_c = \mathbf{f}_c. \tag{5.8}$$

Finally, pre-multiplying both sides by the inverse of the left bracketed term (assuming its exists), we obtain an expression for the coupled response,

$$\mathbf{v}_c = \left(\mathbf{Y}_{Scc}^{-1} + \mathbf{Y}_{Rcc}^{-1}\right)^{-1} \mathbf{f}_c. \tag{5.9}$$

Equation 5.9 relates an external force applied to the interface DoFs c to the coupled response at said DoFs. Hence, the bracketed term must represent the point FRF matrix of the coupled assembly's interface,

$$\mathbf{Y}_{Ccc} = \left(\mathbf{Y}_{Scc}^{-1} + \mathbf{Y}_{Rcc}^{-1}\right)^{-1}. \tag{5.10}$$

From Equation 5.10 we see that the coupled impedance may be expressed simply as the sum of substructure impedances,

$$\mathbf{Z}_{Ccc} = \mathbf{Z}_{Scc} + \mathbf{Z}_{Rcc}. \tag{5.11}$$

This is a well-known but important result; *when two structural components are in rigid contact, the impedance at their connection point is equal to the matrix sum of their respective impedances.* This result was used in 2.3.2 to justify the in situ characterisation of resilient elements, but at the time was not proven.

Although perfectly admissible, Equation 5.10 can be written in a more convenient (at least computationally) form using the mathematical relation,

$$\mathbf{Y}_{Scc}^{-1} + \mathbf{Y}_{Rcc}^{-1} = \mathbf{Y}_{Rcc}^{-1}\left[\mathbf{Y}_{Scc} + \mathbf{Y}_{Rcc}\right]\mathbf{Y}_{Scc}^{-1}. \tag{5.12}$$

Substitution into Equation 5.10 then yields,

$$\mathbf{Y}_{Ccc} = \mathbf{Y}_{Scc}\left[\mathbf{Y}_{Scc} + \mathbf{Y}_{Rcc}\right]^{-1}\mathbf{Y}_{Rcc}. \tag{5.13}$$

In this modified form, the double matrix inversion is avoided. Matrix inversions are not only computationally expensive, but are known to cause an 'amplification' of errors and thus increase overall levels of uncertainty, as discussed in Chapter 8.

Coupled transfer FRF

Let us now consider the transfer FRF of the assembly between the interface c and some remote receiver DoFs r, $\mathbf{Y}_{Crc}$. Note that for the vibro-acoustic FRF $\mathbf{H}_{Crc}$ we simply replace $\mathbf{Y}_{Rrc} \to \mathbf{H}_{Rrc}$ and $\mathbf{v}_{Rr} \to \mathbf{p}_{Rr}$ in what follows.

The remote and interface responses of the uncoupled substructures S and R due to an external interface forcing are given by,

$$\mathbf{v}_{Sc} = \mathbf{Y}_{Scc}\left(\mathbf{f}_{Sc} + \mathbf{g}_{Sc}\right), \quad \mathbf{v}_{Rr} = \mathbf{Y}_{Rrc}\left(\mathbf{f}_{Rc} + \mathbf{g}_{Rc}\right) \tag{5.14}$$

where $\mathbf{f}$ and $\mathbf{g}$ again represent the externally applied force and a set of interface coupling forces, respectively.

Isolating the interface forces,

$$\mathbf{Y}_{Scc}^{-1}\mathbf{v}_{Sc} - \mathbf{f}_{Sc} = \mathbf{g}_{Sc}, \quad \mathbf{Y}_{Rrc}^{-1}\mathbf{v}_{Rr} - \mathbf{f}_{Rc} = \mathbf{g}_{Rc} \tag{5.15}$$

and satisfying equilibrium and continuity ($\mathbf{v}_{Sc} \rightarrow \mathbf{v}_c$, $\mathbf{v}_{Rr} \rightarrow \mathbf{v}_r$) yields,

$$\mathbf{Y}_{Scc}^{-1}\mathbf{v}_c - \mathbf{f}_{Sc} = -\mathbf{Y}_{Rrc}^{-1}\mathbf{v}_r + \mathbf{f}_{Rc} \tag{5.16}$$

where it is noted that the responses now represent coupled responses. Introducing the total external force $\mathbf{f}_c = \mathbf{f}_{Rc} + \mathbf{f}_{Sc}$ and rearranging to isolate the remote receiver response,

$$\mathbf{v}_r = \mathbf{Y}_{Rrc}\mathbf{f}_c - \mathbf{Y}_{Rrc}\mathbf{Y}_{Scc}^{-1}\mathbf{v}_c. \tag{5.17}$$

Substituting the coupled interface response from Equation 5.9 yields,

$$\mathbf{v}_r = \left(\mathbf{Y}_{Rrc} - \mathbf{Y}_{Rrc}\mathbf{Y}_{Scc}^{-1}\left(\mathbf{Y}_{Scc}^{-1} + \mathbf{Y}_{Rcc}^{-1}\right)^{-1}\right)\mathbf{f}_c \tag{5.18}$$

from which we identify the coupled transfer FRF as,

$$\mathbf{Y}_{Crc} = \mathbf{Y}_{Rrc} - \mathbf{Y}_{Rrc}\mathbf{Y}_{Scc}^{-1}\left(\mathbf{Y}_{Scc}^{-1} + \mathbf{Y}_{Rcc}^{-1}\right)^{-1} \tag{5.19}$$

or equivalently using Equation 5.13,

$$\mathbf{Y}_{Crc} = \mathbf{Y}_{Rrc} - \mathbf{Y}_{Rrc}\left[\mathbf{Y}_{Scc} + \mathbf{Y}_{Rcc}\right]^{-1}\mathbf{Y}_{Rcc}. \tag{5.20}$$

A physical interpretation of the above is given as follows. Considering Equation 5.18 we see that the external force generates two distinct response fields,

$$\mathbf{v}_{r_1} = \mathbf{Y}_{Rrc}\mathbf{f}_c \qquad \text{and} \qquad \mathbf{v}_{r_2} = -\mathbf{Y}_{Rrc}\mathbf{Y}_{Scc}^{-1}\left(\mathbf{Y}_{Scc}^{-1} + \mathbf{Y}_{Rcc}^{-1}\right)^{-1}\mathbf{f}_c. \tag{5.21}$$

The response $\mathbf{v}_{r_1}$ is simply that of the uncoupled receiver, i.e. with the source removed. The response $\mathbf{v}_{r_2}$ is then a secondary field arising due to the contact forces generated by the attached source. Focusing on this second term we see that, with $\left(\mathbf{Y}_{Scc}^{-1} + \mathbf{Y}_{Rcc}^{-1}\right)^{-1}$ being the coupled interface FRF, $\left(\mathbf{Y}_{Scc}^{-1} + \mathbf{Y}_{Rcc}^{-1}\right)^{-1}\mathbf{f}_c$ is the coupled interface response $\mathbf{v}_c$. When we pre-multiply this interface response by the inverse source FRF we obtain the contact force imparted on the source by the receiver, $\mathbf{g}_{Rc} = \mathbf{Y}_{Scc}^{-1}\mathbf{v}_c$. Recalling the condition of equilibrium, this same force is applied to the receiver, though with opposite sign (thus the minus term). Finally, the response at r due to this contact force is obtained from the transfer FRF term $\mathbf{Y}_{Rrc}$. This is the interpretation behind Equation 5.19.

For Equation 5.20 the interpretation is identical, though the details differ slightly. We similarly have two response fields,

$$\mathbf{v}_{r_1} = \mathbf{Y}_{Rrc}\mathbf{f}_c \qquad \text{and} \qquad \mathbf{v}_{r_2} = -\mathbf{Y}_{Rrc}\left[\mathbf{Y}_{Scc} + \mathbf{Y}_{Rcc}\right]^{-1}\mathbf{Y}_{Rcc}\mathbf{f}_c \tag{5.22}$$

the latter of which can be understood as follows. The right most term $\mathbf{Y}_{Rcc}\mathbf{f}_c$ represents the free velocity response of the receiver due to the external force $\mathbf{f}_c$. When we pre-multiply this by the inverse sum of the interface FRFs we obtain the contact force $\mathbf{g}_{Rc}$ (see Equation 3.18), which again is negated and propagated by the remaining FRF term $\mathbf{Y}_{Rrc}$.

An alternative form of Equation 5.20 often found in the literature can be obtained by rearranging Equation 5.10 in the form,

$$\mathbf{Y}_{Scc}^{-1} = \mathbf{Y}_{Ccc}^{-1} - \mathbf{Y}_{Rcc}^{-1} \tag{5.23}$$

and substituting this into Equation 5.19,

$$\mathbf{v}_r = \left(\mathbf{Y}_{Rrc} - \mathbf{Y}_{Rrc} \left(\mathbf{Y}_{Ccc}^{-1} - \mathbf{Y}_{Rcc}^{-1} \right) \left(\mathbf{Y}_{Scc}^{-1} + \mathbf{Y}_{Rcc}^{-1} \right)^{-1} \right) \mathbf{f}_c \tag{5.24}$$

which, after some simple rearrangement (in particular the cancellation of $\mathbf{Y}_{Ccc}^{-1}$ and $(\mathbf{Y}_{Scc}^{-1} + \mathbf{Y}_{Rcc}^{-1})^{-1}$) yields,

$$\mathbf{Y}_{Crc} = \mathbf{Y}_{Rrc} \mathbf{Y}_{Rcc}^{-1} \left(\mathbf{Y}_{Scc}^{-1} + \mathbf{Y}_{Rcc}^{-1} \right)^{-1} \tag{5.25}$$

or equivalently,

$$\mathbf{Y}_{Crc} = \mathbf{Y}_{Rrc} \left[\mathbf{Y}_{Rcc} + \mathbf{Y}_{Scc} \right]^{-1} \mathbf{Y}_{Scc}. \tag{5.26}$$

Equations 5.10, 5.13 and 5.19 5.20, 5.25, 5.26 are the direct formulations of the point and transfer FRFs for a rigidly coupled assembly. They provide convenient and easy to implement expressions for the coupled assembly FRFs. They are, however, somewhat limited if more complex systems with multiple components are considered. A more general framework for dynamic substructuring can be derived using the so called three field formulation, as discussed in 5.2.2.

Coupled transfer FRF with resilient connection

When dealing with general coupling elements (see Figure 5.2b), such that no assumptions can be made about their behaviour, it is more convenient to perform substructuring using the three field formulation, as will be detailed in 5.2.2. However, for simple spring-like couplings, which conserve force across their length, the direct method can be used to obtain concise expressions for the coupled FRFs of a resiliently isolated source.

For a resiliently coupled assembly, the interface and remote responses of the (uncoupled) substructures S and R due to interface forcing are given by,

$$\mathbf{v}_{Sc_1} = \mathbf{Y}_{Sc_1c_1} \left(\mathbf{f}_{Sc_1} + \mathbf{g}_{Sc_1} \right), \quad \mathbf{v}_{Rc_2} = \mathbf{Y}_{Rc_2c_2} \mathbf{g}_{Rc_2}, \quad \mathbf{v}_{Rr} = \mathbf{Y}_{Rrc_2} \mathbf{g}_{Rc_2}. \tag{5.27abc}$$

where c_1 represents the source-isolator interface, and c_2 the isolator-receiver interface. Note that external forces are only considered at c_1, where the source is attached. In the presence of a resilient coupling the condition of continuity is no longer satisfied, $\mathbf{v}_{Sc_1} \neq \mathbf{v}_{Rc_2}$. Rather, the source and receiver responses differ and, assuming a simple coupling element, satisfy the equilibrium equation,

$$\mathbf{g}_{Sc_1} = -\mathbf{g}_{Rc_2} = -\mathbf{Z}_{Ic_2c_1} \left(\mathbf{v}_{Sc_1} - \mathbf{v}_{Rc_2} \right) \tag{5.28}$$

where $\mathbf{Z}_{Ic_2c_1}$ (here after denoted simply $\mathbf{Z}_I$) is a diagonal matrix containing the transfer impedance of each coupling element. Recall that Equation 5.28 has already been

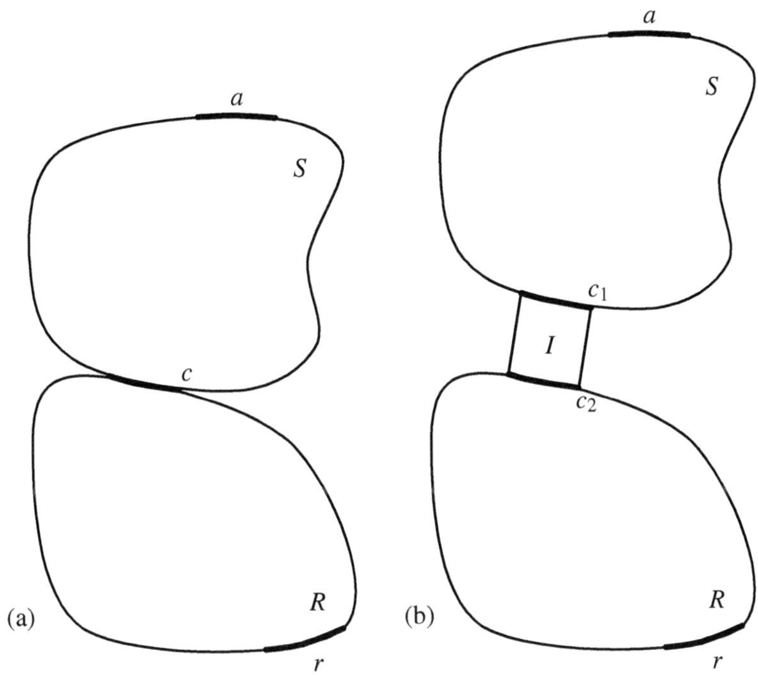

Figure 5.2 Rigid (a) and resiliently (b) coupled assemblies.

introduced in the context of force identification by the so-called mount stiffness method in 3.1.2.

By rearranging Equation 5.27a for the coupling force $\mathbf{g}_{Sc_1}$ and substituting this into Equation 5.28,

$$\mathbf{Y}_{Sc_1c_1}^{-1}\mathbf{v}_{Sc_1} - \mathbf{f}_{Sc_1} = -\mathbf{Z}_I(\mathbf{v}_{Sc_1} - \mathbf{v}_{Rc_2}) \tag{5.29}$$

$$\mathbf{Z}_I^{-1}\mathbf{Y}_{Sc_1c_1}^{-1}\mathbf{v}_{Sc_1} - \mathbf{Z}_I^{-1}\mathbf{f}_{Sc_1} = -(\mathbf{v}_{Sc_1} - \mathbf{v}_{Rc_2}) \tag{5.30}$$

$$(\mathbf{Y}_{Sc_1c_1}\mathbf{Z}_I)^{-1}\mathbf{v}_{Sc_1} - \mathbf{Z}_I^{-1}\mathbf{f}_{Sc_1} = -(\mathbf{v}_{Sc_1} - \mathbf{v}_{Rc_2}) \tag{5.31}$$

$$(\mathbf{Y}_{Sc_1c_1}\mathbf{Z}_I)^{-1}\mathbf{v}_{Sc_1} - \mathbf{Z}_I^{-1}\mathbf{f}_{Sc_1} + \mathbf{v}_{Sc_1} = \mathbf{v}_{Rc_2} \tag{5.32}$$

$$\left((\mathbf{Y}_{Sc_1c_1}\mathbf{Z}_I)^{-1} + \mathbf{I}\right)\mathbf{v}_{Sc_1} - \mathbf{Z}_I^{-1}\mathbf{f}_{Sc_1} = \mathbf{v}_{Rc_2} \tag{5.33}$$

we obtain for $\mathbf{v}_{Sc_1}$,

$$\mathbf{v}_{Sc_1} = \left((\mathbf{Y}_{Sc_1c_1}\mathbf{Z}_I)^{-1} + \mathbf{I}\right)^{-1}(\mathbf{Y}_I\mathbf{f}_{Sc_1} + \mathbf{v}_{Rc_2}). \tag{5.34}$$

Similarly, if we rearrange Equation 5.27b for $\mathbf{g}_{Rc_2}$ and substitute in Equation 5.28, we obtain,

$$\mathbf{v}_{Rc_2} = \left((\mathbf{Y}_{Rc_2c_2}\mathbf{Z}_I)^{-1} + \mathbf{I} \right)^{-1} \mathbf{v}_{Sc_1}. \tag{5.35}$$

Having satisfied equilibrium, the responses $\mathbf{v}_{Sc_1}$ and $\mathbf{v}_{Rc_2}$ now represent coupled assembly responses. We replace them with $\mathbf{v}_{c_1}$ and $\mathbf{v}_{c_2}$, respectively.

After substituting Equation 5.34 into 5.35,

$$\mathbf{v}_{c_2} = \left((\mathbf{Y}_{Sc_1c_1}\mathbf{Z}_I)^{-1} + \mathbf{I} \right)^{-1} \left((\mathbf{Y}_{Rc_2c_2}\mathbf{Z}_I)^{-1} + \mathbf{I} \right)^{-1} (\mathbf{Y}_I\mathbf{f}_{Sc_1} + \mathbf{v}_{c_2}) \tag{5.36}$$

and solving for $\mathbf{v}_{c_2}$ we obtain,

$$\mathbf{v}_{c_2} = \left[\mathbf{Y}_I^{-1} \left\{ \left((\mathbf{Y}_{Rc_2c_2}\mathbf{Z}_I)^{-1} + \mathbf{I} \right) \left((\mathbf{Y}_{Sc_1c_1}\mathbf{Z}_I)^{-1} + \mathbf{I} \right) - \mathbf{I} \right\} \right]^{-1} \mathbf{f}_{c_1} \tag{5.37}$$

where we have replaced $\mathbf{f}_{Sc_1} = \mathbf{f}_{c_1}$, i.e. the external force is now applied to the coupled assembly. From the above we can identify the coupled transfer FRF as,

$$\mathbf{Y}_{Cc_2c_1} = \left[\mathbf{Y}_I^{-1} \left\{ \left((\mathbf{Y}_{Sc_1c_1}\mathbf{Z}_I)^{-1} + \mathbf{I} \right) \left((\mathbf{Y}_{Rc_2c_2}\mathbf{Z}_I)^{-1} + \mathbf{I} \right) - \mathbf{I} \right\} \right]^{-1}. \tag{5.38}$$

Equation 5.38 can be simplified further. If we expand the brackets and rearrange,

$$\mathbf{Y}_{Cc_2c_1} = \left[\mathbf{Y}_I^{-1} \left((\mathbf{Y}_{Sc_1c_1}\mathbf{Z}_I)^{-1} (\mathbf{Y}_{Rc_2c_2}\mathbf{Z}_I)^{-1} + (\mathbf{Y}_{Rc_2c_2}\mathbf{Z}_I)^{-1} + (\mathbf{Y}_{Sc_1c_1}\mathbf{Z}_I)^{-1} \right) \right]^{-1} \tag{5.39}$$

after canceling $\mathbf{Y}_I^{-1}\mathbf{Z}_I^{-1} = \mathbf{I}$ terms, we obtain,

$$\mathbf{Y}_{Cc_2c_1} = \left[\mathbf{Y}_{Sc_1c_1}^{-1} + \mathbf{Y}_{Rc_2c_2}^{-1} + \mathbf{Y}_{Sc_1c_1}^{-1} \mathbf{Y}_I \mathbf{Y}_{Rc_2c_2}^{-1} \right]^{-1}. \tag{5.40}$$

Factoring out the source and receiver FRFs we have, equivalently,

$$\mathbf{Y}_{Cc_2c_1} = \left[\mathbf{Y}_{Sc_1c_1}^{-1} (\mathbf{Y}_{Rc_2c_2} + \mathbf{Y}_{Sc_1c_1} + \mathbf{Y}_I) \mathbf{Y}_{Rc_2c_2}^{-1} \right]^{-1} \tag{5.41}$$

$$= \mathbf{Y}_{Rc_2c_2} (\mathbf{Y}_{Rc_2c_2} + \mathbf{Y}_{Sc_1c_1} + \mathbf{Y}_I)^{-1} \mathbf{Y}_{Sc_1c_1}. \tag{5.42}$$

Note that Equation 5.42 bares a strong resemblance to Equation 5.13, the equation for rigid coupling, the difference being the presence of the isolator FRF term $\mathbf{Y}_I$.

To obtain the remote receiver transfer FRF $\mathbf{Y}_{Crc_1}$, we use Equations 5.27b and 5.42 to solve for the interface force, $\mathbf{g}_{Rc_2}$, due to the external force $\mathbf{f}_{c_1}$,

$$\mathbf{g}_{Rc_2} = \mathbf{Y}_{Rc_2c_2}^{-1} \mathbf{v}_{c_2} = \mathbf{Y}_{Rc_2c_2}^{-1} \mathbf{Y}_{Cc_2c_1} \mathbf{f}_{c_1} = (\mathbf{Y}_{Rc_2c_2} + \mathbf{Y}_{Sc_1c_1} + \mathbf{Y}_I)^{-1} \mathbf{Y}_{Sc_1c_1} \mathbf{f}_{c_1} \tag{5.43}$$

before substituting this into Equation 5.27c,

$$\mathbf{v}_r = \mathbf{Y}_{Rrc_2} (\mathbf{Y}_{Rc_2c_2} + \mathbf{Y}_{Sc_1c_1} + \mathbf{Y}_I)^{-1} \mathbf{Y}_{Sc_1c_1} \mathbf{f}_{c_1}. \tag{5.44}$$

From the above we have,

$$\mathbf{Y}_{Crc_1} = \mathbf{Y}_{Rrc_2} \left(\mathbf{Y}_{Rc_2c_2} + \mathbf{Y}_{Sc_1c_1} + \mathbf{Y}_I \right)^{-1} \mathbf{Y}_{Sc_1c_1}. \tag{5.45}$$

Again, Equation 5.45 bares a strong resemblance to Equation 5.26, the difference again being the additional isolator FRF term.

Equation 5.42 and 5.45 are the direct substructuring equations for a resiliently coupled source. A summary of the direct substructuring equations for both rigid and resilient assemblies can be found in Table 5.1. We now move onto the three field formulation which allows more complex assemblies to be constructed.

5.2.2 THREE FIELD FORMULATION

The direct method introduced above provides convenient expressions for the coupled FRF of simple assemblies composed of just a source and receiver component. However, generalisation to arbitrarily complex assemblies with multiple components becomes impractical and a more general substructuring framework is required. Two such frameworks are available though the so-called three field formulation as introduced below.

We begin by considering the equations of motion of P *uncoupled* substructures in a block diagonal form, as given by,

$$\mathbf{Z}\mathbf{v} = \mathbf{f} + \mathbf{g} \tag{5.46}$$

where, $\mathbf{Z}$ is the block diagonal impedance matrix of the P uncoupled substructures and $\mathbf{v}$, $\mathbf{f}$ and $\mathbf{g}$ are the corresponding partitioned column vectors of responses applied forces, and coupling interface forces, respectively.

$$\mathbf{Z} = \begin{bmatrix} \mathbf{Z}^{(1)} & & & \\ & \mathbf{Z}^{(2)} & & \\ & & \ddots & \\ & & & \mathbf{Z}^{(P)} \end{bmatrix} \tag{5.47}$$

$$\mathbf{v} = \begin{pmatrix} \mathbf{v}^{(1)} \\ \mathbf{v}^{(2)} \\ \vdots \\ \mathbf{v}^{(P)} \end{pmatrix}, \quad \mathbf{f} = \begin{pmatrix} \mathbf{f}^{(1)} \\ \mathbf{f}^{(2)} \\ \vdots \\ \mathbf{f}^{(P)} \end{pmatrix}, \quad \mathbf{g} = \begin{pmatrix} \mathbf{g}^{(1)} \\ \mathbf{g}^{(2)} \\ \vdots \\ \mathbf{g}^{(P)} \end{pmatrix}. \tag{5.48}$$

Note that the vibro-acoustic FRF matrix $\mathbf{H}$ may be included by simply augmenting the appropriate impedance matrix, see for example Equation 2.35 and the surrounding discussion.

As in the direct formulation, to assemble the individual substructures we must enforce the conditions of equilibrium and continuity. To do so we express Equations

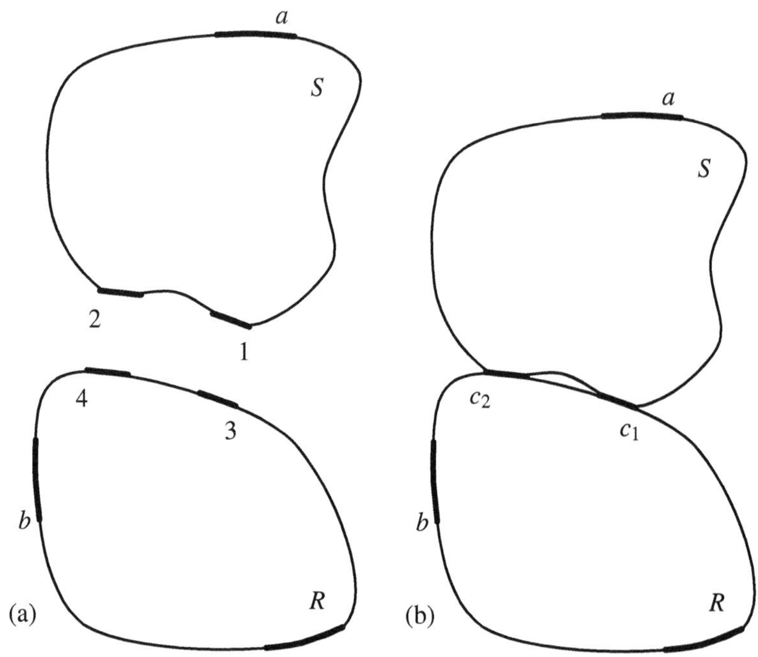

Figure 5.3 Dynamic substructuring example. (a) – uncoupled substructures with remote (a and b) and interface (c) DoFs. (b) – Coupled assembly.

5.3 and 5.4 more generally in the form,

$$\mathbf{Bv} = \mathbf{0} \tag{5.49}$$

and

$$\mathbf{L}^{\mathrm{T}}\mathbf{g} = \mathbf{0} \tag{5.50}$$

respectively, where **B** and **L** represent signed (contain only 0s, 1s and -1s) and unsigned (contain only (0s and 1s) Boolean matrices, respectively. Together, Equations 5.46-5.50 are referred to as the three field formulation [11]. They may be solved in a primal or dual manner, depending on whether Equation 5.49 or 5.50 is satisfied *a priori*.

For the demonstrative example in Figure 5.3, **B** and $\mathbf{L}^{\mathrm{T}}$ are given, respectively, by,

$$\mathbf{B} = \overset{\begin{matrix} a & 1 & 2 & 3 & 4 & b \end{matrix}}{\left[\begin{array}{ccc|ccc} 0 & 1 & 0 & -1 & 0 & 0 \\ 0 & 0 & 1 & 0 & -1 & 0 \end{array}\right]} \begin{matrix} c_1 \\ c_2 \end{matrix} \tag{5.51}$$

and

$$\mathbf{L}^{\mathrm{T}} = \begin{array}{c} \begin{array}{cccccc} a & 1 & 2 & 3 & 4 & b \end{array} \\ \left[\begin{array}{ccc|ccc} 1 & 0 & 0 & 0 & 0 & 0 \\ 0 & 1 & 0 & 1 & 0 & 0 \\ 0 & 0 & 1 & 0 & 1 & 0 \\ 0 & 0 & 0 & 0 & 0 & 1 \end{array}\right] \end{array} \begin{array}{c} a \\ c_1 \\ c_2 \\ b \end{array}. \tag{5.52}$$

Note that for both $\mathbf{B}$ and $\mathbf{L}^{\mathrm{T}}$ the matrix columns correspond to the uncoupled DoFs of the assembly. For $\mathbf{B}$, the rows correspond to the interface DoFs where coupling occurs, and for $\mathbf{L}^{\mathrm{T}}$ the coupled DoFs of the assembly. Their substitution into Equation 5.49 and 5.50 yields,

$$\begin{bmatrix} 0 & 1 & 0 & -1 & 0 & 0 \\ 0 & 0 & 1 & 0 & -1 & 0 \end{bmatrix} \begin{pmatrix} v_a \\ v_1 \\ v_2 \\ v_3 \\ v_4 \\ v_b \end{pmatrix} = \begin{pmatrix} 0 \\ 0 \end{pmatrix} \tag{5.53}$$

and

$$\begin{bmatrix} 1 & 0 & 0 & 0 & 0 & 0 \\ 0 & 1 & 0 & 1 & 0 & 0 \\ 0 & 0 & 1 & 0 & 1 & 0 \\ 0 & 0 & 0 & 0 & 0 & 1 \end{bmatrix} \begin{pmatrix} 0 \\ g_1 \\ g_2 \\ g_3 \\ g_4 \\ 0 \end{pmatrix} = \begin{pmatrix} 0 \\ 0 \\ 0 \\ 0 \end{pmatrix} \tag{5.54}$$

which clearly satisfy the required constraints for rigid coupling.

From the above example we can see that to construct a signed Boolean matrix $\mathbf{B}$, for each connection (i.e. row of $\mathbf{B}$), we assign a 1 and -1 to the columns corresponding to the DoFs that are to be coupled. The size of $\mathbf{B}$ is such that the number of rows is equal to the number of connection points, and the number of columns to the number of uncoupled DoFs.

To construct the unsigned Boolean matrix $\mathbf{L}^{\mathrm{T}}$, for each coupled interface DoF, a pair of 1s are placed in the columns corresponding uncoupled DoF. If a DoF is to be left uncoupled, a 1 is placed in the corresponding column. The size of $\mathbf{L}^{\mathrm{T}}$ is such that the number of rows is equal to the number of coupled DoFs, and the number of columns to the number of uncoupled DoFs.

Primal formulation

The primal formulation begins by defining a unique set of DoFs $\mathbf{v}_C$ representing the *coupled* assembly. Note that the number of coupled DoFs is always less than the number of uncoupled DoFs; for every pair of DoFs that we couple we reduced by one the total number of coupled DoFs, see for example Figure 5.3. As per the condition of continuity, the coupled velocity $\mathbf{v}_C$ can be related to the uncoupled velocity $\mathbf{v}$ by an appropriate unsigned Boolean matrix,

$$\mathbf{v} = \mathbf{L}\mathbf{v}_C. \tag{5.55}$$

The condition of continuity is then given by,

$$\mathbf{B}\mathbf{L}\mathbf{v}_C = \mathbf{0}. \tag{5.56}$$

Note that the above is satisfied irrespective of $\mathbf{v}_C$, hence $\mathbf{B}$ and $\mathbf{L}$ represent the null space of one another. This can provide a convenient way of obtaining one Boolean matrix from the other.

Substituting Equation 5.55 into Equation 5.46 (representing the uncoupled equation of motion) enforces continuity and yields,

$$\mathbf{Z}\mathbf{L}\mathbf{v}_C = \mathbf{f} + \mathbf{g}. \tag{5.57}$$

Pre-multiplying both sides of Equation 5.57 by $\mathbf{L}^{\mathrm{T}}$ we obtain,

$$\mathbf{L}^{\mathrm{T}}\mathbf{Z}\mathbf{L}\mathbf{v}_C = \mathbf{L}^{\mathrm{T}}\mathbf{f} + \mathbf{L}^{\mathrm{T}}\mathbf{g}. \tag{5.58}$$

Recalling Equation 5.50 it is clear that the condition of equilibrium has now also been satisfied. Pre-multiplying both sides of Equation 5.58 by the inverse of the left most bracketed term yields,

$$\mathbf{v}_C = \left(\mathbf{L}^{\mathrm{T}}\mathbf{Z}\mathbf{L}\right)^{-1}\mathbf{f}_C \tag{5.59}$$

where $\mathbf{f}_C = \mathbf{L}^{\mathrm{T}}\mathbf{f}$ represents the total external force applied to the coupled assembly, i.e. the sum of external forces acting on each substructure. Finally, substituting the block diagonal impedance matrix for the inverse mobility matrix,

$$\mathbf{Z} = \mathbf{Y}^{-1}, \qquad \begin{bmatrix} \mathbf{Z}^{(1)} & & & \\ & \mathbf{Z}^{(2)} & & \\ & & \ddots & \\ & & & \mathbf{Z}^{(P)} \end{bmatrix} = \begin{bmatrix} \mathbf{Y}^{(1)} & & & \\ & \mathbf{Y}^{(2)} & & \\ & & \ddots & \\ & & & \mathbf{Y}^{(P)} \end{bmatrix}^{-1} \tag{5.60}$$

where $\mathbf{Y}^{(P)}$ is the free-interface FRF of the Pth component, we arrive at,

$$\mathbf{Y}_C = \left(\mathbf{L}^{\mathrm{T}}\mathbf{Y}^{-1}\mathbf{L}\right)^{-1}. \tag{5.61}$$

where $\mathbf{Y}_C$ is the FRF of the coupled assembly.

Equation 5.61 provides a general framework for determining the FRF matrix (in its entirety) of a coupled assembly. It can be interpreted as a generalisation of the impedance summation obtained by the direct method, as in Equation 5.10. In an experimental context Equation 5.61 has the downside of requiring several matrix inversions (each block matrix of $\mathbf{Y}$ must be inverted in full). These are not only computationally expensive, but prone to error amplification. For this reason, the primal formulation tends to be considered less reliable than its dual counterpart (discussed below). Nevertheless, for relatively small systems of equations, negligible differences are observed.

Dual formulation

The dual formulation begins by satisfying the condition of equilibrium *a priori*. This is done by defining the coupling force $\mathbf{g}$ in the form,

$$\mathbf{g} = -\mathbf{B}^{\mathrm{T}}\boldsymbol{\lambda} \tag{5.62}$$

where $\mathbf{B}$ is the same signed Boolean matrix introduced earlier, and λ a Lagrange multiplier representing physically the unknown coupling interface force intensities. λ is termed a force intensity as it possess no directional information. This is allocated by the Boolean matrix $\mathbf{B}$.

The signed nature of $\mathbf{B}$ ensures co-located coupling forces are equal and opposite in direction. The condition of equilibrium is then given by,

$$\mathbf{L}^{\mathrm{T}}\mathbf{g} = -\mathbf{L}^{\mathrm{T}}\mathbf{B}^{\mathrm{T}}\boldsymbol{\lambda} = \mathbf{0}. \tag{5.63}$$

Recalling that $\mathbf{L}^{\mathrm{T}}$, the Boolean matrix satisfying equilibrium as per Equation 5.50, is in the null space of $\mathbf{B}^{\mathrm{T}}$, the condition of equilibrium is always satisfied.

Substituting Equation 5.62 into Equation 5.46 the equations of motion become,

$$\mathbf{v} = \mathbf{Y}\left(\mathbf{f} - \mathbf{B}^{\mathrm{T}}\boldsymbol{\lambda}\right) \tag{5.64}$$

where $\mathbf{Y}$ again represents the block diagonal matrix of component free-interface FRFs,

$$\mathbf{Y} = \begin{bmatrix} \mathbf{Y}^{(1)} & & & \\ & \mathbf{Y}^{(2)} & & \\ & & \ddots & \\ & & & \mathbf{Y}^{(P)} \end{bmatrix}. \tag{5.65}$$

To determine the coupled assembly response, we must first find λ. To do so we begin by substituting Equation 5.64 into the continuity condition (Equation 5.4),

$$\mathbf{BY}\left(\mathbf{f} - \mathbf{B}^{\mathrm{T}}\boldsymbol{\lambda}\right) = \mathbf{BYf} - \mathbf{BYB}^{\mathrm{T}}\boldsymbol{\lambda} = \mathbf{0}. \tag{5.66}$$

In this sense, λ can be interpreted as the force intensity required to ensure compatibility between connected DoFs. Rearranging the above to find λ,

$$\boldsymbol{\lambda} = \left(\mathbf{BYB}^{\mathrm{T}}\right)^{-1}\mathbf{BYf} \tag{5.67}$$

and substituting this into Equation 5.64 yields the coupled response (since continuity has now be enforced),

$$\mathbf{v}_C = \mathbf{Y}\left(\mathbf{f} - \mathbf{B}^{\mathrm{T}}\left(\mathbf{BYB}^{\mathrm{T}}\right)^{-1}\mathbf{BYf}\right) = \left(\mathbf{Y} - \mathbf{YB}^{\mathrm{T}}\left(\mathbf{BYB}^{\mathrm{T}}\right)^{-1}\mathbf{BY}\right)\mathbf{f}. \tag{5.68}$$

From the above we can identify the coupled mobility $\mathbf{Y}_C$ as,

$$\mathbf{Y}_C = \mathbf{Y} - \mathbf{YB}^{\mathrm{T}}\left(\mathbf{BYB}^{\mathrm{T}}\right)^{-1}\mathbf{BY}. \tag{5.69}$$

Note that Equation 5.69 takes a very similar form to the direct formulation of Equation 5.20. Indeed we can interpret it in the same way. Re-introducing the external force **f**, we have for the coupled response,

$$\mathbf{v}_C = \underbrace{\mathbf{Yf}}_{\mathbf{v}} - \overbrace{\underbrace{\mathbf{Y}\mathbf{B}^{\mathrm{T}}\overbrace{\left(\mathbf{BYB}^{\mathrm{T}}\right)^{-1}\underbrace{\mathbf{BYf}}_{\Delta\mathbf{v}}}^{\lambda}}_{\mathbf{g}}}^{\mathbf{v}_g}. \tag{5.70}$$

As in Equation 5.19, the left most **Yf** term in Equation 5.70 describes the response of each component to the applied force whilst uncoupled; being uncoupled, forces only generate responses within the component to which they are applied. The secondary term, $\mathbf{YB}^{\mathrm{T}}\left(\mathbf{BYB}^{\mathrm{T}}\right)^{-1}\mathbf{BYf}$ describes the additional contribution $\mathbf{v}_g$ to this response that arises due to the coupling forces **g** generated by the connected components. It is through this term that forces applied to one component generate responses in others. Breaking down this second term in detail, starting from the right, **Yf** is again the free response of the uncoupled components due to the applied force **f**, and so pre-multiplication by **B** yields the *relative* response between the coupling DoFs, $\Delta\mathbf{v}$. The product $\mathbf{BYB}^{\mathrm{T}}$ when expanded simply yields the sum of the coupling interface FRFs, and so pre-multiplication by the inverse term $\left(\mathbf{BYB}^{\mathrm{T}}\right)^{-1}$ yields the coupling force intensity (see Equation 3.18). This force is applied to the connecting components with equal magnitude but in opposite directions by $\mathbf{B}^{\mathrm{T}}$, before being propagated through each component by **Y**, whereupon its contribution is combined with that of the force applied directly to the component.

Like Equation 5.61, Equation 5.69 provides a general method for determining the FRF matrix of a coupled assembly. Note that in contrast to the primal formulation, the dual formulation requires only a single matrix inversion. Furthermore, after pre- and post-multiplication by **B** and $\mathbf{B}^{\mathrm{T}}$ the size of the required inversion is reduced to those of the interface DoFs alone. Hence, the dual formulation is a) computationally more efficient, and b) less prone to error amplification.

It is important to note that in deriving Equation 5.69 we did not define a unique set of assembly DoFs. And so in the dual formulation $\mathbf{Y}_C$ retains all uncoupled DoFs (interface DoFs are present twice). Care should be taken to remove/avoid these duplicate rows and columns where necessary.

Including resilient coupling elements

The inclusion of resilient elements, such as vibration isolators, in a substructuring procedure is generally straight forward; the coupling element impedance matrix $\mathbf{Z}_I$ is simply included in the primal formulation as an individual substructure, or for the dual formulation its FRF matrix $\mathbf{Y}_I$ (methods for characterising resilient couplings are described in 2.3.2). However, in the case of an ideal massless spring-like coupling, the coupling element's FRF matrix $\mathbf{Y}_I$ cannot be computed and so an alternative approach is required as described in the following.

Let us consider the simplest resilient coupling, that of an ideal spring-damper. The impedance matrix of such an element is given by,

$$\mathbf{Z}_I = \begin{bmatrix} \frac{K}{i\omega}+C & -(\frac{K}{i\omega}+C) \\ -(\frac{K}{i\omega}+C) & \frac{K}{i\omega}+C \end{bmatrix} \quad (5.71)$$

where K is the stiffness of the spring, and C is its damping (assumed here to be of viscous type). An element of this form can be substituted directly in to the primal formulation as the equilibrium and continuity constraints are applied to the impedance matrices of each substructure (see Equation 5.59). In contrast, the dual formulation requires the mobility matrix of each element (see Equation 5.69), and so the coupling impedance must be inverted. Note however, that $\mathbf{Z}_I$ is singular. Its columns are linearly related by a factor of -1. Physically, this is because a force applied to an unconstrained, mass-less system causes an infinite response. Hence, an element impedance of this form is not compatible with the dual formulation as described above.

There are number of ways to get around this apparent issue, two of which we will discuss here. The simplest approach is to add a small 'virtual mass' M to either end of the ideal spring. By doing so the element's impedance becomes,

$$\mathbf{Z}_I = \begin{bmatrix} \frac{K}{i\omega}+C+i\omega M & -(\frac{K}{i\omega}+C) \\ -(\frac{K}{i\omega}+C) & \frac{K}{i\omega}+C+i\omega M \end{bmatrix} \quad (5.72)$$

which is no longer singular. It can be inverted to obtain the mobility $\mathbf{Y}_I$ and used within the dual formulation, though, care should be taken when adding this virtual mass. If it is made too large, noticeable effects will be introduced to the systems dynamics. In [28] it is recommended that the mass be chosen such that the local non-zero eigen frequencies of the coupling are at least 10 times greater than the frequency range of interest. Alternatively, the effect of the additional mass can be cancelled by coupling a second *negative virtual* mass to the assembly. This way the virtual mass can be chosen arbitrarily.

A second approach is to consider the resilient coupling as a relaxation of the continuity condition. That is, the connecting DoFs are no longer constrained to have the same response; the presence of a resilient element allows for their relative movement. We can then redefine an appropriate continuity condition as,

$$\mathbf{Bv} = \delta \neq \mathbf{0} \quad (5.73)$$

where δ represents the relative response of the interface DoFs. In the presence of an ideal spring-like coupling, this relative velocity can be expressed in terms of the internal force generated by the spring-damper λ,

$$\mathbf{Bv} = \delta = \Gamma\lambda \quad (5.74)$$

where $\Gamma = \text{diag}\left[-\frac{K_1}{i\omega} - C_1, \cdots - \frac{K_N}{i\omega} - C_N\right]^{-1}$ describes the inverse transfer impedance (i.e. 'fexibility') of the N coupling elements. Note that the condition of equilibrium is unchanged. This is because of the massless assumption of the spring-damper coupling, meaning force is conserved across it, and so $\mathbf{g}_1 = -\mathbf{g}_2$. Substituting Equation 5.64 into 5.74 we can rearrange to get λ,

$$\lambda = \left(\mathbf{BYB}^{\mathrm{T}} + \Gamma\right)^{-1} \mathbf{BYf}. \tag{5.75}$$

Finally, substituting Equation 5.75 back into 5.64, we arrive at,

$$\mathbf{Y}_C = \mathbf{Y} - \mathbf{YB}^{\mathrm{T}} \left(\mathbf{BYB}^{\mathrm{T}} + \Gamma\right)^{-1} \mathbf{BY}. \tag{5.76}$$

Equation 5.76 describes the FRF of a resiliently (spring-damper-like) coupled assembly, where Γ characterises the flexibility of the coupling element.

Note that the now permitted relative motion between the two sides of the coupling means that the dual method no longer produces duplicate rows/columns.

Table 5.1
Summary of substructuring equations for direct and three field formulations.

	Method	Formulation
Direct	Rigid coupling, point FRF	$\mathbf{Y}_{Ccc} = \left(\mathbf{Y}_{Scc}^{-1} + \mathbf{Y}_{Rcc}^{-1}\right)^{-1}$ $\mathbf{Y}_{Ccc} = \mathbf{Y}_{Scc}\left[\mathbf{Y}_{Scc} + \mathbf{Y}_{Rcc}\right]^{-1}\mathbf{Y}_{Rcc}$
	Rigid coupling, transfer FRF	$\mathbf{Y}_{Crc} = \mathbf{Y}_{Rrc} - \mathbf{Y}_{Rrc}\left[\mathbf{Y}_{Scc} + \mathbf{Y}_{Rcc}\right]^{-1}\mathbf{Y}_{Rcc}$ $\mathbf{Y}_{Crc} = \mathbf{Y}_{Rrc}\left[\mathbf{Y}_{Rcc} + \mathbf{Y}_{Scc}\right]^{-1}\mathbf{Y}_{Scc}$
	Resilient (spring-like) coupling, transfer FRF	$\mathbf{Y}_{Cc_2c_1} = \mathbf{Y}_{Rc_2c_2}\left(\mathbf{Y}_{Rc_2c_2} + \mathbf{Y}_{Sc_1c_1} + \mathbf{Y}_I\right)^{-1}\mathbf{Y}_{Sc_1c_1}$ $\mathbf{Y}_{Crc_1} = \mathbf{Y}_{Rrc_2}\left(\mathbf{Y}_{Rc_2c_2} + \mathbf{Y}_{Sc_1c_1} + \mathbf{Y}_I\right)^{-1}\mathbf{Y}_{Sc_1c_1}$
Three field	Primal	$\mathbf{Y}_C = \left(\mathbf{L}^{\mathrm{T}}\mathbf{Y}^{-1}\mathbf{L}\right)^{-1}$
	Dual (rigid)	$\mathbf{Y}_C = \mathbf{Y} - \mathbf{YB}^{\mathrm{T}}\left(\mathbf{BYB}^{\mathrm{T}}\right)^{-1}\mathbf{BY}$
	Dual (resilient)	$\mathbf{Y}_C = \mathbf{Y} - \mathbf{YB}^{\mathrm{T}}\left(\mathbf{BYB}^{\mathrm{T}} + \Gamma\right)^{-1}\mathbf{BY}$

The substructuring equations summarised in Table 5.1 amount to different formulations of the same underlying physics. Which approach to use is largely down to personal preference, though for large complex systems, the dual formulation offers advantages in terms of computational efficiency and reduced numerical error.

In the context of virtual prototyping, the purpose of these methods (collectively known as dynamic substructuring) is to take the passive properties (i.e. free-interface FRFs) of an assembly's constituent components, enforce the conditions of equilibrium and continuity, and so obtain a model for the *coupled* assembly's passive properties. With the passive properties established, appropriate loading can be applied to the passive model at the interfaces of the connected vibration sources, so predicting its operational response.

5.2.3 SUBSTRUCTURE DECOUPLING

So far this section has been concerned with the act of *coupling*, i.e. building an assembly model from the free-interface FRFs of its constituent components. In cases where the free-interface conditions necessary for a component's characterisation cannot be adequately approximated by resilient suspension (see 2.3.1), it may be necessary to compensate for, or remove, the dynamics of the mounting structure from the component. This is the opposite problem to substructure coupling described above, namely, *substructure decoupling*. Whilst substructure decoupling relates to the characterisation of individual components, and so could belong in Chapter 2, it follows more naturally now having introduced substructure coupling in detail.

The concept of substructure decoupling was briefly introduced in 2.3.1, though having now introduced the three field formulation, a more general substructuring decoupling framework can be established.

The decoupling of two substructures, say a component A and a test-bench B, can be achieved by considering the primal or dual substructuring formulations, as presented above, in the presence of a negative test-bench FRF; this is equivalent to applying a set of negative interface forces $-\mathbf{g}_B$ that cancel the dynamic contribution of B onto A.

Given the coupled assembly FRF $\mathbf{Y}_C$ (component mounted to the test bench), and the residual FRF $\mathbf{Y}_B$ (test bench), the target component mobility $\mathbf{Y}_A$ can be extracted by first constructing the block diagonal matrix with a negative residual FRF,

$$\mathbf{Y} = \begin{bmatrix} \mathbf{Y}_C & \\ & -\mathbf{Y}_B \end{bmatrix} \tag{5.77}$$

and then following the primal,

$$\mathbf{Y}_A = \left(\mathbf{L}^{\mathrm{T}}\mathbf{Y}^{-1}\mathbf{L}\right)^{-1} \tag{5.78}$$

or dual,

$$\begin{bmatrix} \mathbf{Y}_A & \mathbf{0} \\ \mathbf{0} & -\mathbf{Y}_B \end{bmatrix} = \mathbf{Y} - \mathbf{Y}\mathbf{B}^{\mathrm{T}}\left(\mathbf{B}\mathbf{Y}\mathbf{B}^{\mathrm{T}}\right)^{-1}\mathbf{B}\mathbf{Y} \tag{5.79}$$

substructuring procedures as normal [13].

For a standard decoupling, the Boolean matrix $\mathbf{B}$ is defined identically to the coupling case, i.e. including only interface DoFs. However, unlike substructure coupling, where continuity and equilibrium are satisfied only at the interface, with decoupling these constraints can be enforced at additional remote locations across the test bench; the decoupling enforces the response at measured internal DoFs to be equal to that of the uncoupled assembly. This offers the opportunity to in some way 'over-determine' the decoupling problem. Indeed, there is a great deal of flexibility over how these constraints are imposed. Further details on the families of available decoupling techniques and their respective benefits can be found in [45].

5.2.4 STRUCTURAL MODIFICATION

Substructuring is a procedure used to couple component-level FRFs so as to form a model of a virtual assembly. Often, however, an assembly will exist physically, and instead the question is how best to modify it so as to improve its vibro-acoustic performance. Mathematically, this problem is identical to that of substructuring; rather than coupling component-level FRFs together, we wish to couple a *structural modification* to an already assembled structure. Our uncoupled impedance matrix (Equation 5.47) then becomes,

$$\mathbf{Z} = \begin{bmatrix} \mathbf{Z}_C & \\ & \Delta\mathbf{Z} \end{bmatrix} \tag{5.80}$$

where $\Delta\mathbf{Z}$ represents an arbitrary structural modification. For example, for the addition of a localised mass,

$$\Delta\mathbf{Z} = i\omega m \tag{5.81}$$

whilst for a distributed (i.e. connecting two DoFs) stiffness,

$$\Delta\mathbf{Z} = \begin{bmatrix} \frac{k}{i\omega} & -\frac{k}{i\omega} \\ -\frac{k}{i\omega} & \frac{k}{i\omega} \end{bmatrix}. \tag{5.82}$$

As with resilient elements, care should be taken when using the dual method; the modification impedance must be invertible to obtain the uncoupled mobility matrix $\mathbf{Y}$. Of course, structural modifications are not limited to simple lumped mass or stiffness terms, but can be more complex, for example tuned mass-spring-dampers, etc.

5.2.5 EXPERIMENTAL SUBSTRUCTURING

Whilst the equations that underlie substructuring are relatively straightforward, obtaining good quality predictions from experimental data is far from trivial. But why?

Lots of coupling DoFs

Whilst only a subset of the interface DoFs might strongly couple and contribute to the assembly's dynamics, these DoFs are generally not known *a priori*, and so all DoFs must be included. Take for example a standard vibration source with 4 connections points. Assuming a rigid interface model with 6 DoFs per connection, 24 source-receiver DoF pairs will need to be coupled. DoFs that do not contribute significantly to coupling, but are included for completeness, can introduce experimental error without the benefiting predictive accuracy. This is especially so for hard to measure DoFs, such as rotations. One useful trick to deal with a multitude of DoFs is to perform a coupled measurement using a dummy component that has similar dynamic characteristics to the intended component. From this coupled measurement various tests (e.g. rank and completeness) can be performed to access interface completeness and the contribution of individual DoFs, see 6.4.1.

Example 5.1: A simple substructuring example

In this example we show the result of a simple substructuring calculation. The assembly considered is shown in Figure 5.4; two steel beams are coupled at a single point via a resilient element. In this example we consider only coupling in the vertical z direction, thus simplifying the experimental procedure, analysis and presentation of results.

Figure 5.4 Experimental case study for substructuring; two resiliently coupled beams.

The source and receiver were both measured under approx. free boundary conditions by resting them on foam pads. The isolator was characterised by installing it between a pair of masses (as in Example 2.2). The source was characterised by a single point FRF at its coupling interface, Y_{Scc}. The receiver, which includes an additional remote target DoF r, is characterised by a 2×2 FRF matrix,

$$\mathbf{Y}_R = \begin{bmatrix} Y_{Rcc} & Y_{Rcr} \\ Y_{Rrc} & Y_{Rrr} \end{bmatrix}. \tag{5.83}$$

The point FRFs Y_{Scc} and Y_{Rcc} are shown in Figure 5.5. The isolator is characterised by its transfer impedance $Z_{Ic_1c_2}$ (obtained using the in situ method, see 2.3.2). This can be used either directly, as per the dual and direct methods, or to build a 2×2 impedance matrix $\mathbf{Z}_I$ (as required by the primal method). The isolator stiffness is shown in Figure 5.6. Note that at low frequency the characterisation becomes sensitive to noise, due to rigid body motion of the test assembly. Here we extrapolate the 100 Hz result down to 10 Hz.

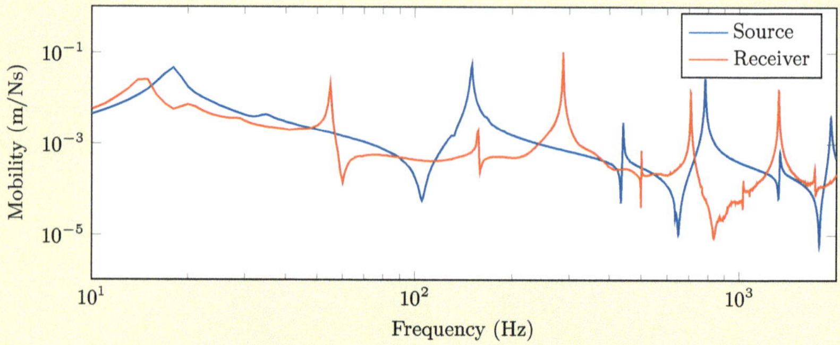

Figure 5.5 Uncoupled source (point) and receiver (transfer) FRFs used for substructuring calculations.

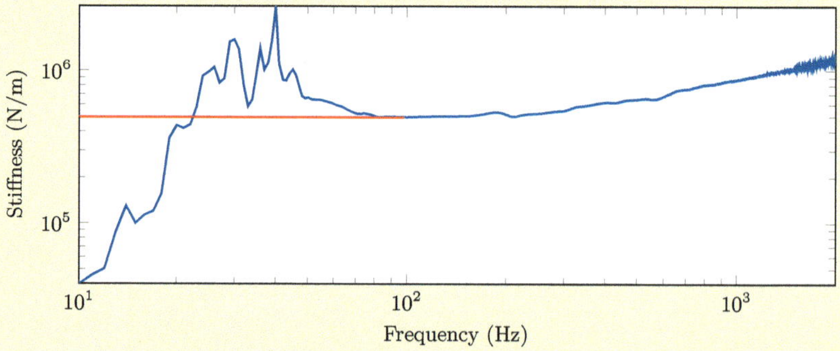

Figure 5.6 Dynamic transfer stiffness of isolator used in substructuring. Red plot shows extrapolated low frequency stiffness.

The source, receiver and isolator FRFs are substructured using primal (Equation 5.61), dual (Equation 5.76) and direct (Equation 5.45) formulations, all giving identical results. The resulting mobility matrix has the form,

$$\mathbf{Y}_C = \begin{bmatrix} Y_{Cc_1c_1} & Y_{Cc_1c_2} & Y_{Cc_1r} \\ Y_{Cc_2c_1} & Y_{Cc_2c_2} & Y_{Cc_2r} \\ Y_{Crc_1} & Y_{Crc_2} & Y_{Crr} \end{bmatrix} \tag{5.84}$$

where c_1 and c_2 are the interfaces above and below the isolator, respectively. Shown in Figure 5.7 is the substructured prediction for the transfer FRF Y_{Crc_1} compared against a direct measurement. Considering the simplifications made (i.e. using only z DoFs) good agreement is obtained across the whole frequency range.

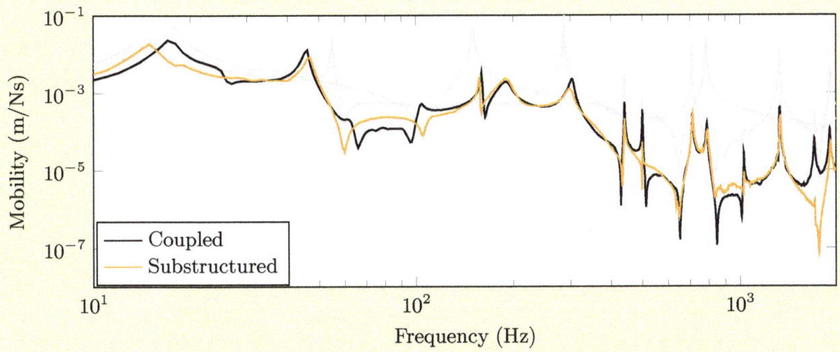

Figure 5.7 Directly measured and substructured transfer FRFs Y_{Crc_1}. Substructured prediction made using primal, dual and direct methods are identical. Also shown in grey are the source and receiver FRFs from Figure 5.5.

Non-rigid interface dynamics

The 'rigid' coupling employed in substructuring is an idealisation, only approximately met in practice. It assumes a) point like (i.e. 6 DoF) interface dynamics, and b) strong continuity across the interface, neither of which are exactly true in practice. In the mid-to high frequency regime, structural wave lengths become comparable in size to the interface connection and the 6 DoF model breaks down (this problem is taken to the extreme when dealing with continuous interfaces). Even in the low-mid frequency regime where the 6 DoF model is valid, rarely do interface connections possess strong continuity; complex frictional phenomena introduce additional stiffness and damping properties to the interface. Methods to characterise and include the dynamics of joints as a means to improve substructuring accuracy is an area of increasing interest. Some recent works include [12, 34, 35].

Ill-conditioning and FRF consistency

Though not strictly an inverse problem, substructuring involves the inversion and manipulation of measured FRF matrices, and so is subject to the same sensitivity and ill conditioning effects. Around component resonances, component FRFs will be close to rank 1, meaning that small errors will be amplified. Similarly, it has been shown that small errors or inconsistencies in the measured FRF matrices can lead to spurious peaks in the substructured estimates [33]. A common method to alleviate these issues is to apply some light regularisation to the matrix inversion (see 7.4.3) or to attempt to improve the consistency of the FRF matrix through various filtering schemes.

Co-located coupling DoFs

It is important that when substructuring two components, their interface DoFs are co-located. This ensures a physically correct coupling is realised. Such co-location can be very difficult to achieve using the Finite Difference (FD) and Equivalent Multi-Point Connection (EMPC) interface models. For this reason, the Virtual Point (VP) transformation is generally recommended if substructuring in multiple DoFs is required.

5.3 TRANSFER PATH ANALYSIS

Having introduced methods for characterising passive (Chapter 2) and active components (Chapter 3), and procedures for coupling their passive properties to form an assembled structure (see 5.2), we are now in a position to consider the construction of a Vibro-Acoustic Virtual Prototype (VAVP). In general, the purpose of such a VAVP is to predict the operational response of an assembly at some set of reference, or target, DoFs r in the receiving part of the structure. The VAVP can then be used for diagnostic (e.g. identifying dominant transmission paths) or predictive (e.g. investigating the effect of structural modification) purposes. These applications constitute the class of methods known widely as Transfer Path Analysis (TPA).

Transfer Path Analysis is broad class of methods that are used for analysing the propagation of noise and vibration in complex built-up structures, for example, ships, vehicles, trains, etc. It has become an essential tool in the development and refinement of structures whose vibro-acoustic response is of interest. A variety of TPA methodologies have been proposed over the years (the reader is referred to [43] for a detailed review of their history and development). Common to all methods is the division of an assembly into active and passive components. Active components, as already mentioned, are regarded as those that contain vibration (or sound) generating mechanisms (such as engines, gearboxes, pumps, motors, etc.), whilst passive components simply transmit, and/or radiate, this vibration (vibration isolators, body panels, etc.). It is the role of TPA to predict the contribution of each active component to the total vibro-acoustic response at some specified positions, e.g. the noise level experienced by a vehicle occupant. To do so the activity, or 'strength', of each active component must first be estimated. Then, the transmission paths through which these components contribute must be identified. Depending on the type of TPA considered, the methods employed and quantities obtained will differ. The focus of this book is on in situ and blocked force-based methods, and so the blocked force and component-based variants of TPA are of most relevance here. However, for completeness we will also introduce some other variants of TPA commonly found in the literature.

Most TPA methods rely on a two part measurement procedure and involve the determination of the acting 'source strengths' (usually forces) by indirect methods. In part one, the passive properties (i.e. FRFs) of the structure of interest are measured between the source-receiver (active-passive) interface and some set of indicator positions, which may or may not include the interface DoFs. Note that this 'source-receiver' interface is somewhat arbitrary. Often passive components, for example vibration isolators, are included in the definition of an active component. In this case the source-receiver interface would be that between the passive side of the isolator and the remaining portion of the assembly. The notion of sources within sources is discussed in more detail in 3.2.2. In part two, the structure is operated and its activity (i.e. operational acceleration, velocity, etc.) is recorded at the same indicator positions. Based on the measured FRFs, the operational responses are used to infer the unknown excitation force at the defined source-receiver interface. Inverse problems of this nature are notoriously sensitive to experimental error, making a rigorous treatment of their uncertainty all the more necessary. This issue is covered in detail in Chapter 8.

5.3.1 CLASSICAL TPA

In classical TPA each active component is characterised by the *contact* force $\mathbf{g}_c \in \mathbb{C}^{N_c}$ that it imparts on its connected receiver structure. This contact force can be measured directly using installed force transducers, though is more typically obtained indirectly using FRFs measured on the *uncoupled* receiver structure (see 3.1.2), following the standard inverse procedure as detailed in 7.4 [40, 44],

$$\mathbf{g}_c = \begin{bmatrix} \mathbf{Y}_{Rcc} \\ \mathbf{Y}_{Rbc} \end{bmatrix}^+ \begin{pmatrix} \mathbf{v}_c \\ \mathbf{v}_b \end{pmatrix} \tag{5.85}$$

where; $\mathbf{v}_c \in \mathbb{C}^{N_c}$ and $\mathbf{v}_b \in \mathbb{C}^{N_b}$ are the operational responses measured at, respectively, the interface (c) and remote (b) DoFs (together termed '*indicator*' DoFs), and $\mathbf{Y}_{Rcc} \in \mathbb{C}^{N_c \times N_c}$ and $\mathbf{Y}_{Rbc} \in \mathbb{C}^{N_b \times N_c}$ are the point and transfer FRFs of the uncoupled receiver that relate the indicator and interface DoFs. In theory it is not necessary to include both c and b points in Equation 5.85. However, by doing so we over-determine the problem and obtain a least squares solution which is usually a better, or at least more robust, estimate (see 7.4.1).

Recall from 3.1.2 that the contact force is *dependent* on the dynamics of the receiver structure. Once obtained, it may be used to predict the operational response of the *same structure* using the forward FRFs of the uncoupled receiver,

$$\begin{pmatrix} \mathbf{v}_r \\ \mathbf{p}_r \end{pmatrix} = \begin{pmatrix} \mathbf{Y}_{Rrc} \\ \mathbf{H}_{Rrc} \end{pmatrix} \mathbf{g}_c \tag{5.86}$$

where $\mathbf{v}_r \in \mathbb{C}^{N_r}$ and $\mathbf{p}_r \in \mathbb{C}^{N_r}$ are, respectively, the predicted structural and sound pressure response at the target (r) DoFs, and $\mathbf{Y}_{Rrc} \in \mathbb{C}^{N_r \times N_c}$ and $\mathbf{H}_{Rrc} \in \mathbb{C}^{N_r \times N_c}$ are the structural and vibro-acoustic FRFs that relate the interface force to the target response.

Note that this prediction is for the same assembly (C) in which the source characterisation (Equation 5.85) is performed. Once the contact force vector has been determined, the contribution of individual forces (or groups thereof) to the assembly's vibro-acoustic response can be calculated,

$$\begin{aligned}\mathbf{v}_r &= \mathbf{Y}_{Rrc_1}\mathbf{g}_{c_1} + \mathbf{Y}_{Rrc_2}\mathbf{g}_{c_2} + \cdots + \mathbf{Y}_{RrcN}\mathbf{g}_{cN} \\ &= \mathbf{v}_r^{c1} \qquad + \mathbf{v}_r^{c2} \qquad + \cdots + \mathbf{v}_r^{cN}\end{aligned} \tag{5.87}$$

where $\mathbf{v}_r^{c_n} \in \mathbb{C}^{N_r}$ represents the contribution to the total response ($\mathbf{v}_r$) due to the contact force at the Nth subset of interface DoFs $c_n \in c$, $\mathbf{g}_{c_n}$. These subsets might describe different source components, connection points, or interface DoFs (i.e. translations or rotations) depending on the level of detail required. These contributions can be rank ordered and used to identify dominant source contributions and help diagnose troublesome noise and vibration issues.

Although well established and widely employed, classical TPA, as described above, is time consuming and often impractical to apply; it requires the assembly to be dismantled for the FRF measurements. Furthermore, the contact forces obtained are dependent upon the dynamics of the receiver structure and not transferable to other installations. This limits classical TPA to diagnostic applications, meaning that the effect of structural modification/redesign cannot be investigated; any change to the receiver structure would require a new set of contact forces to be determined.

5.3.2 MOUNT STIFFNESS TPA

Mount stiffness TPA, like classical TPA described above, represents source activity by the interface contact force. As the name suggests, this contact force is obtained using knowledge of the stiffness (or impedance) of the installed isolators (see 3.1.2 for details),

$$\mathbf{g}_{c_2} = -\mathbf{Z}_{Ic_1c_2}[\mathbf{v}_{c_1} - \mathbf{v}_{c_2}]. \tag{5.88}$$

For this reason, the approach is only applicable to resiliently mounted machines.

Once the contact force has been obtained, the assembly's vibro-acoustic response is predicted in the same way as for classical TPA (see Equation 5.86) and can similarly be subject to a contribution analysis (see Equation 5.87). Like classical TPA, mount stiffness TPA is based on the use of contact force, and so is limited to diagnostic, rather than predictive, applications.

5.3.3 BLOCKED FORCE (IN SITU) TPA

An alternative procedure, known widely as blocked force TPA (also in situ TPA), relies on the *independent* characterisation of each active component by their *blocked force* [15]. These blocked forces are obtained using the in situ approach presented in 3.2. The steps followed are identical to classical TPA except that the source and receiver are not separated during the passive test, so that the coupled FRFs of the assembly are obtained,

$$\bar{\mathbf{f}}_c = \begin{bmatrix} \mathbf{Y}_{Ccc} \\ \mathbf{Y}_{Cbc} \end{bmatrix}^+ \begin{pmatrix} \mathbf{v}_c \\ \mathbf{v}_b \end{pmatrix}. \tag{5.89}$$

Avoiding the cost (i.e. time) of dismantling an assembly is crucial, especially in the automotive sector, where disassembly is non-trivial (especially with electric vehicles), the time to market is short, and the time to troubleshoot NVH problems even shorter.

Once obtained the blocked force can be used to predict the operational response of the structure using the forward FRFs of the coupled assembly,

$$\begin{pmatrix} \mathbf{v}_r \\ \mathbf{p}_r \end{pmatrix} = \begin{pmatrix} \mathbf{Y}_{Crc} \\ \mathbf{H}_{Crc} \end{pmatrix} \bar{\mathbf{f}}_c. \tag{5.90}$$

Note that this prediction is for the same assembly (C) in which the characterisation is performed. As in classical TPA, the contribution of individual blocked forces can be used to identify troublesome sources by rank ordering [15, 36].

Blocked force TPA offers two distinct advantages over classical TPA:

1) The characterisation procedure is performed entirely *in situ*. This means that the active components need not to be removed from their intended installation, unlike classical TPA which requires the assembly to be dismantled.
2) Perhaps of greater importance, the active components are characterised *independently* [24]. This means that the same set of blocked forces can be used to excite a range of different assemblies, or be applied in the presence of a structural modification (i.e. they are transferable) [14, 30]. Consequently, a contribution analysis can be repeated for different versions of the passive receiver. This is in contrast to classical TPA, where the contact forces obtained are valid only for the assembly in which they were characterised.

Since its development, (in situ) blocked force TPA has seen considerable interest due to its straightforward implementation, transferable source properties, related standardisation [1] (ISO 20270: 2019) and integration into commercially available software packages.

It is worth noting that quite often in the literature, blocked force TPA is referred to as component(-based) TPA, a further variant that also uses the blocked force to characterise its active components. This 'component-based' pre-fix is attributed to the independent nature of the blocked force; it is a component level quantity. However, the forward FRFs used in blocked force TPA are assembly properties (typically measured directly), and not component level. For this reason, we reserve the prefix *component-based* for the particular variant of TPA whereby the forward FRFs are also based on component-level quantities, i.e. obtained using substructuring, as discussed in 5.3.6.

5.3.4 AIRBORNE TPA

With the in situ blocked force method forming the basis of blocked force TPA, it should be no surprise that an acoustic equivalent exists based on the in situ free volume velocity method described in 3.4.4.

The free volume velocity of an acoustic source describes the discretised (hence volume) velocity profile over an enveloping surface, when the connecting domain is removed. It is simply the acoustic equivalent to the free velocity. Based the equivalent field theorem (see 3.4.4), the free volume velocity can be obtained using the inverse relation,

$$\tilde{\mathbf{q}}_c = \mathbf{H}^+_{Cbc}\mathbf{p}_b \tag{5.91}$$

where $\mathbf{H}_{Cbc} \in \mathbb{C}^{N_b \times N_c}$ is the acoustic transfer FRF between a set of points/DoFs c that discretise the continuous surface interface and a set of remote measurement positions b, and $\mathbf{p}_b \in \mathbb{C}^{N_b}$ is the operational sound pressure at said remote points. Note that unlike structural TPA, for practical reasons it is not typical to measure the *interface* response $\mathbf{p}_c$.

Once the free volume velocities are obtained, response predictions can be made at the target (r) DoFs using,

$$\mathbf{p}_r \approx \mathbf{H}_{Crc}\tilde{\mathbf{q}}_c. \tag{5.92}$$

In the same way as for classical and blocked force TPA, contributions of individual sources can be separated and used to perform a rank ordering.

5.3.5 TRANSMISSIBILITY-BASED TPA

The variants of TPA described above (classical, mount stiffness, and blocked force) share two common traits; 1) source activity is represented by a force loading applied to the structure at the source-receiver interface, and 2) to identify these loads, and relate them to target responses, a series of force-response FRFs (e.g. mobility) must be measured as part of a passive test campaign. Transmissibility-based TPA provides an alternative approach that avoids one or both of these force-based requirements. For a detailed discussion on the concept of transmissibility, see 2.4.2. Like classical and mount stiffness TPA, transmissibility-based TPA has limited applications in the context of virtual prototyping, as will be discussed shortly.

The central idea behind transmissibility-based TPA (of which there are multiple variants), is to avoid the determination of operational loads by instead decomposing the response at the target DoFs in terms of the responses at a series of DoFs located elsewhere, typically at the source-receiver interface c. These DoFs are then interpreted as the operational inputs to the system. Whilst placing these DoFs at the interface helps with the interpretation of results, as we are not considering force loading, these DoFs can in theory be positioned arbitrarily.

Mathematically, we have for a combined structural and acoustic target responses,

$$\begin{pmatrix} \mathbf{v}_r \\ \mathbf{p}_r \end{pmatrix} = \begin{bmatrix} \mathbf{T}^{(v)}_{Rrc} \\ \mathbf{T}^{(p,v)}_{Rrc} \end{bmatrix} \mathbf{v}_c \tag{5.93}$$

where $\mathbf{v}_c \in \mathbb{C}^{N_c}$ is the operational response at the interface DoFs c, $\mathbf{v}_r \in \mathbb{C}^{N_{r1}}$ and $\mathbf{p}_r \in \mathbb{C}^{N_{r2}}$ are the operational responses at the target DoFs r (such that $N_{r1} + N_{r2} = N_r$ is the total number of responses), and $\mathbf{T}^{(v)}_{Rrc} \in \mathbb{C}^{N_{r1} \times N_c}$ and $\mathbf{T}^{(p,v)}_{Rrc} \in \mathbb{C}^{N_{r2} \times N_c}$ are the FRFs relating the above. Note that $\mathbf{T}^{(v)}_{Rrc}$ relates the structural interface response to the structural target response, and so represents a response transmissibility (see 2.4.2) [17].

As with force-based TPA, Equation 5.93 can be broken down into a series of contributions,

$$\begin{pmatrix} \mathbf{v}_r \\ \mathbf{p}_r \end{pmatrix} = \begin{bmatrix} \mathbf{T}^{(v)}_{Rrc_1} \\ \mathbf{T}^{(p,v)}_{Rrc_1} \end{bmatrix} \mathbf{v}_{c_1} + \begin{bmatrix} \mathbf{T}^{(v)}_{Rrc_2} \\ \mathbf{T}^{(p,v)}_{Rrc_2} \end{bmatrix} \mathbf{v}_{c_2} + \cdots + \begin{bmatrix} \mathbf{T}^{(v)}_{Rrc_N} \\ \mathbf{T}^{(p,v)}_{Rrc_N} \end{bmatrix} \mathbf{v}_{c_N} \tag{5.94}$$
$$= \begin{pmatrix} \mathbf{v}_r \\ \mathbf{p}_r \end{pmatrix}^{c_1} + \begin{pmatrix} \mathbf{v}_r \\ \mathbf{p}_r \end{pmatrix}^{c_2} + \cdots + \begin{pmatrix} \mathbf{v}_r \\ \mathbf{p}_r \end{pmatrix}^{c_N}$$

where $c_n \in c$ represents a subset of the interface DoFs, and their relative importance ranked to identify possible noise and vibration issues.

In 2.4.2 we showed that transmissibilities can be obtained from standard FRF measurements, and that they are invariant to the nature of the loading applied for their measurement. In [18], as part of their so-called Global Transmissibility Direct Transmissibility (GTDT) TPA methodology (later termed Advanced TPA [27]), these transmissibilities are determined from conventional transmissibilities (measured using external excitations, e.g. by impact hammer), by means of some simple linear algebraic manipulations.

It is important to note that $\mathbf{v}_{c_n}$ used in Equation 5.94 describes the local response which is itself a function of all the loads ($\mathbf{g}_c$) applied across the interface. Whilst the local response is likely to be dominated by the local forces, this is not guaranteed and strong modal response could see the significant contributions from remote forces. For this reason care should be taken when interpreting the results of transmissibility-based TPA.

Furthermore, with $\mathbf{v}_{c_n}$ being a property of the coupled assembly (and not an invariant source property as with the blocked force), if any structural modifications are made, $\mathbf{v}_{c_n}$ will change accordingly. And so transmissibility-based TPA does not provide the necessary invariance required for virtual prototyping applications. Its application is thus limited to diagnostic investigations. Even within its diagnostic remit, transmissibility-based TPA should be used with care. The contributions that it predicts can be quite different to those obtained from a force-based TPA. This shouldn't be a surprise, after all, transmissibility-based TPA does not include a source identification step; it is entirely based on measured responses.

Operational-TPA

In a standard transmissibility-based TPA (e.g. ATPA), the transmissibilities are determined as part of a passive test campaign using external excitations, much like a conventional force-based TPA. In general, this passive test is the most expensive or time consuming part of a TPA. Hence, there is an interest in alternative methods that avoid the need to perform any passive testing, and thus rely only on operational measurements. The most well established method for doing so is known as Operational TPA [10, 25].

The operational identification of the transmissibility FRFs follows the output-only procedure outlined in 2.4.2. We consider P different operational states of the assembly and write an equation in the form of Equation 5.93 for each one. These different states might include: stationary states with different relative contributions of each source, states with different source loading, time windows from time-varying operations, e.g. run up/downs, etc. These equations can be combined by placing (synchronised) responses corresponding to each operational state in the columns of a pair of response *matrices*,

$$\left[\begin{pmatrix} \mathbf{v}_r \\ \mathbf{p}_r \end{pmatrix}^{(1)} \cdots \begin{pmatrix} \mathbf{v}_r \\ \mathbf{p}_r \end{pmatrix}^{(P)} \right] = \left[\begin{array}{c} \mathbf{T}_{Rrc}^{(v)} \\ \mathbf{T}_{Rrc}^{(p,v)} \end{array} \right] \left[(\mathbf{v}_c)^{(1)} \quad \cdots \quad (\mathbf{v}_c)^{(P)} \right] \tag{5.95}$$

or in matrix notation,

$$\mathbf{VP}_r = \mathbf{T}_{Rrc}\mathbf{V}_c \tag{5.96}$$

where $\mathbf{VP}_r \in \mathbb{C}^{N_r \times P}$ and $\mathbf{V}_c \in \mathbb{C}^{N_c \times P}$ are matrices of complex Fourier spectra, and $\mathbf{T}_{Rrc} \in \mathbb{C}^{N_r \times N_c}$ is the transmissibility matrix relating them. At this point we can solve for $\mathbf{T}_{Rrc}$ by post-multiplying both sides by the (pseduo-)inverse interface response matrix,

$$\mathbf{T}_{Rrc} = \mathbf{VP}_r\mathbf{V}_c^+. \tag{5.97}$$

In the case that the number of operational states P is equal to the number of interface DoFs c, the above reduces to a standard matrix inverse. In practice, however, it is generally beneficial to include additional states such that $P > N_c$, thus over-determining the problem and returning the least squares estimate. Furthermore, careful regularisation through singular value discarding (see 7.4.3) has been found to improve results [10]. A requirement when solving Equation 5.97 is that the interface response matrix is full rank, i.e. the number of independent operational states should at least equal the number of interface DoFs N.

Once the transfer function matrix $\mathbf{T}_{Rrc}$ has been estimated, as per Equation 5.97, it can be used to break down contributions of new operational states as per Equation 5.94.

Whilst it was shown in 2.4.2 that the transmissibility is invariant to the specifics of the source distribution, meaning the same transmissibility is obtained through operational and FRF measurements, this was based on the assumption that the underlying dynamics of the assembly do not change whilst it is operational. This might indeed not be the case in a practical scenarios, for example a vehicle accelerating under load.

Relation to MIMO FRF estimation

As already mentioned, the above output only FRF estimation should be familiar to the reader, it is identical to the output only transmissibility procedure presented in 2.4.2. On closer inspection, noting that the matrix (right) pseduo-inverse can be written as,

$$\mathbf{A}^+ = \mathbf{A}^{\mathrm{H}}(\mathbf{A}\mathbf{A}^{\mathrm{H}})^{-1}. \tag{5.98}$$

Equation 5.97 also bares a strong resemblance to the general MIMO FRF estimation described in 7.3.1. From Equation 5.98 and Equation 5.97 we have,

$$\mathbf{T}_{Rrc} = \mathbf{VP}_r\mathbf{V}_c^{\mathrm{H}}(\mathbf{V}_c\mathbf{V}_c^{\mathrm{H}})^{-1} \tag{5.99}$$

where $\mathbf{VP}_r\mathbf{V}_c^{\mathrm{H}} = \mathbf{S}_{VP,V}$ represents the averaged cross-power spectrum between interface and target responses, and $\mathbf{V}_c\mathbf{V}_c^{\mathrm{H}} = \mathbf{S}_{V,V}$ averaged the auto-power spectrum of the interface response. Note that the averaging is taken care of by the matrix multiplication, which adds together all the operational states. Equation 5.99 clearly takes the same form as the MIMO FRF estimation described in 7.3.1 (specifically Equations 7.58 and 7.59).

Path interpretation in TPA

The use of generalised transmissibilities in the context of TPA, as discussed above, introduces an important subtlety to the interpretation of what we term a 'transfer path' or a 'source contribution'.

As discussed in 2.4.2, the generalised (response) transmissibility is defined in such a way that a blocking constraint is present on all inputs except the one through which a response is being applied. In contrast, the forward FRFs used in a conventional TPA are unconstrained. This difference is illustrated in Figure 5.8.

Whilst there is no agreed upon definition, it is argued in the GTDT TPA literature that the use of transmissibilities provides a more 'path-like' contribution, as all others have been blocked. In contrast, the contributions obtained through a force-based

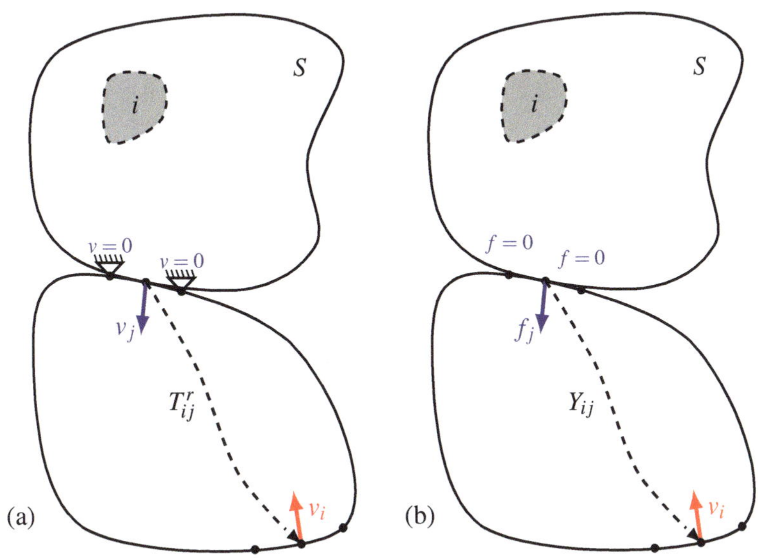

Figure 5.8 Difference in 'path' interpretation between (a) transmissibility-based and (b) force-based TPA.

TPA describe the response at a target location that has been transmitted through all paths connecting the input with the target. As such, these might be better described as *source* or *input contributions*.

Not only do we see an interesting difference in the interpretation of contributions, but when comparing transmissibility and blocked force TPA, we see an intriguing duality. The transmissibility, when defined using the complete set of interface DoFs c, becomes an invariant property of the *receiver* (see 2.4.2), owing to the blocking constraints in its definition. The operational response, however is a property of the *assembly*. Compare this with a blocked force TPA, where the operational blocked force is by definition an invariant *source property*, whilst the forward FRFs are properties of the *coupled assembly*. The invariance thus switches between the FRF and the operational activity.

With the above discussion in mind, one might argue that the name 'Transfer Path Analysis' is somewhat of a misnomer, when describing force-based methods since the contributions to the receiver response which it quantifies are not strictly via transmission paths. An alternative, and perhaps more accurate, name would be 'Source Contribution Analysis' which although appearing in some publications appears now to have lost favour. Indeed, the name 'TPA' though misleading, has now become almost universal and so has been adopted here.

5.3.6 COMPONENT-BASED TPA

So far, none of the TPA methods described above answer question '*how do we predict the response of a virtual assembly?*'. Answering this question will be the focus of the remainder of this chapter.

The advent of blocked force TPA opened up the possibility of applying the results of a TPA analysis in a far wider form than had previously been possible. In particular, because the blocked forces remains valid when receiver components are modified, TPA can now serve a predictive role, allowing for design optimisation and 'what if' analysis of virtual structures [29, 31, 42]. Key to this predictive capability, in addition the blocked force, is to replace measured forward FRFs with ones predicted by substructuring. The combination of blocked force TPA with substructuring has widely become known as 'component-based TPA', a name that we will begrudgingly use here. We stress begrudgingly because the authors do not feel that this name is representative of the method's main purpose; to predict the overall response of an assembly that does not physically exist. A more appropriate name might be *component-based virtual modelling*, or as we have adopted through this book, *Vibro-Acoustic Virtual Prototyping*. Regardless of its name, it should be noted that fundamentally, there is no technical difference between component-based TPA and blocked force TPAm but the latter term typically implies usage in a diagnostic role and the former in a predictive role.

In a component-based TPA, the blocked force of each active component is obtained either in situ, or by using an appropriate test bench, as per the standard inverse formulation,

$$\bar{\mathbf{f}}_c = \left[\begin{array}{c} \mathbf{Y}_{Ccc} \\ \mathbf{Y}_{Cbc} \end{array} \right]^+ \left(\begin{array}{c} \mathbf{v}_c \\ \mathbf{v}_b \end{array} \right). \tag{5.100}$$

Once obtained, the blocked force is used to predict the vibro-acoustic response ($\mathbf{v}_r$ and $\mathbf{p}_r$) of the now virtual assembly X using its forward FRFs,

$$\left(\begin{array}{c} \mathbf{v}_r \\ \mathbf{p}_r \end{array} \right) = \left(\begin{array}{c} \mathbf{Y}_{Xrc} \\ \mathbf{H}_{Xrc} \end{array} \right) \bar{\mathbf{f}}_c \tag{5.101}$$

where $\mathbf{Y}_{Xrc}$ and $\mathbf{H}_{Xrc}$ are the output of some substructuring (DS) calculation,

$$\left(\begin{array}{c} \mathbf{Y}_{Xrc} \\ \mathbf{H}_{Xrc} \end{array} \right) = DS\left(\mathbf{Y}_S, \mathbf{Y}_I, \mathbf{Y}_R, \mathbf{H}_R\right). \tag{5.102}$$

With the forward FRFs $\mathbf{Y}_{Xrc}$ and $\mathbf{H}_{Xrc}$ obtained by substructuring (direct, primal or dual formulations), they need not correspond to an assembly that exists physically. Alternatively, they might describe an assembly that does exists physically, but has been subject to a structural modification, as described in 5.2.4.

Clearly, component-based TPA (or VAVP) presents a highly flexible framework in which components can be interchanged and modified for the purpose of design optimisation and target setting. Whats more, this framework allows for seamless integration of experimental and modelled data. This topic is discussed further in 5.4.

5.4 HYBRID MODELS

In the preceding sections, we have perhaps implicitly assumed that all the required FRFs are obtained experimentally. Whilst for diagnostic applications of TPA this is almost certainly the case, predictive applications (such as component-based TPA) are not so restricted. Indeed, it is possible to combine both measured and modelled component FRFs to form so-called *hybrid models*. This unlocks the ability not only to model complex structures that have not been previously assembled, but also to include components that do not exist physically.

5.4.1 FINITE ELEMENT – EXPERIMENTAL (FE-X) MODELS

Component-based TPA provides a general framework that allows for the virtual interchange of components in a physically consistent manner. It provides design engineers with a means to investigate the effect of structural modification/redesign in the presence of a realistic operational excitation (in the form of a measured blocked force). Importantly, the introduction of dynamic substructuring facilitates the coupling of experimental and *numerically modelled* components (say by Finite Elements) to form hybrid FE-X assemblies. This extends our virtual prototyping capabilities to systems whereby one or more of its components do not yet exist physically. Furthermore, it provides a means of optimising component level models to with respect to whole assembly targets.

Though more capable, the requirements of FE-X modelling are no different to that of component-based modelling with exclusively experimental subsystems provided that the connecting interfaces of the FE model are represented in a compatible form, congruent with the components to be connected. This is readily achieved using, for example, the virtual point method (see 4.2.2). It is also necessary that the FE model's FRFs are exported at the same frequency resolution as any experimental FRFs.

In theory, substructuring procedures apply equally to experimental or modelled data. However, in practice it is surprisingly challenging to combine experimental descriptions of component FRFs into a reliable prediction of the assembly FRF (as mentioned in 5.2.5). These difficulties are perhaps even greater when attempting to combine measured and modelled data. The sources of error are, broadly speaking, the same as those mentioned in 5.2.5: disparity between the measured or modelled data and the real-life components and incorrect or insufficiently detailed interface descriptions.

For an active component, there is an additional consideration if its free-interface FRF is to be modelled, as opposed to measured. This is because the active (i.e. blocked force) and passive (i.e. FRF) properties of an active component share underlying terms. As discussed in 8.3.2, lack of *consistency* between these two descriptions leads to incomplete cancellation and potentially large errors in the forward response prediction. Thus, for a source characterisation, the mixing of measured and modelled data is not recommended. In practice this means that source FRFs will ideally need to be measured.

5.4.2 HYBRID FINITE ELEMENT – STATISTICAL ENERGY ANALYSIS – EXPERIMENTAL (FE-SEA-X) MODELS

The component-based TPA framework enables the construction of hybrid FE-X models, where Finite Element and eXperimental component representations are combined using standard dynamic substructuring procedures (direct, primal or dual). However, complex engineering structures often contain components whose dynamics are too complex to be modelled using conventional deterministic means (such as FE), which are typically limited to low frequency, or long wave-length regimes.

Consider the following example. The engine of a vehicle (characterised by its blocked force and free-interface FRF) is resiliently connected to a subframe. Being relatively stiff, the subframe can be modelled with reasonable accuracy using FE. However, attached to the subframe are a series of body panels that radiate into a closed cabin. These body panels are thin and highly flexible (at least relative to the subframe) and so have much shorter wavelengths propagating. Should we model these panels as additional FE components? Probably not... This is because a) the number of DoFs required to represent such components at short wave-lengths may become prohibitively large, and b) the sensitivities of such components to small imperfections (e.g. due to manufacturing tolerances) introduce uncertainties that can limit the interpretation of a deterministic analysis.

So how then should we model these components? The above limitations are largely overcome by a method known as Statistical Energy Analysis (SEA), an approach that considers the energetic response of a component, as opposed to kinematic [26].

It would appear that the solution to our multi-scale (long and short wavelength components) VAVP problem would be to combine experimental, Finite Element and SEA representations, as illustrated in Figure 5.9. That is the focus of this section. But first, for the unfamiliar reader, we introduce the concept of SEA.

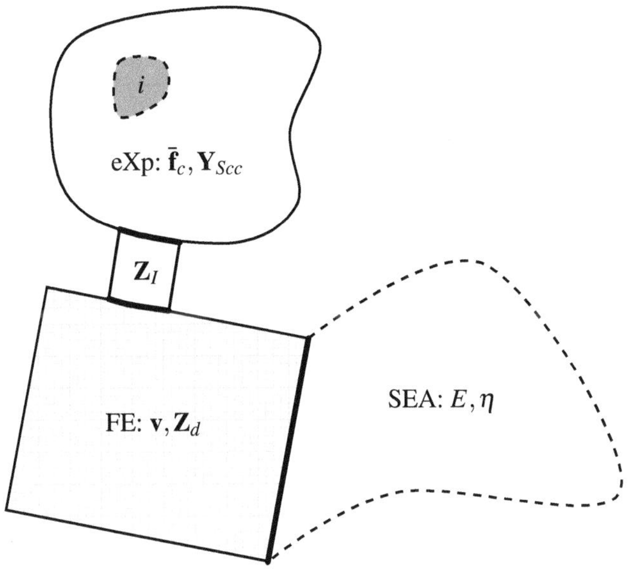

Figure 5.9 Example of Hybrid FE-SEA-X model. An eXperimental source component (characterised by its free interface FRF matrix and blocked force) is coupled to a FE component (characterised by its mass, stiffness and damping matrices, or by its numerical impedance FRF), itself coupled to an SEA component (characterised by its material properties such as Young's modulus and loss factor).

Statistical Energy Analysis

In SEA a complex system is subdivided into a collection of subsystems, each of which is characterised by a single quantity, its vibrational (or acoustical) energy. The theory of SEA is well developed, and detailed expositions are available in numerous books and articles [4, 16, 19, 20, 26, 46], to which the reader is referred for more detail.

In accordance with the conservation of energy, the sum of all power flows within a steady state system should equal 0. With reference to Figure 5.10, for a subsystem i, a power balance of this form can be written as,

$$P_{i,in} - P_{i,diss} + \sum_{j}^{N} \overbrace{P_{i,j} - P_{j,i}}^{\Pi_{i,j}} = 0 \tag{5.103}$$

where $P_{i,in}$ is the power input directly to the ith subsystem, $P_{i,diss}$ is the power dissipated internally within the ith subsystem, e.g. due to damping or friction, $P_{i,j}$ represents the power delivered to subsystem i from j, with $P_{j,i}$ being the power lost from

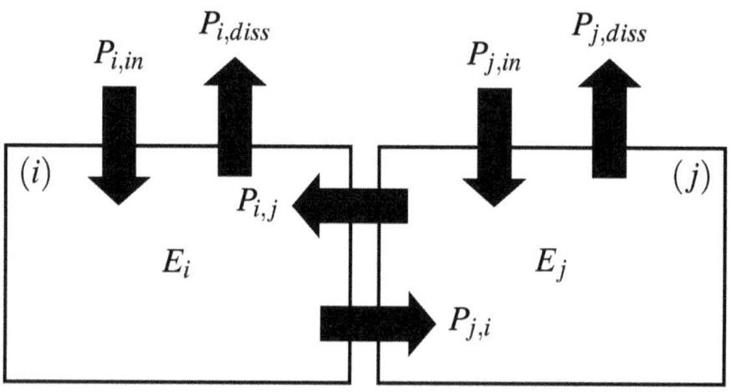

Figure 5.10 Simple SEA model of two connected subsystems.

from i to j. The summation accounts for the power transfer between N connected subsystems. The net power flow between subsystem i and j is denoted by $\Pi_{i,j}$.

If the subsystems are linearly coupled, then this power flow is directly proportional to the difference in subsystem energies and is given by,

$$\Pi_{i,j} = \omega\left(\eta_{i,j}E_i - \eta_{j,i}E_j\right) \tag{5.104}$$

where E_i and E_j are the subsystem energies, and $\eta_{i,j}$ is the so-called coupling loss factor.

Similarly, the dissipated power can be related to the subsystem energy by,

$$P_{i,diss} = \omega\eta_i E_i \tag{5.105}$$

where η_i is the subsystem damping loss factor. With the above, the power balance of subsystem i can be rewritten in the form,

$$P_{i,in} = \omega\eta_i E_i + \omega\sum_j\left(\eta_{i,j}E_i - \eta_{j,i}E_j\right) \tag{5.106}$$

Writing a power balance of this form for each subsystem yields a set of simultaneous equations,

$$\omega\begin{bmatrix} \eta_1 + \sum_{i\neq 1}\eta_{1,i} & -\eta_{2,1} & \cdots & -\eta_{N,1} \\ -\eta_{2,1} & \eta_2 + \sum_{i\neq 2}\eta_{2,i} & & -\eta_{N,2} \\ \vdots & & \ddots & \vdots \\ -\eta_{1,N} & -\eta_{2,N} & \cdots & \eta_N + \sum_{i\neq N}\eta_{N,i} \end{bmatrix}\begin{pmatrix} E_1 \\ E_2 \\ \vdots \\ E_N \end{pmatrix} = \begin{pmatrix} P_{1,in} \\ P_{2,in} \\ \vdots \\ P_{N,in} \end{pmatrix} \tag{5.107}$$

or equivalently in terms of modal energies $\hat{E}_i = E_i/n_i$,

$$\omega\begin{bmatrix} n_1\left(\eta_1 + \sum_{i\neq 1}\eta_{1,i}\right) & \cdots & -n_N\eta_{N,1} \\ \vdots & \ddots & \vdots \\ -n_1\eta_{1,N} & \cdots & n_N\left(\eta_N + \sum_{i\neq N}\eta_{N,i}\right) \end{bmatrix}\begin{pmatrix} \frac{E_1}{n_1} \\ \frac{E_2}{n_2} \\ \vdots \\ \frac{E_N}{n_N} \end{pmatrix} = \begin{pmatrix} P_{1,in} \\ P_{2,in} \\ \vdots \\ P_{N,in} \end{pmatrix} \tag{5.108}$$

where n_i is the modal density of subsystem i. This latter form is often preferred, as it yields a symmetric matrix of loss factors, since,

$$\frac{\eta_{i,j}}{n_j} = \frac{\eta_{j,i}}{n_i}. \tag{5.109}$$

Whatever form is chosen, expressed in matrix form we obtain,

$$\omega \mathbf{C}\mathbf{E} = \mathbf{P} \tag{5.110}$$

where $\mathbf{C} \in \mathbb{R}^{N\times N}$ is a matrix of loss factors, $\mathbf{E} \in \mathbb{R}^{N}$ is a vector of subsystem (modal) energies, and $\mathbf{P} \in \mathbb{R}^{N}$ is a vector of input powers. Solving an SEA system involves determining the subsystem energies given a specified input power,

$$\mathbf{E} = \frac{1}{\omega}\mathbf{C}^{-1}\mathbf{P}. \tag{5.111}$$

The obtained energies can subsequently be related to subsystem specific quantities, for example the mean structural velocity or sound pressure level.

This SEA approach to analysing complex systems can greatly simplify the modelling process, whilst retaining a ‘feel’ for the physics of the problem. A complex model with many thousands of DoFs can be replaced by a relatively simple power balance between just a few subsystems. All that is required are a few subsystem properties, including the modal density and loss factor, and the coupling loss factors that describe the separating junctions. For simple junctions, these coupling loss factors are available directly from the literature [26] or can be derived analytically [23]. For more complex junctions they can be determined from FE models [39, 41], or obtained experimentally [3, 7].

The SEA approach, as described above, is built on two important assumptions: 1) the subsystems are weakly coupled and 2) the vibration field within the analysis bandwidth constitutes a good approximation of a diffuse field. These assumptions are largely satisfied in the analysis of high frequency (short wavelength) vibration. At low frequencies (long wavelengths), low modal density means that the statistical assumptions are not satisfied and significant errors can result.

Hybrid FE-SEA-X

Finite Elements and Statistical Energy Analysis are suitable, respectively, for the analysis of long and short wavelength (sub)systems. However, the substructures that comprise complex engineering structures often have a wide range of wavelengths. In this case neither FE nor SEA are appropriate to model the complete system. Based on a diffuse field reciprocity relation [21, 37], a general wave-based approach for coupling both theories into a single model has been proposed by Shorter and Langley [38]. Termed the ‘Hybrid FE-SEA’ method, this approach has been validated both numerically and experimentally [9], and has proven itself as a wide-band vibro-acoustic prediction tool. It has been further extended to predict the variance of a response due to both non-parametric (i.e. arising from SEA subsystems) [22] and parametric (i.e. arising from FE components) uncertainties [5, 6] .

The hybrid FE-SEA method as described above provides a powerful tool for modelling complex, multi-scale, structures. Nevertheless, there exist notable barriers that restrict its more general application. To achieve an accurate response prediction the assembled model must 1) correctly describe the assembly's passive properties (i.e. its response to a unit excitation) and 2) correctly model the excitation induced by its active components. Whilst the hybrid FE-SEA method is able to predict the passive properties of complex structures with impressive accuracy (including descriptions of uncertainty), the mechanisms that lead to vibratory excitation lay outside its capabilities. They are however, amenable to characterisation using the blocked force. The use of experimental data (both active and passive) to represent complex subsystems that cannot be modelled directly was proposed in [8], where complex vibration sources were characterised by their experimental dynamic stiffness and blocked force [32], and coupled to a statistical subsystem as part of a hybrid SEA-eXperimental model.

A diagrammatic example of a hybrid FE-SEA-X model is shown in Figure 5.9. A vibration source S and accompanying resilient element I (characterised experimentally) are coupled to an FE subsystem, which itself is connected to an SEA subsystem. This simple example exemplifies many practical engineering structures. As an automotive example, we might consider an engine (experimental) resiliently coupled to a subframe (FE). Connected to the subframe are a variety of body panels (SEA) which themselves might radiated into the cabin (SEA). As a building acoustics example, some mechanical plant (experimental) might be attached to the building structure (FEA), which surrounds and radiates into a number of acoustic spaces (SEA).

A broad class of vibro-acoustic problems can, in principle, be tackled using the hybrid FE-SEA-X method. In what follows we present the key underlying theory and accompanying equations. The reader is referred to the many published works on the hybrid FE-SEA method for further details.

In the literature, the hybrid method is typically described in terms of displacement and dynamic stiffness ($\mathbf{x}$ and $\mathbf{D}$), but for consistency with the notation adopted in this book, we introduce the hybrid method in terms of velocity and impedance ($\mathbf{v} = i\omega\mathbf{x}$ and $\mathbf{Z} = \mathbf{D}/i\omega$).

The hybrid FE-SEA method requires each component of the vibro-acoustic system under study to be identified as either deterministic or statistical. The deterministic part is represented by a finite set of DoFs $\mathbf{v}$, and the statistical part by a set of subsystems, each of which possesses a single DoF, their vibrational (or acoustical) energy E.

Treatment of the deterministic part follows standard numerical procedures, e.g. by Finite Element Analysis. For a statistical subsystem, we consider the contained wave field as a combination of two separate fields: the response due to the initially generated waves (direct field), and the contribution from all the waves reflected by the subsystem's boundaries (reverberant field). These boundaries are considered unknown or uncertain, and so the reverberant field is considered random. Based on this separation a direct field impedance matrix $\mathbf{Z}_{dir}$ is defined for each statistical subsystem. This is the impedance of the subsystem in the absence of any reflections, i.e. the infinite subsystem response. Analytical solutions for $\mathbf{Z}_{dir}$ are available for many simple cases (e.g. plates, cavities, etc.). Alternatively, a boundary element analysis can be used for complex structures.

The contribution of the kth statistical subsystem to the system response $\mathbf{v}$ is included by first adding its direct field impedance to that of the deterministic stiffness matrix $\mathbf{Z}_d$, and then applying an appropriate force $\mathbf{f}_{rev}^{(k)}$, arising due to the reverberant field in the statistical subsystem. The governing equations of motion are then,

$$\mathbf{Z}_{tot}\mathbf{v} = \mathbf{f}_d + \sum_k \mathbf{f}_{rev}^{(k)} \tag{5.112}$$

where $\mathbf{f}_d$ represents the forcing applied directly to the deterministic part, and

$$\mathbf{Z}_{tot} = \mathbf{Z}_d + \sum_k \mathbf{Z}_{dir}^{(k)} \tag{5.113}$$

is the total impedance matrix. This assemblage of subsystems is equivalent to the primal formulation presented in 5.2.2, and can equivalently be expressed as,

$$\mathbf{Z}_{tot} = \mathbf{L}^{\mathrm{T}}\left[\mathbf{Z}\right]\mathbf{L} \tag{5.114}$$

with the block diagonal impedance matrix $[\mathbf{Z}]$. As written in Equation 5.113, each impedance matrix must be zero padded to ensure consistent dimensions and that connecting DoFs align.

In the presence of j experimental subsystems (i.e. an FE-SEA-X model) the deterministic impedance matrix $\mathbf{Z}_d$ is modified as so,

$$\mathbf{Z}_d = \mathbf{Z}_{d,FE} + \sum_j \mathbf{Z}_{d,exp}^{(j)} \tag{5.115}$$

where $\mathbf{Z}_{d,FE}$ and $\mathbf{Z}_{d,exp}^{(j)}$ represent the impedance matrices of the numerical part and the jth experimental subsystem, respectively. In the case that the experimental subsystems are active (i.e. they operate as vibration sources) their activity must also be accounted for. To do so the deterministic force vector $\mathbf{f}_d$ is modified as so,

$$\mathbf{f}_d = \mathbf{f}_{ext} + \sum_j \bar{\mathbf{f}}_{exp}^{(j)} \tag{5.116}$$

where $\mathbf{f}_{ext}$ represents an external forcing, and $\bar{\mathbf{f}}_{exp}^{(j)}$ the *blocked force* of the jth active experimental subsystem. If the jth experimental subsystem is purely passive, then $\bar{\mathbf{f}}_{exp}^{(j)} = \mathbf{0}$.

To obtain a Hybrid FE-SEA-X model the above equations are not enough. The energetic properties of a statistical subsystem must be related to its reverberant force and direct field impedance. This key relation was derived by Shorter and Langley [21, 37] and is known as diffuse field reciprocity. It is an identity that relates the cross-spectral matrix of a subsystem's reverberant force, denoted as $\mathbf{S}_{ff}^{(k),rev} = \mathrm{E}[\mathbf{f}_{rev}^{(k)}\mathbf{f}_{rev}^{(k)*T}]$, where $\mathrm{E}[\square]$ denotes the ensemble average, to its energy E_k and direct field impedance

$\mathbf{Z}_{dir}$. This relationship, valid when the ensemble response constitutes a diffuse wave field [21], can be expressed as

$$\mathbf{S}_{ff}^{(k),rev} = \frac{4a_k E_k}{\pi n_k} \mathrm{Re}\{\mathbf{Z}_{dir}^{(k)}\} \tag{5.117}$$

where E_k and n_k are, respectively, the ensemble averaged vibrational energy and modal density of the subsystem, and a_k is a term to take into account local concentrations in the wave field [21, 22].

Converting Equation 5.112 into a quadratic form such that $\mathbf{S}_{vv} = \mathrm{E}[\mathbf{v}\mathbf{v}^{*T}]$, and substituting the diffuse field reciprocity relation (Equation 5.117) leads to,

$$\mathbf{S}_{vv} = \mathbf{Y}_{tot} \left[\mathbf{S}_{ff} + \sum_k \frac{4a_k E_k}{\pi n_k} \mathrm{Re}\{\mathbf{Z}_{dir}^{(k)}\} \right] \mathbf{Y}_{tot}^{*T} \tag{5.118}$$

where $\mathbf{Y}_{tot} = \mathbf{Z}_{tot}^{-1}$ and in the presence of active experimental subsystems,

$$\mathbf{S}_{ff} = \mathbf{S}_{ff}^{ext} + \sum_j \mathbf{S}_{\bar{f}\bar{f}}^{(j),exp}. \tag{5.119}$$

In the above $\mathbf{S}_{vv}$ represents the cross-spectral matrix of the system response $\mathbf{v}$, where $\mathbf{S}_{ff}^{ext}$ and $\mathbf{S}_{\bar{f}\bar{f}}^{(j),exp}$ are the cross-spectral matrices corresponding to the external force and the blocked force of the jth experimental subsystem, respectively.

To solve Equation 5.118 the energy E_k of each subsystem must first be determined. This is done by solving a power balance equation of the form [8],

$$\omega\left(\eta_j + \eta_{d,j}\right) E_j + \sum_k \omega \eta_{jk} n_j \left(\frac{E_j}{n_j} - \frac{E_k}{n_k} \right) = P_j + P_{in,j}^{ext} \tag{5.120}$$

where P_j and $P_{in,j}^{ext}$ are the power inputs from external sources, applied on the subsystem and deterministic part, repetitively, η_j is the loss factor of subsystem j, η_{jk} is the coupling loss factor between subsystems k and j, and $\eta_{d,j}$ is the loss factor associated with the deterministic system. These terms can be related to the total and direct field impedance matrices as so:

$$\omega \eta_{d,j} = -\frac{2a_j}{\pi n_j} \sum_{r,s} \mathrm{Re}\{\mathbf{Z}_d\}_{r,s} \left(\mathbf{Y}_{tot} \mathrm{Re}\{\mathbf{Z}_{dir}^{(j)}\} \mathbf{Y}_{tot}^{*T} \right)_{r,s} \tag{5.121}$$

$$\omega \eta_{jk} n_j = -\frac{2a_j}{\pi} \sum_{r,s} \mathrm{Re}\left\{\mathbf{Z}_{dir}^{(j)}\right\}_{r,s} \left(\mathbf{Y}_{tot} \mathrm{Re}\{\mathbf{Z}_{dir}^{(j)}\} \mathbf{Y}_{tot}^{*\mathrm{T}} \right)_{r,s} \tag{5.122}$$

and

$$P_{in,j}^{ext} = -\frac{1}{2} \sum_{r,s} \mathrm{Re}\left\{\mathbf{Z}_{dir}^{(j)}\right\}_{r,s} \left(\mathbf{Y}_{tot} \mathbf{S}_{ff} \mathbf{Y}_{tot}^{*\mathrm{T}} \right)_{r,s}. \tag{5.123}$$

To solve the above, Equation 5.120 is rewritten in matrix form,

$$\mathbf{C}_0 \hat{\mathbf{E}} = \mathbf{P} + \mathbf{P}_{in}^{ext} \tag{5.124}$$

where $\hat{\mathbf{E}}$ is a vector of ensemble averaged modal energies such that $\hat{E}_k = E_k/n_k$, $\mathbf{P}$ and $\mathbf{P}_{in}^{ext}$ are vectors of input powers due to direct excitation of the statistical component and external forces (including experimental blocked forces) applied to the deterministic part, respectively, and $\mathbf{C}_0$ is a matrix of coupling loss factors (obtained by assembling Equations 5.121 and 5.122. On solving Equation 5.124, one obtains the energetic response of the statistical subsystems. Substituting the modal energies into Equation 5.118 yields the response of the deterministic system.

Equations 5.118 and 5.124 let us predict the vibro-acoustic response of complex built-up structures, possessing both active and passive subsystems, with long and/or short wavelengths, represented numerically, statistically, and/or experimentally. Equations are also available that provide estimates for the variance of both subsystem energies and assembly cross-spectra due to parametric (i.e. FE) [5, 6] and non-parametric (i.e. SEA) [22] uncertainty. The reader is referred to [8] and references within for further details.

Example 5.2: Hybrid SEA-X model

In this example we show the results obtained from a simple hybrid SEA-X model. Full details of the example can be found in [8]. The assembly considered is shown in Figure 5.11. It consists of a 4 footed vibration source connected to a large aluminium plate ($1 \times 0.8 \times 0.003$m) through 4 resilient elements.

Figure 5.11 Photo of the hybrid SEA-X test assembly. The receiver plate, to modelled as an SEA subsystem, is randomised using small point masses to create an ensemble of 20 nominally identical plates.

To model this assembly, the vibration source is characterised in the usual way by its free-interface FRF matrix (see 2.3.1) and blocked force (obtained using in situ method, see 3.2). The resilient elements are characterised by their dynamic transfer stiffness, obtained using the in situ method (see 2.3.2). The receiver plate has been designed to be representative of a typical short-wave length subsystem, displaying a high sensitivity to perturbations in its geometry/material properties. For this reason, it is best treated as an SEA subsystem. As an SEA subsystem, the receiver plate is characterised by its material properties (density 2700 kg/m^3, Young's modulus 70 Gpa, Poisson ratio 0.33, and loss factor 2.3%) and thickness (3 mm).

With the receiver being modelled as an SEA subsystem, the purpose of our SEA-X model is not to predict the precise response at a single position for a single receiver, but rather to predict the average energy or velocity magnitude squared across an ensemble of nominally identical receivers. To provide a validation of this result, an ensemble of 20 receivers was generated artificially by randomly positioning a series of small point masses across its surface (see small metal disks in Figure 5.11). The total mass added was 650 g, approximately 10% of the initial mass of the plate. From this ensemble we can obtain averaged energy and velocity results using the 6 accelerometers attached to the plate.

The loss factor required for the receiver's SEA description was obtained experimentally from its response to a point force without the source and isolators attached. Shown in Figure 5.12a) in red is the ensemble average energy response of the plate considering a single point force. This average is obtained from 20 individual ensemble members, the energies of which are shown in grey. In blue is the response obtained by modelling the plate as an SEA subsystem. Also shown are the upper 95% and 99% confidence bounds on the energy. In Figure 5.12b) the relative variance of the energy is shown for the physical ensemble (red) and the SEA model.

We can see from Figure 5.12a) that the ensemble average and SEA result are in good agreement above approx 200 Hz. In this region, the added masses cause significant shifts in the natural frequencies such that they begin to overlap. When averaged this overlap smooths the response, tending towards that of an infinite plate, represented by the SEA result in blue. Below 200 Hz, the behaviour of the plate is insensitive to the added masses, and distinct modes can be observed. It is in this frequency range that a standard component-based TPA approach (using a measured or FE modelled receiver) would be more appropriate.

Satisfied that the receiver plate can be adequately represented as an SEA subsystem, we can now build an SEA-X model of the assembly. With no FE components, the deterministic impedance matrix $\mathbf{Z}_d$ is obtained by summing the free-interface source impedance $\mathbf{Z}_S = \mathbf{Y}_S^{-1}$ and the combined isolator impedance matrix $\mathbf{Z}_I$,

$$\mathbf{Z}_d = \begin{bmatrix} \mathbf{Z}_S & \mathbf{0} \\ \mathbf{0} & \mathbf{0} \end{bmatrix} + \begin{bmatrix} \mathbf{Z}_I & -\mathbf{Z}_I \\ -\mathbf{Z}_I & \mathbf{Z}_I \end{bmatrix}. \tag{5.125}$$

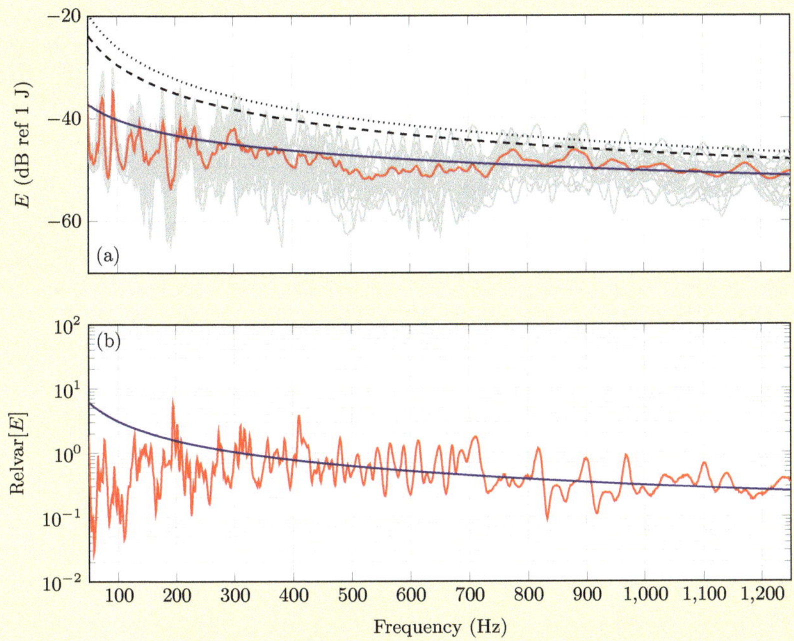

Figure 5.12 Average (a) and relative variance (b) of the plate energy due to a point force. Grey – energy of ensemble members, red – ensemble average, blue – SEA prediction, dashed/dotted black – 95/99% confidence bounds.

For the direct field impedance matrix of the plate, we compute the analytical response of a thin infinite plate using,

$$Y_{Rij} = Y_R(r_{ij}) = \frac{H_0^{(2)}(k_b r_{ij}) - 2i/\pi K_0(k_b r_{ij})}{8i\omega\sqrt{D\rho h}} \tag{5.126}$$

where r_{ij} is the distance between the ith and jth connections points, $H_0^{(2)}$ is the Hankel function of the second kind and zeroth order, K_0 is the modified Bessel function of the second kind and zeroth order, D is the flexural rigidity of the plate, ρ is its mass density, h is its thickness and $k_b = (\rho h\omega^2/D)^{1/4}$ is the plate bending wave number. Equation 5.126 can be used to build the direct field impedance matrix as,

$$\mathbf{Z}_{dir} = \begin{bmatrix} \mathbf{0} & \mathbf{0} \\ \mathbf{0} & \mathbf{Y}_R^{-1} \end{bmatrix}. \tag{5.127}$$

Finally, the deterministic force vector is built from the blocked force,

$$\mathbf{f}_d = \begin{pmatrix} \bar{\mathbf{f}}_c \\ \mathbf{0} \end{pmatrix}. \tag{5.128}$$

Using the above, Equation 5.121-5.122 are computed and used to solve Equation 5.124 for the modal energy E. This energy can then be substituted into Equation 5.118 to compute the cross-spectral response of the assembly. This result is shown in Figure 5.13 for the SEA-X assembly considered here.

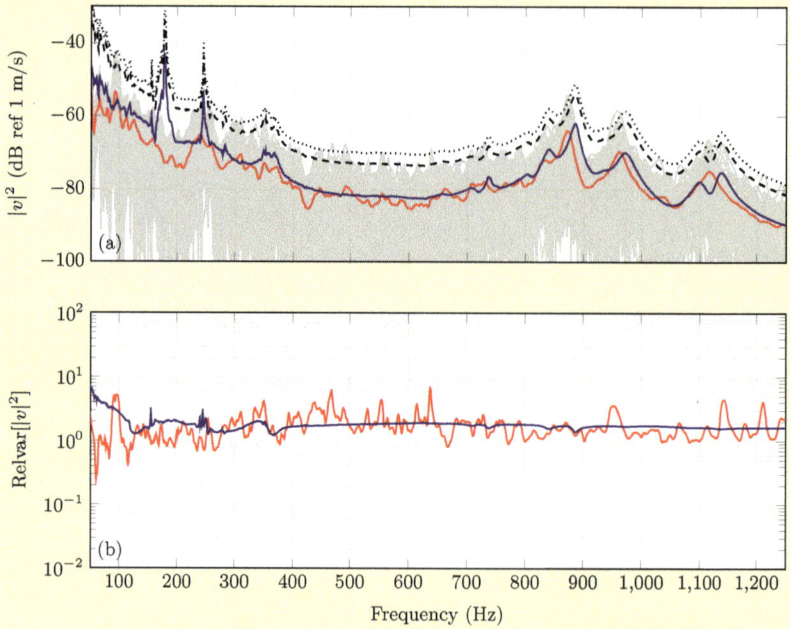

Figure 5.13 Magnitude squared velocity response of the assembly (a) and its relative variance (b). Grey – magnitude squared velocity response of individual ensemble members, red – ensemble average, blue – SEA-X prediction, dashed/dotted black – 95/99% confidence bounds.

As before, we use an artificially generated ensemble to provide validation. The individual members of this ensemble are shown in grey in Figure 5.13a).

The average of this ensemble is shown in red. The hybrid SEA-X prediction is in blue. 95/99% confidence bounds are again shown by dashed/dotted lines. Figure 5.13b) shows this uncertainty estimate as a relative variance.

We see that the SEA-X results are in good agreement with the experimental ensemble. The widely fluctuating plate behaviour, illustrated by the spread across the ensemble in grey, is reduced to a smoothly varying mean response, whilst the deterministic features of the source are clearly evident by the pronounced resonant peaks above approx 800 Hz. We further see that the confidence bounds accurately capture the bounds of the ensemble response.

Note that if a standard CB-TPA simulation were performed, i.e. by treating the receiver as a deterministic component, one would obtain an estimate for just one of the grey ensemble members. Clearly, if the assembly being considered is subject to any manufacturing uncertainty/tolerances, this single member estimate is of limited use. In this case a hybrid SEA-X model might provide more worthwhile, as it captures the average response across an ensemble, including uncertainty bounds. Of course, Finite Element components can be readily incorporated alongside experimental ones within this hybrid framework.

5.5 SUMMARY

The aim of this chapter was to present methods for coupling together the components of a VAVP to form an *assembly*. In addition, we have introduced TPA as a class of diagnostic, and now predictive, methods that can be used to identify trouble noise and vibration issues in existing or entirely virtual assemblies.

First, the dynamic substructuring equations for assembling components are presented: the *direct* method followed by *three field formulations* known as *primal* and *dual*. Connection of components with and without resilient connectors is considered in all cases. The direct method is perhaps simpler to follow but the three field formulations are more easily extended to larger systems. The equations for substructure *decoupling* are also provided. This is a variation on the coupling theme and is relevant to characterisation of components by allowing the dynamic influence of attached structures, such as a test bench, to be removed. Another variation is *component modification* which allows components or assemblies to be modified, for example by adding mass or stiffness.

We then move on to discuss the class of methods known as Transfer Path Analysis (TPA), a number of variations of which exist. One grouping of these methods, comprising *classical* and *mount stiffness TPA* are based on indirectly measured contact force at the interface. Transmissibility-based methods work with measured responses. The aim is to quantify the contributions of given connections or paths to the overall response of the assembly. These variants of TPA are restricted to diagnostic applications since the source properties are not transferable. A comparatively new variant, blocked force TPA, however, is based the invariant blocked force. This, as well as serving a similar diagnostic purpose, opens up significant new possibilities for prediction. Indeed, the blocked forces can be transferred into a VAVP and combined with component-level data with complete flexibility. This variant is often known as 'component-based TPA'.

The last part of the chapter describes an extension of the component-based framework to include components defined only by statistical properties. The conventional VAVP framework allows measured and FE modelled components to be combined, provided of course that the interfaces and frequency resolution are compatible. However, at high frequency an FE model is not appropriate for components with short wavelengths and a statistical energy analysis (SEA) description is more likely to be successful. A hybrid framework has been presented within which FEM and SEA modelled components can be combined with experimental components. As yet, this approach does not allow auralisation, but hybrid modelling of this sort seems likely to play an important role in the future.

Bibliography

1. ISO 20270:2019 Acoustics — characterization of sources of structure-borne sound and vibration — indirect measurement of blocked forces. 2019.
2. MS Allen, DJ Rixen, M Van der Seijs, P Tiso, T Abrahamsson, and RL Mayes. *Substructuring in engineering dynamics*. Springer, 2020.
3. DA Bies and S Hamid. In situ determination of loss and coupling loss factors by the power injection method. *Journal of Sound and Vibration*, 70(2):187–204, 1980.
4. CB Burroughs, RW Fischer, and FR Kern. An introduction to statistical energy analysis. *The Journal of the Acoustical Society of America*, 101(4):1779–1789, 1997.
5. A Cicirello and RS Langley. The vibro-acoustic analysis of built-up systems using a hybrid method with parametric and non-parametric uncertainties. *Journal of Sound and Vibration*, 332(9):2165–2178, 2013.
6. A Cicirello and RS Langley. Efficient parametric uncertainty analysis within the hybrid finite element/statistical energy analysis method. *Journal of Sound and Vibration*, 333(6):1698–1717, 2014.
7. BL Clarkson and MF Ranky. On the measurement of the coupling loss factor of structural connections. *Journal of Sound and Vibration*, 94(2):249–261, 1984.
8. A Clot, JWR Meggitt, RS Langley, AS Elliott, and AT Moorhouse. Development of a hybrid fe-sea-experimental model. *Journal of Sound and Vibration*, 452:112–131, 2019.
9. V Cotoni, P Shorter, and RS Langley. Numerical and experimental validation of a hybrid finite element-statistical energy analysis method. *The Journal of the Acoustical Society of America*, 122(1):259–270, 2007.
10. D de Klerk and A Ossipov. Operational Transfer Path Analysis: theory, guidelines and tire noise application. *Mechanical Systems and Signal Processing*, 24(7):1950–1962, 2010.
11. D de Klerk, DJ Rixen, and SN Voormeeren. General framework for dynamic substructuring: history, review and classification of techniques. *AIAA journal*, 46(5):1169–1181, 2008.
12. M Di Manno, J Brunetti, W D'Ambrogio, and A Fregolent. A critical evaluation of semm-based joint identification procedure to reduce the error propagation effects. *Mechanical Systems and Signal Processing*, 208:111063, 2024.
13. W D'Ambrogio and A Fregolent. Direct decoupling of substructures using primal and dual formulation. In *Linking Models and Experiments, Volume 2: Proceedings of the 29th IMAC, A Conference on Structural Dynamics, 2011*, pages 47–76. Springer, 2011.
14. AS Elliott, N Sanei, and M Glesser. Combining structural modification with in-situ Transfer Path Analysis to solve noise and vibration problems. In *Proceedings of the 23rd International Congress on Acoustics*, 2019.

15. AS Elliott, AT Moorhouse, T Huntley, and S Tate. In-situ source path contribution analysis of structure borne road noise. *Journal of Sound and Vibration*, 332(24):6276–6295, 2013.
16. FJ Fahy. Statistical energy analysis: a critical overview. *Philosophical Transactions of the Royal Society of London. Series A: Physical and Engineering Sciences*, 346(1681):431–447, 1994.
17. P Gajdatsy, K Janssens, W Desmet, and H Van Der Auweraer. Application of the transmissibility concept in Transfer Path Analysis. *Mechanical Systems and Signal Processing*, 24(7):1963–1976, 2010.
18. O Guasch, C García, J Jové, and P Artís. Experimental validation of the direct transmissibility approach to classical Transfer Path Analysis on a mechanical setup. *Mechanical Systems and Signal Processing*, 37(1-2):353–369, 2013.
19. KH Heron. Advanced statistical energy analysis. *Philosophical Transactions of the Royal Society of London. Series A: Physical and Engineering Sciences*, 346(1681):501–510, 1994.
20. AJ Keane and WG Price. Statistical energy analysis: an overview, with applications in structural dynamics. Cambridge University Press. 1997.
21. RS Langley. On the diffuse field reciprocity relationship and vibrational energy variance in a random subsystem at high frequencies. *The Journal of the Acoustical Society of America*, 121(2):913–921, 2007.
22. RS Langley and V Cotoni. Response variance prediction for uncertain vibro-acoustic systems using a hybrid deterministic-statistical method. *The Journal of the Acoustical Society of America*, 122(6):3445–3463, 2007.
23. RS Langley. A derivation of the coupling loss factors used in statistical energy analysis. *Journal of Sound and Vibration*, 141(2):207–219, 1990.
24. D Lennström, M Olsson, F Wullens, and A Nykänen. Validation of the blocked force method for various boundary conditions for automotive source characterization. *Applied Acoustics*, 102:108–119, 2016.
25. M Lohrmann. Operational Transfer Path Analysis: comparison with conventional methods. In *International Congress on Sound and Vibration*, pages 2671–2676, 2008.
26. RH Lyon. *Theory and Application of Statistical Energy Analysis*. Butterworth-Heinemann, 2014.
27. FX Magrans, PV Rodriguez, and GC Cousin. Low and mid-high frequency advanced transmission path analysis. In *Proceedings of the Twelfth International Congress on Sound and Vibration ICSV12*, 2005.
28. A El Mahmoudi, DJ Rixen, and CH Meyer. Comparison of different approaches to include connection elements into frequency-based substructuring. *Experimental Techniques*, 44(4):425–433, 2020.
29. JWR Meggitt, AS Elliott, and AT Moorhouse. Virtual assemblies and their use in the prediction of vibro-acoustic responses. In *Proceedings of the Institute of Acoustics, Warickshire*, 2016.
30. JWR Meggitt, AS Elliott, AT Moorhouse, A Jalibert, and G Franks. Component replacement tpa: A transmissibility-based structural modification method for in-situ Transfer Path Analysis. *Journal of Sound and Vibration*, 499:115991, 2021.
31. JWR Meggitt. *On in-situ methodologies for the characterisation and simulation of vibro-acoustic assemblies*. PhD thesis, University of Salford, 2017.
32. AT Moorhouse, AS Elliott, and TA Evans. In situ measurement of the blocked force of structure-borne sound sources. *Journal of Sound and Vibration*, 325(4-5):679–685, 2009.
33. Daniel J Rixen. How measurement inaccuracies induce spurious peaks in frequency based substructuring. In *International modal analysis conference IMAC-XXVI*, pages 1–15, 2008.
34. Zeeshan Saeed, Christian M Firrone, and Teresa M Berruti. Joint identification through hybrid models improved by correlations. *Journal of Sound and Vibration*, 494:115889, 2021.
35. Zeeshan Saeed, Steven WB Klaassen, Christian M Firrone, Teresa M Berruti, and Daniel J Rixen. Experimental joint identification using system equivalent model mixing in a bladed disk. *Journal of Vibration and Acoustics*, 142(5):051001, 2020.
36. Taejin Shin, Yeon Soo Kim, Kanghyun An, and Sang Kwon Lee. Transfer path analysis of rumbling noise in a passenger car based on in-situ blocked force measurement. *Applied Acoustics*, 149:1–14, 2019.
37. Phil J Shorter and Robin S Langley. On the reciprocity relationship between direct field radiation and diffuse reverberant loading. *The Journal of the Acoustical Society of America*, 117(1):85–95, 2005.
38. Phil J Shorter and Robin S Langley. Vibro-acoustic analysis of complex systems. *Journal of Sound and Vibration*, 288(3):669–699, 2005.

39. C Simmons. Structure-borne sound transmission through plate junctions and estimates of sea coupling loss factors using the finite element method. *Journal of Sound and Vibration*, 144(2):215–227, 1991.
40. JM Starkey, GL Merrill, et al. On the ill-conditioned nature of indirect force-measurement techniques. *Journal of Modal Analysis*, 4(3):103–108, 1989.
41. AN Thite and BR Mace. Robust estimation of coupling loss factors from finite element analysis. *Journal of Sound and Vibration*, 303(3-5):814–831, 2007.
42. Maarten van der Seijs. *Experimental dynamic substructuring: Analysis and design strategies for vehicle development*. PhD thesis, TU Delft, 2016.
43. Maarten V van der Seijs, Dennis de Klerk, and Daniel J Rixen. General framework for Transfer Path Analysis: History, theory and classification of techniques. *Mechanical Systems and Signal Processing*, 68:217–244, 2016.
44. Jan W Verheij. Experimental procedures for quantifying sound paths to the interior of road vehicles. In *Technical papers 2nd international conference vehicle comfort: ergonomic, vibrational, noise and thermal spects: October 14-16, 1992, Palazzo della Cultura e dei Congressi Bologna, Italy*, pages 483–491, 1992.
45. SN Voormeeren and DJ Rixen. A family of substructure decoupling techniques based on a dual assembly approach. *Mechanical Systems and Signal Processing*, 27:379–396, 2012.
46. J Woodhouse. An approach to the theoretical background of statistical energy analysis applied to structural vibration. *The Journal of the Acoustical Society of America*, 69(6):1695–1709, 1981.

Part 2: Practical Implementation

6 Measurement

6.1 INTRODUCTION

We now move on to Part 2 of the book where we consider how to implement the theoretical descriptions from Part 1 into a practical component-based simulation, or VAVP. In this chapter we consider *Measurement*, followed by *Processing* and *Uncertainties* in Chapters 8 and 9. Chapter 10 then provides a walk-through example, pulling together all the various elements covered so far.

Considering *measurement*, we described *what* and *where* to measure in the characterisation of VAVP components in Chapter 2. These topics will now be looked at in a little more detail in 6.2. The main aim of *this* chapter however, is to consider the question of '*how*' to conduct the necessary measurements. While the measurement procedures themselves are relatively straightforward, acquiring a reliable representation of a component as a whole will turn out to be far more challenging. Actually *obtaining* the data then is only a small part of 'measurement' for a VAVP; a more significant aspect is the *evaluation* of that data's suitability to represent a component in a VAVP. To this end the main focus of the chapter is in describing a range of checks and validation procedures for this purpose, both at the level of *individual measurements* (6.3) and at *component-level* (6.4).

At the level of individual measurements, even a single noisy or inaccurate data point can have a disastrous effect on a subsequent inverse solution, so careful checking is required. *Coherence* checks will help to identify unreliable data and are considered essential for every measurement. For multi-point measurements, additional checks are advised on the consistency and the physicality of the data set as a whole, including *reciprocity* and *passivity*. *Condition number* is also introduced since it provides an important measure of the *sensitivity* of a set of FRFs to subsequent inversion errors.

Even the best measurements though will not yield good results if, for example, sensors are located in the wrong place or if insufficient DoFs are included. In short, the previous checks tell us if the measurements are right, but not if they are the right measurements! To obtain the right measurements we need to consider the whole campaign at the *component level*. Recall that components are characterised almost exclusively at their interfaces; accurately representing the component's dynamics with the *interface model* is arguably the most important and difficult stage in building a VAVP. In 6.4, we describe a range of procedures and checks that have evolved to assess the reliability of this model. These are of two types: *interface* model tests and *validation* checks. A measurement campaign will typically involve application of interface checks followed by several stages of validation tests. In 6.4.2 we describe how a carefully staged validation programme can help to separate out and minimise errors from different causes.

DOI: 10.1201/9781003399131-6

Finally, our treatment of measurement would not be complete without some consideration of measurement hardware and instrumentation which is provided in 6.5. Readers without a particular need may wish to skip this section.

6.2 PROBLEM DEFINITION

Before setting up for a measurement, it is of course necessary to know *what* needs to be measured and *where* to place sensors. Recall from Chapter 2 that characterisation of *receiver* components requires matrices of FRFs at their interfaces and at any target locations. In the simplest cases, these may be measured directly with the component freely suspended as shown in Figure 6.1a. For *source* components on the other hand, a full characterisation requires blocked forces at the interface DoFs for which *in situ* measurements are required (Figure 6.1b). The *in situ* setup may also be required for receiver components if free conditions cannot be adequately met or if validation is required. The *in situ* test arrangement is therefore more complicated but more universal in its application and for this reason it forms the main focus of this chapter.

The set up phase for an *in situ* measurement includes the following:

a) Decide on an interface model i.e. identify the connection points to be considered part of the interface and the DoFs to be included at each (Chapter 4). These DoFs are labelled c.
b) For each interface point decide on a representation to capture all DoF, e.g. by Finite Difference, as a Virtual Point, or Equivalent Multi-Point Connection (Chapter 4). Locate sensors and excitation positions to allow for this representation.
c) Identify target locations and DoF. These points are labelled r.
d) If the problem is to be overdetermined (recommended), select additional indicator points on the receiver (6.4.2). These points are labelled b.
e) If validation with artificial excitation is to be conducted (recommended), select artificial excitation points on the source (6.4.2). These points are labelled a.

The location of sensors around the interface, c, is determined by a combination of the interface model and the arrangement chosen to capture all the necessary interface DoFs (e.g. virtual point etc). For target DoFs, r, the sensor positions are typically predetermined according to some specification or supplier agreement and no choice is involved. For indicator, b, and artificial excitation, a, DoFs on the other hand there is often scope for selecting optimum positions. Detailed guidelines for selecting these points are presented later (6.4.2).

Considering the above sets of DoFs, all or some of the FRFs detailed in Table 6.1 will require measurement. In addition for *source* components, *operational* measurements will be required at c and, if the problem is to be over-determined, also at b. These measurements will employ the same sensor locations as already used for FRF measurement. Indeed, ideally the sensors should be left in place for the operational phase to avoid uncertainties relating to detachment and reattachment. Some of the

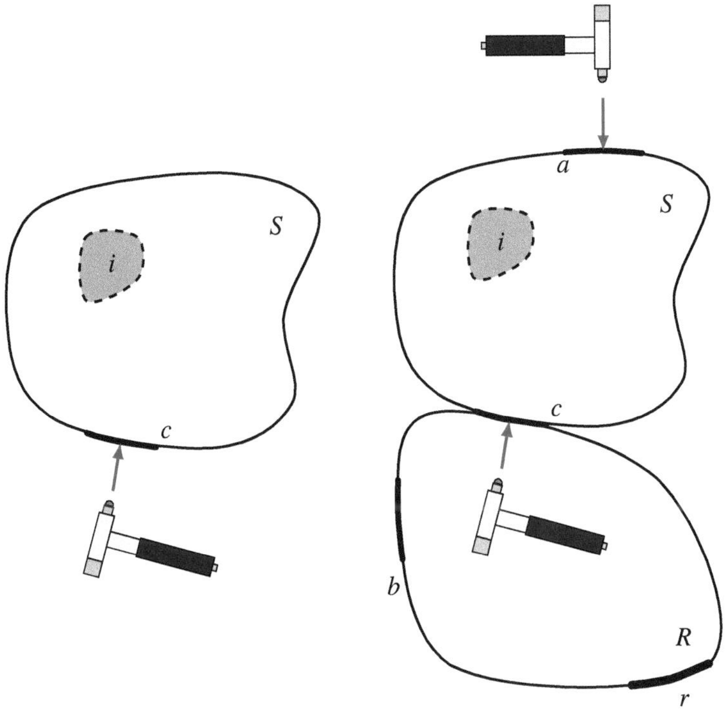

(a) Free interface condition - passive component characterisation.

(b) Installed source - in situ blocked force characterisation.

Figure 6.1 Measurement configurations for passive (a) and active (b) source characterisation.

Table 6.1
Main FRFs and their role in source characterisation.

FRF	Purpose
$\mathbf{Y}_{Ccc}$	Interface completeness, blocked force characterisation
$\mathbf{Y}_{Cbc}$	Interface completeness, blocked force characterisation (if over-determined)
$\mathbf{Y}_{Cca}$	Rank check, interface completeness, artificial excitation
$\mathbf{Y}_{Cba}$	Rank check, interface completeness, artificial excitation
$\mathbf{Y}_{Crc}$	Target response prediction

above measurements are required only as an intermediate step and will not directly form part of the characterisation.

The receiver structure, *R*, employed for *in situ* testing is often predetermined, for example it may be part of an existing installation. However, there is sometimes the opportunity to design a receiver specifically for the measurements, a test bench for example. Guidelines on test bench design are given in 6.4.3.

6.3 INDIVIDUAL MEASUREMENTS

Having defined the FRFs and operational quantities to be measured and determined the location of sensors and excitations required to capture them, we are now almost ready to begin measurements. First, there are some preliminary decisions to make such as *averaging* and *frequency resolution* which are discussed in 6.3.1. In 6.3.2 we go on to present a number of checks that should be conducted for every individual FRF measurement. Operational measurements are considered in 6.3.3.

6.3.1 PRELIMINARY REQUIREMENTS AND PRE-TEST CHECKS

Before any actual measurements are performed, some preliminary decisions need to be made as described in the following.

Averaging

It is standard practice when measuring FRFs to perform repeated excitations and to average the auto and cross-spectra prior to computing an estimate of the FRF. Doing so has the advantage of improving the signal to noise ratio (SNR) and reducing the influence of any random errors (as detailed in 7.2.3).

However with impact-based FRF measurements, too much averaging can cause more harm than good. In particular, it can lead to a smearing of the FRF anti-resonances. This is because anti-resonance frequencies depend on the precise excitation position, so any inconsistency of excitation location or orientation causes them to shift in frequency from one measurement to the next. When averaged, the FRFs become smeared around anti-resonances giving an apparent increase in damping (resonances are not affected in this way). Whilst it might be tempting to think that these frequency regions are unimportant because of the low FRF amplitude, it should be remembered that an error in the anti-resonance of a single FRF term is typically propagated across the entire FRF matrix when a matrix inversion is performed. In practice, the authors have found between 3 and 6 averages to be a good compromise between SNR and FRF quality. Of course, this value does depend on the structure – stiff structures are often less sensitive to excitation inaccuracy compared with flexible ones.

Whilst averaging has been standard practice since the advent of FRF measurements, with modern computing it has become feasible to store every individual FRF measurement. This opens the possibility to select the best combination of individual measurements so as to maximise the physicality of the FRF matrix (i.e. its reciprocity and passivity). Furthermore, as will be discussed in Chapter 8, it also has the advantage of allowing a full uncertainty analysis. The disadvantage is of course that the data storage requirements are significantly increased, but if this can be accommodated, it is recommended to retain the results of every hammer hit or shaker window. Similarly, for operational measurements, unless data storage is an insurmountable problem, it is recommended to retain all time history data rather than just the averages. As discussed in 7.5, this will facilitate the auralisation of VAVP predictions if later required.

Frequency resolution and sample rate

Construction of a VAVP requires combining both active and passive quantities and it is essential that they share a common frequency resolution. The same requirement applies to any numerically modelled components that are to be included.

As discussed in Chapter 7, the frequency resolution is dictated by the sample rate and measurement/FFT length (according to Equation 7.14). For a passive characterisation, the latter should be long enough that the structure's response to an impact excitation has reduced sufficiently by the end of the measurement period. To ensure the same frequency resolution is obtained, the measurement period for both active and passive measurements should be equal. For an operational measurement though, it is preferred to record (at the consistent sample rate) for an extended period of time and use the overlap-add window approach detailed in 7.5 to capture the full time structure of the signal. In this case, the length of each window must equal that used for the FRFs so as to ensure consistent frequency resolution.

Windows and overlap

A traditional aspect of FFT measurements has been to apply windows of various types to each time window as the data is captured, prior to any analysis, to reduce the effects of spectral leakage. We provide some discussion of popular window types for VAVP measurement in Chapter 7.

For impact-based FRF measurements, a rectangular window is generally preferred. One issue to be aware of is that for highly resonant structures with minimal damping, the response may not have decayed sufficiently within the FFT length. The sudden truncation of this response can introduce undesirable spectral effects. In this case one could a) extend the FFT length (thus altering the frequency resolution) to caputre the full signal decay or b) apply a single sided window (e.g. exponential) to bring the signal to zero by the end of the window. This latter option is generally discouraged as it amounts to adding artificial damping to the measurement. It is better to adjust the window length so as to capture the full decay.

For operational response data, the use of a window depends on whether or not the aim is to reconstruct a time series response. If that is the aim, we recommend using a rectangular window with a 50% overlap. These windows can later be stitched back together using a Hanning window as a cross-fade (see 7.5). If the aim is to simply predict a spectrum, more flexibility is available, a safe option being a Hanning window with a 50% overlap.

Uncertainties

It is generally good scientific practice to estimate the uncertainty of any measured or derived quantity. Whilst some error metrics are computed as standard during experimental campaigns, for example coherence, these are not sufficient to provide an estimation of uncertainty in any derived parameters, such as blocked forces and substructured FRFs. In order to estimate uncertainties for a VAVP then, it is necessary to extend standard practice. A detailed discussion of this topic is the subject of Chapter 8. It will be seen that the FRFs obtained from individual excitations must be retained such that their uncertainty can be quantified correctly, i.e. including all possible correlations. These correlations are necessary to compute the overall uncertainty in the measured FRF matrices and accurately estimate the propagation of uncertainty through any subsequent calculations. Therefore, if a full uncertainty analysis is to be conducted, it is essential to make this decision prior to starting measurements so that data for every FRF excitation and operational window is retained.

6.3.2 FRF QUALITY CHECKS

To indicate the presence of possible errors in FRFs, a number of simple data checks are available. These are based on the principles of coherence, linearity, reciprocity and passivity. In addition, the condition number provides an important indicator of the sensitivity to errors from inversion.

Coherence

As discussed in detail in Chapter 7, the coherence γ^2_{XY} is a measure of linear dependence between two signals $x(t)$ and $y(t)$, as a function of frequency [16]. In the context of an FRF measurement, it is the coherence between an applied excitation force $f(t)$ and the measured response $a(t)$ (typically an acceleration). It is calculated according to,

$$\gamma^2_{fa}(\omega) = \frac{|S_{fa}(\omega)|^2}{S_{ff}(\omega)S_{aa}(\omega)} \tag{6.1}$$

where $S_{fa}(\omega)$ is the (single sided) averaged cross-spectrum between a and f, at frequency ω, and $S_{aa}(\omega)$ and $S_{ff}(\omega)$ are their averaged auto-spectra, respectively.

The value of coherence is bound between 0 and 1. A value of one indicates an ideal linear system; the input and output are exactly linearly related and ‘phase locked’. A value less that one can indicate: the presence of uncorrelated noise in a or f, a nonlinearity in the system, or a response a that is influenced by some other unknown input. Coherence is a standard output calculated by most multi-channel analysers during the measurement of FRFs and should be carefully reviewed alongside FRFs during the measurement process so as to provide an early indication of errors.

Some examples of good vs. poor coherence can be found in Figure 6.4, with the corresponding FRFs in Figure 6.3. Some typical causes of poor coherence are described below.

A worsening of coherence across a broad frequency band may indicate a loose connection or faulty sensor. Any such results should be eliminated from the data set.

Poor coherence at a minimum in the spectrum may indicate that the signal is contaminated by background noise at those frequencies. However, if the minimum is an anti-resonance then there is a more subtle possibility due to the fact that the anti-resonance frequencies of an FRF depend on excitation position. This is in contrast to resonant frequencies (i.e. peaks in the FRF curve) which do not depend on excitation positions (they are global properties of the structure) and are inherently stable. If an FRF is measured repeatedly with slightly different excitation position (for example due to inaccuracy with an impact hammer), the shift in anti-resonance frequency will cause a sharp notch in the coherence function due to lack of repeatability at those frequencies. These notches are often unavoidable, although care should be taken to minimise them as best as possible. For well damped systems, this effect is much less severe.

Poor coherence at resonance frequencies is unlikely to be due to background noise, because the signal is higher there, but may be due to nonlinearity resulting from high response amplitudes. A further nonlinearity check (6.3.2) may be helpful in this case.

Points that are separated by distance or by structural features may suffer from poor coherence over a broad frequency band, even when background noise is low. This is likely to be due to loss of consistency in their phase and hence will typically be worse if they are located in a reverberant field. Here, many reflections of waves from nearby boundaries cause phase to vary rapidly in space so that it does not repeat exactly each measurement. Acoustic and vibro-acoustic FRFs often suffer this problem.

Generally, frequency ranges with poor coherence should be considered with care since blocked force or substructuring predictions may be more susceptible to errors in these ranges.

Linearity

The entire VAVP framework is built upon the assumption of linear and time invariant component behaviour. However, certain types of construction can introduce non-linearity, for example bolted joints and parts with small clearances and so it is important to check whether the component under study satisfies the linear assumption. A first check of linearity can be conducted simply by listening to the response of the component when struck; any rattling or buzzing is a likely indication of non-linearity.

A more rigorous check involves measuring an FRF with a given level of excitation, whether from a hammer or shaker, and then repeating the measurement with a significantly higher or lower level of excitation. Because FRFs are normalised to the input force, the FRF should be the same in both cases and therefore any differences between the two results are potentially due to non-linearity and should be investigated further.

Reciprocity

Discussed in detail in 2.2.7, the principle of reciprocity states that the transfer FRF of a linear system between two DoFs, say i and j, remains unchanged if the excitation and response positions are interchanged. That is, $Y_{ij} = Y_{ji}$. This principle can be used to check the quality of measured FRFs.

For example, if measurements are made at the interface of a component (as required for substructuring) we can invoke reciprocity as a check on the quality of the measured FRF matrix $\mathbf{Y}_{Ccc}$. According to the reciprocity principle we should have that,

$$\mathbf{Y}_{Ccc} = \mathbf{Y}_{Ccc}^{\mathrm{T}} \quad \text{or} \quad Y_{Cc_ic_j} = Y_{Cc_jc_i} \tag{6.2}$$

Reciprocal agreement is most easily checked visually, by plotting the two FRFs together, though a quantitative comparison can be achieved using various methods (see 6.4.3). Poor reciprocal agreement can indicate a non-linearity in the system, or that the excitation applied at one or both DoFs is not consistent with the measured response (for example a force is applied in a different direction or at a different location to the response measurement). An example of good vs. poor reciprocity is shown in Figure 6.7.

If response and excitations are measured at c and b we can similarly compare $\mathbf{Y}_{Cbc} = \mathbf{Y}_{Ccb}^{\mathrm{T}}$. However, typically when the remote DoFs b are used, for example when over-determining the blocked force, they are used as response-only positions, i.e. only $\mathbf{Y}_{Cbc}$ is measured, and so reciprocity cannot be inspected. For this reason, if possible, it is always good practice to measure both the direct and reciprocal FRFs.

Whilst it is straightforward to 'correct' for reciprocity by averaging the FRF matrix and its transpose, this does not generally tend to improve calculations, in fact rather the opposite. Some researchers have investigated alternative methods of enforcing reciprocity as a means of improving prediction accuracy [6].

Passivity

A fourth useful data check is that of passivity. Recalling from 3.3 that the power injected into a passive structure due to an applied point force must always be positive we reiterate Equation 3.55 [8],

$$P = \frac{1}{2}\mathbf{g}^{\mathrm{H}}\mathrm{Re}\left(\mathbf{Y}\right)\mathbf{g}. \tag{6.3}$$

For the single point case, this implies that the *real part* of a point *mobility* must always be *positive* and this condition can be used as a check on the measurement quality. Note that for displacement and acceleration based FRFs the power is related to the *imaginary* part and the checks should be adjusted accordingly (see 3.3). Negative measured values indicate some lack of physicality, for example if the excitation and response sensors are not exactly co-located so that the intended point mobility measurement is in reality that of a transfer mobility. Passivity checks can also help to identify errors or inconsistency in the sign convention adopted if, for example the real parts are positive at some points and negative at others.

For the multi-point case there is no restriction on the sign of the transfer mobilities but Equation 6.3 must hold for any excitation vector, which means that $\text{Re}(\mathbf{Y})$ must be a *positive semi-definite* matrix. The properties of such matrices provide another way to check the data for the multi-point case. In addition to having positive diagonal elements (point mobilities), as already mentioned, a positive semi-definite matrix has purely non-negative eigenvalues. This may be seen by expressing the power as an eigenvalue expansion of Equation 6.3,

$$P = \mathbf{g}^{\mathrm{H}}\mathbf{X}^{\mathrm{H}}\Lambda\mathbf{X}\mathbf{g} = \sum_{i} |a_i|^2 \lambda_i \tag{6.4}$$

where $\mathbf{X}$ is a matrix whose columns are the eigenvectors of $\text{Re}(\mathbf{Y})$, $\Lambda = \text{diag}[\cdots \lambda_i \cdots]$ is a diagonal matrix containing the corresponding eigenvalues, and $|a_i|$ is a constant that indicates the contribution of each eigenmode. We see from Equation 6.4 that all *eigenvalues* must be positive to ensure that the power remains positive for any excitation vector.

The eigenvalues of the measured FRF matrix are frequency-dependent mobility-type functions as illustrated in Figure 6.2. This plot is from an 8×8 FRF matrix, although it is seen that of the 8 eigenvalues, only the largest 4 or 5 are positive over most of the frequency range. The lower eigenvalues fall into noise and are partly negative. This is a fairly typical situation for measured data. Also of note is that there is typically one dominant eigenvalue at resonances [13]. This is related to the rank of the matrix as will be discussed later in the chapter in relation to interface checks (6.4.1).

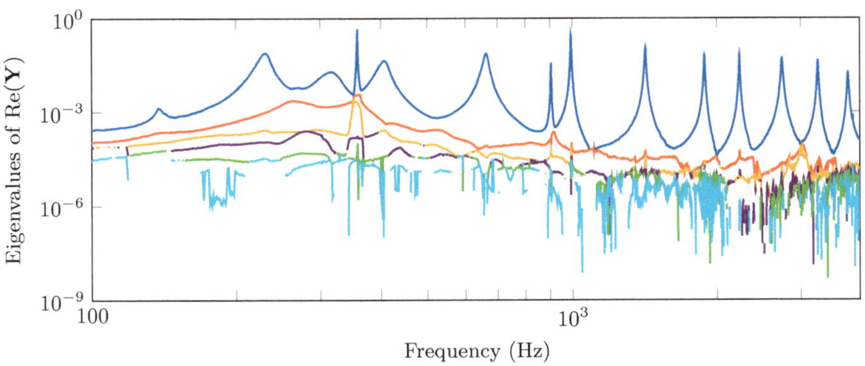

Figure 6.2 First 6 eigenvalues of the real part of an 8×8 mobility matrix of a resiliently mounted steel beam. Missing parts of some curves indicate negative values.

The eigenvalue test can be applied to any submatrix formed from a subset of points since Equation 6.7 must hold for every such case. This can be used to check off-diagonal terms one by one or in groups. Consider the simple case of a 2×2

matrix from any pair of points,

$$\mathbf{Y} = \begin{bmatrix} Y_{ii} & Y_{ij} \\ Y_{ji} & Y_{jj} \end{bmatrix} \tag{6.5}$$

Assuming the diagonal terms (point mobilities) have already been checked for positivity, then the eigenvalues effectively provide a way to check the off-diagonal element, i.e. the transfer mobility $\mathrm{Re}(Y_{ij})$. By applying this repeatedly for every pair of points, each off-diagonal element may be checked one by one. As an illustrative example, consider the special case where two point mobilities are equal, $\mathrm{Re}(Y_{ii}) = \mathrm{Re}(Y_{jj})$, in which case the two eigenvalues are given by,

$$\lambda_{max}, \lambda_{min} = \mathrm{Re}(Y_{ii}) \pm \mathrm{Re}(Y_{ij}) \tag{6.6}$$

which implies a bound on the magnitude of the off-diagonal element,

$$|\mathrm{Re}(Y_{ij})| \leq \mathrm{Re}(Y_{ii}). \tag{6.7}$$

Condition number

It is common practice when analysing dynamic systems to associate the so-called condition number of the system matrix (in our case the FRF matrix $\mathbf{Y}$) with the level of uncertainty, or error, present in the solution of an inverse problem (i.e. a blocked force estimate).

Considering the inverse problem,

$$\mathbf{v} = \mathbf{Y}\bar{\mathbf{f}}, \tag{6.8}$$

the condition number $\kappa(\mathbf{Y})$ can defined as the (maximum) ratio of the relative error on the output (i.e. $\bar{\mathbf{f}}$) for a relative error on the input (i.e. $\mathbf{v}$) or the system matrix (i.e. $\mathbf{Y}$), in a normed sense [19],

$$\kappa(\mathbf{Y}) = \max \left\{ \frac{\frac{||\Delta\bar{\mathbf{f}}||}{||\bar{\mathbf{f}}||}}{\frac{||\Delta\mathbf{v}||}{||\mathbf{v}||}} \right\} \quad \text{or} \quad \max \left\{ \frac{\frac{||\Delta\bar{\mathbf{f}}||}{||\bar{\mathbf{f}}||}}{\frac{||\Delta\mathbf{Y}||}{||\mathbf{Y}||}} \right\} \tag{6.9}$$

where $\frac{||\Delta\bar{\mathbf{f}}||}{||\bar{\mathbf{f}}||}$ represents the normed relative error of the blocked force, $\frac{||\Delta\mathbf{v}||}{||\mathbf{v}||}$ is the same for the velocity, and $||\square||$ is any consistent norm. Rewriting Equation 6.9 as,

$$\kappa(\mathbf{Y}) = \max \left(\frac{||\Delta\bar{\mathbf{f}}|| \cdot ||\mathbf{v}||}{||\bar{\mathbf{f}}|| \cdot ||\Delta\mathbf{v}||} \right) = \max \left(\frac{||\mathbf{Y}^{-1}\Delta\mathbf{v}|| \cdot ||\mathbf{Y}\bar{\mathbf{f}}||}{||\bar{\mathbf{f}}|| \cdot ||\Delta\mathbf{v}||} \right) \tag{6.10}$$

and using the identity $||\mathbf{A}\mathbf{x}|| \leq ||\mathbf{A}|| \cdot ||\mathbf{x}||$ to evaluate its maximum we obtain,

$$\kappa(\mathbf{Y}) = \frac{||\mathbf{Y}^{-1}|| \cdot ||\Delta\mathbf{v}|| \cdot ||\mathbf{Y}|| \cdot ||\bar{\mathbf{f}}||}{||\bar{\mathbf{f}}|| \cdot ||\Delta\mathbf{v}||} = ||\mathbf{Y}^{-1}|| \cdot ||\mathbf{Y}|| \tag{6.11}$$

Noting that, according to the min-max theorem, the largest singular value σ_1 of a matrix is equal to the norm of said matrix, and that the singular values of a matrix inverse are the reciprocal of those of the original matrix (and so the largest being $1/\sigma_N$), the condition number can also be expressed as,

$$\kappa(\mathbf{Y}) = \frac{\sigma_1}{\sigma_N} \tag{6.12}$$

where σ_1 and σ_N are, respectively, the first (largest) and last (smallest) singular values of the FRF matrix $\mathbf{Y}$.

A large condition number implies that a small error on the input or system matrix, will be amplified considerably and become a large error on the output. Large condition numbers are often therefore associated with increased levels of uncertainty.

By inspecting the condition number of a measured FRF matrix it is possible to identify frequency ranges over which the blocked force may be particularly sensitive to errors on the FRF and operational response. In the case of substructuring, since this also involves the inversion of a measured FRF matrix, a similar inference can be made regarding frequency ranges that are sensitive to errors.

Note that condition number is indicative of the *sensitivity* to error, but that the error magnitude on the solution (e.g. a blocked force or substructured FRF) also depends on the error present on the *input* (i.e. the measured FRFs and/or operational responses). A large condition number may thus be counteracted by a low level of measurement error on the input. For example, large condition numbers are generally expected at the resonant frequencies of a structure, where all DoFs are constrained to move according to the mode shape. This can lead to an effective rank deficiency in the FRF matrix, and thus a very small minimum singular value σ_N and high condition numbers. However, resonant frequencies also have the largest signal to noise ratio and are inherently stable, so tend to have low levels of uncertainty on the input (in comparison to anti-resonances, for example).

This effect is illustrated in Figure 6.6 where the peaks in condition number do not necessarily coincide with drops in coherence (i.e. increased experimental uncertainty).

In summary, whilst the condition number provides a convenient single number indication of which frequencies the inverse problem is most sensitive to error, it does not take into account the experimental error present in the measurements themselves, nor does it provide uncertainty estimates for predicted quantities. This is the role of uncertainty propagation, to be discussed in more detail in Chapter 8.

Example 6.1: FRF estimation and data checks

In this example we will show a comparison of some of the input-output relations (FRF estimates and coherence) and data checks for a set of 'good' and 'poor' measurements. The structure under study is the steel beam (dimensions $1515 \times 30 \times 10$mm) first used in Example 2.1. Using the same set-up (two accelerometers spaced roughly 30cm apart) we measure a 2×2 FRF matrix. At each accelerometer position, a total of 20 excitations are applied. Note that this is for illustration and is many more than would be typically applied in practice. This measurement was performed twice. Initially, excitations are applied with reasonable care to yield a 'good' measurement set. In the second case, a more 'relaxed' approach was taken, giving us a 'poor' measurement.

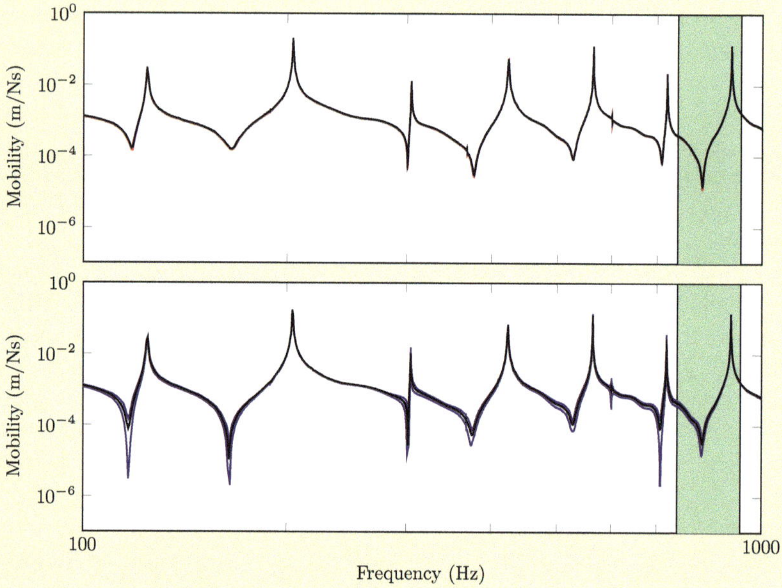

Figure 6.3 Comparison of 'good' and 'poor' point FRF measurements. Black – H_1 estimate, red – $\pm\sigma_{|Y|}$ for good measurement, blue – $\pm\sigma_{|Y|}$ for poor measurement. Zoom of highlighted range shown in Figure 6.5.

Shown in Figure 6.3 are the H_1 estimates of point mobility Y_{11} for the good and poor measurements (see Chapter 7 for more detail on different FRF estimates). Also shown for each case are the estimated $\pm\sigma_{|Y|}$ bounds (from Equation 7.67) to provide an indication of the uncertainty/spread in the individual measurements. It is clear that the good measurement set has a much tighter $\pm\sigma_{|Y|}$ bound across the entire frequency range. It can also be seen that the greatest uncertainty occurs around the anti-resonances. Again, this is an expected result as the resonance frequencies of the beam are global and do not depend on measurement position, whilst anti-resonances vary with excitation and response position.

The $\pm\sigma_{|Y|}$ bounds are estimated from the coherences shown in Figure 6.4 for the good and poor measurements. These coherence results clearly illustrate the effect of poor measurement accuracy and its emphasis around anti-resonances.

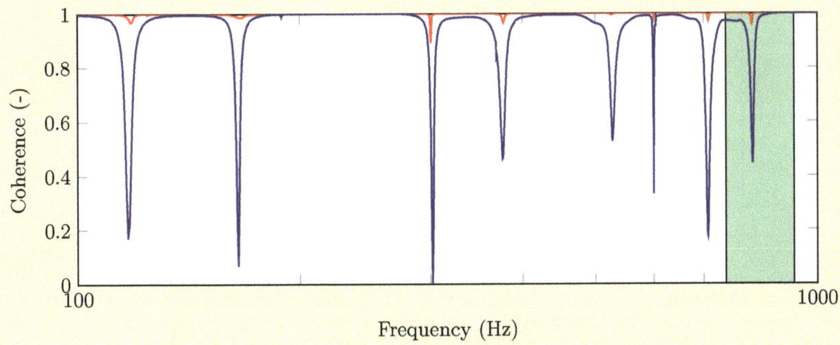

Figure 6.4 Comparison of coherence estimates for 'good' and 'poor' point FRF measurements. Red – good measurements, blue – poor measurements.

The various FRF estimates available are compared in Figure 6.5 over the frequency range 750 – 930 Hz highlighted in Figure 6.3 and 6.4. The mobilities obtained from each individual excitation are shown in grey. From Figure 6.5 we clearly see the effect of anti-resonance smearing due to excessive averaging.

Shown in Figure 6.6 are the condition numbers of the two measurement sets. We see that that frequencies of maximum condition number largely coincide with system resonances, and that there does not appear to be any correlation with measurement error; in fact, the 'good' measurement has the largest condition number.

Shown in Figure 6.7 is the reciprocal agreement obtained from the two measurement sets. It is clear that careful measurement is required to obtain good quality reciprocal agreement. If the 'poor' measurements shown in Figure 6.7 were to be used in a matrix inversion, the reciprocal disagreements would likely lead to large artefacts in the resulting impedance FRF matrix (and so also in any subsequent calculations). An example of such an artefact can be seen in the dynamic stiffness estimation in Figure 2.20 of Example 2.2.

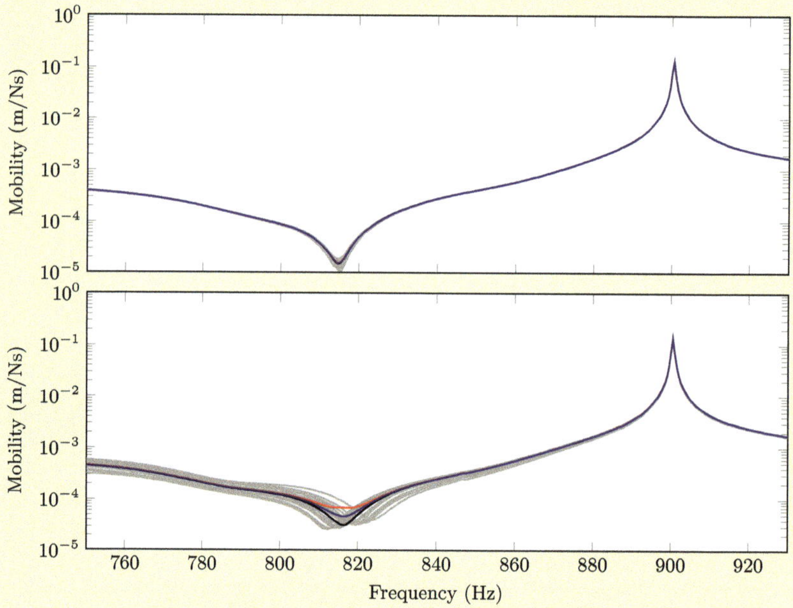

Figure 6.5 Comparison of FRF estimates for 'good' and 'poor' measurements over highlighted frequency range. Grey – individual FRF measurements, black - H_1, red - H_2, blue - H_3.

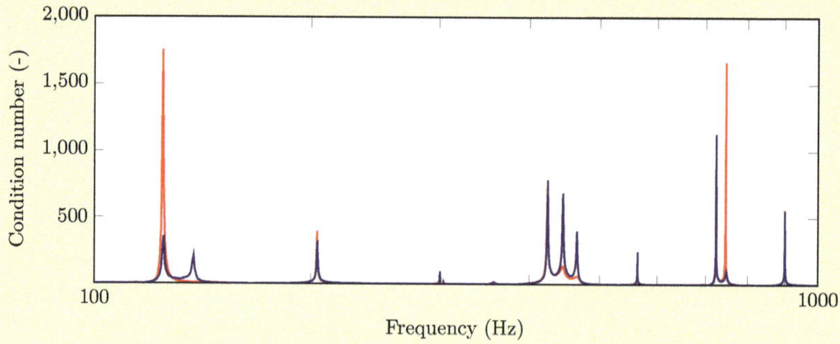

Figure 6.6 Comparison of matrix condition number estimates for 'good' and 'poor' FRF measurements. Red – good measurements, blue – poor measurements.

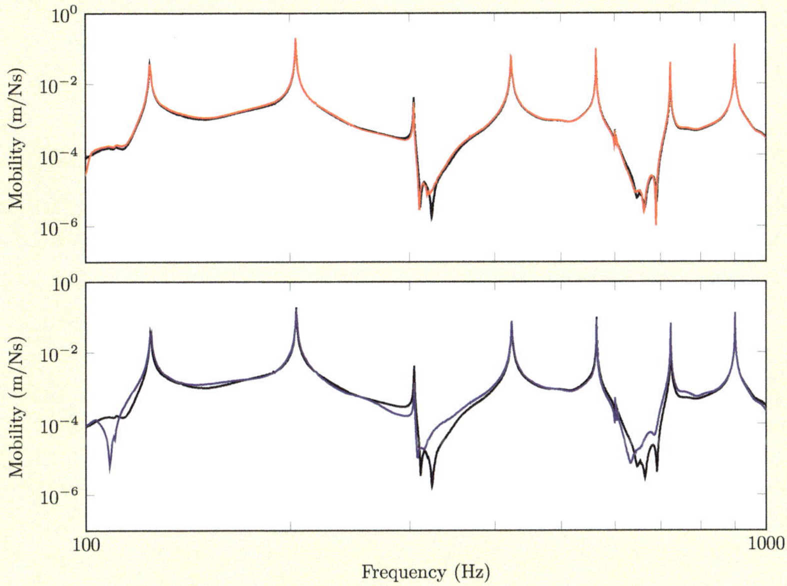

Figure 6.7 Comparison of reciprocity (Y_{12} vs. Y_{21}) for 'good' and 'poor' FRFs measurements. Examples taken from the 20 individual excitations. Red – good measurements, blue – poor measurements.

6.3.3 OPERATIONAL QUALITY CHECKS

Operational response measurements typically use the same sensor array as for the FRF measurements (unless reciprocal FRFs are being used). As far as possible the sensors should remain in place for both types of tests to minimise reattachment errors. The operational response measurement is influenced much less by user error, and so the necessary checks are relatively simple compared with those for FRFs.

Signal to Noise Ratio

An important quality check for operational data is the *signal-to-noise ratio* (SNR) of the operational responses. When solving the inverse problem so as to obtain the blocked force, it is important that noise is kept to a minimum. Since each response measurement contributes towards all elements of the blocked force, a single noisy channel has the potential to corrupt the entire blocked force vector. For this reason it is often recommended that the measured responses be at least 10 dB above the background noise levels.

To check that this requirement is satisfied, a background noise measurement should be done using the same sensors but with the source switched off. The SNR can then be computed, for each indicator position, by taking the log magnitude ratio of the measured operational and noise signals,

$$\mathrm{SNR}_{\mathrm{dB},c_i} = 10\log_{10}\left(\frac{|v_{c_i}|^2}{|v_{c_i,bgn}|^2}\right). \tag{6.13}$$

A poor SNR (< 10 dB) could be obtained for several reasons, including: 1) secondary sources contributing to raised background noise, 2) the indicator DoFs are poorly located, e.g. at a very stiff part of the structure, or 3) if the source is simply too weak or the sensor sensitivity too low.

6.4 BLOCKED FORCE AND COMPONENT CHARACTERISATION

The checks presented so far are considered an essential part of any measurement campaign. However, the most likely place for things to go wrong in building a VAVP is at the level of the component description, particularly the interface model, and we have so far done nothing to check the suitability of the data to actually represent the component. In short, we are now able to check whether the measurements are right but not whether they are the right measurements! In this section, we consider measurements at the *component-level*, aiming to ensure a reliable characterisation of a component for use in a VAVP. We first describe checks on the interface model (6.4.1), then procedures for measuring component-level quantities, particularly blocked forces (6.4). We move on to discuss the various levels of validation available and how they can be staged to isolate specific sources of error (6.4.3. In almost all cases, these checks are conducted with the component attached to another component, which may be part of an installation, a test bench, or a 'dummy' component simply used for the test. Whatever the test setup, a systematic programme of checks and validation should be designed into any measurement campaign.

6.4.1 INTERFACE AND COMPLETENESS CHECKS

We now present procedures to check the validity of the interface description, namely tests for: rigidity, rank and completeness.

Rigidity

Interface rigidity is an important assumption in the development of a VAVP (both for substructuring and blocked force characterisation); assuming point-like connections, the local interface dynamics of a component (i.e. those in the vicinity of a contact point) should be described completely by the 6 rigid body DoFs (x, y and z, along with their corresponding rotations). This requirement is usually readily met in the low-mid frequency range where structural wavelengths are longer than the contact patch, but as wavelength reduces it is important to check the degree to which the interface behaves rigidly. To this end, in 4.2.3 we introduced two indicators for sensor and force consistency.

The general idea is to take the measured interface FRFs ($\hat{\mathbf{Y}}$), apply an appropriate interface transformation (say to a Virtual Point 4.2.2) to obtain the sought after rigid body DoFs, before applying the same transformation in reverse to return to the original measured DoFs. This act of applying and then reversing the transformation effectively filters out any non-rigid interface dynamics. A comparison of the filtered FRF to the original FRF thus provides a measure of flexibility (or lack of rigidity) inherent in the original FRFs. This can be done for the interface sensors and forces separately, or in a combined manner as below,

$$\rho = \text{Corr}\left(\underset{\rightarrow}{\hat{\mathbf{Y}}}, \underset{\rightarrow}{\hat{\mathbf{Y}}'''}\right) \tag{6.14}$$

where Corr ($\square$) is a correlation based comparison (see 6.4.3) and,

$$\hat{\mathbf{Y}}''' = \mathbf{T}_v^+ \left(\mathbf{T}_v \hat{\mathbf{Y}} \mathbf{T}_f^{\mathrm{T}}\right) \mathbf{T}_f^{+\mathrm{T}} \tag{6.15}$$

is the force and sensor filtered FRF. A value of $\rho < 1$ would indicate the presence of some flexible motion across the measured DoFs that is not captured by the transformed interface representation. Of course, ρ is a frequency dependent value, which will typically decrease with increasing frequency as a result of shortening wavelength. For more detail the reader is referred back to 4.2.3.

Note that if an extended (i.e. flexible) interface representation is used (e.g. Virtual Point with flexible deformation modes, 4.2.5) interface rigidity is no longer a strict requirement. However, the interface representation should be such that any flexibility present is captured.

Rank

It has been emphasised throughout that identifying interface DoFs is a critical step in building a VAVP. Here we present a *rank* test to determine the *number* of important DoFs at an interface. Identifying *which* DoFs requires a slightly more involved *completeness* test which is described in the next section.

The rank test is based on the idea that the interface acts as a *bottleneck*, only allowing through excitation in a limited number of independent patterns. As discussed in 4.5.1, this can be interpreted in terms of the Round Trip identity,

$$\text{rank}(\mathbf{Y}_{Cba}) = \text{rank}(\mathbf{Y}_{Cac}\mathbf{Y}_{Ccc}^{-1}\mathbf{Y}_{Ccb}) \leq \text{rank}(\mathbf{Y}_{Ccc}) \tag{6.16}$$

where c are interface DoFs and a and b DoFs either side (see Figure 6.8).

In words, *the rank of any FRF matrix that traverses the interface is constrained to be equal to or less than the number of physical DoFs present at said interface.* This fact can be used to estimate the number of contributing interface DoFs.

The experimental procedure is quite simple, and illustrated in Figure 6.8a. We measure the transfer FRF $\mathbf{Y}_{Cba}$ by exciting a number of points one side of the interface, say a, and measuring the response at some points on the other side, say b.

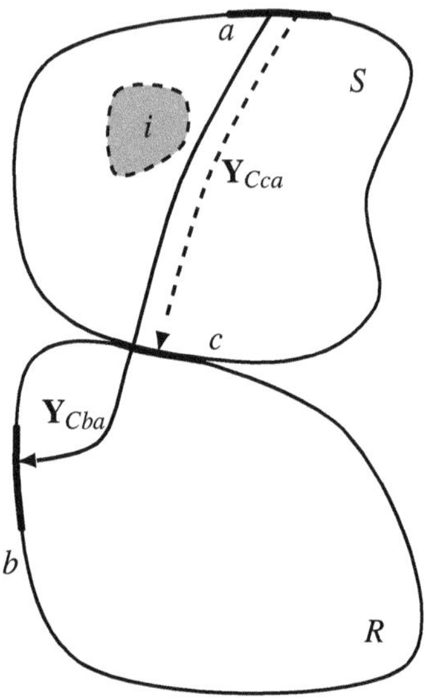

(a) Measurements required for interface rank test. Dashed FRF is optional.

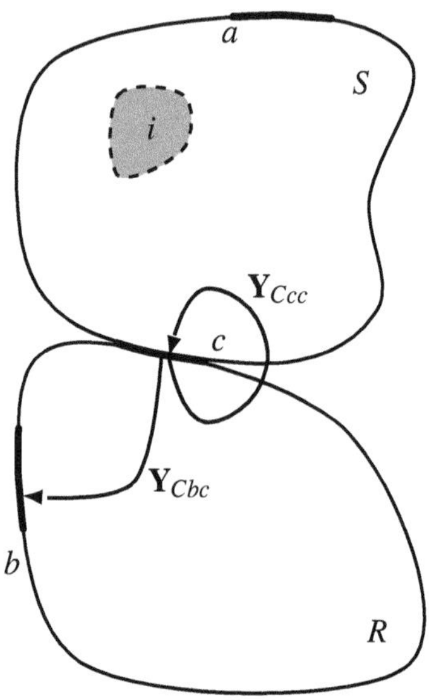

(b) Additional measurements required for interface completeness check.

Figure 6.8 Passive tests required for interface rank and completeness checks.

Provided there are enough a and b points ($\min[a,b] > c$) then the number of interface DoFs can be determined without exciting the interface or even measuring the response there. Furthermore, it is not necessary to know where the a or b points are provided their number is sufficient. In practice, the interface DoFs c are usually instrumented anyway as part of the source characterisation, and so it is convenient to include these in our assessment of the rank. To do so, we simply combine our FRF matrices, $[\mathbf{Y}_{Cba}^{\mathrm{T}} \mathbf{Y}_{Cca}^{\mathrm{T}}]^{\mathrm{T}}$, and inspect the rank of the resulting matrix.

A minor issue is to establish the matrix rank which is not precisely defined for measured data. The problem of establishing the rank of a noisy matrix is a widely studied problem for example in economics and seismology. For our purposes it is most easily solved by evaluating the singular values of the transfer FRF matrix. While mathematically, the matrix rank is the number of non-zero singular values, with noisy data it can be considered equal to the number of singular values above the noise floor or the number before the 'elbow' of a scree plot, as illustrated in Figure 6.11.

Since we do not need to know where the a points are, in principle the excitation can also come from an *operational excitation*. If one substructure, say A, is a source, then the response vectors are recorded for a number of operating states i and are stacked side by side to form a matrix of responses,

$$\begin{pmatrix} \mathbf{v}_c^{(i)} \\ \mathbf{v}_b^{(i)} \end{pmatrix} \rightarrow \begin{bmatrix} \mathbf{V}_c \\ \mathbf{V}_b \end{bmatrix}. \tag{6.17}$$

The rank of this matrix can be determined in the same way as before.

Note that whilst it is not necessary to know the precise location of excitations a (hence they can be internal and operational), it is necessary that they are varied enough to generate more independent excitations than interface DoFs. This issue was discussed in 2.4.2.

Example 6.2: Assessing interface rank with singular values

In this example we show how the singular values of a measured FRF matrix can be used to assess the physical rank of an interface. A transfer FRF matrix is measured between a series of points, a and b, either side of an interface c. The test assembly is that of two steel beams coupled at a single point, as illustrated in Figure 6.9.

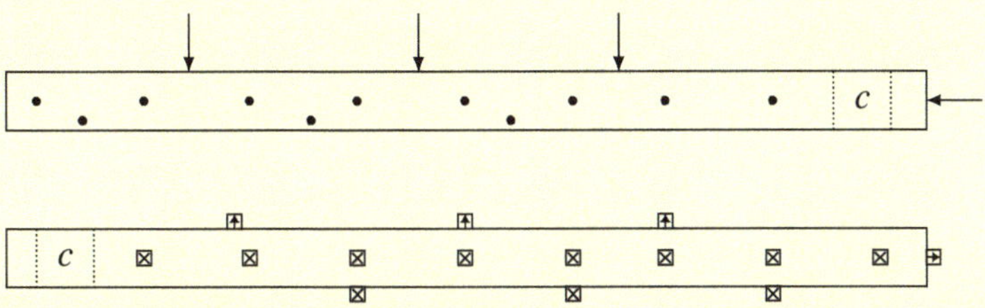

Figure 6.9 Diagram of rank test excitation (top) and response (bottom) positions. Note that excitations and responses are in different directions and deliberately offset from centre line. Dashed region c is where the two beams were overlapped and attached with glue.

A total of 15 excitations are applied, spanning x, y and z directions. Similarly, 15 responses are measured, also in the x, y and z directions. The FRF matrix $\mathbf{Y}_{Cba} \in \mathbb{C}^{15\times 15}$ is assembled and its SVD computed. Shown in Figure 6.10a) is the spatially averaged FRF magnitude. In Figure 6.10b) are the FRF matrix's singular values obtained at each frequency. Figure 6.11a) and b) shows the scree plots of the singular values at the specific frequencies of 30 Hz and 300 Hz (indicated by the dashed vertical line in Figure 6.10).

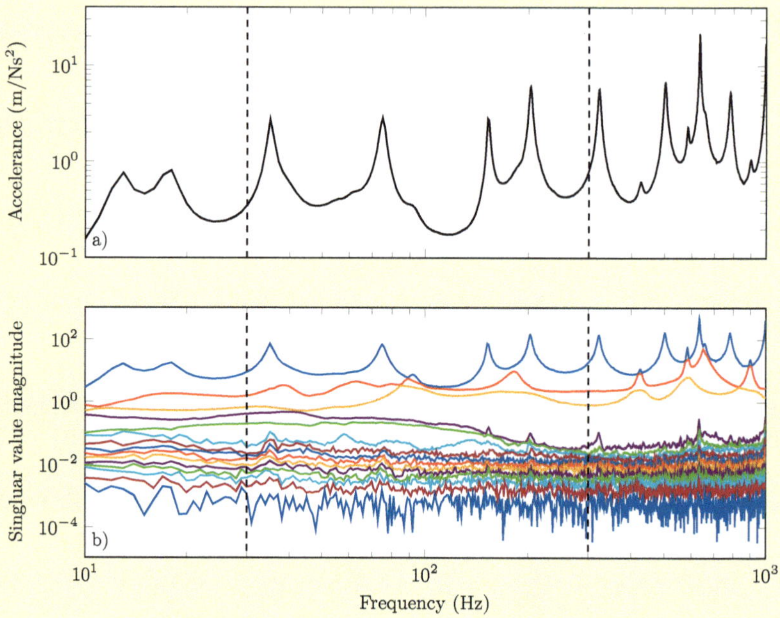

Figure 6.10 Singular values of a 15×15 transfer FRF $\mathbf{Y}_{Cba}$ (note that the accelerance FRF is plot rather than mobility) measured across the interface of two rigidly connected beams. a) Averaged FRF across all excitation and response positions. b) Singular values vs. frequency.

We see from Figure 6.11b) that at 300 Hz there is a clear 'knee' in the plot. Here, we can say with some confidence that the highest 3 singular values are of significance, i.e. the system is rank 3 at this frequency. From Figure 6.11a), the situation is less clear cut at 30 Hz, with no obvious point at which the singular values drop into noise. This situation is common in practice and one may need to draw on an understanding of the physics to help decide the matrix rank. For example, in this case rigid-body motion is expected below the first beam resonance at 34Hz which suggests a rank of at most 6, corresponding to 6 rigid body modes.

Whilst Figures 6.10 and 6.11 indicate that there are e.g. 3 significant contributions at 300 Hz and above, they do not tell us *which* DoFs these are. Indeed, at two different frequencies, these 3 singular values could be made up of different physical DoFs. What the results do tell us though, is that to characterise this interface we should include *at least* 3 DoFs. Deciding which DoFs these should be would of course require some engineering judgement which could be supplemented with a test of completeness (6.4.1).

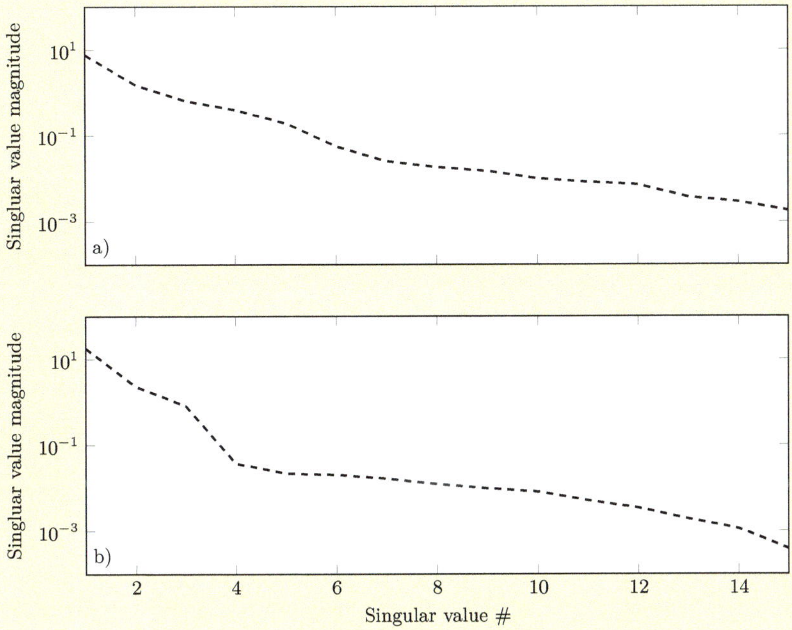

Figure 6.11 Scree-plots of singular values at a) 30 Hz and b) 300 Hz.

Completeness

The interface rank test described above can provide an indication of how many significant DoFs are contributing at a particular frequency. It does not however, provide any indication as to which DoFs those are (which will of course change with frequency), or how well the chosen interface representation is able to capture them. To achieve this, a more involved *completeness* test is required.

Interface completeness is a measure of how well an interface representation (i.e. the measured FRF matrix $\mathbf{Y}_{Ccc}$ after interface transformations are applied) is able to capture the full set of interface dynamics present. A *complete* interface model is essential if transferable blocked forces are to be obtained (see Example 4.1) and if substructuring is to be successful.

To quantify interface completeness, an FRF-based metric has been proposed, termed the Interface Completeness Criterion (ICC) (see 8.3.1 for a full discussion). In short, the ICC assesses the interface completeness by reconstructing a hypothetical transfer FRF matrix between a and b using only those interface DoFs c_i included in the measurement interface representation and excluding unknown and/or purposefully neglected DoFs. This reconstruction is achieved using the Round Trip identity (see 2.5),

$$\mathbf{Y}_{Cba}^{(c_i)} = \mathbf{Y}_{Cbc_i}\mathbf{Y}_{Cc_ic_i}^{-1}\mathbf{Y}_{Cc_ia} \tag{6.18}$$

where the superscript c_i indicates that the resulting FRF includes only the contributions of the c_i DoFs. Hence, by comparing $\mathbf{Y}_{Cba}^{(c_i)}$ against the directly measured FRF $\mathbf{Y}_{Cba}^{(c)}$, which by definition includes contributions through all interface DoFs, the ICC is able to quantify the level of completeness. This comparison can be done in a variety of ways, though the most common is using a correlation-based approach, as described in 6.4.3,

$$\text{ICC} = \text{Corr}\left(\underset{\rightarrow}{\mathbf{Y}}_{Cba}^{(c_i)}, \underset{\rightarrow}{\mathbf{Y}}_{Cba}^{(c)}\right). \tag{6.19}$$

Values of ICC $<< 1$ indicate that some important DoFs have not been captured by the chosen representation.

A test of completeness requires some additional measurements over and above those required for a test of rank. Specifically, the FRFs $\mathbf{Y}_{Ccc}$ and $\mathbf{Y}_{Cbc}$, both of which require excitations at the interface, must now be measured as illustrated in Figure 6.8b.

An important theoretical condition for the completeness test is that the excitations at the remote DoFs a and the responses at b should be able to fully *control* and *observe* the interface, respectively. The number of a and b points must therefore at least equal the number of interface DoFs c and in practise will typically exceed the number by a factor of between 2 and 3 so as to provide some degree of over-determination. The aim is to maximise the number of linearly independent excitations and responses. In practice, this can be achieved by exciting and measuring responses in different directions so as to excite a wide range of structural modes. This is the reason for the excitation and response points in x, y and z directions and located off the centre line in Example 6.2.

Note that the requirements for response DoFs b are effectively the same as will be considered later for selection of indicator DoFs (6.4.2). Similarly, those for excitation points a are the same as for artificial excitation points, also considered in 6.4.2. The interested reader may refer to those sections for further discussion.

Clearly, to conduct a test of interface completeness, the component must be installed, or at least attached to another component. In practice, one uses the same sensor arrays as for other tests including blocked force determination and subsequent validation tests.

6.4.2 IN SITU BLOCKED FORCE CHARACTERISATION

In this section, we recap the in situ characterisation of the blocked force and describe the experimental procedure in more detail.

Key equations

The *in situ* blocked force method is based on the *in situ* relation (reproduced from Equation 3.29 for convenience),

$$\begin{pmatrix} \mathbf{v}_c \\ \mathbf{v}_b \end{pmatrix} = \begin{bmatrix} \mathbf{Y}_{Ccc} \\ \mathbf{Y}_{Cbc} \end{bmatrix} \bar{\mathbf{f}}_c \tag{6.20}$$

where $\mathbf{v}_c$ and $\mathbf{v}_b$ are the responses measured at the interface and remote DoFs c and b whilst the source is in operation (together we term c and b the *indicator* DoFs), and $\mathbf{Y}_{Ccc}$ and $\mathbf{Y}_{Cbc}$ are the FRF matrices relating these DoFs to the interface c.

Equation 6.20 represents the so-called *forward* problem. The blocked force is obtained indirectly by solving the *inverse* problem, that is, taking the inverse, or more generally the pseudo-inverse (see 7.4), of the FRF matrix giving,

$$\bar{\mathbf{f}}_c = \begin{bmatrix} \mathbf{Y}_{Ccc} \\ \mathbf{Y}_{Cbc} \end{bmatrix}^+ \begin{pmatrix} \mathbf{v}_c \\ \mathbf{v}_b \end{pmatrix}. \tag{6.21}$$

For simplicity this is written in a generic form in Figure 6.12 as,

$$\bar{\mathbf{f}}_c = \mathbf{Y}_C^+ \mathbf{v} \tag{6.22}$$

where we have omitted reference to any specific DoFs b or c.

It is important to recall that Equation 6.22 assumes that the measured interface DoFs have been transformed appropriately, for example using one of the methods described in Chapter 4. For an FRF matrix, this transformation takes the form,

$$\mathbf{Y} = \mathbf{T}_v \hat{\mathbf{Y}} \mathbf{T}_f^{\mathrm{T}} \tag{6.23}$$

where $\hat{\mathbf{Y}}$ is the directly measured FRF matrix. Note that in this form both the measured force and response DoFs are transformed. For Equation 6.22 to remain consistent, the operational responses must also be transformed in the same way,

$$\mathbf{v} = \mathbf{T}_v \hat{\mathbf{v}} \tag{6.24}$$

where $\hat{\mathbf{v}}$ are the directly measured responses.

Alternatively, the response transformation can be left out all-together, after all it is ultimately the blocked force that we want

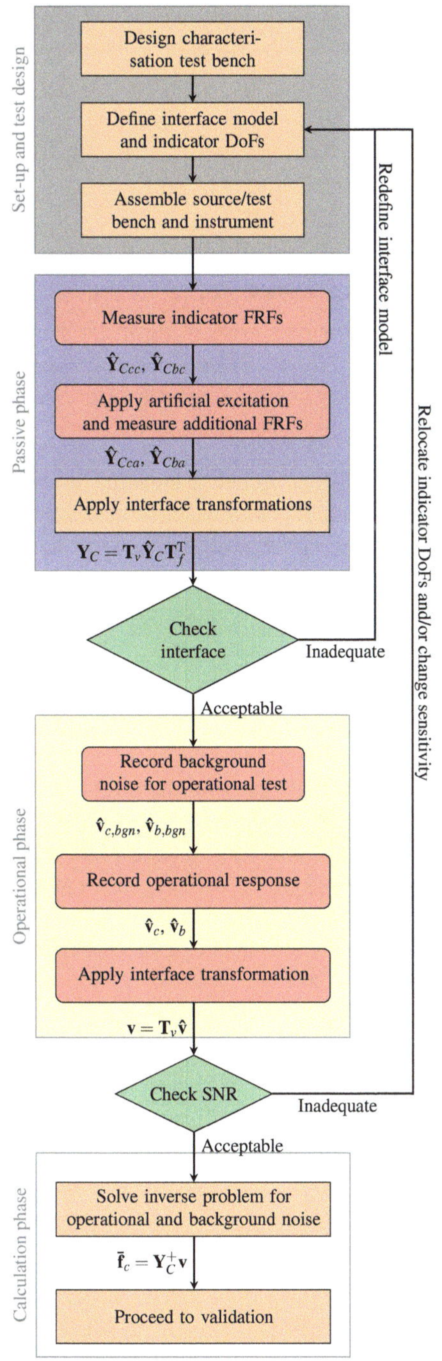

Figure 6.12 Blocked force characterisation flow chart.

to project onto our interface model. In this case we leave the operational response untransformed and only transform the FRF forces,

$$\bar{\mathbf{f}}_c = \left(\hat{\mathbf{Y}}\mathbf{T}_f^{\mathrm{T}}\right)^+ \hat{\mathbf{v}}. \tag{6.25}$$

Note that this reduced transformation can only be done for a blocked force characterisation. Any substructuring calculations will require the FRF matrix to be fully transformed.

Procedure

Though simple in form, implementation of Equation 6.22, including its validation, is a relatively involved process. Figure 6.12 illustrates the general procedure graphically and the steps are described in more detail below.

Set-up phase

The set-up phase includes test bench design, interface definition, indicator point selection, and physical construction. See 6.4.2 and 6.4.3 for further discussion of indicator point selection and test bench design.

Passive test phase

For the passive test phase the source is turned off, and the FRFs needed for characterisation ($\mathbf{Y}_{Ccc}$ or $\mathbf{Y}_{Cbc}$ or both) are measured (see Figure 6.13a).

This typically involves excitation of the interface c with an instrumented force hammer, or shaker, whilst simultaneously recording responses at all indicator DoFs. In some cases, the FRFs for remote indicator positions might be measured reciprocally by applying the force excitation directly to the remote indicator points.

Next, a series of *artificial excitations* are applied to the surface of the source, and the FRFs to the indicator DoFs measured, $\mathbf{Y}_{Cca}$ and $\mathbf{Y}_{Cba}$. These FRFs are used to assess the completeness of the interface representation according to the ICC described above in 6.4.1. They can also be used in their own right to inspect the inverse problem in the presence of a simpler (non-operational) excitation, as discussed below.

Once all FRFs have been measured, appropriate interface transformations (e.g. Finite Difference or Virtual Point) are applied to extract translational and rotational quantities of interest.

Before proceeding to the operational phase, the interface representation should be checked, specifically completeness and rigidity should be assessed and the interface model revised if necessary. If the interface representation is deemed sufficient, we can proceed to the operational phase.

Operational phase

Before turning on the source, a background noise measurement should be taken. Ideally, sensors will already be in place from the passive test phase. Once completed, the source can be turned on and its operational response recorded (see Figure 6.13b). During this measurement it is typical to record for a period of 30 seconds or so, or at least long enough to capture the manoeuvre being performed. The responses should ideally be processed using an overlap-add window approach, as described in 7.5.

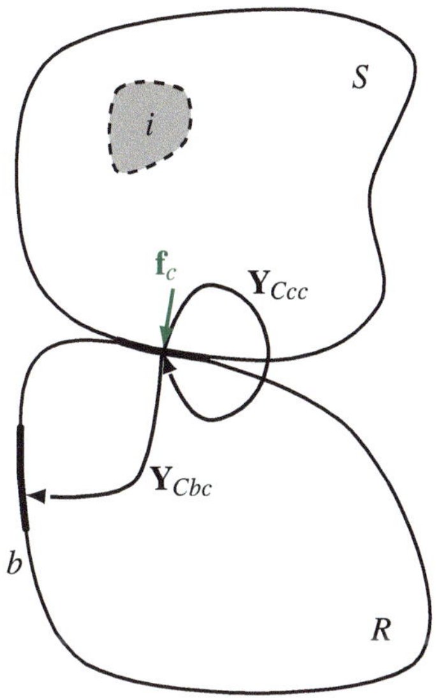

(a) Passive phase - measurement of FRF matrices.

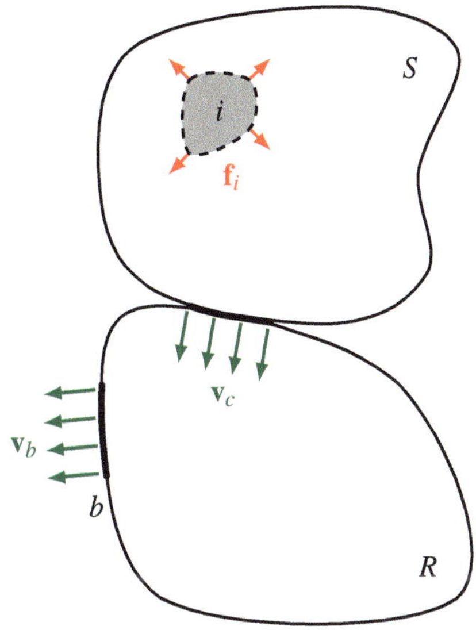

(b) Operational phase - measurement of the operational responses.

Figure 6.13 Two part measurement required for a blocked force characterisation.

If a full interface transformation is applied to the measured FRF (i.e. forces and responses are both transformed), then the operational response should also be transformed in the same way to ensure consistency with the FRF matrix.

Before proceeding to calculate the blocked force, Signal to Noise Ratio (SNR) should be checked and, if not sufficient, adjustments made. This might involve relocating the remote b indicators closer to the interface, or perhaps using higher sensitivity accelerometers. Once an acceptable SNR is achieved, we can proceed to compute the blocked force.

Calculation phase

Finally, we are in a position to calculate our blocked forces, that is, to implement Equation 6.22. We do this for both background noise measurement and operational responses. The former allows us to propagate the noise floor of our measurements through a VAVP prediction, as described in more detail below.

Of course, calculating the blocked force isn't the final step. Before we can use the blocked force within a VAVP, we must ensure it is fit for purpose. This requires a validation, usually a multi-stage validation which will be described in 6.4.3. A coded example of a blocked force characterisation and validation is provided in Chapter 9.

Indicator point selection

The indicator points used to obtain a blocked force may include *interface* DoFs (labelled c) plus additional *remote* points on the receiver (labelled b) as illustrated

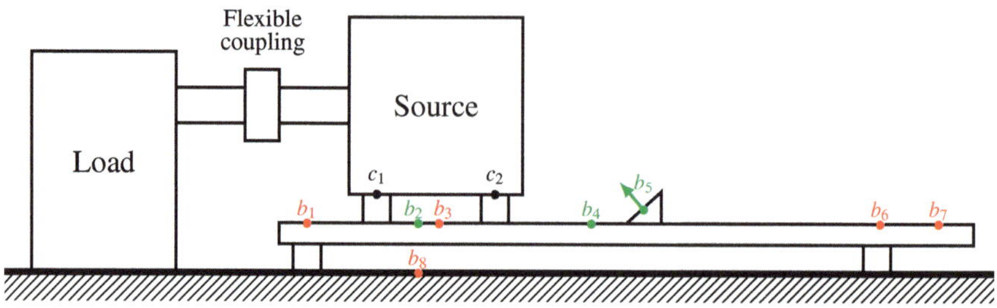

Figure 6.14 Guide for selection of indicator and validation points.

in Figure 6.14. This gives three broad options for determining blocked forces:

$$\text{From interface DoFs only } (c)\text{:} \quad \bar{\mathbf{f}}_c = \mathbf{Y}_{Ccc}^{-1}\mathbf{v}_c \tag{6.26}$$

$$\text{From remote points only } (b)\text{:} \quad \bar{\mathbf{f}}_c = \mathbf{Y}_{Cbc}^{+}\mathbf{v}_b \tag{6.27}$$

$$\text{From a combination of } b \text{ and } c\text{:} \quad \bar{\mathbf{f}}_c = \begin{bmatrix} \mathbf{Y}_{Ccc} \\ \mathbf{Y}_{Cbc} \end{bmatrix}^{+} \begin{pmatrix} \mathbf{v}_c \\ \mathbf{v}_b \end{pmatrix} \tag{6.28}$$

Any of the above options can in theory provide a valid solution. However, we may expect differences in the quality of the results depending on the choice of indicator points as discussed in the following.

Interface DoFs (c) have the advantage that they directly represent the blocked forces at the point of action. However, as was discussed in relation to the Round Trip method in 2.5, obtaining reliable measured interface FRFs can be challenging, particularly those involving certain directions, such as rotational or in-plane DoFs. Measurement of response is not usually the main problem, rather, the difficulty is to apply excitation in these DoFs at the interface, often made worse by limited access. Consequently interface FRFs, particularly those involving difficult DoFs, tend to be more prone to errors.

Remote DoFs (b) on the other hand, being more or less removed from the interface, are less directly linked to the blocked forces. However, their advantage is that by invoking reciprocity ($\mathbf{Y}_{Ccb} = \mathbf{Y}_{Cbc}^{\mathrm{T}}$) one may avoid any difficulties of applying excitation to the interface. Instead, the excitation is applied at the remote DoFs whose location and direction may be selected to make this as easy as possible. FRFs at remote points are therefore likely to be more reliable. A further advantage is that, unlike interface DoFs, one may arbitrarily add extra remote points to increase over-determination of the problem (see 7.4.1 for a discussion on the benefits of over-determination). It should be understood that it is necessary to measure the *responses* at *all* interface DoFs (assuming reciprocity is invoked for remote points); the distinction between them concerns whether the *excitation* is applied at potentially awkward interface DoFs or at specially selected remote points.

From the above, *interface DoFs* should always be included if the corresponding FRFs $\mathbf{Y}_{Ccc}$ can be reliably measured. However, given that FRF uncertainty is liable to be amplified during inversion, one should try to exclude unreliable FRFs which, as described above, is more likely to occur for interface DoFs in difficult directions. In this case, it may be preferable not to include these DoFs but to add additional remote points instead. Guidelines for judging the quality of measured FRFs were discussed in 6.3 including coherence, reciprocity, passivity and condition number and these checks can inform decisions about which interface DoFs to include.

However, not all remote points are equally useful. Figure 6.14, illustrates some points that would be a poor choice because they are either: a) too distant, b) too noisy, or c) too close to other indicator points. Expressing these criteria in more conceptual terms, the responses at indicator locations should ideally be:

- *dependent* on the interface DoFs;
- free from *background noise*;
- *independent* of other indicator responses.

The first point above is another way of saying that the interface should be fully *observable* from the indicator points. For this, we need to select points with a good 'view' of the interface – the more distant the point, the less prospect of observing the interface dynamics from there [23]. Point b_7 in Figure 6.14, for example, is not only distant from the interface but is located the other side of a rigid block which will tend to obscure the signal from the interface even more. The same comments apply to point b_6 and b_8, the former being located at the rigid attachment point, and the latter below it on the test bench foundation. In practice, good coherence of the elements in $\mathbf{Y}_{Cbc}$ indicates strong dependence (or good observability).

Generally, coherence will tend to decrease as the distance between excitation and response increases [23]. Partly, this is because relative error in phase becomes more significant as the wavelength distance increases; phase is therefore less repeatable resulting in a lowering of coherence. Low coherence may also be caused by background noise. For example, point b_1 in Figure 6.14 is less distant than b_4 but is located at a rigid point on the structure so the signal is likely to be weak and consequently contamination by noise is more likely. The same can be said about point b_6. These points are thus unlikely to be a good choice for indicator locations.

The remaining points in Figure 6.14 that are neither too distant, nor subject to noise contamination are b_2, b_3, b_4, and b_5. However, selection of all of these points is unlikely to be a good choice because of the third bullet point in the list above i.e. the need for *linear independence*.

Note that b_2 and b_3 are close together which means that the associated rows and columns of $\mathbf{Y}_{Cbc}$ are likely to be very similar. Theoretically, when rows or columns of a matrix are equal the matrix becomes singular and the inversion blows up. Exact singularity does not occur with experimental data, but one can expect an increase the condition number making the inverse solution more sensitive to errors in the input data. Any potential advantage of over-determination from adding extra points may be more than negated if it leads to a higher condition number. Practically, linear independence can be achieved by selecting points that are not too close together and that, as far as possible, provide excitation in different directions so as to excite different structural modes within the assembly. The wedge-shaped attachment illustrated in Figure 6.14 is one way to achieve this.

Having discussed the best *locations* for indicator points, there is also a question of *how many* points to include. In other words, the extent of *over-determination* needs to be decided. Briefly, a factor of between 2 and 3 is often recommended (number of indicators is 2-3 times the number of interface DoFs, or $2c \leq c+b \leq 3c$). The reader seeking a deeper understanding is referred to 7.4.1 where this issue is discussed in relation to solving of the inverse problem.

Artificial excitation

Artificial excitation is useful technique in which the real operating forces of the source are replaced with a series of impact or shaker excitations applied to its housing, $\mathbf{f}_a$. The resulting responses can be subject to the same treatment as described above so as to yield a set of blocked forces for the 'artificial' source.

Artificial excitation has the advantage that, unlike real operational forces, it is repeatable and controllable. It can also provide excitation over a wide frequency band, which is not always the case for real sources, and without features such as tones which may complicate validation comparisons. For these reasons, even for real sources, it is recommended to employ artificial excitation prior to a real operational test [1] as will be discussed later in relation to validation in 6.4.2.

Only minor modification to the previous analysis is required in that, whereas for real excitation one measures absolute responses, with artificial excitation they are normalised to the input force, in other words we use FRFs in place of the operational response. This normalisation is of course not possible for real excitation since the internal forces are unknown.

Thus, Equation 6.21 becomes

$$\bar{\mathbf{F}}'_c = \begin{bmatrix} \mathbf{Y}_{Ccc} \\ \mathbf{Y}_{Ccb} \end{bmatrix}^+ \begin{pmatrix} \mathbf{Y}_{Cca} \\ \mathbf{Y}_{Cba} \end{pmatrix} \tag{6.29}$$

where the prime indicates that the blocked force $\bar{\mathbf{F}}'_c$ is normalised to the input force. A second difference is that, whereas we have shown operational responses as a single vector, we now have matrices $\mathbf{Y}_{Cca}$ and $\mathbf{Y}_{Cba}$, the columns of which correspond to different artificial excitation positions. Collecting the blocked force due to each artificial excitation we obtain the matrix of blocked forces $\bar{\mathbf{F}}'_c$. In fact, it is common to use this matrix form for real excitation too, where the columns correspond to different operating conditions, or simply to different time windows.

It will come as no surprise that the *excitation points* for artificial excitation must be carefully selected to give the best results. Fortunately, the same criteria apply as have already been discussed for indicator points in 6.4.2. The only adjustment to the previous discussion is that rather than requiring the interface to be fully *observable* as it was from the indicator (response) points, from the point of view of excitation positions it should be fully *controllable*. This is purely a semantic distinction however, and in practice if the role of source and receiver were interchanged, points that make good indicators would also make good locations for artificial excitation. For both indicator and artificial excitation points then, we need points with a good 'view' of the interface which can be confirmed from the coherence of the transfer FRF terms $\mathbf{Y}_{Cca}$ and $\mathbf{Y}_{Ccb}$. Points should also not be too close together as they would induce very similar responses across the interface and provide little additional information.

It is worth noting that the normalised blocked force due to an artificial excitation is closely related to the Interface Completeness Criterion (ICC) described above. In fact, the ICC is simply a correlation-based comparison (see below) between a) the predicted response at b due to the normalised blocked forces (due to excitation at a) and b) the directly measured response at b due to artificial excitation at a. The ICC simply boils down multiple excitation and response positions into an easy to digest number scaled between 0 and 1. Of course, the artificial excitation can also be considered individually if a more detailed analysis is required.

Background noise

When measuring the operational response it is good practice to also record a background noise measurement $\mathbf{v}_{bgn}$, i.e. the operational response whilst the source is turned off. Importantly, this measurement should be done with the exact same settings as the operational measurement. This result can then be compared against the operational measurement and used to identify frequency ranges where the measured response falls below the background noise. It can also be used to identify anomalous non-physical tones, for example due to ground loops or mains hum. Moreover, the background noise can be subject to the same post-processing as applied to the operational response, and used to compute the background noise on the blocked force estimate as well as any response predictions,

$$\bar{\mathbf{f}}_{c,bgn} = \mathbf{Y}_{Ccc}^{-1}\mathbf{v}_{c,bgn} \qquad \longrightarrow \qquad \mathbf{v}_{r,bgn} = \mathbf{Y}_{Crc}\bar{\mathbf{f}}_{c,bgn}. \tag{6.30}$$

From the above, $\mathbf{v}_{r,bgn}$ provides a lower limit on the VAVP's prediction capability.

6.4.3 VALIDATION

In any measurement, the evaluation and reduction of errors is extremely important. This is particularly so for VAVPs, not least because any measurement errors may be amplified, potentially enormously, in the solution of the inverse problem. Any measurement campaign therefore needs to integrate an effective validation strategy so that the dominant sources of error may be first identified and secondly, where possible, minimised. Note that for simplicity the term 'errors' is taken here to include both bias errors and random uncertainties; the distinction between these terms will be discussed in Chapter 8.

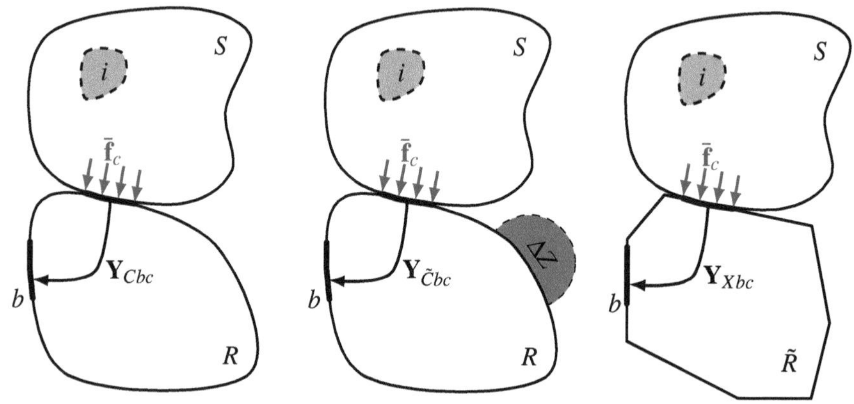

(a) On-board. **(b)** Modified receiver. **(c)** Transferability.

Figure 6.15 Validation methods. (a) On-board validation – prediction in the same substructure, C. (b) Modified receiver validation – prediction in the same substructure but modified, $\tilde{C}$. (c) Transferability validation – prediction in a different substructure, X. For a full VAVP validation the FRF in (c) is obtained by substrucutring.

A staged validation approach is therefore recommended in which the complexity of tests is increased gradually so as to introduce additional sources of error, as far as possible, one at a time. Four broad stages are summarised in Figure 6.15 and are described in more detail in the following, although variations are of course possible. In every case, the validation revolves around the comparison of a predicted response with a directly measured response, as illustrated in flow chart of Figure 6.16. The prediction uses blocked forces obtained *in situ* with the source attached to a receiver (possibly a test bench). The response is predicted either in the same receiver, or in the same receiver but modified in some way, or in a completely different receiver. This gives three levels of validation known as *on-board validation*, *modified receiver validation* and *transferability validation*. If required, separate validation of the substructuring step can also be incorporated into this last phase as per Figure 6.16 to give a *full VAVP validation*. As we have seen in 6.4.2, it is advantageous to conduct each stage first with artificial excitation and then with real excitation of the source.

These four levels of validation, each with real or artificial excitation, introduce different sources of error as summarised in Table 6.2. These will be described in more detail in the following, but taking an overview if we proceed from left to right and top to bottom through this table we will introduce additional sources of error at each stage such that each can be to some extent isolated.

The overall errors naturally include the accumulated effect of errors from each stage in the prediction, including those in the *input data*, i.e. the measured FRFs and operational responses. Hence, it is important to apply all the measurement checks described in 6.3 so that the component-level validation is built on an input data set already cleaned of any 'poor' data. For similar reasons, it is important to check the suit-

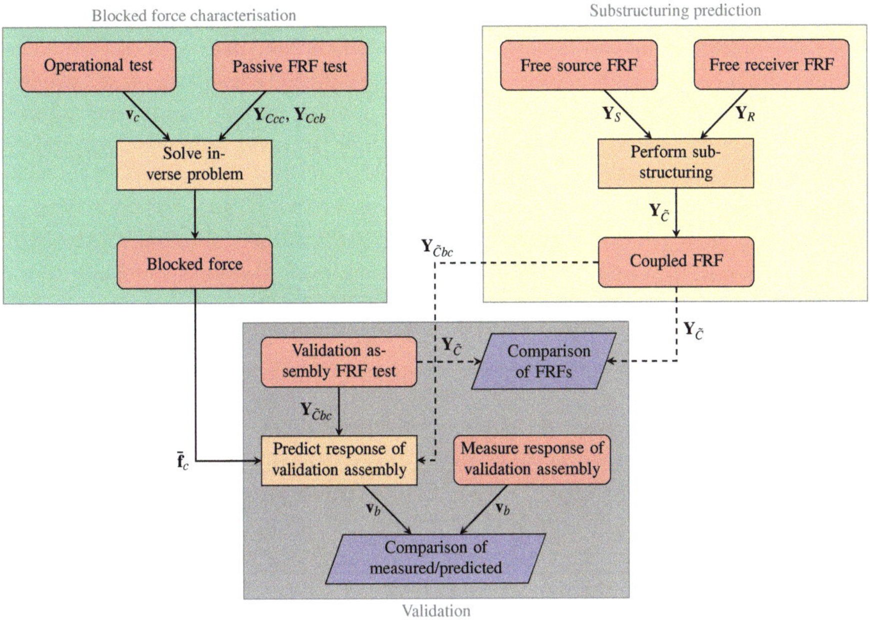

Figure 6.16 Flow chart for validation. (a) The blocked force is first obtained; (b) the response is then predicted from the blocked force in the same or a different receiver and compared with direct measurement; (c) the assembly FRFs can also be obtained from those of the individual substructures to test the substructuring step.

ability of the interface model as described in 6.4.1 before embarking on component-level validation tests.

One of the main new sources of error to be introduced at the component-level validation stage is that associated with *solving the inverse problem*. It could be argued that this is not a source of error in itself but rather a process in which existing errors may become amplified. Whichever way we view it though, testing its effect is an important part of the component-level validation to be described in the following.

On-board validation

The simplest approach to validation is to predict the operational response of the same assembly from which the blocked force was determined (Figure 6.15a). We term this an 'on-board validation'. In equation form it reads,

$$\overbrace{\bar{\mathbf{f}}_c = \mathbf{Y}_C^{+}\mathbf{v}^{(\text{meas})}}^{\text{Characterisation}} \quad \rightarrow \quad \overbrace{\mathbf{v}^{(\text{pred})} = \mathbf{Y}_C\bar{\mathbf{f}}_c}^{\text{Prediction}} \quad \rightarrow \quad \overbrace{\mathbf{v}^{(\text{meas})} \stackrel{?}{=} \mathbf{v}^{(\text{pred})}}^{\text{Comparison}} \qquad (6.31)$$

where it is noted that the same $\mathbf{Y}_C$ is used in both the characterisation and prediction stages. We can distinguish between two forms of on-board validation. The first uses the response $\mathbf{v}^{(\text{pred})}$ predicted at the same DoFs as used in the characterisation step, i.e. $\mathbf{v}^{(\text{meas})}$. We call this an 'indicator validation'. This validation provides an indication of whether the blocked force is able to reproduce the responses from which it was determined. To some extent this provides a check on both the inverse solution and the completeness of the interface description. However, it is not a particularly good test of the blocked force's predictive capabilities. The second form of on-board validation, termed here as 'cross-validation' uses a new set of DoFs (say $\tilde{b}$) that are not included in the characterisation, though they are measured simultaneously. In the literature, the term on-board validation is usually used to refer specifically to this cross-validation approach. In practice, one can simply omit one or more of the b DoFs used in the characterisation, saving these as validation points. In this case, the cross-validation requires no additional measurements beyond those of the characterisation itself.

Once a prediction is made the simplest, and most often used, method of comparison against a measured response is to plot them against one another. Examples of such a comparison can be found in several of the invited case studies through Chapter 10 and also the coded example of Chapter 9. In addition to a visual comparison, more quantitative options are available but discussion of these will be deferred until we have described the remaining validation setups.

On-board validation is relatively simple to implement since the only setup needed is that for the blocked force characterisation. It avoids uncertainties related to the (dis)assembly of the source and receiver and any variations in source operation. Of course, errors in the input data (particularly FRFs) are inherently included as part of the overall error and cannot be separated out at this stage. This is why we have previously emphasised careful checking of the FRFs at every stage so that the inverse procedure to obtain blocked forces is based on carefully checked data. The additional sources of error introduced in on-board validation (see Table 6.2) are those due to the interface representation and the inversion process.

Table 6.2

Types of validation and additional errors introduced at each stage.

Validation procedure	Artificial excitation	Operational excitation
On-board	Inverse procedure, interface completeness	Source repeatability, consistency
Modified receiver	Interface completeness under new constraints	Blocked force invariance under different constraints
Transferability	Attachment repeatability	Blocked force invariance after reattachment
Full VAVP	Substructuring calculations	All sources of error and uncertainty

The table should be read left to right and top to bottom, then each additional step will introduce the additional potential source of error indicated.

In the case of real excitation, there is an additional possible source of error that the FRF of the source is different in the passive and operational phases. This could occur, for example, because of a temperature difference between the operational and FRF tests or if there is stiffening due to the operating load. If this occurs then the blocked forces obtained are *inconsistent* with the measured FRFs and may affect the predicted responses in ways that are not at all obvious. It is assumed that this inconsistency would not occur for artificial excitation since the 'operational' test and the FRF test are usually conducted at the same time, or at least under similar conditions. It has been described that the correlation coefficient for an on-board validation with artificial excitation gives a measure of interface *completeness*; if essentially the same on-board validation is repeated with real excitation and the correlation is found to be worse, then a likely explanation is *inconsistency*. Inconsistency is explained in more detail in 8.3.2.

Modified receiver validation

Whilst the on-board validation provides a convenient (low effort) means of validation, it is not a particularly rigorous test. A truer and more challenging test is to predict the response in a different receiver structure. In a staged validation approach, one ideally wants to change as little as possible at each stage and this can be achieved by creating a new receiver through modification of the existing (i.e. characterisation) receiver (Figure 6.15b). This modification might be the addition or removal of a point mass, or perhaps a stiffening element such a beam or flange. The exact type of modification is not important, so long as it provides a sufficient change in the receiver's dynamics (this can be checked by comparing the FRFs measured on the original and modified assembly).

In equation form the modified receiver validation reads,

$$\bar{\mathbf{f}}_c = \mathbf{Y}_C^+ \mathbf{v}^{(\text{meas})} \quad \rightarrow \quad \mathbf{v}^{(\text{pred},\tilde{C})} = \mathbf{Y}_{\tilde{C}} \bar{\mathbf{f}}_c \quad \rightarrow \quad \mathbf{v}^{(\text{meas},\tilde{C})} \stackrel{?}{=} \mathbf{v}^{(\text{pred},\tilde{C})} \tag{6.32}$$

where it is noted that $\tilde{C}$ represents the modified assembly. Clearly, $\mathbf{v}^{(\text{meas},\tilde{C})}$ requires that the source is operated a second time with the receiver modification in place. Similarly, it is necessary to remeasure the FRFs $\mathbf{Y}_{\tilde{C}}$ for the validation step. With the modification in place, it would be fair to use any of the measured DoFs for comparison (as they are all different to those used in the characterisation).

Importantly with this form of validation, while the receiver itself is modified the source remains attached throughout. In practice, any modification of the contact has the potential to introduce some variation in the interface dynamics; for example, if the source and receiver were unbolted it is difficult to reattach the bolts in precisely the same way. This method therefore allows validation whilst avoiding the additional complication of variations due to dis(assembly) of the source and receiver. With the source having to be run a second time, the validation also includes the effect of any variation/repeatability issues in the operation of the source.

Transferability validation

A transferability validation, like the modified receiver validation, aims to predict the response in a new assembly. It does this though, by physically replacing the receiver. That is, the source is detached from the characterisation receiver and transferred to a completely new receiver (see Figure 6.15c). Thus, unlike on-board and modified-receiver validation, the transferability validation includes any error or variance due to reproducibility, i.e. detachment and reattachment (see Table 6.2). In this sense, it provides an error estimate more representative of what might be obtained in an actual VAVP prediction.

In equation form, the transferability validation reads,

$$\bar{\mathbf{f}}_c = \mathbf{Y}_C^+ \mathbf{v}^{(\mathrm{meas})} \quad \rightarrow \quad \mathbf{v}^{(\mathrm{pred},X)} = \mathbf{Y}_X \bar{\mathbf{f}}_c \quad \rightarrow \quad \mathbf{v}^{(\mathrm{meas},X)} \stackrel{?}{=} \mathbf{v}^{(\mathrm{pred},X)} \tag{6.33}$$

where X represents the assembly with a new receiver. As with the modified receiver, one must rerun the source to measure $\mathbf{v}^{(\mathrm{meas},X)}$ and remeasure the FRFs $\mathbf{Y}_X$.

Full VAVP validation

A final level of validation can be provided by replacing the measured assembly FRF $\mathbf{Y}_X$, as used in the transferability validation, with one obtained by a substructuring prediction from the separate components (see Figure 6.16). This form of validation moves us on from component-level to the level of the whole VAVP and combines all sources of error that would be encountered in a full VAVP prediction.

Substructuring validation

The new element in the full VAVP validation is the prediction of the assembly FRFs by substructuring. This procedure will validate the interface FRFs of both components as well as the substructuring computations. It will also provide a check on the interface model similar to that in a completeness test. Ideally, the substructuring will be done as a separate step prior to applying the blocked force to the assembly and comparing the predicted output (see Figure 6.16).

In a substructuring validation, the component is coupled to a matching component both physically and by substructuring computations and the measured and predicted assembly FRFs compared. The matching component may be a real component or a dummy component designed to replicate the dynamics of the intended installation. The procedure is outlined in Figure 6.16: assuming previously measured free-interface FRFs are available for the component in question, $\mathbf{Y}_S$, the steps in more detail are:

- a suitable 'dummy' component is selected and its free-interface FRFs obtained, $\mathbf{Y}_R$;
- the assembly FRF, $\mathbf{Y}_C$ is predicted using any of the substructuring methods from Chapter 5;

- the two components are physically coupled and the assembly FRFs measured;
- directly measured and predicted FRFs, $\mathbf{Y}_X$ and $\mathbf{Y}_C$ respectively, are compared.

Whatever method is used in the substructuring calculation, it will require a full set of interface FRFs for both components (see Chapter 5). Ideally, assembly FRFs will also be measured for all of these DoFs for comparison, however, the comparison can be conducted at remote points on the matching substructure if necessary, albeit with reduced confidence in the conclusions, for example if access to the assembly interface is impeded.

Whilst in the case of blocked forces, their measurement and validation can both be done in the same assembly (using on-board validation), this is not the case for substructuring. Recall from Chapter 2 that independent component FRFs can only be directly measured with the component in isolation, usually freely suspended and any validation will require a separate setup with the component attached to a 'dummy' component.

Test rig design

In some cases, it is advantageous to characterise the blocked force on a specially designed test rig. The question then arises as to how it should be designed and specifically, how similar or different it should be to the real installation. Intuitively two receivers that are dynamically very different will provide a more rigorous test in a validation test, particularly of the assumption of blocked force invariance. The extreme case is where one approximates a free, and the other a fixed boundary; if one of these extremes can be predicted from the other then one would expect also to be able to predict any intermediate state and thus one might argue that the characterisation could be considered validated for any installation. For example in [29] a free response was successfully predicted from a fixed one and vice versa.

However, whilst this may provide a reassuring test of blocked force invariance it may not be necessary, or advisable, to go to these extremes if they lie outside the normal range for the given source. Recall from the discussion of mobility matching in Chapter 3 that source-receiver interaction tends to be more complicated when their mobilities are closely matched. Source components that are designed for a narrow range of receiver structures, for example for resilient mounts or installation on a vehicle subframe, would not necessarily expect to encounter a full range of mobility ratios from free to fixed. The most important requirement in these cases is that the blocked force and validation assemblies are representative of the range of dynamic behaviour expected from real installations.

Importantly, 'representative' implies simultaneously in all DoFs. Thus, one approach to achieving representative boundary conditions is to design a test rig of a similar construction to the eventual installation, e.g. a beam or plate of similar material and thickness, or an isolator of e.g. low or high bending stiffness. In this way, one can be confident that the source operation is representative of what can be expected in the real installation. Other considerations for design of test rigs are:

- ease of access - the better the access, the better the quality of FRFs that can be expected;
- low background noise, ideally the rig should be isolated from other noise sources;
- the need for loading the source to achieve normal operation.

Most source component suppliers have well-developed expertise in to how to load their own components realistically for testing. The design challenge then becomes how to apply these techniques in a way such that the load does not become an interfering source of background noise for the operational tests (see Figure 6.14).

Methods of comparison

Having obtained some predicted responses (or FRFs) there are numerous ways in which their agreement with direct measurement can be evaluated. The simplest way is by visual inspection of the measured and predicted responses when plotted on the same axes (examples of which can be found throughout Chapters 9 and 10). This is the most common form of comparison and has the advantage that peaks and troughs in the predicted response are visible so one can make a judgement as to the overall importance of any discrepancies in a given frequency range. Perhaps for this reason, this method is recommended in the blocked force standard [14]. However, it is often necessary to provide a *quantitative* measure of agreement, and so below we discuss various methods for quantifying the similarity or difference between a measurement and prediction. Specifically, we discuss the *mean squared error* (MSE), *correlation*, and *conjugate*-based approaches. We also expand on the idea of *cross-validation* which was introduced earlier in relation to the on-board validation.

It should be noted that whilst the following methods of comparison are described with respect to measured and predicted *responses*, they can easily be applied to FRFs by simply transforming them into vectors by vectorisation ($\underset{\sim}{\mathbf{Y}}_C$, see Equation 8.28).

Mean squared error

In general terms, one may evaluate the quality of a model from its *residual* which, adopting the notation from above, is given by,

$$\Delta \mathbf{v} = \mathbf{v}^{(\text{meas})} - \mathbf{v}^{(\text{pred})} = \mathbf{v}^{(\text{meas})} - \mathbf{Y}_C \bar{\mathbf{f}}_c. \tag{6.34}$$

To compare more than one DoF simultaneously, it is more convenient to look at the mean of the squared residual (termed the Mean Squared Error, or *MSE*),

$$MSE = \frac{1}{N} ||\mathbf{v}^{(\text{meas})} - \mathbf{Y}_C \bar{\mathbf{f}}_c||_2^2 \tag{6.35}$$

where $||\square||_2^2$ denotes the squared L_2 norm and N is the number of DoFs being averaged.

The MSE as defined above is a frequency-dependent quantity with units of velocity squared which gives a measure of the error averaged over a number of DoFs. It is also possible to average over frequency so as to give a single number indicator.

When applied to the results of an on-board validation, it is important to distinguish between:

a) predicting the response at the characterisation DoFs and,
b) predicting the response at some new (unused) DoFs.

Note that this is not required for modified receiver or transferability validations, as the predicted DoFs are always different to those of the characterisation.

In the former case, the MSE gives an indication of how well the acquired blocked force is able to reproduce the responses from which it was determined. Intuitively, a more genuine test would require prediction of an *unknown* response, i.e. the latter case. We might refer to this as *cross-validation.* Here, more responses are measured than are needed for characterisation, one or more of which are not used in the calculation of blocked forces but are reserved for later comparison. The advantage is that circularity is to some extent avoided by predicting a response that was 'unknown' at the stage of determining blocked forces.

In a cross-validation, the measured response vector is in effect partitioned into a *training set* (including c and b) and a *validation set* (say $\tilde{b}$). The blocked force is characterised using the training set according to Equation 6.21 and then used to predict the response at the validation set according to,

$$\mathbf{v}_{\tilde{b}}^{(\text{pred})} = \mathbf{Y}_{C\tilde{b}c}\bar{\mathbf{f}}_c. \tag{6.36}$$

The MSE is then evaluated between the measured and predicted responses at the validation DoFs $\tilde{b}$.

Cross-validation is a standard method of validation within the statistics and machine learning community, and there are many ways to select the training and validation sets. For blocked force characterisation, with relatively small data sets a *'leave one out'* cross validation is often employed where a single response is held back for validation. Clearly, any remote response can be left out so that multiple cross checks can be performed with the same data by leaving out a different responses each time. If more than one, say p, responses are held back each time this is known as a *'leave p out'* cross-validation.

Correlation-based approach

An alternative to the mean squared error is to compute the *correlation* between measured and predicted response vectors,

$$\rho = \text{Corr}\left(\mathbf{v}^{(\text{meas})}, \mathbf{v}^{(\text{pred})}\right) \qquad \text{with} \qquad \text{Corr}(\mathbf{a}, \mathbf{b}) = \frac{\left|\mathbf{a}^{\text{H}}\mathbf{b}\right|^2}{\mathbf{a}^{\text{H}}\mathbf{a} \cdot \mathbf{b}^{\text{H}}\mathbf{b}} \tag{6.37}$$

where $\text{Corr}(\mathbf{a}, \mathbf{b})$ is based on a reformulation of the Cauchy–Schwarz inequality $\left|\mathbf{a}^{\text{H}}\mathbf{b}\right|^2 \leq \mathbf{a}^{\text{H}}\mathbf{a} \cdot \mathbf{b}^{\text{H}}\mathbf{b}$. The correlation provides a single (frequency dependent) number between 0 and 1. A value less than 1 indicates that the measured responses are not perfectly reproduced by the prediction. Note this is the same correlation-based metric as used by the Modal and Frequency Response Assurance Criterion (MAC, FRAC) [25].

The value of $\text{Corr}(\mathbf{v}^{(\text{meas})}, \mathbf{v}^{(\text{pred})})$ can be interpreted as the cos of the 'angle' between the two complex vectors $\mathbf{v}^{(\text{meas})}$ and $\mathbf{v}^{(\text{pred})}$. This interpretation reveals two important features of the correlation-based comparison. First, to compute their correlation, $\mathbf{v}^{(\text{meas})}$ and $\mathbf{v}^{(\text{pred})}$ *must be vectors*; a response prediction is required for at least 2 positions. If a single response prediction is used, the correlation will simply return 1 regardless of the agreement obtained. Secondly, the angle between two vectors is independent of their respective magnitudes, and so the correlation is not affected by any constant off-sets in the measured or predicted responses. That is, $\mathbf{v}^{(\text{meas})}$ and $\mathbf{v}^{(\text{pred})}$ would have the same correlation as $\mathbf{v}^{(\text{meas})}$ and $10 \times \mathbf{v}^{(\text{pred})}$.

With the above in mind, a correlation-based comparison is best applied to the whole set of measured responses, rather than a selected few as is done for cross validation. If a single value across the entire frequency range is required, the responses obtained for each frequency can simply be stacked to form a single, much larger, vector, and the correlation function used as before.

Conjugate-based

An alternative to the correlation-based comparison, that is sensitive to constant changes in magnitude, can be formulated by noting that given $a, b \in \mathbb{C}$,

$$(a+b)(a+b)^* = aa^* + bb^* + a^*b + ab^* \overset{a=b}{\rightarrow} 2(aa^* + bb^*) \tag{6.38}$$

where $\square^*$ denotes the complex conjugate. Based on the above, a conjugate-based comparison is given by [22],

$$\kappa = \text{Conj}\left(\mathbf{v}^{(\text{meas})}, \mathbf{v}^{(\text{pred})}\right) \qquad \text{with} \qquad \text{Conj}(a,b) = \frac{(a+b)\cdot(a+b)^*}{2(aa^* + bb^*)} \tag{6.39}$$

where it is noted that $\text{Conj}(a,b)$ is applied element wise to $\mathbf{v}^{(\text{meas})}$ and $\mathbf{v}^{(\text{pred})}$ (such that κ is a vector), and so does not require multiple response positions. If multiple responses are available, their individual conjugate comparisons can be averaged to obtain a single value. Like the correlation-based comparison, $\text{Conj}(a,b)$ is bound between 0 and 1, with 1 indicating perfect agreement.

6.5 INSTRUMENTATION

The component-based approach is entirely based around the idea that the vibro-acoustic behaviour of a physical prototype can be represented in digital form. However, we have so far said little about the hardware by which conversion, or *transduction* from physical to digital takes place. These topics are addressed in this section although the treatment is inevitably brief since (a) transduction is a large subject and (b) the hardware itself is constantly changing and evolving so that any detailed description would quickly go out of date. The interested reader is encouraged to refer to the many excellent sources of detailed and up-to-date information, often provided by manufacturers themselves. Readers without a particular interest in the hardware may skip this section.

The main hardware required to build a VAVP consists of *sensors* and *means of excitation* which are addressed below. Data acquisition systems and associated electronics are also required, though not discussed here.

6.5.1 SENSORS

The vast majority of vibro-acoustic measurements are conducted using piezoelectric accelerometers and force transducers combined with condenser microphones. Other sensing technologies may be required in specific situations, such as to measure down to low frequencies for tactile vibration, and are described later.

Sensor characteristics

Whatever sensing technology is employed, it is important to understand some basic properties of sensors as follows.

Sensitivity

The *sensitivity* is the ratio of electrical output to mechanical or acoustic input and is strictly valid only at a single reference frequency. For typical vibro-acoustic sensors it has units of millivolts per physical unit (acceleration, velocity, displacement, force or sound pressure). Too low, and the measurements will be noisy, but too high a sensitivity may result in overloads.

Frequency response

The *frequency response* is effectively the frequency-dependent sensitivity of the sensor normalised to the stated sensitivity $G(\omega) = s(\omega)/s(\omega_{ref})$ or in dB as $20log_{10}(G(\omega))$ where $s(\omega)$ and $s(\omega_{ref})$ are the frequency dependent sensitivity and that at the reference frequency, respectively. Frequency response is often only defined for amplitude although phase variation is also important.

Linearity

For an ideal linear transducer a change in input corresponds to a proportional (linear) change in output. However, in practice the output typically becomes non-linear at high amplitudes, for example due to large displacements or overloading the electronics. This provides an upper limit to the amplitude that can be reliably measured.

Noise floor

All sensors have a level of self noise, either in the transducer itself or the associated electronics, so that they generate an output even when no genuine input signal is present. The noise signal is typically random, producing a flat broadband output spectrum known as the *noise floor*. The true signal should be sufficiently above the noise floor to avoid contamination, usually taken as at least 10 dB and preferably at least 20 dB.

Dynamic range

Dynamic range is the working range of amplitudes for the combination of the sensor and associated electronics. It is the ratio between the smallest amplitude that can be sensed without noise contamination to the largest amplitude that can be accepted without distortion or overload. With older technology, the dynamic range of the measurement chain as a whole was often limited by that of the electronics, but a much wider dynamic range is now available [10].

Frequency range

The working frequency range of a sensor is that over which its sensitivity is within a given percentage of its rated sensitivity. Piezo-electric devices have an inherent low frequency limit and are unable to measure down to 0 Hz, hence alternatives may need to be considered for low frequency measurement. At high frequencies the dynamics of the sensor itself become apparent, for example, mass-spring resonances of accelerometers and diaphragm resonances of microphones (both typically occur above the frequency range for VAVPs). Frequency range is normally defined in terms of amplitude, but phase may also be important.

Piezo-electric accelerometers

Piezoelectric accelerometers are ideal for most vibro-acoustic applications. A piezoelectric material/crystal is sandwiched between a small mass, m, and the sensor housing, as illustrated in Figure 6.17. Movement of the housing causes an inertia force proportional to the acceleration according to Newton's 2nd law, $f = ma$. This in turn causes an electrical charge in the crystal proportional to the force and so the acceleration. The charge is then converted to a voltage signal by a charge amplifier.

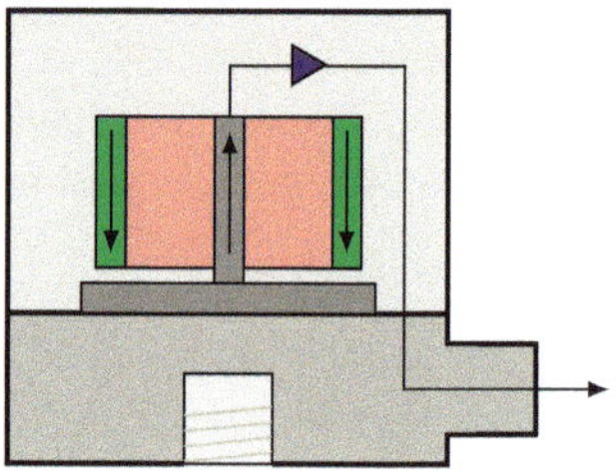

Figure 6.17 Schematic of conventional uni-axial IEPE accelerometer. Piezoelectric crystal is sandwiched between the rigid housing (gray) and an inertial mass (green).

'Piezoelectric' means 'electricity resulting from pressure' and piezoelectric transduction is hence said to be *force sensitive*. In principle, the forces can be applied by compression, by shear or by flexion (bending) which provides a range of options for building the elements into transducers.

Sensitivity typically ranges from 0.005 (for shock applications) to 1000 mV/ms^{-2} (for seismic applications). For most VAVP applications, sensitivity of the order of 10 mV/ms^{-2} is appropriate, although higher might be required for measurements on heavy-weight components or through well isolated transmission paths.

Accelerometers are designed to exhibit a maximum sensitivity along a particular axis. However, there will always be some *transverse sensitivity* such that there will be some output even when presented with a purely transverse signal. Tri-axial accelerometers allow simultaneous data capture in 3 orthogonal x, y and z directions. Although there is more flexibility with placement of uni-axial types, tri-axial models have the advantage of compactness, especially for interfaces where space is limited.

Most modern sensors are of the *Integrated Electronics Piezo-Electric* (IEPE) type with an integrated charge amplifier. These require a constant current power supply via the signal cable but are less susceptible to interfering noise and often equipped with a *Transducer Electronic Data Sheet* (TEDS) allowing them to be setup in the data acquisition system automatically (an important benefit for larger measurement arrays). Older models are often passive *charge mode* accelerometers which require an external charge amplifier and manual inputting of sensitivity.

An accelerometer must be rigidly fixed to the test object so as to faithfully capture its acceleration. The most common method is to use cyanoacrylate glue, although screwed studs can be used for more permanent fixings [12]. Only a small drop of glue is required for a secure fix suitable for up to typically 10 kHz. It is important that between each test, any remaining glue is removed from the sensor base, as this can build up and lead to imperfect contact in future tests.

Once calibrated, piezo devices are extremely stable and their sensitivity does not normally change significantly unless there is some damage.

Piezo-electric force transducers

Piezoelectric force transducers operate on a similar principle to accelerometers but no mass is required because the crystal is compressed by the force between two substructures (see Figures 6.18 and 3.3). Force transducers and 'force ring' sensors usually include an additional pre-load bolt in their construction, as in Figure 3.3 (not shown in Figure 6.18) that may also fix the test objects together. If so, the calibration must account for the portion of the contact force that passes through the bolt and is not sensed by the piezo element (as described in 3.1.2).

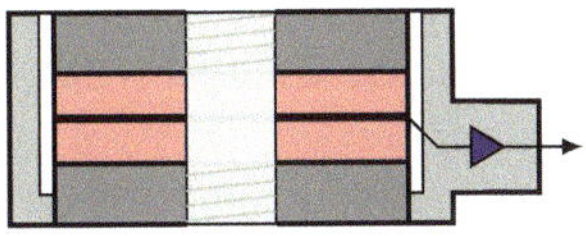

Figure 6.18 Schematic of conventional piezoelectric force transducer. The central hole is for a pre-load bolt (not shown).

Like accelerometers, tri-axial, or more specifically, 3-component, force transducers are available. These produce signals proportional to the out-of-plane (z) compression/tension forces, but also the in-plane (x and y) shear forces applied across the sensor.

Most modern piezoelectric force transducers, like accelerometers, are available with built in electronics (IEPE) and with their sensitivity on a chip (TEDS) for automatic setting up.

It is important when installing a force transducer that it is correctly aligned with the test objects such that it is loaded uniformly across its surface, minimising any edge loading. Where the surfaces are not flat or parallel, this can be difficult and an intermediate surface may be needed to improve alignment.

Condenser microphones

The *condenser* (or capacitor) microphone employs a thin membrane, coated in a conductive material and held a short distance away (though electrically isolated) from a conducting back plate. The membrane and back-plate are polarised with an applied voltage so that a capacitor is formed. Movement of the diaphragm caused by incident sound pressure then causes the capacitance to vary and this change is converted to a fluctuating voltage proportional to the sound pressure. The polarisation may be achieved with an external supply, often 200V, delivered via the signal cable. *Pre-polarisation* can also be achieved (such as in the *electret microphone*) without external supply, by coating the back-plate with a polymer with an embedded charge.

Condenser microphones designed for measurement (as opposed to sound recording) provide a flat frequency response over a wide range. The diaphragm is however delicate and its sensitivity can be affected by dust or water droplets. Hence, careful handling is important and the calibration should ideally be checked before every measurement.

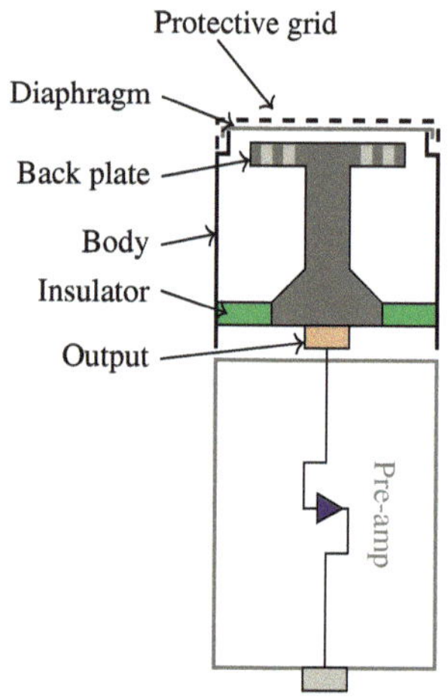

Figure 6.19 Condenser microphone.

For high frequency measurements (above about 10 kHz) there is a subtle difference in the operation of free-field, diffuse field and pressure microphones. *Free-field* microphones are designed to correct for their own presence on the sound field, and output a level that would be measured if the microphone was not present. They are designed to measure sound at normal incidence, and so should be pointed directly towards the source. *Diffuse-field* microphones are designed to respond uniformly to signals arriving simultaneously from all angles. *Pressure* microphones output the pressure on the surface of their diaphragm, making no corrections for its physical presence. As such, pressure microphones should be flush mounted on a boundary (e.g. a wall) to avoid diffraction effects. For most VAVP applications a diffuse field microphone is appropriate. The response for other types can be corrected if necessary [18].

Alternative sensing methods

The combination of piezoelectric accelerometers and force transducers with condenser microphones as described above is sufficient for most VAVP applications. However, other sensing technologies are available and may be advantageous in some situations, such as to extend to low frequencies for tactile vibration or for non-contact measurements. Some alternatives are introduced below, some of which may need special hardware in the data acquisition system.

Laser doppler vibrometry

For light structures at high frequencies, the mass loading of accelerometers can be significant. Whilst this can be corrected through procedures akin to substructure decoupling [4, 5, 15], there exist alternative *non-contact* measurement methods that avoid this issue entirely, whilst also providing additional benefits. The most well establish of these methods is Laser Doppler Vibrometry (LDV) in which the velocity of a surface is sensed from the Doppler shift of reflected laser light [17].

LDV provides an accurate measure of velocity [21] and can be applied in situations where standard accelerometers cannot function, including over distances up to hundreds of metres, at frequencies above 20kHz, on high temperature surfaces and even on rotating surfaces such as axles, shafts or spindles.

Perhaps the main disadvantage of LDV, other than cost (relative to the humble accelerometer) is that it provides only a single point measurement. To overcome this issue, *scanning* LDVs are available, employing servo controlled mirrors to redirect the laser without having to move the whole device. Scanning LDVs can provide high resolution measurements of surface velocity fields, ideal for visualising and extracting mode shapes and other quantities such as structural intensity. Nevertheless, the measurements are still taken sequentially, as opposed to simultaneously as with an accelerometer array. Use of LDV for multi-DoF FRF measurements therefore relies on precise repetition of the input. Further limitations are that LDVs measure only out-of-plane velocity and require direct line of sight, so cannot be used for many enclosed geometries. It is also essential that the measurement unit itself is isolated from unwanted vibration.

Laser displacement sensors

In contrast to LDVs, Laser Displacement Sensors (LDSs) are far simpler in their operation, though more limited in terms of measurement capabilities. LDSs have a narrower working distance range than LDV, with an optimal standoff distance typically a few cm from the test object. Frequency range is also more limited, partly because LDSs sense displacement whose amplitude reduces at high frequency compared with velocity. Nevertheless, for their relative cost, LDSs offer some advantages over their more costly LDV cousins.

Strain gauges

Strain gauges use the principle that when a conductor is stretched or compressed, its electrical resistance changes in a measurable way. This effect is known as *piezoresistivity*. Strain gauges typically consist of thin printed conductors of metal or semi-conductors on a strip that is rigidly glued to the surface of a solid structure. The change in resistance is proportional to the strain in a given direction.

Strain gauges are often built into load cells which can be calibrated and used to measure low frequency forces down to d.c. Compared with piezoelectric force transducers however, their dynamic range is more limited. Furthermore, they require an external power supply and the sensing element must be built into a Wheatstone bridge arrangement to enable its resistance to be measured. Temperature compensation may also be required in some applications, although these functions are built into some modern analysers.

Electro-dynamic sensors

For measurement of a low frequency structural response, it may be necessary to employ *geophones*, which use the *electro-dynamic* principle with a coil and a magnet. Either the coil or the magnet is attached to the housing and the other acts as an inertial mass whose relative movement causes a current to flow in the coil. The output signal is directly proportional to *velocity* giving them improved low frequency sensitivity compared with accelerometers.

Electro-dynamic 'moving coil' microphones are also available but do not generally have significant advantages over condenser microphones for measurement. The same principle can however be used with a loudspeaker used 'in reverse' to act as a microphone. This technique has proved useful for reciprocal measurement of volume velocity [20].

MEMS sensors

Micro-Electro Mechanical Systems (MEMS) sensors are now available, such as those used in mobile phones. While their dynamic range and frequency response is typically inferior to that of piezoelectric sensors, their reduced size means they can be embedded into structures to make measurements at inaccessible or 'buried' interfaces. For example in [24], Zabel et al. obtained blocked forces of a closely-coupled electric motor using embedded MEMS accelerometers that would not otherwise have been possible.

6.5.2 EXCITATION METHODS

For measurement of FRFs it is necessary to excite the component or assembly with a known input. This requires a means both of applying the excitation and sensors to register its strength. Structural excitation is applied either with a force hammer or a shaker. For acoustic excitation, a calibrated volume-velocity source is used.

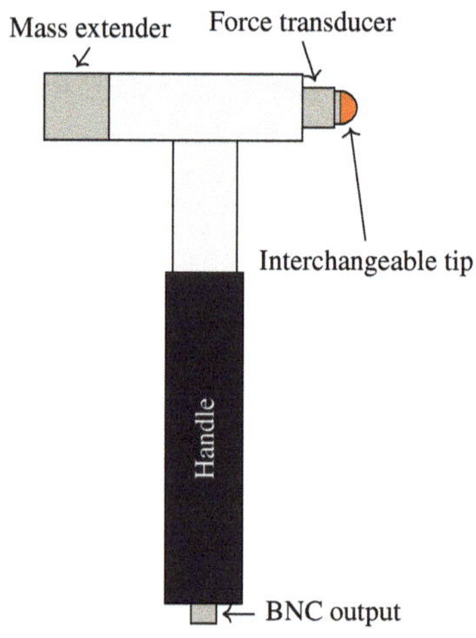

Figure 6.20 Typical instrumented force hammer.

Instrumented Hammer

Instrumented force hammers (impact hammers) include integrated piezoelectric force transducers to measure the force with which the test object is struck. Some control can be exerted over the frequency range of the excitation by changing the hammer tip and/or mass: stiff tips (like metal) provide short impulses, containing greater high frequency content, whereas softer materials provide longer impulse duration, with greater low frequency energy. The maximum frequency is typically up to 5 kHz for large hammers and 20 kHz for small ones.

Impact hammers have been an integral part of modal testing for decades and are available in a range of sizes, ranging from miniature designs for exciting lightweight objects such as circuit boards, to large sledgehammer-style designs, for heavyweight structures like ships and buildings. As their size increases, their sensitivity generally decreases to cater for the high force levels being applied. For most typical problems and VAVP applications, a sensitivity in the range 1-10 mV/N would be well suited.

Impact hammers have the advantage of simplicity in that, unlike shakers, nothing is attached to the test object that could change its response. They are also able to access relatively inaccessible locations. The main disadvantage is that obtaining accurate and consistent results requires skill from the operator: inconsistency in the position and direction of the applied force increases uncertainty, particularly at anti-resonances as discussed in 6.3. Accidental 'double hits' are also difficult to avoid for lightweight/flexible structures.

Shakers

Shakers apply a force to the test object, usually through a thin 'stinger' and a piezoelectric force transducer. The high axial stiffness of the stinger transmits the shaker-generated force normal to the surface, whilst its lateral flexibility minimises any in-plane loads. Compared with impact hammers, they have the advantages of allowing long averaging times without inconsistency and the excitation signal can be shaped as desired. The main disadvantages are that they require setting up for each excitation position and that they may alter the test object dynamics by being attached to it.

Shakers were originally the preferred choice for modal testing but have largely been replaced by hammer excitation for VAVP applications because of the convenience when exciting at a number of positions. However, a new breed of mini-shakers is now available that can be fixed directly to the test object. The shaker itself acts as a reaction mass so that an inertial force is generated when the test object is excited. Such mini-shakers are generally limited to lower input forces but greatly simplify the attachment compared with the traditional, larger design.

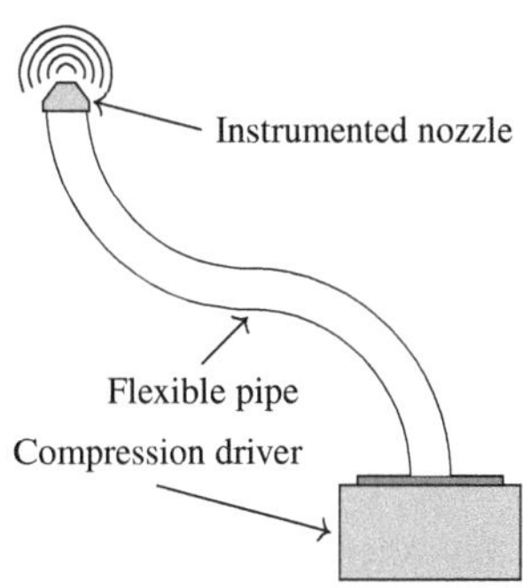

Figure 6.21 Typical volume velocity source design. Nozzle is made with relatively narrow opening to optimise omni-directionality.

Volume Velocity Source

The fundamental acoustic point source is the monopole, sometimes described as a small *breathing sphere*. It displaces fluid at a rate quantified by its *volume velocity* with units of m^3s^{-1}. An ideal monopole is *omni-directional*, meaning it radiates sound equally in all directions. Reproducing this omni-directionality is relatively easy for small sources at low frequency (much less than an acoustic wavelength in size) but becomes much more challenging as frequency increases.

A typical design uses a compression driver attached to a length of flexible pipe so as to facilitate precise placement. Small pipe diameters increase the omni-directional frequency range but have high internal impedance, and therefore lower output levels. Larger diameter pipes have higher output levels, but a narrower frequency range. An acceptable compromise can typically only be found over a limited frequency range. To circumvent this issue, manufactures sometimes offer two or three devices, each targeting a particular frequency band, together covering the entire range. The obvious disadvantage is that repeating tests with a different device lengthens the measurement time. Perhaps more importantly though, combining overlapping results from tests over different frequency ranges is rarely straightforward.

To measure the volume velocity generated, most devices employ a microphone embedded near the pipe opening. By measuring the radiated sound pressure level in free field (in an anechoic chamber), a calibration factor can be derived that converts the measured internal pressure to a volume velocity [3, 11]. This calibration factor also helps correct for the effect of standing waves within the pipe. Some designs incorporate two spaced microphones at the opening, which can be used to decompose the pipe's internal wave field to obtain the particle velocity and thus the volume velocity [9, 11]. This approach does not require any free field calibration measurements.

If vibro-acoustic FRFs are to be measured reciprocally then the volume velocity source must produce a measurable response in the solid structure; for heavyweight structures, e.g. engine blocks, etc. this can be challenging.

6.5.3 CALIBRATION

Proper calibration is of vital importance to ensure reliability and consistency of measurements. In principle, every measured quantity, for example acceleration in ms^{-2}, should ultimately be referred back to the *primary standards* for that quantity. Manufacturers and calibration laboratories are typically set up to perform *traceable* calibrations of this type according to procedures defined in International standards [1, 2]. Here, we focus on checks that can be performed in-house.

For piezoelectric sensors, the sensitivity generally changes very little over time unless there is some damage. In-house checks can be done by comparing the output with that of a reference accelerometer when both are subject to the same input, for example when mounted side by side on a shaker table. These *back-to-back* or *comparison* checks should show up any change in frequency response that might indicate damage.

Condenser microphones, on the other hand, are more prone to drift and so should have their calibration checked for every measurement. This is typically done with a portable calibrator that fits over the microphone and provides a *known input*, usually at a single frequency. Of course, the calibrator itself requires calibration. As well as these regular checks, condenser microphones will usually be sent to a calibration laboratory to be checked against primary references, often every 1 or 2 years.

Force transducer calibration is more difficult to check in-house. The most commonly used method is in conjunction with an accelerometer to measure the FRF of a rigid mass, m, whose mobility is known: $Y_{mass} = 1/i\omega m$ [7]. In effect this performs a back-to-back test but against an accelerometer rather than another force transducer. Although simple in principle, in practise it is difficult not to strike the mass at an angle. The mass is typically suspended from a frame, or rested on a table or simply in the operator's hand for the test, but none of these methods ever works perfectly. Nevertheless, it is usually possible to obtain a frequency response sufficient to check for changes in sensitivity or damage.

6.6 SUMMARY

Checking and validation of data at every stage of measurement is of paramount importance in building a VAVP. The most important check, recommended for every measurement is *coherence*, which will identify when the output of an FRF is not linearly related to its input. *Linearity* checks are possible using different levels of excitation. *Reciprocity* should be checked where reciprocal FRFs are available for comparison and *passivity* checks may help to identify noisy data and errors in sign convention. The *sensitivity* of a measured FRF matrix to error amplification on inversion is an important property and is quantified by the *condition number*. It is recommended when measuring FRFs that they are not immediately averaged, as is standard practice, but that data from each excitation, typically a hammer hit, is retained for later uncertainty analysis.

In addition to these individual measurement checks, *component-level* checks and validation are required to ensure that the measured data adequately represents the component. The most important is the *interface completeness* check which identifies the important DoFs at the interfaces. A simpler test of *interface rank* can be used to determine the *number* of interface DoFs but does not identify which ones they are. If the virtual point approximation is being used, then a check on *interface rigidity* is also advisable.

Blocked force measurement involves a two-phase test with a passive and an operational phase. Typically, sensors are required at *indicator* points in addition to at the interface. These do not form part of the characterisation but are needed as an intermediate step to over-determine the inverse solution. Good placement of indicator positions is important: they should have a good 'view' of the interface and should excite as many modes as possible but should not be too close together. In the operational phase, background noise should be measured with the source switched off prior to an operational test

Having obtained the blocked forces it is essential to carry out a staged validation. Four stages are outlined, each using the same procedures but with the source attached to a different receiver so as to introduce additional errors at each stage. In *on-board* validation the response is predicted in the same receiver as used for blocked force determination. *Modified receiver* validation uses the same receiver but modified and *transferability* validation uses a different receiver. The final stage, *full VAVP* validation, also includes predicting the assembly FRFs from those of the components so as to validate the substructuring step.

It is recommended that all stages of validation be conducted with *artificial excitation* as well as with operation excitation. Here, the same procedures are followed as an operational test but with a more controllable excitation provided by hammer or shaker on the source housing rather than by the real source operation.

For most VAVP measurements, piezoelectric accelerometers and force transducers combined with condenser microphones are sufficient. Alternative sensing methods such as lasers and strain gauges may be required for specific situations such as low frequencies and non-contact measurements. Excitation can be provided by hammer or shaker. Acoustic excitation, if required, uses a volume velocity source. The importance of calibration to international standard quantities is emphasised. Full calibration is usually only done by specialist laboratories but *back to back* comparison tests can usefully be performed in-house.

Bibliography

1. IEC 61094: multiple parts. Electroacoustics - measurement microphones.
2. ISO 16063: multiple parts. Methods for the calibration of vibration and shock transducers.
3. DK Anthony and SJ Elliott. A comparison of three methods of measuring the volume velocity of an acoustic source. *Journal of the Audio Engineering Society*, 39(5):355–366, 1991.
4. S Bi, J Ren, W Wang, and G Zong. Elimination of transducer mass loading effects in shaker modal testing. *Mechanical Systems and Signal Processing*, 38(2):265–275, 2013.
5. O Cakar and KY Sanliturk. Elimination of transducer mass loading effects from frequency response functions. *Mechanical Systems and Signal Processing*, 19(1):87–104, 2005.
6. N Contartese, E Nijman, and W Desmet. A procedure to restore measurement induced violations of reciprocity and passivity for frf-based substructuring. *Mechanical Systems and Signal Processing*, 167:108556, 2022.
7. DJ Ewins. *Modal testing: theory, practice and application.* John Wiley & Sons, 2009.
8. F Fahy and P Gardonio. Sound and structural vibration: Radiation, transmission and response. 2007.
9. S Gade, N Møller, J Hald, and L Alkestrup. The use of volume velocity source in transfer measurements. In *Proceedings of the 2004 International Conference on Modal Analysis Noise and Vibration Engineering (ISMA)*, pages 2641–2648, 2004.
10. NJ Jacobsen and OT Andersen. Dyn-x technology: 160 DB in one input range, Technical Review no. 1. Technical report, Brüel & Kjær, 2006.
11. A Lindberg and G Pavic. Experimental characterisation of a small compression driver using an internal microphone. In *Proc. of the International Conference on Noise and Vibration Engineering (ISMA)*, pages 1111–1119, 2014.
12. J Mathews. Guide to accelerometer installation. *ENDEVCO Corporation, Technical Paper*, 319.
13. AT Moorhouse and BM Gibbs. Simplified characterisation of multiple point excited structures using mobility matrix eigenvalues and eigenvectors. *Acta Acustica United with Acustica*, 84(5):843–853, 1998.
14. ISO 20270:2019 Acoustics — characterization of sources of structure-borne sound and vibration — indirect measurement of blocked forces, 2019.
15. O Özsahin, HN Özgüven, and E Budak. Analysis and compensation of mass loading effect of accelerometers on tool point frf measurements for chatter stability predictions. *International Journal of Machine Tools and Manufacture*, 50(6):585–589, 2010.
16. RB Randall. Frequency analysis. *Naerum (DK): Bruel & Kjaer*, 1987.

17. SJ Rothberg, MS Allen, P Castellini, D Di Maio, JJJ Dirckx, DJ Ewins, BJ Halkon, P Muyshondt, N Paone, T Ryan, et al. An international review of laser doppler vibrometry: Making light work of vibration measurement. *Optics and Lasers in Engineering*, 99:11–22, 2017.
18. GRAS Sound and Vibration. What are the free-field and random incidence corrections? how should i use them? `https://www.grasacoustics.com/component/dc/?dc_id=83#:~:text=Page%201-,What%20are%20the%20free%2Dfield%20and%20random%20incidence%20corrections%3F,due%20to%20the%20diffraction%20effect`. Accessed: 10-01-2024.
19. G Strang. *Linear algebra and its applications*. Academic Press, 2012.
20. T Ten Wolde. Reciprocity measurements in acoustical and mechano-acoustical systems. review of theory and applications. *Acta Acustica United with Acustica*, 96(1):1–13, 2010.
21. F Trainotti, TFC Berninger, and DJ Rixen. Using laser vibrometry for precise frf measurements in experimental substructuring. In *Dynamic Substructures, Volume 4: Proceedings of the 37th IMAC, A Conference and Exposition on Structural Dynamics 2019*, pages 1–11. Springer, 2020.
22. MV van der Seijs, D van den Bosch, DJ Rixen, and D de Klerk. An improved methodology for the virtual point transformation of measured frequency response functions in dynamic substructuring. In *4th ECCOMAS thematic conference on computational methods in structural dynamics and earthquake engineering*, volume 4, 2013.
23. MWF Wernsen, M Van Der Seijs, and D de Klerk. An indicator sensor criterion for in-situ characterisation of source vibrations. In *Sensors and Instrumentation, Volume 5: Proceedings of the 35th IMAC, A Conference and Exposition on Structural Dynamics 2017*, pages 55–69. Springer, 2017.
24. D Zabel, M Sturm, T Alber, and AT Moorhouse. Embedded mems accelerometers for the in-situ measurement of blocked forces in coupled structures. In *DAGA*, volume 1, pages 7–10, 2017.
25. RJ Allemang. The modal assurance criterion–twenty years of use and abuse. *Sound and vibration*, volume 37(8), pages 14–23, 2003.

7 Processing

7.1 INTRODUCTION

In this chapter we focus on the processing techniques needed to transform raw measured data into a complete working VAVP. In outline, the steps in the whole process are as follows: raw analogue *time series* are captured, digitised and transformed to the frequency domain; the *frequency domain* data is then organised into matrix-vector form and independent component characteristics obtained, including by *solving the inverse problem* to obtain the blocked forces; the VAVP is then assembled by combining component properties and the output is calculated, potentially requiring transformation back to the *time domain* for auralisation. The main topics in this chapter follow this sequence, starting with *raw data* processing applied to operational (single channel) data in 7.2, leading on to the calculation, organisation and manipulation of *FRFs* in 7.3. We then consider *solving the inverse problem* in 7.4 and transformation back to the time domain in 7.5. A final section (7.6) considers the conversion to and from octave band data which raises the prospect of synthesising missing phase.

Many of the processing operations presented in this chapter are well known and described in more detail in standard texts [4, 16, 16]. Furthermore, most standard operations are typically built into data acquisition systems so that component-based simulation can often be achieved perfectly well without a detailed understanding of their inner workings. Nevertheless, an appreciation of the principles is important and thus the aim of the chapter is to provide a complete, albeit brief, presentation of the signal processing principles required for building a VAVP.

Recall that both operational and passive properties (FRFs) may be required to characterise a component, but in both cases the processing begins with the capture of analogue time series data. Thus, the first topic considered is the *discretisation* of these analogue signals into digital form. Windowing operations are also discussed here as a prelude to transformation to the frequency domain. The primary processing tool for time-to-frequency transformation is then introduced, i.e. the *Discrete Fourier Transform* (DFT). The DFT transforms a block of discrete time data into a *Fourier spectrum* also called a DFT (the same term being used both for the process of transforming and for the resulting spectrum). The DFT is the fundamental property on which all other frequency-domain properties are based. The first of these derived properties to be considered are the auto- and cross-spectra which, unlike Fourier spectra, have a constant phase reference that allows averaging.

To characterise a component's passive properties we require *FRFs*, i.e. *relative* quantities relating a measured output to an applied input. In theory, FRFs can be computed from the Fourier spectrum of the output normalised to that of the input. However, in the real world, there is always noise and averaging is required to improve signal-to-noise ratio. Thus in practice, FRFs are estimated based on combinations of auto- and cross-spectra. Application to single and multiple-output measurements is

DOI: 10.1201/9781003399131-7

discussed and various forms of *coherence* function, which quantify the linear dependence of the output on the input, are presented.

One of the most important aspects of processing is in *solving the inverse problem* to obtain blocked forces. This is a delicate process that can go badly wrong so an understanding of the inner workings is important. One technique for improving reliability is *over-determination*. A further important technique, often required in combination with over-determination, is *regularisation*. It must be however used with appropriate caution to avoid 'cheating'. As well as relating to blocked force determination, the important principles of matrix inversion presented here are also relevant for substructuring calculations.

Having presented the processing needed to obtain a frequency-domain output from a VAVP, the final step is to transform back to the time domain. However, this is not simply a matter of inverse transformation because the time structure of most operational signals is typically longer than a single DFT block. An overlap-add method is therefore presented which effectively allows time sequences of unlimited length to be propagated through the VAVP for subsequent auralisation. The traditional cross-spectral approach, which gives time-averaged responses, is also presented for completeness and comparison.

Finally, given that many standard measurements in acoustics employ processing in third octave and octave bands, we consider their relation to the narrow band analysis used for VAVPs.

7.2 RAW DATA PROCESSING

For most readers, the data processing described through 7.2 and 7.3 will be performed automatically by the data acquisition system in use. In this case, the primary role of the engineer is to: a) interpret and inspect the data as it is being collected to ensure good quality measurements are obtained, b) export the processed FRFs and operational responses and load this data into their preferred programming environment, and c) construct the necessary matrices and vectors for blocked force and substructuring calculations. Nevertheless, it is sometimes necessary to work directly with the raw time data, and so here we describe the key processing steps.

7.2.1 TIME SERIES DATA

There are generally two 'types' of quantity in a VAVP: relative and absolute. Relative quantities are the FRFs (i.e. the response output relative to some input) that characterise the passive properties of the component or assembly. Absolute quantities are typically the operational responses (e.g. velocities) of the component or assembly when the source is active. Whilst the interpretation and purpose of these quantities are quite different, both begin with the capture of time series data from one or more sensors installed on the component/assembly under study.

Signal discretisation

In general, without reference to any particular quantity or sensor type, a time varying voltage or current is output from the installed sensor(s) and input to an acquisition system (see 6.5 for some details on the common types of sensor used in a VAVP). The first main step in the data processing pipeline is to discretise the continuous input signal such that it can be stored on a computer and further processed. This discretisation can be thought of as a multiplication between our continuous input signal, and a pulse train made up of a series of (unit valued) delta functions separated by the small time interval Δt. This time period is inversely related to the so-called 'sampling rate' of the recorded signal, $f_s = 1/\Delta t$. The sample rate controls the theoretical maximum frequency that can be captured by the discretisation, $f_{\max} = f_s/2$, according to the Nyquist-Shannon sampling theorem [23]. This discretisation is illustrated in Figure 7.1.

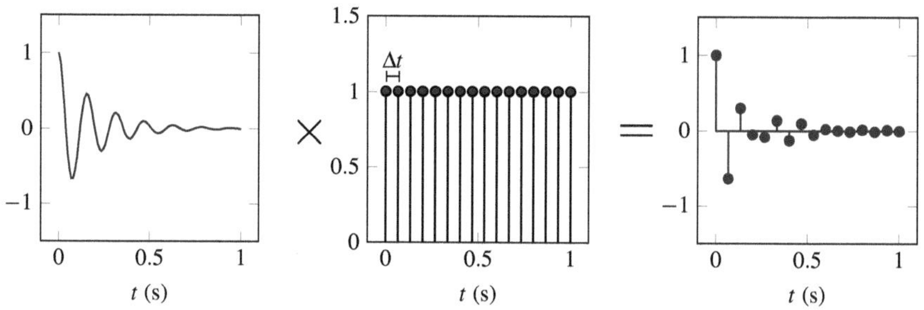

Figure 7.1 Sampling process – multiplication of continuous signal by pulse train.

A continuous signal of length L seconds, will be discretised into $N = L/\Delta t$ samples. The window length L (or equivalently the number of samples N) and sampling period Δt (or equivalently the sample rate f_s) are important measurement parameters, to be specified by the test engineer. They control the resolution obtained once measurements have been transformed into the frequency domain, for example whether we are able to make a prediction at every 1 Hz or 10 Hz.

Windowing

Before we transform our signal into the frequency domain, which is where all major calculations are performed, we may need to apply appropriate windowing functions. It will be seen later that transformation into the frequency domain is done with *blocks* of time data of specific length, N samples or $N\Delta t$ seconds in length. Block lengths in powers of 2 are typically used since this allows more efficient processing (e.g. 1024 or 2048 samples). If the time sequence is longer than the desired block length (chosen to achieve the desired frequency resolution), then it will need to be split into a number of blocks each of the desired length, N. However, simply truncating the signal after N samples may leave an abrupt cutoff which causes a phenomenon known as 'spectral leakage' upon transformation to the frequency domain. A range of windowing functions have therefore been developed to aid the smooth and efficient transformation to the frequency domain e.g. by tapering the beginnings and ends of each block. Naturally, these modifications introduce artefacts which must be minimised. Another reason to apply a window is to reduce noise in a portion of the time sequence that contains no signal (for example after a transient signal has finished). Whatever the reason for windowing, the windows are applied prior to transformation to the frequency domain.

Windowing a signal is achieved by multiplying the samples of the discretised signal with the appropriate window function,

$$x_n = w(n)\hat{x}_n \tag{7.1}$$

where $\hat{x}_n$ is the nth sample of the initial signal, $w(n)$ is the window function evaluated for sample n, and x_n is the windowed signal.

A variety of window functions $w(n)$ are available, each with their own particular benefits in terms of their spectral characteristics (main lobe width, side lobe level, fall off rate, etc). A comprehensive discussion of window types and their spectral characteristics is beyond the scope of this book (see [15, 21]) but the three most common windows used in the context of VAVPs are briefly described below and illustrated in Figure 7.2.

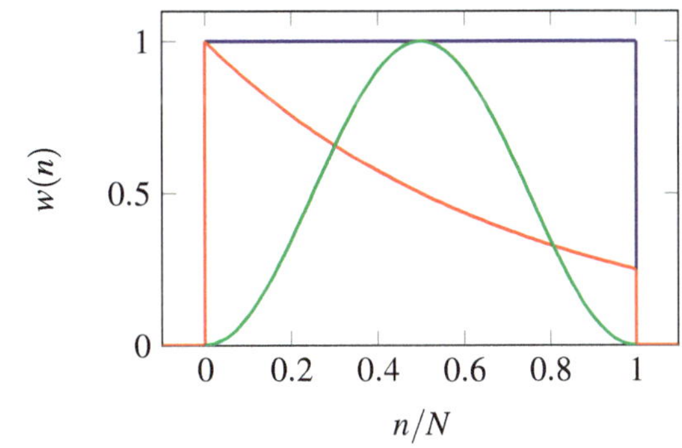

Figure 7.2 Rectangular, Hanning and exponential ($d = 0.25$) windows.

The *rectangular window* is in essence the 'default' window for any measurement. Any signal of length N can be thought of as the application of a rectangular window of length N to a signal of length $> N$. It is defined simply as,

$$w_r(n) = 1 \qquad 0 \leq n \leq N-1 \tag{7.2}$$

$$w_r(n) = 0 \qquad n > N-1. \tag{7.3}$$

It multiplies all signal values within the window by 1, and those outside by 0. The rectangular window is the window of choice for standard impact-based FRF measurements, for which the measured signals naturally start at zero, and providing there exists sufficient damping, decay to zero within the measurement period N.

The *Hanning window* (also Hann) is a continuous cos-based window that attenuates the beginning and end of a signal to zero. It is defined as,

$$w_h(n) = 0.5\left(1-\cos\left(2\pi\frac{n}{N-1}\right)\right) \qquad 0 \leq n \leq N-1 \tag{7.4}$$

$$w_h(n) = 0 \qquad 0 \leq n > N-1. \tag{7.5}$$

By tapering both ends of the signal to zero, the Hanning window enforces periodicity, minimising abrupt changes in the signal and thus reducing spectral leakage. It is usually a good choice for continuous signals (such as operational responses, or shaker-based FRF measurements). It should not be used for transient signals like impact based-FRF measurements, as the initial transient will be severely attenuated by the window function.

The *exponential window*, unlike the Hanning, is designed to attenuate primarily the end of a signal. Its principle application is to hasten the decay of a measured transient signal which would otherwise extend beyond the measurement period N. It is defined as,

$$w_e(n,d) = e^{\ln(d)\frac{n}{N-1}} \qquad 0 \leq n \leq N-1. \tag{7.6}$$

The attenuation is exponential, starting with a weighting value of 1, and finishing with a value of d at sample $N-1$. Different values of d can be used to provide a more or less rapid decay. Similar *asymmetric* windows can be built, for example using the half cos function etc. Their purpose is generally the same; to maintain the signal during the initial transient phase, whilst smoothly reducing the amplitude for later times.

The type of window used depends on the nature of the signal being measured. Generally, continuous signals (for example, those recorded during the operational response of a vibration source, or a shaker-based FRF measurement) should be processed using a symmetric window, for example the rectangular or Hanning. Transient signals (for example, those recorded during an impact-based FRF measurement) should be processed with a rectangular window or, if necessary, by a window that does not attenuate the beginning of the signal where the transient occurs (rectangular or exponential). The rectangular window is generally preferred, though in some cases where low damping is present, the signal decay can extend beyond the length of the measurement period. In this case an exponential (or similar) window can be applied to hasten the decay to zero. This should, however, be used with caution, as multiplication by an exponential window is equivalent to adding artificial damping to the system. It would be better practice to alter the measurement duration (L or N) such that the signal decays sufficiently in that time period (though this would of course alter the frequency resolution obtained).

Clearly, applying a window attenuates the signal amplitude, and therefore reduces the amount of energy in the signal. This reduction must be compensated for in order to obtain a correct level when transformed into the frequency domain. It is convenient to apply the scaling factor α along with the window itself as,

$$x_n = \frac{w(n)}{\alpha}\hat{x}_n \tag{7.7}$$

where $\hat{x}_n$ is the initial discretised signal and x_n is the windowed and scaled signal.

By scaling the signal whilst it remains in the time domain, the compensation will correct for the average amplitude or level of the signal. The amplitude correction factor is obtained from the (arithmetic) average level of the window,

$$\alpha = \frac{1}{N-1}\int_0^{N-1} w(n)dn. \tag{7.8}$$

For the rectangular window we get $\alpha_r = 1$. For the Hanning window, $\alpha_h = 0.5$. For the exponential window it depends on the parameter d and is given by $\alpha_e = \frac{d-1}{\log d}$.

Overlap

Quite often, one must make a measurement over an extended period of time, for example to capture the operational response of a particular manoeuvre like a run-up or run-down. It is sometimes useful to work with such sequences in their entirety, for example to plot against RPM or some other quantity. However, transforming lengthy sequences to the frequency domain results in spectra with narrow resolution that may not be compatible with other data. The solution is to adopt a sequential windowing method, as illustrated in Figure 7.3, where the measured signal is split into N_w, potentially overlapping, blocks each of length N.

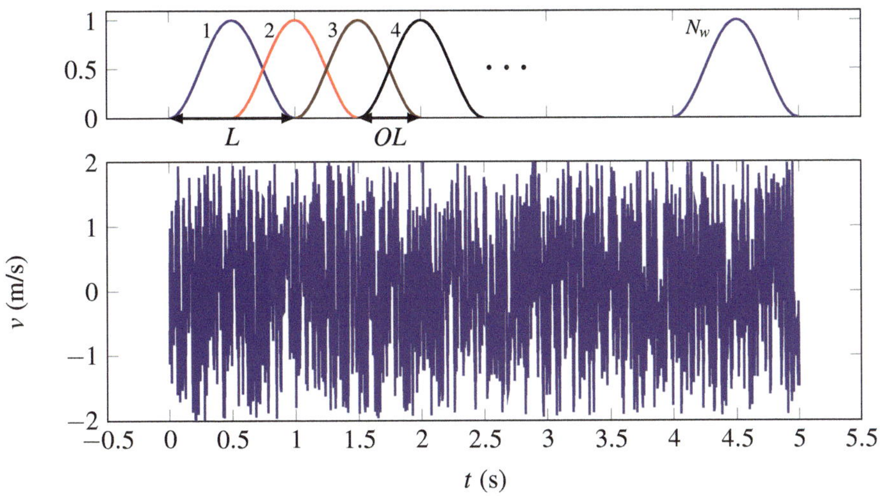

Figure 7.3 Windowing of time domain signals.

By splitting the signal in this way, we can chose to either 1) capture time varying behaviour by retaining all the individual blocks, or 2) obtain the time averaged response by computing a single average spectra from all the blocks. In either case, if a windowing function (e.g. a Hanning) is applied to each individual block, important information can be lost in the transition from one block to the next. For example if transients occur towards the beginning or end of the block, they will be attenuated by the tapered window function and their influence underestimated. A simple solution is to overlap the windows by an appropriate percentage, such that information lost by one block, is picked up its neighbours and thus correctly represented in any subsequent analysis.

If no windowing function is applied (i.e. a rectangular window is used), overlapping may still offer some benefit, as it would enable more averages to be performed over a given time period. In this way, the effect of overlapping is to provide a smoothing effect in the time averaged result. This can help make trends easier to identify.

If the intention is to produce an auralisation of a VAVP, it is important that we retain all time windows and that no information is lost in the transition from one block to the next. This enables the final result (i.e. a frequency domain VAVP prediction) to be converted back to the time domain and auralised, e.g. for subjective evaluation. Specific details on how this is most easily achieved are given in 7.5.

After discretising, windowing, amplitude compensating, and overlapping our measured signals, we are ready to transform them into the frequency domain.

7.2.2 FREQUENCY DOMAIN REPRESENTATION

The vast majority of component-based simulation and virtual prototyping is carried out in the frequency domain. This simplifies the underlying equations and reduces computational cost as each frequency component is dealt with separately. The frequency domain approach of course applies only to linear and time-invariant systems and if these assumptions are not met then more complicated time domain approaches will generally be required. However, these are beyond the scope of the book and the following discussion relates purely to frequency domain analysis.

With all our calculations being performed in the frequency domain, it is necessary to convert our discretised time domain signals (representing accelerations, forces, etc.) into their frequency domain representations. This is done by means of the Discrete Fourier Transform (DFT).

Discrete Fourier Transform

Before discussing the DFT, it is worthwhile considering, at least briefly, the (continuous) Fourier Transform.

A continuous time domain signal $x(t)$ can be transformed into an equivalent frequency domain representation by means of the Fourier Transform (FT). This transform is based on the notion that a real signal can be written as the summation of an infinite and continuous series of complex sinusoids,

$$x(t) = \int_{-\infty}^{\infty} \hat{x}(f) e^{i2\pi ft} df \tag{7.9}$$

where $\hat{x}(f)$ is a complex number that conveys both the amplitude and phase of the sinusoid at frequency f, $\hat{x}(f) = |\hat{x}(f)|e^{i\phi_{\hat{x}}(f)}$. The function $\hat{x}(f)$ for all values of f is said to be the frequency domain representation of $x(t)$. Alternatively, Equation 7.9 can be written as,

$$x(t) = \frac{1}{2\pi} \int_{-\infty}^{\infty} \hat{x}(f) e^{i\omega t} d\omega \tag{7.10}$$

where the integration is now over the radian frequency $\omega = 2\pi f$.

The frequency domain representation $\hat{x}(f)$ is obtained by multiplying the time signal $x(t)$ with a complex exponential of the desired frequency and integrating,

$$\hat{x}(f) = \int_{-\infty}^{\infty} x(t)e^{-i2\pi ft}dt. \tag{7.11}$$

Equation 7.11 is referred to as the Fourier Transform (FT), and Equation 7.9 as the Inverse Fourier Transform.

To transform a discrete signal into the frequency domain we use the aptly named *Discrete* Fourier Transform (DFT). For more detailed discussion on the interpretation and mathematical treatment of the FT and DFT the reader is referred to the previously mentioned textbooks [15, 21]. Here, we limit ourselves to the details necessary to understand the processing required to build a VAVP.

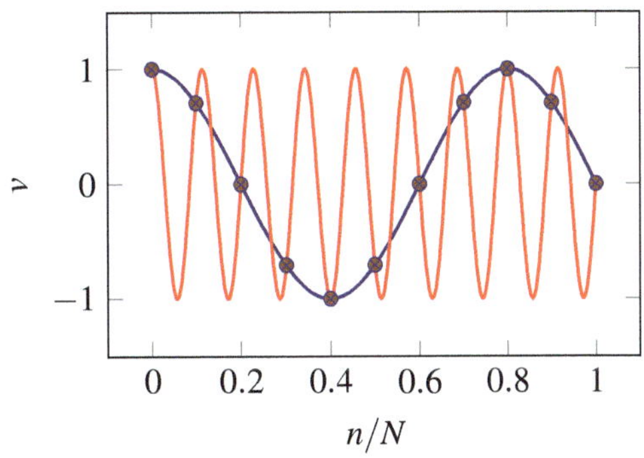

Figure 7.4 Illustration of aliasing – if the red signal is sampled at too low a rate the samples appear as a lower frequency 'alias' shown in blue.

In the current context, where our measured signal is discrete, of finite duration (N), and real, the DFT transforms the sequence of N sample values x_n (time domain representation) into a sequence of N complex numbers X_k (frequency domain representation), according to,

$$X_k = \frac{1}{N}\sum_{n=0}^{N-1} x_n e^{-\frac{i2\pi}{N}f_k n}, \qquad \mathbf{X} = \mathscr{F}(\mathbf{x}) \tag{7.12}$$

where f_k represents the kth frequency component, and the summation runs over the N samples of the signal block. Similar to the continuous case, X_k is a complex coefficient that conveys the amplitude and phase of the sinusoid at frequency f_k. Given the full set of complex coefficients X_k, the original signal is obtained from the weighted sum of N sinusoids as,

$$x_n = \sum_{k=0}^{N-1} X_k e^{\frac{i2\pi}{N}f_k n}, \qquad \mathbf{x} = \mathscr{F}^{-1}(\mathbf{X}). \tag{7.13}$$

With reference to Equation 7.12 and 7.13, we refer to $\mathbf{X} = [X_0, X_1, \cdots, X_{N-1}]$ as the complex Fourier spectrum of the discrete time signal $\mathbf{x} = [x_0, x_1, \cdots, x_{N-1}]$, and use the notation $\mathscr{F}(\square)$ and $\mathscr{F}^{-1}(\square)$ to denote the (discrete) Fourier transform and its inverse, respectively.

Compared with the continuous Fourier transform (Equations 7.9-7.11), the DFT has two important differences, each with significant consequences that should be understood.

First, the signal is now of finite duration, say L. As a result, the signal can only be decomposed into frequencies f that are periodic across the interval L (block length), i.e. those whose period $T = 1/f$ is an integer ratio of the block length, $T = L/k$ or $f = k/L$, for $k = 0, 1, 2, 3, \cdots$. Note that one way to view the function of windows like the Hanning, is that they reduce frequency components that are not L/k periodic by tapering the signal to zero at the beginning and end of the block.

Second, the signal is no longer continuous. As elucidated by the Nyquist-Shannon sampling theorem, we are unable to determine the coefficient of any frequency greater than half the sample rate, the so-called Nyquist limit. That is, $f_{max} = f_s/2$ is the highest frequency we can capture using the DFT. Moreover, if the original signal contains information at frequencies above the Nyquist limit, these will be sampled in such a way that they appear as lower frequency components due to the effect of *aliasing*, as illustrated in Figure 7.4. To avoid this issue, it is essential that anti-aliasing filters are applied prior to sampling the signal. These are built in to most data acquisition systems but aliasing problems can still sometimes occur for example when resampling.

Recalling that measurement length L is related to the number of samples and the sample rate by, $L = N/f_s$, the allowable DFT frequencies are $f_k = kf_s/N$. From this we obtain the frequency resolution (i.e. spacing between frequency bins),

$$\Delta f = \frac{f_s}{N}. \tag{7.14}$$

Equation 7.14 is a useful one; it is used to determine the length of each time block, N, required to achieve the desired frequency resolution Δf knowing the sample frequency f_s.

Negative frequencies and the single-sided spectrum

Above we have noted that the highest frequency that the DFT can capture is $f_{max} = f_s/2$ (the Nyquist limit), and that the DFT returns N equally spaced frequency components with spacing $\Delta f = f_s/N$. At first, these two statements appear contradictory. Dividing f_{max} by the frequency resolution Δf indicates that there $N/2$ frequency components between 0 and f_{max}. The DFT, however returns N complex coefficients.

It turns out that the DFT produces a *double-sided* spectrum, only one half of which is useful for our purposes. Frequency bins between 0 Hz and just below the Nyquist frequency (i.e. $N/2-1$) constitute the normal *single-sided* spectrum that is used almost exclusively for representation and computation in the frequency-domain. Above the Nyquist limit, an *alias* spectrum is produced which can be dropped for subsequent operations.

To see this in a bit more detail, consider the $(N-k)$th frequency component; following some simple algebraic manipulations, and noting that $e^{-i2\pi n} = 1$ for all integer values of n, we have that,

$$\begin{aligned} X_{N-k} &= \sum_{n=0}^{N-1} x_n e^{-\frac{i2\pi}{N}(N-k)n} \\ &= \sum_{n=0}^{N-1} x_n e^{-i2\pi n} e^{\frac{i2\pi}{N}kn} \\ &= \sum_{n=0}^{N-1} x_n e^{\frac{i2\pi}{N}kn} = X_k^*. \end{aligned} \tag{7.15}$$

We see that the $(N-k)$th frequency component is equal to the complex conjugate of the kth frequency component. Equation 7.15 shows that above the $N/2$ Nyquist frequency component, the complex coefficients are *conjugate-even*. In other words the conjugate of the single-sided spectrum is reflected or 'flipped' around the Nyquist frequency. Because the function is periodic we can equally consider the reflected spectrum as *above* the Nyquist frequency ($N/2+1 \leq k \leq N$), or *below* zero, ($-(N/2+1) \leq k \leq -1$). Hence, the reflected half of the spectrum is often considered to occur at *negative frequencies*.

The reflected spectrum contains no additional information and its meaning is somewhat obscure. Hence, to reduce computational effort, it is typical to deal only with the positive ($0 < k < N/2$) frequencies, i.e with the *single-sided spectrum*. The dual-sided spectrum can be reconstructed if there is a need to convert back into the time domain as will be described shortly. As we will deal almost exclusively with the single-sided spectrum from now on, it will be denoted as simply X_k, the additional $\hat{\square}$ accent will denote the dual-sided spectrum where required.

When converting to the single-sided spectrum, it is necessary to account for the missing energy contained by the negative frequencies. To do so we use the scaling,

$$X_k = \hat{X}_k, \qquad \text{for} \quad k = 0, \frac{N}{2} \tag{7.16}$$

$$X_k = \sqrt{2}\hat{X}_k, \qquad \text{for} \quad 1 < k < \frac{N}{2} - 1 \tag{7.17}$$

The remainder of the spectrum is discarded, i.e. $\frac{N}{2} < k \leq N-1$.

The above scaling ensures that the Parseval's theorem is satisfied, i.e. that the original time domain signal and the single sided spectrum have the same total energy,

$$\sum_{n=0}^{N} |x_n|^2 = \frac{1}{N}\sum_{k=0}^{N-1} |\hat{X}_k^* \hat{X}_k|^2 = \frac{1}{N}\sum_{k=0}^{N/2} |X_k^* X_k|^2. \tag{7.18}$$

where the instantaneous single-sided complex Fourier spectrum, X_k, of the discretised signal x_n will form the basis of all operational response and FRF calculations in the frequency domain.

Conversion to double-sided spectrum

It is sometimes necessary to perform the conversion in the other direction, i.e. to convert from single-sided to double-sided spectra. For example, to auralise the output of a VAVP we will typically start with a single-sided spectrum that needs to be transformed to the time-domain and for this we need to convert to a *double-sided* spectrum prior to applying an inverse DFT. There are other ways to achieve this but we will not describe them here to avoid confusion.

There are two steps prior to applying the inverse DFT (Equation 7.13): a) the above scaling must be reversed and b) the negative frequency components must be recreated. The scaled, double-sided spectrum can therefore be regenerated as follows,

$$\hat{X}_k = \frac{1}{\sqrt{2}} X_k, \qquad \text{for} \quad 1 < k < \frac{N}{2} - 1 \tag{7.19}$$

$$\hat{X}_k = 0, \qquad \text{for} \quad k = 0, \frac{N}{2} \tag{7.20}$$

$$\hat{X}_k = \frac{1}{\sqrt{2}} X^*_{-k}, \qquad \text{for} \quad -\left(\frac{N}{2} - 1\right) \leq k \leq -1 \tag{7.21}$$

which can then be inverse transformed to time domain.

Note that the lines at 0 Hz ($k = 0$) and at the Nyquist frequency ($k = N/2$) are set to zero to avoid a DC offset in the time sequence; this would not be audible but may create numerical problems which are best avoided.

7.2.3 AUTO AND CROSS-SPECTRA

The single-sided Fourier spectrum X_k specifies the amplitude and phase of the kth harmonic component of the discrete signal x_n. Often, interest lies not in the amplitude of a signal's harmonics, rather in their energy or power, which are time-averaged quantities. To this end, we consider the signal's auto-spectrum. Omitting the frequency index k for clarity, the instantaneous (as opposed to average) auto-spectrum of X is defined (for each frequency bin) as,

$$S_{XX} = X^* X = |X|^2 \tag{7.22}$$

the final equality of which follows from the polar form of $X = |X| e^{i\phi_X}$. In the context of a VAVP, the auto-spectrum alone is not of much use. By definition it is strictly real and therefore does not contain any phase information. This means we cannot use the auto-spectrum in any calculations where the relative phase between signals is important (for example, in determining the blocked force), nor are we able to reconstruct the original time domain signal.

To capture relative phase information between pairs of signals X and Y we introduce the (instantaneous) cross-spectrum, defined as,

$$S_{XY} = X^* Y = |X| \, |Y| e^{i(\phi_Y - \phi_X)} \tag{7.23}$$

where again, the final equality follows from the polar representation of $X = |X| e^{i\phi_X}$ and $Y = |Y| e^{i\phi_Y}$. We read S_{XY} as the cross-spectrum from X to Y; the first subscripted index being the conjugated quantity.

From Equation 7.23 it is clear that the phase of the cross-spectrum is indeed the relative phase between the two signals X and Y, $\phi_{YX} = \phi_Y - \phi_X$. As such, the cross-spectrum S_{YX} (i.e. in the opposite direction) has the same magnitude, but opposite phase,

$$S_{YX} = Y^*X = |X|\,|Y|e^{i(\phi_X - \phi_Y)} = S_{XY}^*. \tag{7.24}$$

The combination of the cross-spectrum and auto-spectrum will allow us to compute FRFs using estimators that are robust against corrupting noise, as will be discussed shortly.

Averaged spectra and phase

The auto and cross-spectrum, as defined in Equations 7.22 and 7.23, are 'instantaneous', that is, they apply to a single time window. We now consider what is involved in *averaging* over several measurements. We will only consider averaging in the *frequency domain*, usually referred to as *spectral averaging*. Essentially the same operation is sometimes also referred to, perhaps confusingly, as *time-averaging* (see later) but this should not be confused with *time-domain averaging* where the time signals themselves are averaged and which will not be considered.

First, we note that direct averaging of the complex Fourier spectra X is not possible because of the lack of a common *phase reference*. With a DFT, each repeated measurement of X is phase-referenced to the beginning of its time window, and whilst the phase is consistent across multiple channels captured simultaneously, the phase of the next measurement is arbitrary depending on the moment that data capture starts. Thus, each subsequent window will have a different phase reference and a meaningful average cannot be obtained. The solution is to make use of the auto and cross-spectra, the former having no phase component, and the latter having a *relative* phase component, i.e. the phase is referenced to one of the signals and is therefore consistent not just across channels but from one measurement to the next.

In the case of FRF measurements with a hammer, one simply averages the instantaneous auto/cross-spectra captured after each hit whenever they occur. For operational or shaker measurements, the process is essentially the same, except that the time blocks will typically be contiguous or overlapping rather than spaced out in time. This may give the impression of a 'continuous' average and may hence be referred to as *time-averaging*, although it still consists of the averaging of *spectra* and is hence done in blocks.

Having justified the use of auto and cross-spectra for averaging, let us now consider the role of *noise*, which is present in any measurement. In this case, using $\tilde{\square}$ to denote a *measured* quantity, the measured signals $\tilde{x}$ and $\tilde{y}$ take the form,

$$\tilde{x}(t) = x(t) + e_x(t), \qquad \tilde{y}(t) = y(t) + e_y(t) \tag{7.25}$$

where $x(t)$ and $y(t)$ are the true underlying signals, and $e_x(t)$ and $e_y(t)$ are the corrupting (noise) signals on x and y, respectively. Discretising, scaling, and transforming to the frequency domain we have,

$$\tilde{X} = X + \varepsilon_X, \quad \tilde{Y} = Y + \varepsilon_Y \tag{7.26}$$

where ε_X and ε_Y are the complex Fourier spectra of the corrupting noise. Owing to the linearity of the DFT, ε_X and ε_Y remain as additive terms onto the true underlying DFT spectra X and Y. Assuming that both ε_X and ε_Y constitute uncorrelated, stationary, zero-mean noise, we can readily compute the time averaged auto and cross-spectra.

Considering first the *auto-spectrum*, the arithmetic average over N repeated measurements (time windows, hammer impacts, etc.) is given by,

$$\begin{aligned}\langle S_{\tilde{X}\tilde{X}}\rangle &= \frac{1}{N}\sum_n^N S_{\tilde{X}_n\tilde{X}_n} \\ &= \frac{1}{N}\sum_n^N |X_n|^2 + |\varepsilon_{X_n}|^2 + X_n\varepsilon_{X_n}^* + X_n^*\varepsilon_{X_n} \xrightarrow{\text{as } N\to\infty} = \frac{1}{N}\sum_n^N |X_n|^2 + |\varepsilon_{X_n}|^2 \\ &= \langle S_{XX}\rangle + \langle S_{\varepsilon_X\varepsilon_X}\rangle \end{aligned} \tag{7.27}$$

Being uncorrelated and having zero mean, the terms involving both X_n and ε_{X_n} tend towards 0 as the number of averages is increased. Note that $|\varepsilon_{X_n}|^2$ is no longer zero mean, and so the resulting auto-spectrum is the sum of the underlying signal and noise auto-spectra. In other words, the averaged auto-spectrum of the measured signal $\tilde{X}$ is a *biased* estimate, $\langle S_{\varepsilon_X\varepsilon_X}\rangle$ representing the bias error.

Let us now consider the *cross-spectrum*. Assuming ε_X and ε_Y are uncorrelated with X, Y and each other,

$$\begin{aligned}\langle S_{\tilde{Y}\tilde{X}}\rangle &= \frac{1}{N}\sum_n^N X_n{}^*Y_n + \varepsilon_{X_n}^*Y_n + X_n^*\varepsilon_{Y_n} + \varepsilon_{X_n}^*\varepsilon_{Y_x} \xrightarrow{\text{as } N\to\infty} = \frac{1}{N}\sum_n^N X_n^*Y_n \\ &= \langle S_{XY}\rangle \end{aligned} \tag{7.28}$$

where, again, all terms containing either ε_X or ε_Y tend towards 0 as the number of averages is increased. Unlike the auto-spectrum, the obtained cross-spectrum is an *unbiased* estimate that approaches the true cross-spectrum as the number of averages increases. This property can be leveraged to obtain improved FRF estimates that are robust against noise on the input or output, as discussed in 7.3.1.

Hereafter we will omit the notation $\langle\square\rangle$ to denote an average. It will be assumed that any auto or cross-spectra are averaged, unless otherwise specified.

So far we have dealt with the auto and cross-spectra of single signals or pairs of signals. In the context of a VAVP, quite often we might wish to compute the auto-spectrum for a number of signals, and also the cross-spectrum between each pair. This can be done in one-shot using the so-called cross-spectral matrix $\mathbf{S_{XX}}$, defined for the averaged case as,

$$\mathbf{S_{XX}} = \frac{1}{N}\mathbf{XX}^{\mathrm{H}} \tag{7.29}$$

where the signal matrix $\mathbf{X}$ is constructed as,

$$\mathbf{X} = \begin{bmatrix} \mathbf{X}_1 & \mathbf{X}_2 & \cdots & \mathbf{X}_N \end{bmatrix} = \begin{bmatrix} X_{1,1} & X_{1,2} & \cdots & X_{1,N} \\ X_{2,1} & X_{2,2} & \cdots & X_{2,N} \\ \vdots & \vdots & \ddots & \vdots \\ X_{N,1} & X_{N,2} & \cdots & X_{N,N} \end{bmatrix} \tag{7.30}$$

with $X_{i,j}$ being the jth repeated measurement of the ith signal. The resulting cross-spectral matrix takes the form,

$$\mathbf{S_{XX}} = \begin{bmatrix} S_{X_1X_1} & S_{X_1X_2} & \cdots & S_{X_1X_N} \\ S_{X_2X_1} & S_{X_2X_2} & \cdots & S_{X_2X_N} \\ \vdots & \vdots & \ddots & \vdots \\ S_{X_1X_N} & S_{X_NX_2} & \cdots & S_{X_NX_N} \end{bmatrix} \tag{7.31}$$

where the matrix multiplication takes care of the spectral averaging. The averaged auto-spectra are found along the diagonal and all pairs of cross-spectra populate the off-diagonals. Note that the auto-spectral matrix satisfies $\mathbf{S}_{\mathbf{XX}}^{\mathrm{H}} = \mathbf{S_{XX}}$.

We can similarly define the cross-spectral matrix between two different sets of spectra as,

$$\mathbf{S_{YX}} = \frac{1}{N}\mathbf{X}\mathbf{Y}^{\mathrm{H}} \tag{7.32}$$

where $\mathbf{Y}$ is defined similarly to $\mathbf{X}$. Note that the cross-spectral matrix satisfies $\mathbf{S}_{\mathbf{YX}}^{\mathrm{H}} = \mathbf{S_{XY}}$.

To conclude this section we comment that, whilst it is usual practice to discard the individual spectra once their average has been obtained, the authors advocate that all spectra should be retained for subsequent analysis. Whilst this clearly requires more data storage, it enables both rigorous uncertainty analysis, as discussed in Chapter 8, and the reconstruction of long time sequences for auralisation, as outlined in 7.5.

7.3 INPUT-OUTPUT RELATIONS

The data processing described so far, i.e. the DFT and calculation of auto and cross-spectra, is sufficient to represent the *operational* response signals required for VAVP. It remains to consider the component's *passive* properties, that is, their FRFs. Measurement and manipulation of FRFs is of central importance to all aspects of virtual prototyping.

7.3.1 FREQUENCY RESPONSE FUNCTIONS

FRFs are the primary building block of a VAVP and provide a complete (black box) characterisation of linear, time invariant systems between a set of specified inputs and outputs. For a single-input, single-output (SISO) system at a given frequency f, the FRF is defined as the ratio of the complex Fourier spectrum of the output, Y, to that of the input, X,

$$H(f) = \frac{Y(f)}{X(f)}. \tag{7.33}$$

Its magnitude $|H|$ represents the change in level between input and output and its phase ϕ_H the delay incurred as signal passes through the system.

For multi-input multi-output systems the FRF is extended to a matrix form $\mathbf{H}$, relating the vector of inputs $\mathbf{X}$ to the vector of outputs $\mathbf{Y}$,

$$\mathbf{Y} = \mathbf{HX} \quad \text{or} \quad \begin{pmatrix} Y_1 \\ Y_2 \\ \vdots \\ Y_{N_Y} \end{pmatrix} = \begin{bmatrix} H_{11} & H_{12} & \cdots & H_{1N_X} \\ H_{21} & H_{22} & \cdots & H_{2N_X} \\ \vdots & \vdots & \ddots & \vdots \\ H_{N_Y1} & H_{N_Y2} & \cdots & H_{N_YN_X} \end{bmatrix} \begin{pmatrix} X_1 \\ X_2 \\ \vdots \\ X_{N_X} \end{pmatrix}. \tag{7.34}$$

The definition of a single entry of the FRF matrix $\mathbf{H}$ now comes with the requirement that all other inputs are held to 0 (see 2.2.8),

$$H_{ij} = \left.\frac{Y_i}{X_j}\right|_{X_{k\neq j}=0}. \tag{7.35}$$

Though defined as a simple ratio of complex Fourier spectra, in practice are FRFs not calculated in this way. Rather, they are estimated using auto and cross-spectra, which provide robust estimates in the presence of noise, as discussed below.

For a complex system with multiple inputs and outputs (as for most structural components), a *matrix* of FRFs is required to characterise the system. The FRFs that make up this matrix can be measured in one of three ways; with a single input and single output, a single input and multiple outputs, or multiple inputs and multiple outputs. These different estimation procedures are also detailed below.

Single-input single-output (SISO)

The *single-input single-output* approach measures the FRF matrix entries one at a time, by applying a single input and observing a single output. In practice this method is rarely used in favour of a single-input multiple-output approach. However, the process of FRF estimation is virtually identical and so we describe it here.

When measuring the FRF of a system, external disturbances or noise are unavoidable, at least to some extent. We can consider the three scenarios depicted in Figure 7.5: 1) noise is present on the output, but the input is noise free, 2) noise is present on the input, but the output is noise free, and 3) noise is present on both the input and output. Depending on which scenario is considered, the optimum FRF estimation differs. In short, to minimise the influence of noise we want to avoid computing the auto-spectrum for any quantities that are subject to an added noise, as this leads to a bias error, as discussed in 7.2.3.

Considering only *noise on the output* Y (Case 1, Figure 7.5), we want to avoid computing the auto-spectrum S_{YY} which will be biased by the output noise. Multiplying Equation 7.33 above and below by the conjugate Fourier spectra of the input,

$$H_1 = \frac{YX^*}{XX^*} \xrightarrow{\text{Averaging}} \frac{S_{XY}}{S_{XX}} \tag{7.36}$$

we obtain an unbiased estimate of the FRF, which tends towards the true FRF, H, as the number of averages is increased. This is because the influence of noise on the output is reduced by averaging the cross-spectrum (which incurs no bias error). This particular FRF estimate is referred to as the H_1 estimate.

Now considering only *noise on the input* X (Case 2), we want to avoid computing the auto-spectrum S_{XX}, which will be biased by the input noise, and rely instead on S_{YY} and S_{YX}. Multiplying Equation 7.33 above and below by the conjugate Fourier spectra of the output,

$$H_2 = \frac{YY^*}{XY^*} \xrightarrow{\text{Averaging}} \frac{S_{YY}}{S_{YX}} \tag{7.37}$$

we obtain an unbiased estimate of the FRF, which tends towards the true FRF H as the number of averages is increased. This particular FRF estimate is referred to as the H_2 estimate.

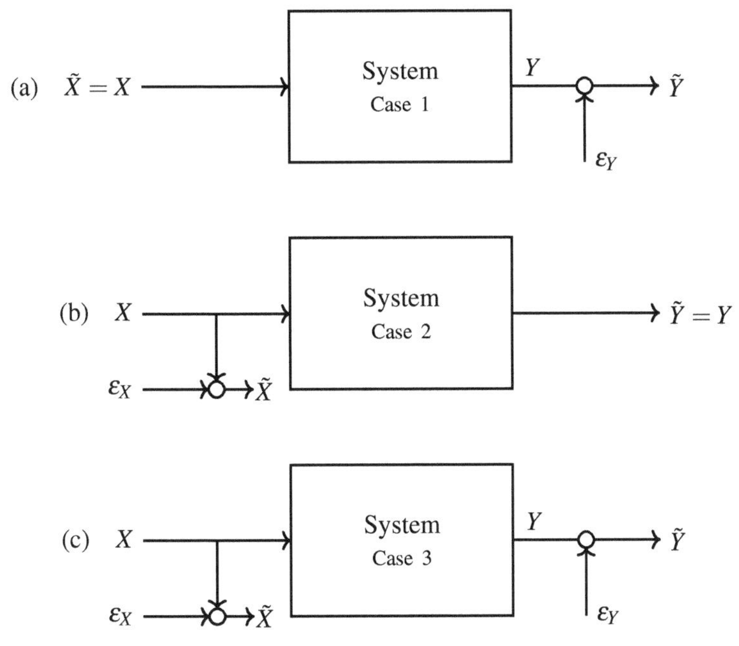

Figure 7.5 Illustration of: noise on the (a) output, (b) input, and (c) both input and output.

In practice, when FRFs are measured using an instrumented force hammer (or shaker), the input signal (i.e. the force signal) is generally recorded with a very good

SNR. The responses on the other hand might be located at some distance away, and therefore have reduced SNR. Similarly, other factors such as external disturbances, inaccurate hammer excitations, and weak non-linearities tend to effect the output response much more than in the input. For this reason, in most scenarios, it is the H_1 FRF estimate that should be used.

In the above, we have implied, though not proved, that the H_1 and H_2 are the optimal estimates of H in the presence of output and input noise, respectively. To add some rigour, it is straightforward to show that these estimates are in fact the least squares solutions for H, given their respective noise sources. Starting with H_1, for the output noise problem we have,

$$Y_i + \varepsilon_{Yi} = HX_i \tag{7.38}$$

where i denotes the ith measurement. Recalling that our X and Y represent *complex* Fourier spectra, we express the squared error of the ith measurement using the conjugate product,

$$|\varepsilon_{Yi}|_2^2 = (HX_i - Y_i)(HX_i - Y_i)^* \tag{7.39}$$

The average squared error then becomes,

$$\begin{aligned} J &= \frac{1}{N}\sum_i^N |\varepsilon_{Yi}|_2^2 \\ &= \sum_i (HX_i - Y_i)(HX_i - Y_i)^* \\ &= \sum_i Y_iY_i^* - Y_i^*HX_i - Y_iH^*X_i^* + HH^*X_iX_i^* \end{aligned} \tag{7.40}$$

To obtain the least squares solution we look to find the H that minimises J. We do this in the usual way by taking the derivative of J with respect to H and setting this to zero,

$$\frac{dJ}{dH} = \frac{1}{N}\sum_i^N -Y_i^*X_i + H^*X_iX_i^* = 0. \tag{7.41}$$

Note that when dealing with complex derivatives, we can treat H and H^* as independent variables [40]. Substituting for the auto and cross-spectrum,

$$S_{YX} = H^*S_{XX} \tag{7.42}$$

we obtain for the conjugate FRF,

$$H_1^* = \frac{S_{YX}}{S_{XX}} \quad \text{or} \quad H_1 = \frac{S_{XY}}{S_{XX}}. \tag{7.43}$$

Equivalently, taking instead the derivative of J with respect to H^*,

$$\begin{aligned} \frac{dJ}{dH^*} &= \frac{1}{N}\sum_i^N -Y_iX_i^* + HX_iX_i^* = 0 \\ &\rightarrow S_{XY} = HS_{XX} \end{aligned} \tag{7.44}$$

we obtain for the FRF,

$$H_1 = \frac{S_{XY}}{S_{XX}}. \tag{7.45}$$

Hence, the H_1 FRF estimate is the least squares solution given noise on the output.

Now with H_2 for the input noise problem we have,

$$Y_i = H(X_i + \varepsilon_{Xi}). \tag{7.46}$$

Isolating the error term,

$$\frac{Y_i}{H} - X_i = \varepsilon_{Xi} \tag{7.47}$$

the average squared error becomes,

$$\begin{aligned} J &= \frac{1}{N}\sum_i^N |\varepsilon_{Xi}|_2^2 \\ &= \sum_i \left|\frac{Y_i}{H} - X_i\right|_2^2 \\ &= \sum_i \frac{Y_i Y_i^*}{HH^*} - \frac{Y_i^*}{H^*}X_i - \frac{Y_i}{H}X_i^* + X_i X_i^*. \end{aligned} \tag{7.48}$$

As before, we take the derivative of J with respect to H and set this to 0,

$$\frac{dJ}{dH} = \frac{1}{N}\sum_i^N - \frac{Y_i Y_i^*}{H^2 H^*} + \frac{Y_i}{H^2}X_i^* = 0. \tag{7.49}$$

Multiplying both sides by H^2, and substituting for the auto and cross-spectra we get that,

$$\frac{1}{N}\sum_i^N \frac{Y_i Y_i^*}{H^*} = \frac{1}{N}\sum_i^N Y_i X_i^* \rightarrow S_{YY} = H^* S_{XY} \tag{7.50}$$

and so,

$$H_2^* = \frac{S_{YY}}{S_{XY}} \quad \text{or} \quad H_2 = \frac{S_{YY}}{S_{YX}} \tag{7.51}$$

Hence, the H_2 FRF estimate is the least squares solution for the case of input noise.

From inspection of Equation 7.36 and 7.37 we can make the following observation on the H_1 and H_2 estimates. At resonance $|H|$ becomes a maximum. For H_1, this implies that $|S_{XX}|$ becomes a minimum and/or $|S_{XY}|$ becomes a maximum. $|S_{XY}|$ being at a maximum implies that it is well above the noise level and so provides an accurate estimate of H. In contrast, at an anti-resonance, $|H|$ becomes a minimum. For H_1, this implies that $|S_{XX}|$ becomes a maximum and/or $|S_{XY}|$ becomes a minimum. As $|S_{XY}|$ approaches a small value, the relative contribution of noise becomes greater, and so a less accurate estimate is obtained. Let us now consider H_2. Resonance implies that $|S_{YY}|$ is a maximum, and/or $|S_{YX}|$ is at a minimum. With $|S_{YX}|$ being close to the noise level, we get a poor estimate of H. At an anti-resonance, the opposite is true, $|S_{YX}|$ becomes a maximum and is well separated from the noise, thus giving an accurate estimate. And so, the H_1 estimate is most accurate at resonant frequencies, whilst H_2 is most accurate at anti-resonant frequencies.

Based on the H_1 and H_2 estimates described above it is possible to define an estimate that offers a compromise between the two by simply considering their geometric average. This estimate is typically denoted H_3 and is given by,

$$H_3 = \sqrt{H_1 H_2}. \tag{7.52}$$

Let us now consider the more complicated Case 3, with *noise on the input and output*,

$$Y_i + \varepsilon_{Yi} = H(X_i + \varepsilon_{Xi}). \tag{7.53}$$

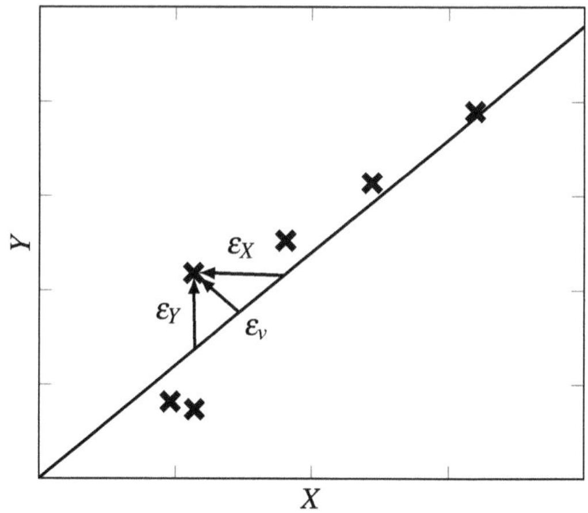

Figure 7.6 Illustration of input, output and combined error. For input error the deviation of the measurement point from the least squares solution is considered to be along the X axis. For output error, deviation is considered to be along the Y axis. For combined error it is perpendicular to the least squares solution.

Solving for the least squares optimum H is now equivalent to minimising both the input and output error simultaneously (in the case of equal error magnitudes $|\varepsilon_{Yi}| = |\varepsilon_{Xi}|$ this can be interpreted visually as the error ε_v illustrated in Figure 7.6). This is also known as total least squares. In general, the total least squares problem can only be solved with some *a priori* information about the input and output errors, for example the ratio of their auto-spectra. General descriptions of total least squares can be found in [9, 12, 26]. An application of total least squares to FRF estimation was first presented by Vold in the 1980's [28, 30]. We omit any derivations here as a) they are rather lengthy and b) the resulting estimator is less commonly used than the H_1 and H_2 estimators introduced above.

Following the total least squares procedure, we obtain the so-called H_s FRF estimate,

$$H_s = \frac{S_{YY} - s^2 S_{XX} + \sqrt{(S_{YY} - s^2 S_{XX})^2 + 4s^2 S_{XY} S_{YX}}}{2S_{YX}}. \tag{7.54}$$

Equation 7.54 provides the total least squares FRF estimate under the assumption that $S_{\varepsilon_Y \varepsilon_Y} = s^2 S_{\varepsilon_X \varepsilon_X}$, where s is a specified parameter that balances the error in the auto-spectra for the input and output. It can be shown that varying s between 0 and ∞, the H_s estimator takes values between H_2 and H_1 [11, 25], and that the H_s estimator corresponds to the Maximum Likelihood estimate for H [29].

In the not uncommon case where the input and output noise are of the same origin (e.g. thermal amplifier noise, etc.), such that we can assume $s = 1$, the H_s estimator reduces to the so-called H_v estimate,

$$H_v = \frac{S_{YY} - S_{XX} + \sqrt{(S_{YY} - S_{XX})^2 + 4S_{XY} S_{YX}}}{2S_{YX}}. \tag{7.55}$$

The factor s can be interpreted as placing an emphasis on minimising the input or output error ε_X or ε_Y. With $s = 1$ an equal weighting is applied.

Single-input multiple-output (SIMO)

In the above we considered the estimation of a single FRF using a single-input single-output measurement approach. In practice, a single FRF is not usually sufficient to characterise a component. Rather, it is necessary to build *FRF matrices* that relate the input-output at and between all (important) interface DoFs, as well as any remote target locations. These matrices can regularly include upwards of 30+ DoFs, giving over 900 individual FRF elements at each frequency. Even exploiting reciprocity to reduce the number of individual FRFs, measuring such matrices element by element is not advisable! Instead, it is general practice to measure such matrices *column by column*, i.e. using a single-input multiple-output approach.

By applying a single input force, we know that the measured outputs (now a vector of Fourier spectra $Y_i \rightarrow \mathbf{Y}_i$) are solely the result of the applied force (as well as any external disturbances ε_{Xi} or ε_{Yi}); there is no need to separate the contributions of multiple inputs. As such, the H estimators described above can be applied directly to multiple simultaneous outputs with no need for any modification.

Multiple-input multiple-output (MIMO)

The single-input multiple-output (SIMO) approach is by far the most commonly used method in the context of component characterisation for VAVPs. In the wider context of structural dynamics, particularly experimental modal analysis however, the SIMO approach is sometimes abandoned in favour of multiple-input multiple-output (MIMO) strategies [18]. Reasons for this include:

1) A single excitation might struggle to sufficiently excite a large structure. Simply increasing the force could push the structure, at least locally, into a non-linear regime. Using multiple excitations reduces the force required by each one individually.
2) Structures with strong directional behaviour require excitation in multiple directions to sufficiently excite all their modes.
3) Symmetric structures with degenerate modes (i.e. multiple modes occurring at the same frequency) require multiple inputs to correctly identify these modes.
4) All excitations are performed simultaneously hence reducing experimentation time.

Nevertheless, there is no reason why such MIMO methods cannot be applied to the characterisation of components. Indeed, this may be necessary if the VAVP approach were to be applied, for example, to a large heavyweight structure.

To introduce the MIMO estimation methods, we must extend our complex Fourier spectra to multiple input and output quantities, such that,

$$X \to \mathbf{X} = \begin{pmatrix} X_1 \\ X_2 \\ \vdots \\ X_{N_X} \end{pmatrix} \quad \text{and} \quad Y \to \mathbf{Y} = \begin{pmatrix} Y_1 \\ Y_2 \\ \vdots \\ Y_{N_Y} \end{pmatrix} \tag{7.56}$$

and the MIMO system Equation becomes,

$$\mathbf{Y} = \mathbf{HX} \quad \text{or} \quad \begin{pmatrix} Y_1 \\ Y_2 \\ \vdots \\ Y_{N_Y} \end{pmatrix} = \begin{bmatrix} H_{11} & H_{12} & \cdots & H_{1N_X} \\ H_{21} & H_{22} & \cdots & H_{2N_X} \\ \vdots & \vdots & \ddots & \vdots \\ H_{N_Y1} & H_{N_Y2} & \cdots & H_{N_YN_X} \end{bmatrix} \begin{pmatrix} X_1 \\ X_2 \\ \vdots \\ X_{N_X} \end{pmatrix}. \tag{7.57}$$

To determine the system FRF matrix **H**, the input vector **X** must be replaced by some form of matrix, such that it can be inverted and thus **H** isolated. Depending on how this is done, one can obtain a H_1 or H_2 estimate of the MIMO FRF.

Considering first the H_1 estimate. Post-multiplying both sides of Equation 7.57 by the conjugate transpose of **X**, and averaging over N repeated measurements of **X** and **Y**,

$$\mathbf{YX}^{\mathrm{H}} = \mathbf{HXX}^{\mathrm{H}} \xrightarrow{\text{Averaging}} \mathbf{S_{XY}} = \mathbf{HS_{XX}} \tag{7.58}$$

we obtain a cross-spectral form of Equation 7.57. Assuming it exists, we post-multiply both sides of Equation 7.58 by the inverse auto-spectrum matrix $\mathbf{S}_{\mathbf{XX}}^{-1}$,

$$\mathbf{H}_1 = \mathbf{S_{XY}}\mathbf{S}_{\mathbf{XX}}^{-1}. \tag{7.59}$$

Recall that for noise on the output only, $\mathbf{S_{XY}}$ provides an unbiased estimate of the true cross-spectrum. As such, Equation 7.59 represents a MIMO H_1 FRF estimate.

A MIMO H_2 estimate can be obtained similarly. Post-multiplying Equation 7.57 by the conjugate transpose of **Y**, and averaging over N repeated measurements,

$$\mathbf{YY}^{\mathrm{H}} = \mathbf{HXY}^{\mathrm{H}} \xrightarrow{\text{Averaging}} \mathbf{S_{YY}} = \mathbf{HS_{YX}} \tag{7.60}$$

we obtain an alternative cross-spectral form of Equation 7.57.

Assuming the number of inputs and outputs are equal, $N_Y = N_X$, the cross-spectral matrix $\mathbf{S_{YX}}$ is square, and its inverse can be computed. Post-multiplying both sides of Equation 7.60 by $\mathbf{S}_{\mathbf{YX}}^{-1}$ we obtain,

$$\mathbf{H}_2 = \mathbf{S_{YY}}\mathbf{S}_{\mathbf{YX}}^{-1}. \tag{7.61}$$

Noting that for noise on the input only, $\mathbf{S_{YX}}$ provides an unbiased estimate of the true cross-spectrum, Equation 7.61 thus represents a MIMO H_2 FRF estimate.

Where the number of input and output spectra differ, $N_Y \neq N_X$ Equation 7.61 can be generalised using the pseudo-inverse,

$$\mathbf{H}_2 = \mathbf{S_{YY}}\mathbf{S}^{+}_{\mathbf{YX}} \tag{7.62}$$

which provides a least squares estimate of the H_2 FRF. Note that the case of $N_Y \neq N_X$ has no bearing on the H_1 estimate, as the matrix inversion is applied to the auto-spectral matrix, $\mathbf{S_{XX}}$, which is by definition square.

Using an eigenvector approach, a MIMO estimate of the FRF in the presence of both input and output noise can be obtained, H_v. Details on this method can be found in [14].

A successful MIMO FRF estimation is dependent on a number of factors and can be quite sensitive to details such as input signal type, averaging method employed, and other signal processing options. An important requirement is that no two inputs are perfectly correlated; this would yield a rank deficiency in the auto and cross-spectral matrices and cause numerical issues with their inversion. A more detailed overview of the nuances in MIMO FRF estimation can be found in [19].

7.3.2 COHERENCE

In the development of a VAVP, we assume all components behave linearly and are time invariant. In practice, these are idealisations and all components will exhibit non-linear behaviour to some extent, for example if excited with sufficient energy or if corrupting noise or experimental error appears. Clearly, it would be of great practical use to have a measure of the linear dependence between our measured inputs and outputs. This is provided by the so-called *coherence* function.

In general, coherence describes the linear dependence, or correlation, between two signals X and Y (typically taken to be a system's input and output, respectively). In the context of VAVP development and component characterisation, the most common application of coherence is as a test of the quality of measured FRFs. Indeed, the coherence can be interpreted as a measure of how well the obtained X/Y measurements fit the estimated FRF H. Examples of good and poor fits are shown in Figure 7.7. Note that in this illustrative example, input X and output Y are represented as real valued (or magnitude only). This gives the illusion that it is only magnitude that effects coherence. This is of course not the case. In general, input X and output Y are complex quantities, and so the coherence is also sensitive to phase. Nevertheless, Figure 7.7 illustrates the main points to be made. For the 'good' coherence case we see that the measured data points are grouped closely around the estimated FRF (represented by the gradient of the solid line), displaying a strong linear dependence. In this case the variation around the FRF estimate could be due to noise. For the poor coherence case, we see that, in addition to the scatter due to noise, above a certain input amplitude the output begins to drop off, indicative of a non-linearity. Consequently, the data points do not appear to fit the estimated FRF very well.

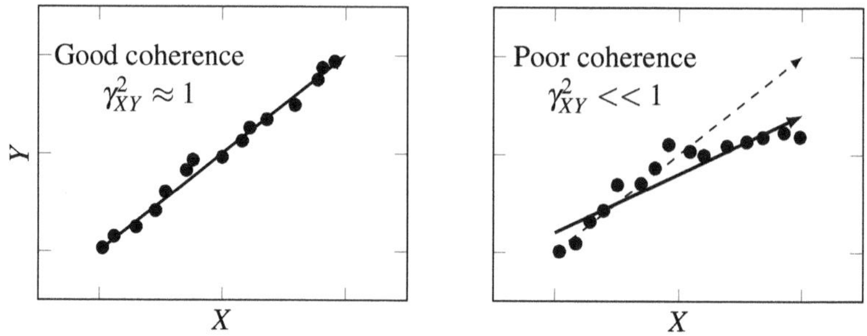

Figure 7.7 Example of good and poor coherence. Dashed arrow represents FRF obtained from low amplitude (i.e. linear) inputs.

Depending on the number of inputs being considered, a number of different coherence metrics are available. We summarise a couple of the main ones below. A more detailed exposition can be found in [1].

Ordinary coherence

To differentiate from the others, we can refer to the single-input coherence as the *ordinary* coherence. It is defined as,

$$\gamma^2_{XY} = \frac{|S_{XY}|^2}{S_{XX}S_{YY}} = \frac{H_1}{H_2}. \tag{7.63}$$

The ordinary coherence is applicable to any single input system. For multiple outputs the coherence is simply calculated for each output individually. It describes the linear dependence between each output Y_i, and the single input X. Poor coherence is generally found around anti-resonances, where weak input-output correlation is present and noise becomes significant, and occasionally at resonances, a feature attributed to signal processing effects such as leakage. Importantly, the coherence must be calculated using time-averaged auto- and cross-spectra. Using the instantaneous auto- and cross-spectra simply yields a value of 1 regardless of the underlying linear dependency.

The coherence, as defined above, can be interpreted much like a squared correlation coefficient, where the auto-spectra represent the variance of signals X and Y, and the cross-spectrum the covariance between them. Like the correlation coefficient, the coherence is a real number bound between 0 and 1. A value of 1 indicates perfect correlation (i.e. linear dependence, zero error). A value less than 1 indicates some divergence from linear dependence; this could be due to non-linearity, or perhaps noise. With this cross-correlation interpretation, it should not be surprising that the coherence function can be used, under certain assumptions, to estimate the uncertainty (i.e. variance) of auto and cross-spectra, as well as FRF estimates. This application will be detailed later in this section.

Aside from the FRF estimate itself, the ordinary coherence is one of the most important quantities to compute and inspect when performing FRF experiments in order to ensure good quality data. This application of the ordinary coherence has been discussed in more detail in Chapter 6.

Multiple coherence

In the presence of multiple inputs, a coherence metric should describe the linear dependence of the output Y_i, to *all* of the simultaneous inputs $\mathbf{X}$.

Perhaps naively, one might expect an extension of the coherence metric to a multiple input scenario to involve a simple summation of the ordinary coherence obtained between the output and each input. However, in general this is not the case (unless all inputs are perfectly uncorrelated, which is rarely the case in practice); the sum of multiple ordinary coherences is not bound between 0 and 1, and therefore fails to serve as a coherence metric under this bounding definition.

A consistent definition of coherence that remains bound in the presence of multiple partially correlated inputs, the so-called *multiple coherence*, is given by,

$$\gamma^2_{Y_i\mathbf{X}} = \frac{\mathbf{S}_{Y_i\mathbf{X}}\mathbf{S}^{-1}_{\mathbf{XX}}\mathbf{S}_{\mathbf{X}Y_i}}{S_{Y_iY_i}} \tag{7.64}$$

where $\mathbf{S}^{-1}_{\mathbf{XX}}$ is the inverse auto-spectral matrix of the input $\mathbf{X}$, $\mathbf{S}_{Y_i\mathbf{X}} = \mathbf{S}^{\mathrm{H}}_{\mathbf{X}Y_i}$ is a cross-spectral vector between inputs $\mathbf{X}$ and the single output Y_i, and $S_{Y_iY_i}$ is the auto-spectrum of the output. Clearly, a requirement of the multiple coherence is that the input auto-spectral matrix is full rank and can be inverted. This requires that no two inputs are perfectly correlated.

The multiple coherence metric falls below 1 for largely the same reasons as the ordinary coherence, i.e. non-linearity, signal processing effects and noise. Unlike the ordinary coherence however, the multiple coherence is not generally expected to drop around anti-resonances, the reason being that the frequencies at which anti-resonances occur is dependent on the position of the excitation. As such, the anti-resonances due to each input will occur at different frequencies, and thus the output will likely remain coherent with at least 1 input. A basic application of the multiple coherence might be to determine if any unknown inputs are present that contribute towards a measured output.

Conditioned coherence

The multiple coherence describes the linear dependence between the output Y_i and *all* simultaneous inputs $\mathbf{X}$. Often, it is necessary to understand the dependence between a given output, and each particular input. Such information would allow us, for example, to investigate the contribution of individual sources, or perhaps infer their relative strengths or input energies.

To overcome the issues associated with (partially) correlated inputs, coherence metrics are typically based on *conditioned* input auto-spectral matrices. Over the years a variety of such methods have been proposed, including the *partial coherence*, which is based on conditioning against a physical input; and the *virtual coherence*, which is based on conditioning against a *virtual* input (i.e. a singular value of the auto-spectrum). Whilst conditioned coherences can be useful for the analysis of MIMO systems, their use is more focused towards signal processing applications (e.g. active noise control). They are rarely used in the context of VAVP development. For this reason, for detailed discussion of conditioned coherence metrics the reader is referred to the recent review paper [1], and the references therein.

Coherence-based error relations and uncertainty estimates

When dealing with auto and cross-spectra in the presence of corrupting noise it is important to have an understanding of the potential error in our estimates and also a means of quantifying it.

In general, an error can be described as a combination of random and bias error contributions, as illustrated in Figure 7.8. A detailed treatment and discussion of uncertainty in the context of VAVP, including sources of uncertainty, their quantification, and propagation, is presented in Chapter 8. For a more detailed discussion focusing on input-output relations (cross-spectra, FRFs, etc.) see the classic text by Bendat and Piersol [4]. For now it is sufficient to understand bias error as a systematic error, or off-set, that leads to an inaccurate estimate of the sought after quantity (e.g. auto or cross-spectrum). Random errors on the other hand are caused by unknown and unpredictable changes in the experiment, and are characterised using standard statistical quantities such as variance and covariance.

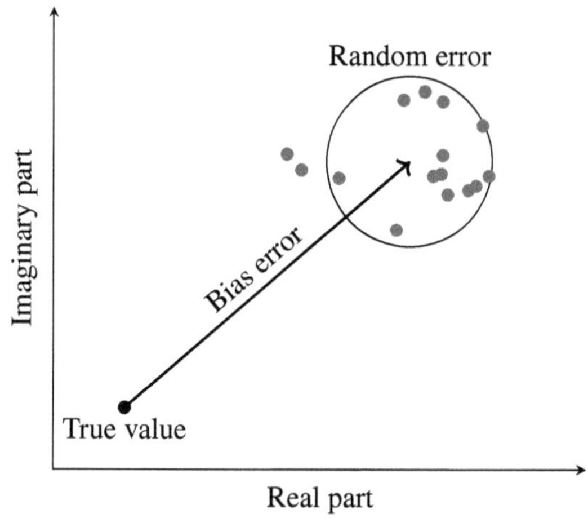

Figure 7.8 Illustration of bias and random error for the measurement of a complex value.

In the presence of corrupting signals, both the auto and cross-spectra are subject to an error or uncertainty. In both cases the random component can be minimised by conducting multiple averages. However, in the case of the auto-spectrum, a bias error term was shown to remain, i.e. the noise auto-spectrum $S_{\varepsilon_X \varepsilon_X}$.

Under fair assumptions on the signals $x(t)$ and $y(t)$, i.e. that they are stationary, ergodic and subject to small levels of noise/uncertainty, simple formulae are available to estimate the random error component described above [4]. Perhaps surprisingly, it can be shown that the variance in the estimate of a Fourier spectrum from a single measurement, is equal to the spectrum itself. Consequently, the variance of the averaged auto-spectrum is given by,

$$\text{Var}(S_{XX}) = \frac{S_{XX}^2}{N} \tag{7.65}$$

where N is the number of measurements used in the average. Similarly, for the magnitude of the cross-spectrum,

$$\text{Var}(|S_{XY}|) = \frac{|S_{XY}|}{\sqrt{\gamma_{XY}^2 N}} \tag{7.66}$$

where γ_{XY}^2 is the ordinary coherence function, introduced above. For both spectra, we see that the random error is reduced by increasing the number of measurements taken.

Based on the above, it is possible to obtain an estimate of the variance of the FRF magnitude and phase [3]. For the FRF magnitude we have,

$$\text{Var}(|H|) \approx \frac{(1-\gamma_{XY}^2)|H|^2}{\gamma_{XY}^2 2N}. \tag{7.67}$$

For the phase angle, assuming small levels of uncertainty, the small angle approximation yields,

$$\text{Var}(\angle H) \approx \frac{\text{Var}(|H|)}{|H|^2}. \tag{7.68}$$

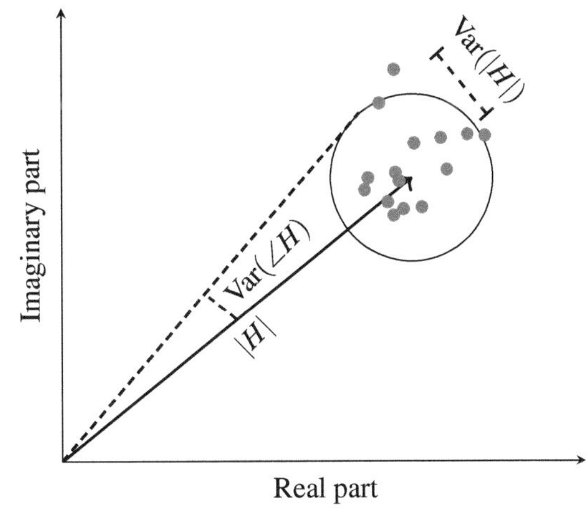

Figure 7.9 Illustration of variance on the FRF magnitude [2].

Assuming that the real and imaginary parts are uncorrelated with the same variance (which would typically be the case for interfering noise), their variances are equal to that of the magnitude,

$$\text{Var}(\text{Re}(H)) = \text{Var}(\text{Im}(H)) = \text{Var}(|H|) \tag{7.69}$$

Though simple and of some practical use, the above variance estimates should be used with care. As will be shown in Chapter 8, for an FRF measured using an impact hammer (the most common method for VAVP component characterisation), strong correlations are introduced not only between real and imaginary parts of each FRF, but also between all simultaneously measured FRFs. These correlations can have a considerable effect on the propagation of uncertainty from a VAVP and must be included for an accurate estimate. The coherence-based estimates described above are not able to capture these correlations.

7.4 SOLVING THE INVERSE PROBLEM

In the previous sections we detailed the key signal processing steps to calculate the operational responses (complex Fourier spectra) and FRFs required to characterise components and build a VAVP. In this section we will focus specifically on methods for solving the so-called inverse problem, which lies at the heart of a blocked force characterisation.

Obtaining a reliable characterisation of blocked forces is a crucial step in development of a VAVP. According to the in situ method, (see 6.4.2), this requires solving an inverse problem, i.e. inferring the blocked force from a set of measured responses and FRFs. Whilst mathematically this can be expressed in simple terms as,

$$\begin{pmatrix} \mathbf{v}_c \\ \mathbf{v}_b \end{pmatrix} = \begin{bmatrix} \mathbf{Y}_{Ccc} \\ \mathbf{Y}_{Cbc} \end{bmatrix} \bar{\mathbf{f}}_c \xrightarrow{\text{Solve inverse problem}} \bar{\mathbf{f}}_c = \begin{bmatrix} \mathbf{Y}_{Ccc} \\ \mathbf{Y}_{Cbc} \end{bmatrix}^+ \begin{pmatrix} \mathbf{v}_{Cc} \\ \mathbf{v}_{Cb} \end{pmatrix} \tag{7.70}$$

achieving a reliable solution with experimental data is often fraught with difficulty. In short, inverse problems are notoriously sensitive to small errors in the measured input data, that is they are typically *ill-posed* or *ill-conditioned*, presenting one or more of the following characteristics:

- they might lack a unique solution (non-uniqueness), e.g. two or more blocked force vectors might reproduce the observed response;
- the observed responses are highly sensitive to small perturbations in the model, e.g. errors in a measured FRF matrix;
- they might lack a solution altogether. Whilst this would generally not be the case for a VAVP, it might occur in other contexts.

To overcome the challenges associated with an ill-conditioned inverse problem, it is necessary to have an intuitive understanding of the underlying problem. In some scenarios, one might also require mitigation strategies to reduce the effects of ill-conditioning. This is most commonly done through processes known as *over-determination* and *regularisation*. In 7.4.1 we discuss the concept of over-determination and its relation to the so-called *least squares* solution. In 7.4.3 we discuss the two most common regularisation methods used for inverse force identification; Tikhinov and the Truncated Singular Value Decomposition.

7.4.1 OVER-DETERMINATION AND THE LEAST SQUARES SOLUTION

As discussed before, the in situ blocked force relation can be written in several ways, depending on which DoFs are used for the indicator positions; c, b or both c and b. In the first instance, using the interface DoFs c (Equation 6.26), the number of measured responses is equal to the number of unknown blocked forces. We call this a *determined* problem, for which the is solution is obtained by inverting the (square) FRF matrix. Providing the FRF matrix is full rank, a unique solution is obtained i.e. there is only one blocked force vector that satisfies this equation.

In the second and third instance (Equation 6.27 and 6.28, respectively), we might have more measured responses than unknown blocked forces. We call this an *over-determined* problem, for which the solution is obtained in a similar way, by *pseudo-inverting* the (now rectangular) FRF matrix. If the number of indicator positions is greater than the number of unknown blocked forces, then there is no unique solutions to Equation 6.27 and 6.28. Instead, one obtains the so-called least squares solution (described in more detail below).

As we will see shortly, it is generally beneficial to solve the over-determined problem, particularly that of Equation 6.28, which includes both interface and remote indicator DoFs. Doing so can help reduce the influence of experimental errors and provide a more reliable and robust blocked force estimate. We will demonstrate this by examining the equivalence of Equation 7.70 to a conventional linear regression.

Linear regression is a popular statistical method for modelling the relationship between an observed response (termed the dependent variable) and one or more predictors (so-called independent variables). It is the role of linear regression to find the optimal set of coefficients that relate these two variables. In matrix notation, a linear regression model takes the form,

$$\mathbf{y} = \mathbf{X}\beta \tag{7.71}$$

where $\mathbf{y}$ is a vector of measurements, $\mathbf{X}$ is a matrix of predictor variables, and β is the vector of unknown weighting coefficients to be determined. Clearly, the above is identical in form to the lefthand side of Equation 7.70, with the blocked force vector $\bar{\mathbf{f}}_c$ taking on the role of the unknown weighting coefficient vector β. In a typical linear regression, the predictor variables are set as increasing powers of the independent variable $(x^0, x^1, x^2, \cdots, x^N)$, or similar. In the context of a blocked force characterisation they are the measured FRFs.

Linear regression is a very well developed topic and is routinely applied across a wide range of disciplines. A detailed accounting of its development and state-of-the-art is beyond the scope of this book. The interested reader is referred to the many available textbooks on this topic, for example [33, 34].

To demonstrate the concept, and potential advantages, of over-determination in the context of a blocked force characterisation lets consider a simple example. Suppose we wish to determine the scalar blocked force $\bar{f}$. If we were to measure a scalar operational response v we would have the equation,

$$v = Y\bar{f}. \tag{7.72}$$

This equation describes a straight line whose intercept passes through the origin, with the FRF being interpreted as the independent variable (x-axis) and the operational response as the dependent variable (y-axis), as illustrated in Figure 7.10a. The measured FRF/response pair are represented by the orange cross. Note that for illustrative purposes Figure 7.10. presents the response and FRF as real valued. In reality they are complex variables, though the concept remains very much the same.

Determining the blocked force $\bar{f}$ is equivalent to solving for the gradient of the straight line. A unique solution is obtained by inverting the FRF to obtain the impedance Z,

$$\bar{f} = \frac{1}{Y} v = Zv. \tag{7.73}$$

Now suppose we measure N responses and their corresponding FRFs. In this case we have the equation,

$$\begin{pmatrix} v_1 \\ \vdots \\ v_N \end{pmatrix} = \begin{pmatrix} Y_1 \\ \vdots \\ Y_N \end{pmatrix} \bar{f}. \tag{7.74}$$

Note that each response/FRF pair represents a data point in Figure 7.10b. So, we now have N equations, but only 1 unknown, the blocked force. This is what we mean by an over-determined problem. Noting that any individual data point could be used to determine a blocked force, it is clear that there is no longer a *unique* solution to the problem.

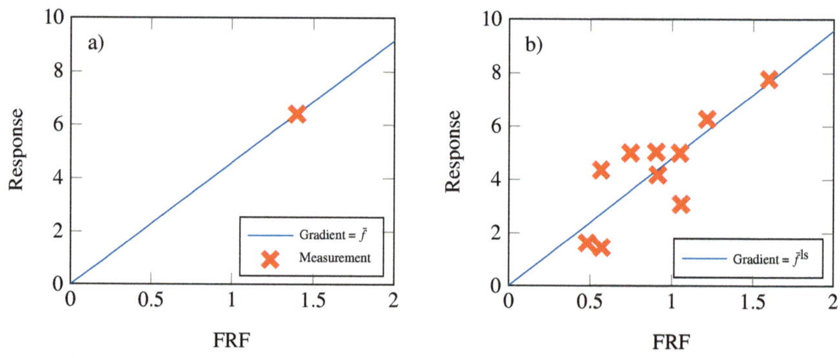

Figure 7.10 Regression interpretation of blocked force calculation for: a) a single measured response and b) 10 measured responses. The blocked force is represented by the gradient of the linear model.

To overcome this non-uniqueness, we can look to find an *optimum* blocked force that satisfies some criterion, for example the minimum squared error. That is, the optimum blocked force estimate is the one that when used to predict the observed responses, produces the smallest sum of squared errors,

$$\bar{f}^{\text{ls}} = \arg\min_{\bar{f}} \sum_i ||v_i - Y_i \bar{f}||_2^2. \tag{7.75}$$

The least squares estimate of the blocked force $\bar{f}^{\text{ls}}$ defines the gradient of the line of best fit through the set of data points in Figure 7.10b.

Fortunately, for linear problems in the form of Equation 7.71 the least square solution can be obtained directly in closed form. For the simple case considered here its solution is obtained by the following equation,

$$\bar{f}^{\mathrm{ls}} = \begin{pmatrix} Z_1 & \cdots & Z_N \end{pmatrix} \begin{pmatrix} v_1 \\ \vdots \\ v_N \end{pmatrix} = \mathbf{Z}\mathbf{v} \tag{7.76}$$

where the impedance matrix $\mathbf{Z}$ is given by,

$$\mathbf{Z} = \left[\begin{pmatrix} Y_1^* & \cdots & Y_N^* \end{pmatrix} \begin{pmatrix} Y_1 \\ \vdots \\ Y_N \end{pmatrix} \right]^{-1} \begin{pmatrix} Y_1^* & \cdots & Y_N^* \end{pmatrix} \tag{7.77}$$

or more concisely as,

$$\mathbf{Z} = \left(\mathbf{Y}^{\mathrm{H}}\mathbf{Y}\right)^{-1} \mathbf{Y}^{\mathrm{H}} = \mathbf{Y}^{+}. \tag{7.78}$$

The operator $\square^{+}$ is called the pseudo-inverse. For the over-determined problem described above it is more specifically the *left* pseudo-inverse. It provides the least squares solution to the over-determined problem. It is simple enough to derive Equation 7.78 for the general case whereby multiple blocked forces are being determined, such that $\bar{\mathbf{f}}$ is a vector. We begin by expressing the least squares error in matrix form as,

$$\Delta\mathbf{v}^2 = ||\mathbf{Y}\bar{\mathbf{f}} - \mathbf{v}||_2^2 = (\mathbf{Y}\bar{\mathbf{f}} - \mathbf{v})^{\mathrm{H}}(\mathbf{Y}\bar{\mathbf{f}} - \mathbf{v}). \tag{7.79}$$

and taking the (complex) differential of both sides,

$$\mathrm{d}\left(\Delta\mathbf{v}^2\right) = \mathrm{d}\left(\bar{\mathbf{f}}^{\mathrm{H}}\mathbf{Y}^{\mathrm{H}}\mathbf{Y}\bar{\mathbf{f}} - \mathbf{v}^{\mathrm{H}}\mathbf{Y}\bar{\mathbf{f}} - \bar{\mathbf{f}}^{\mathrm{H}}\mathbf{Y}^{\mathrm{H}}\mathbf{v} + \mathbf{b}^{\mathrm{H}}\mathbf{b}\right). \tag{7.80}$$

Note that when dealing with complex differentials, $\bar{\mathbf{f}}$ and $\bar{\mathbf{f}}^{\mathrm{H}}$ are treated as separate variables. Considering the differential with respect to $\bar{\mathbf{f}}$, the above reduces to,

$$\mathrm{d}\left(\Delta\mathbf{v}^2\right) = \bar{\mathbf{f}}^{\mathrm{H}}\mathbf{Y}^{\mathrm{H}}\mathbf{Y}\,\mathrm{d}\left(\bar{\mathbf{f}}\right) - \mathbf{v}^{\mathrm{H}}\mathbf{Y}\,\mathrm{d}\left(\bar{\mathbf{f}}\right) \tag{7.81}$$

where the vector derivative can now be identified as,

$$\frac{\partial \Delta\mathbf{v}^2}{\partial \bar{\mathbf{f}}} = \left(\bar{\mathbf{f}}^{\mathrm{H}}\mathbf{Y}^{\mathrm{H}}\mathbf{Y} - \mathbf{v}^{\mathrm{H}}\mathbf{Y}\right). \tag{7.82}$$

To find the force vector with the smallest squared error, we set this derivative to zero,

$$\frac{\partial \Delta\mathbf{v}^2}{\partial \bar{\mathbf{f}}} = \bar{\mathbf{f}}^{\mathrm{H,ls}}\mathbf{Y}^{\mathrm{H}}\mathbf{Y} - \mathbf{v}^{\mathrm{H}}\mathbf{Y} = \mathbf{0} \tag{7.83}$$

take the Hermitian of both sides,

$$\mathbf{Y}^{\mathrm{H}}\mathbf{Y}\bar{\mathbf{f}}^{\mathrm{ls}} = \mathbf{Y}^{\mathrm{H}}\mathbf{v} \tag{7.84}$$

and solve for $\bar{\mathbf{f}}^{\text{ls}}$,

$$\bar{\mathbf{f}}^{\text{ls}} = \left(\mathbf{Y}^{\text{H}}\mathbf{Y}\right)^{-1}\mathbf{Y}^{\text{H}}\mathbf{v}. \tag{7.85}$$

The resulting matrix product is the left pseudo-inverse, as used in Equation 7.78.

Solving the over-determined problem has clear advantages over the determined problem. Bearing in mind that any measurement (operational response or FRF) will be affected to some extent by experimental error, one might expect the over-determined solution to provide a more reliable estimate. Indeed this is the case.

For the determined solution a measurement error on the response (i.e. a vertical shifting of the data point in Figure 7.10a) will clearly affect the blocked force (i.e. the gradient) by an amount inversely proportional to the FRF (i.e. small FRF, large change in gradient). Conversely, an error on the FRF will shift the gradient by an amount proportional to the response.

For the over-determined solution, the error in any single measurement has much less influence over the gradient, which is governed by the collection of all measurements. Assuming that errors are distributed with a zero mean, the least squares solution will provide the best un-biased estimate of the blocked force.

The above interpretation is readily extended to multiple blocked forces; the blocked force vector $\bar{\mathbf{f}}$ can be interpreted as a vector containing the 'best fit' gradients of a hyper-plane in N dimensional space.

In a practical context, it is generally agreed that over-determination is beneficial, and that this is best done by including additional remote indicator points to supplement those at the interface (as opposed to using entirely remote indicators). However, there is no general consensus on what level of over-determination is required. In [41], significant improvements were found with a factor of 1.25 or 1.5 although nowadays a factor of 2 or more is often proposed. Published evidence is not easily found although there is undoubtedly much unpublished experience which possibly informs the higher numbers. The degree of over-determination therefore should be considered on a case-by-case basis.

It is important to note that whilst over-determination is generally beneficial, adding too many sensors can in-fact worsen the quality of a characterisation. If the additional sensors are poorly positioned and do not provide useful information they will simply add more noise and increase the overall level of uncertainty. The guiding principle is to select indicator DoFs with as much independence from existing indicators as possible, for example by orientating them in different directions (see 6.4.2).

7.4.2 UNDER-DETERMINED PROBLEM

As emphasised above, it is important to over-determine the inverse problem where possible. However, in rare circumstances, such as those involving acoustic interfaces, it may be necessary to solve the under-determined problem, i.e. where the number of unknown sources (e.g. blocked forces) exceeds the number of measured responses [6]. Fortunately, this can be achieved using essentially the same formulation as before,

$$\bar{\mathbf{f}}^{\mathrm{ls}} = \mathbf{Y}^{+}\mathbf{v} \tag{7.86}$$

using instead the so-called *right* pseudo-inverse,

$$\mathbf{Y}^{+} = \mathbf{Y}^{\mathrm{H}}\left(\mathbf{Y}\mathbf{Y}^{\mathrm{H}}\right)^{-1}. \tag{7.87}$$

Whilst in the context of VAVPs under-determined problems have received far less attention than over-determined ones, they have proven successful for some problems, for example in [6] where the equivalent field representation was solved as an under-determined problem and used to estimate radiated power. Nevertheless, under-determined solutions are seldom used, and should always be treated with caution.

7.4.3 REGULARISATION

Our discussion of inverse problems would not be complete without mention of regularisation techniques. Before we proceed however, we sound a note of warning: at its worst, regularisation can 'trick' an apparently good solution from a fundamentally flawed model that is difficult to distinguish from a genuinely good fit. This is the reason we have emphasised the need for careful checking and validation of the input data *at every stage* using methods described in Chapter 6 *prior* to any regularisation. If this produces a satisfactory result then there is no need for regularisation and this section can be skipped. Often though, even with good input data it will be necessary or desirable to employ some regularisation to stabilise the result.

An issue that often occurs when solving an inverse problem, such as determining the blocked force, is over-fitting. An over-fitted model is one that contains more parameters than can be justified by the data. For example at an interface instrumented to capture 6 DoFs at each connection point, the vibration source may impart forces only in some directions (which change with frequency). This would lead to redundancy in the interface representation, i.e. more parameters than independent blocked forces, leading to an ill-conditioned problem and an over-fitted model. Regularisation is often used to relieve the issues of over-fitting. More generally, the concept of regularisation refers to a process whereby additional information is introduced so as to solve an ill-posed problem.

Although numerous regularisation techniques exist [35], in the context of source characterisation two particular methods tend to be in favour; Tikhonov regularisation [36] and Truncated Singular Value Decomposition (TSVD, also referred to as singular value rejection) [37].

With many articles and textbooks already devoted to the subject of regularisation our touch will be light, providing only the key equations with some interpretation. The reader will be referred to the literature for more detail.

Tikhinov regularisation

Tikhinov regularisation is the most widely applied form of regularisation. It attempts to deal with the problem of ill-conditioning by introducing an additional constraint onto the least squares solution. In its most general form the Tikhinov solution is given by,

$$\bar{\mathbf{f}}^{\text{tik}} = \arg\min_{\bar{\mathbf{f}}} \quad ||\mathbf{v} - \mathbf{Y}\bar{\mathbf{f}}||_2^2 \quad \text{s.t.} \quad ||\Gamma\bar{\mathbf{f}}||_2^2 < a. \tag{7.88}$$

where Γ is the so-called Tikhinov matrix that acts on the blocked force to describe the constraint imposed. Various matrices can be used depending on the expected nature of the solution. Perhaps the most common Tikhinov matrix is the scaled identity $\Gamma = \alpha\mathbf{I}$, which is equivalent to requiring that the 2-norm of the blocked force solution remains less than some user specified value,

$$\bar{\mathbf{f}}^{\text{tik}} = \arg\min_{\bar{\mathbf{f}}} \quad ||\mathbf{v} - \mathbf{Y}\bar{\mathbf{f}}||_2^2 \quad \text{s.t.} \quad ||\bar{\mathbf{f}}||_2^2 < b \tag{7.89}$$

with $b = a/\alpha$. Within the statistics community, Equation 7.89 is also referred to as *ridge regression.*

A graphical representation of Equation 7.89 is shown in Figure 7.11. The least squares solution represents a specific point in the 2D (or more generally ND) plane where the residual norm $||\mathbf{v} - \mathbf{Y}\bar{\mathbf{f}}||_2^2$ reaches a minimum. Surrounding this point are a series of curves that each represent different possible solutions (i.e. blocked force vectors) that have the same residual norm. Noting that the constraint $||\bar{\mathbf{f}}||_2^2 < b$ represents the area enclosed by a circle (or more generally an ND sphere) of radius b, the solution to Equation 7.89 is the point of intersection between the boundary of the constraint and the tangential residual error curve. Clearly, this constraint can be used to find an optimal solution with a reduced 2-norm.

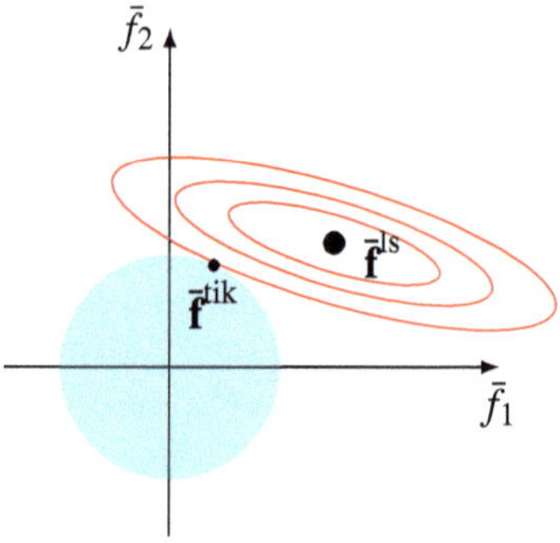

Figure 7.11 Tikhinov regularisation.

Equation 7.89 can written in an alternative Lagrangian form as,

$$\bar{\mathbf{f}}^{\text{tik}} = \arg\min_{\bar{\mathbf{f}}} \quad ||\mathbf{v} - \mathbf{Y}\bar{\mathbf{f}}||_2^2 + \beta||\bar{\mathbf{f}}||_2^2 \tag{7.90}$$

where the constraint is now included directly within the cost function to be minimised.

The additional constraint terms in Equation 7.89 and 7.90 promote blocked force solutions that have smaller 2-norms, that is solutions with less widely varying and inflated values.

One of the main advantages of Tikhinov regularisation is that a closed form solution is available, thus avoiding the need to solve Equations 7.89 or 7.90 by optimisation. This solution is given by,

$$\bar{\mathbf{f}}^{\text{tik}} = \left(\mathbf{Y}^{\text{H}}\mathbf{Y} + \lambda\mathbf{I}\right)^{-1}\mathbf{Y}^{\text{H}}\mathbf{v} \tag{7.91}$$

where $\lambda = \beta^2 \geq 0$ is a real valued scalar term referred to as the regularisation parameter.

Note that Equation 7.91 looks is very similar to the least squares solution using the left pseudo-inverse (see Equation 7.78). The difference is the small additional factor of λ along the diagonal of $\mathbf{Y}^{\mathrm{H}}\mathbf{Y}$; it is this additional diagonal term that gives rise to the name *ridge* regression. Its introduction has the effect of improving the conditioning of the problem, i.e. $\mathbf{Y}^{\mathrm{H}}\mathbf{Y} + \lambda\mathbf{I}$ has a smaller condition number than $\mathbf{Y}^{\mathrm{H}}\mathbf{Y}$ alone and is therefore less susceptible to inversion issues. For $\lambda = 0$, Equation 7.91 reduces to the standard least squares solution.

Key to the successful application of Tikhinov regularisation is the appropriate choice of λ. This is typically done according to some heuristic criterion, often with little or no physical interpretation.

A popular approach involves using the so-called L-curve criterion [5, 7]. With this approach, solutions are obtained for increasing levels of λ and curve is generated that plots the *residual norm* $||\mathbf{v} - \mathbf{Y}\bar{\mathbf{f}}||_2^2$ against the *solution norm* $||\bar{\mathbf{f}}||_2^2$. This curve is said to display an L-shaped trend, but in practice this is rarely clear cut for structural problems. The optimal value of λ is taken as the value that minimises both the residual and solution norm, i.e. the 'elbow' of the L-curve.

An alternative approach is based on a form of cross-validation, where the optimal value of λ is found though a leave-p-out (LPO) cross-validation [7]. In this case a solution is obtained using the measured data set with p responses excluded. A residual is then computed using the estimated solution (obtained using a particular value of λ) and the excluded p responses. This process is repeated for several different subsets of p responses providing a MSE for a given value of λ. The value of λ that produces the smallest LPO MSE is then chosen as the optimal value.

Note that whilst the regularisation parameter λ has little physical meaning, when viewed from a Bayesian perspective, λ is in fact related (under the assumptions of zero mean and uncorrelated Gaussian variables) to the variance of the corrupting noise on $\mathbf{v}$ and $\mathbf{Y}$ and the resulting signal-to-noise ratio [17, 20], and the Tikhinov solution represents the minimal unbiased linear estimator.

Truncated Singular Value Decomposition (TSVD)

Whilst Tikhinov regularisation is widely employed within the statistics and machine learning communities, a lack of physical justification for the parameter λ is somewhat disconcerting. An alternative approach, to which some physical interpretation can be applied, is achieved by utilising the Singular Value Decomposition (SVD) to compute the pseudo-inverse of the FRF matrix $\mathbf{Y}$.

The SVD of a complex rectangular matrix $\mathbf{Y} \in \mathbb{C}^{M\times N}$ takes the form,

$$\mathbf{Y} = \mathbf{U}\boldsymbol{\Sigma}\mathbf{V}^{\mathrm{H}} \tag{7.92}$$

where $\mathbf{U} \in \mathbb{C}^{M\times M}$ and $\mathbf{V} \in \mathbb{C}^{N\times N}$ are unitary matrices (meaning $\mathbf{U}^{\mathrm{H}}\mathbf{U} = \mathbf{V}^{\mathrm{H}}\mathbf{V} = \mathbf{I}$) that contain as their columns the left and right singular vectors, $\mathbf{u}_i$ and $\mathbf{v}_i$, respectively.

$$\mathbf{U} = \begin{bmatrix} \mathbf{u}_1 & \cdots & \mathbf{u}_M \end{bmatrix} \qquad \mathbf{V} = \begin{bmatrix} \mathbf{v}_1 & \cdots & \mathbf{v}_N \end{bmatrix} \tag{7.93}$$

$\Sigma \in \mathbb{R}^{M\times N}$ is a non-negative real rectangular matrix containing the so-called *singular values* along its diagonal, $\Sigma_{ii} = \sigma_i$, with zeros elsewhere. The number of non-zero diagonal elements/singular values is equal to the rank of $\mathbf{Y}$. It is typical to order the singular values (and so also the singular vectors) in descending order such that $\sigma_1 > \sigma_2 > \sigma_3 > \cdots$.

The SVD can be interpreted in a number of ways. Geometrically, we can think of the SVD as breaking down a general linear transformation, as represented by the matrix $\mathbf{Y}$, into a sequence of simpler transformations, specifically: a rotation or reflection (represented by the matrix $\mathbf{V}^{\mathrm{H}}$), followed by a scaling (Σ), and then another a rotation or reflection ($\mathbf{U}$). Another interpretation of the SVD is that it decomposes a matrix into an ordered sum of simpler rank 1 matrices $\mathbf{A}_i$, each formed by the outer product of singular vectors and weighted by their corresponding singular value,

$$\mathbf{Y} = \sum_i \mathbf{A}_i = \sum_i \sigma_i \mathbf{u}_i \mathbf{v}_i^{\mathrm{H}}. \tag{7.94}$$

In this sense, the matrices $\mathbf{A}_i$ associated with the larger singular values represent more important contributions to the original matrix $\mathbf{Y}$. Those matrices associated with smaller singular values contribute only the small fluctuations that are present across $\mathbf{Y}$. These are typically associated with corrupting noise. With this interpretation we see that a low rank approximation of $\mathbf{Y}$ can be obtained by discarding the contribution of small singular values, i.e. truncating the summation early.

One of the main applications of the SVD is the efficient computation of the matrix (pseudo-)inverse. By taking the pseudo-inverse of both sides of Equation 7.96 it can be shown that,

$$\mathbf{Y}^+ = \left(\mathbf{U}\Sigma\mathbf{V}^{\mathrm{H}}\right)^+ = \mathbf{V}^{-\mathrm{H}}\Sigma^+\mathbf{U}^{-1} \tag{7.95}$$

and recalling that $\mathbf{U}$ and $\mathbf{V}$ are unitary, we have that,

$$\mathbf{Y}^+ = \mathbf{V}\Sigma^+\mathbf{U}^{\mathrm{H}} \tag{7.96}$$

where it is noted that, being diagonal, Σ^+ is obtained by simply by taking the reciprocal of its non-zero elements, i.e.,

$$\Sigma^+ = \begin{bmatrix} \frac{1}{\sigma_1} & 0 & 0 \\ 0 & \ddots & 0 \\ 0 & 0 & \frac{1}{\sigma_N} \\ 0 & 0 & 0 \end{bmatrix}. \tag{7.97}$$

Note that the smallest singular values now provide the greatest contribution to the inverse! As already mentioned, these smaller singular values represent low level noise-like contributions to $\mathbf{Y}$. Consequently, its inverse becomes dominated by the effects of noise. To alleviate this issue, one can perform a Truncated Singular Value Decomposition (TSVD) on the matrix inverse. This is done quite simply by retaining the reciprocal singular values only if a certain threshold α is reached,

$$\Sigma_{ii}^+ = \begin{cases} \frac{1}{\sigma_i} & \text{if } \sigma_i \geq \alpha \\ 0 & \text{if } \sigma_i < \alpha. \end{cases} \tag{7.98}$$

By discarding the singular values associated with noise-like contributions, a better conditioned matrix inverse can be obtained. This is simply a form of regularisation; the additional information being introduced is the threshold α. It is worth mentioning that truncated SVD regularisation is included as part of most pseudo-inverse implementations, for example by MATLAB, NumPy and SciPy, requiring specification of the threshold value.

For Tikhinov regularisation, determining the optimum threshold for TSVD can be done using an LPO cross-validation. However, one of the main advantages of the TSVD scheme, is that some physical interpretation can be used to determine, not α, but the number of retained singular values. Consider for example the measurement and inversion of an interface FRF matrix $\mathbf{Y}_{Ccc}$. In the absence of any experimental noise, and assuming rigid interface behaviour, the rank of $\mathbf{Y}$ will be equal to the number of physical DoF present at the interface, i.e. 6 for each contact point. If said interface is represented by > 6 measurement DoF per contact, this will result in a rank deficient matrix, i.e. its smallest singular values will be exactly 0. In the presence of experimental error, the FRF matrix will not be exactly rank deficient, rather its smallest singular values will represent the contribution of noise-like fluctuations. These singular values can be discarded on the basis that they represent 'non-physical' contributions. In this sense, the number of retained singular values can be related to the number of physical DoFs that are contributing at a given frequency.

Having now introduced the SVD, it is worth noting that Tikhinov regularisation can similarly be implemented as a weighting of the singular values according to,

$$\Sigma^{+}_{ii} = \frac{\sigma_i}{\sigma_i^2 + \lambda} \tag{7.99}$$

where λ is the previously mentioned regularisation parameter.

7.4.4 PARTIALLY CORRELATED SOURCES

In this section, we consider the analysis needed when operational responses are partly correlated or uncorrelated and have been averaged prior to solving the inverse problem. In this case, a cross spectral approach is required as described below. The treatment is brief because, as described in the following section, the authors' preferred approach is to *avoid* averaging, at least until the last stage of presentation, and instead to use an *overlap-add* method which retains all time variation. Nevertheless, in the literature we occasionally see the problem expressed in cross-spectral form so the main equations are recounted here for completeness.

We begin by considering the case where all responses are fully correlated such that their relative phases are all well defined. In this case, all relative phase information can be extracted from a single column of the cross-spectral matrix $\mathbf{S_{vv}}$. To do this we must choose a *reference signal* v_i, to which we reference the phase of all other responses. This relative phase $\angle \mathbf{S}_{\mathbf{v}v_i}$ can then be assigned to the square root of the time averaged auto-spectrum to produce a complex (time averaged) response spectrum,

$$\mathbf{v} = \sqrt{\text{diag}(\mathbf{S_{vv}})}\, e^{-i\angle \mathbf{S}_{\mathbf{v}v_i}}. \tag{7.100}$$

The resulting complex response vector $\mathbf{v}$ can subsequently be used in the estimation of the blocked force.

This approach assumes all responses are fully correlated such that a single column of $\mathbf{S_{vv}}$ is sufficient to describe all the relative phases. In the presence of *partially correlated* responses, this will not be the case, and the entire cross-spectrum matrix must be used. To this end, Equation 6.22 is post-multiplied by its own conjugate transpose before a time average is taken,

$$\overline{\mathbf{f}\mathbf{f}^{\mathrm{H}}} = \mathbf{Y}_C^{-1}\mathbf{v}\mathbf{v}^{\mathrm{H}}\mathbf{Y}_C^{-\mathrm{H}} \longrightarrow \mathbf{S}_{\overline{\mathbf{ff}}} = \mathbf{Y}_C^{-1}\mathbf{S}_{\mathbf{vv}}\mathbf{Y}_C^{-\mathrm{H}} \tag{7.101}$$

where $\mathbf{S}_{\overline{\mathbf{ff}}}$ and $\mathbf{S_{vv}}$ are the time-averaged cross-spectral blocked force and response matrices, respectively. We can similarly represent the forward response prediction in a quadratic form as,

$$\mathbf{S}_{\mathbf{v}_r\mathbf{v}_r} = \mathbf{Y}_X\mathbf{S}_{\overline{\mathbf{ff}}}\mathbf{Y}_X^{\mathrm{H}} \tag{7.102}$$

where $\mathbf{Y}_X$ is the forward FRF of the VAVP and $\mathbf{v}_r$ is the response at the target DoFs r.

The off-diagonal terms of the cross-spectral matrices represent the interaction between the various DoFs. If two DoFs are completely *uncorrelated*, in other words if their relative phase $\phi_{YX} = \phi_Y - \phi_X$ (see Equation 7.24) is random, then their cross-spectrum S_{XY} will be negative as often as it is positive and will average to zero. Thus, an uncorrelated source is characterised by the (real) *diagonal* elements of the cross spectrum only. On the other hand, if two DoFs are completely *correlated*, i.e. they are 'phase locked' so that $\phi_Y - \phi_X$ is constant, then the averaging does not change the matrix. Thus, correlated sources are characterised by a *complex vector* of blocked forces and the inverse problem can be solved as described previously.

A likely scenario in practice however is that the source is neither completely uncorrelated nor correlated but is said to be *partially correlated*. In this case, the averaging preserves the correlated part of the off-diagonals and the uncorrelated part averages to zero. Thus, no assumptions can be made about the off-diagonal elements and the *entire cross-spectral matrix* is required to characterise partially correlated sources. Prediction of time-averaged output responses, and their correlations, in the form of a cross spectral matrix is then possible. The cross-spectral approach has the advantage of computational simplicity and efficiency, avoiding calculations for lengthy time series. The disadvantage is that any time-varying characteristics of the source behaviour are lost. Hence, the method is not suitable for auralisation or the evaluation of sound quality metrics. The overlap-add method on the other hand, described shortly, retains all time information and is therefore equally suitable for correlated, partially correlated or uncorrelated time sequences.

In the presence of partially correlated responses, the quadratic form can be further developed using Principle Component Analysis (PCA) to establish a set of uncorrelated and independent *virtual sources* which can be used for contribution analysis, etc. (see for example [8, 27]).

7.5 TIME SERIES RECONSTRUCTION

In the early days of NVH, targets were often set in terms of time-averaged energy quantities such as A-weighted sound pressure level. Increasingly though, perceptual and sound quality targets are being employed and in some cases listening tests are required with human listeners. If the output of a VAVP is to be used within this perceptually-driven framework, it is necessary to synthesise its response in terms of a *time series* suitable for auralisation.

Time series data can of course be generated by the inverse DFT of a response spectrum, for example a sound pressure spectrum computed from the VAVP. The length of the resulting time block is however limited to the size of the initial response spectrum, i.e. to N samples. For example, at a relatively narrow frequency resolution of 1 Hz, a DFT spectrum contains enough samples to synthesise only 1 second of time series, which is not enough for most listening tests or for many perception-based metrics [31]. Therefore, we need to find a way to synthesise time histories longer than a single block from our frequency domain description. Ideally, we need sufficient length to accurately represent the *time structure* of the sound which may have a strong influence on its perception. There are several ways this can be achieved but the authors have found an *overlap-add* method to the simplest and most effective.

Figure 7.13 illustrates the overlap-add process for VAVP time series reconstruction [16]. First, a full length operational response is recorded and split into a series of overlapping blocks/windows, each of N samples in length. The sample frequency f_s and window length N must of course yield a frequency resolution consistent with the FRFs used within the VAVP.

For a typical VAVP the initial time sequence $\mathbf{v}(t)$ will be the operational response measured at some set of indicator DoFs. After windowing, each time block's spectrum is used to compute a blocked force in the usual way, i.e. for the nth time block, $(\bar{\mathbf{f}}_c)_n = \mathbf{Y}_C^+ \mathbf{v}_n$. By repeating this for all time blocks, an array of blocked force vectors are obtained,

$$\left[\begin{array}{ccc} (\bar{\mathbf{f}}_c)_1 & \cdots & (\bar{\mathbf{f}}_c)_n \end{array}\right] = \mathbf{Y}_C^+ \left[\begin{array}{ccc} (\mathbf{v})_1 & \cdots & (\mathbf{v})_n \end{array}\right] \longleftarrow (\mathbf{v})_n = \mathscr{F}\left[\mathbf{v}(\Delta T_n)\right] \quad (7.103)$$

that implicitly captures the time varying behaviour of the source.

The output of the VAVP, typically in the form of a sound pressure spectrum p, is then calculated from the blocked forces for each block in the usual way to give an ordered sequence of output spectra.

$$\left[\begin{array}{ccc} (\mathbf{p})_1 & \cdots & (\mathbf{p})_n \end{array}\right] = \mathbf{H}\left[\begin{array}{ccc} (\bar{\mathbf{f}}_c)_1 & \cdots & (\bar{\mathbf{f}}_c)_n \end{array}\right] \quad (7.104)$$

It remains to transform each spectrum back to the time domain and to add them together to create a long output sequence corresponding to the initial input sequence.

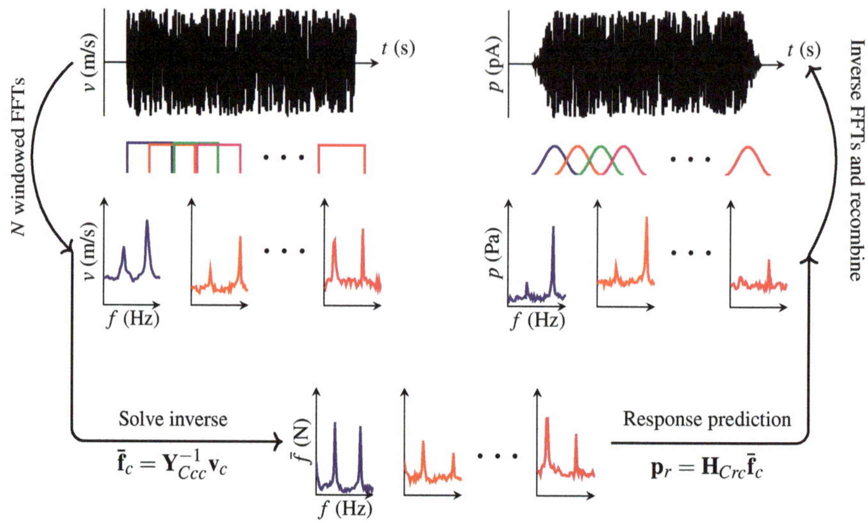

Figure 7.12 Overlap-add scheme for time series reconstruction using blocked forces.

A few points must be noted for the inverse transformation:

- if the calculation has been conducted using single-sided spectra then they must be converted to double-sided spectra prior to inverse transformation (as described in 7.2.2);
- the same overlap must be applied on the output as was used on the input – a 50% overlap is most convenient;
- any windowing should be applied after inverse transformation but before adding.

The authors have found the best implementation of the overlap add method to be with a rectangular window on the input, a 50% overlap, and a Hanning window on the output. This output window effectively 'cross-fades' the overlapped signals and prevents transient artefacts between blocks. If the intention is not to auralise, but to produce time averaged plots a Hanning window could be used on the input to reduce effects of spectral leakage.

Even if a time series reconstruction is not the intention, the above over-lap add scheme is still recommended for processing operational response data. Time-averaged results can be computed from the windowed response predictions as the last calculation step prior to plotting.

Importantly, by retaining the time series information, the above overlap-add method implicitly handles correlated, uncorrelated and partially correlated responses, which can otherwise present challenges when attempting to solve inverse problems.

7.6 OCTAVE BANDS

In component-based simulation we will generally be working with complex, narrow band data. However, there may be occasions when it is useful to export data as *constant percentage band* (CPB) values, particularly third octave or octave band values but also potentially extending to narrower bands such as $\frac{1}{12}$th octaves.

It should be noted that the processing leading to CPB values is fundamentally different to that used for the narrow band spectra assumed throughout the rest of the book. *Narrow band spectra*, as just described, are obtained by discrete Fourier transformation of a recorded time series leading to a *complex* spectrum with *equally spaced lines* on a linear axis, i.e. equally spaced in Hz. *CPB* spectra on the other hand, are obtained by *filtering* of time signals so as to extract the energy in a frequency band whose bandwidth is a constant percentage of the centre frequency and therefore increases, in terms of Hz, proportionally to its centre frequency (see Figure 7.13). Whilst there are fundamental differences between these spectral types, it is possible to approximately construct CPB spectra from narrow band data (and to some extent *vice versa* as will be discussed in 7.6.2.).

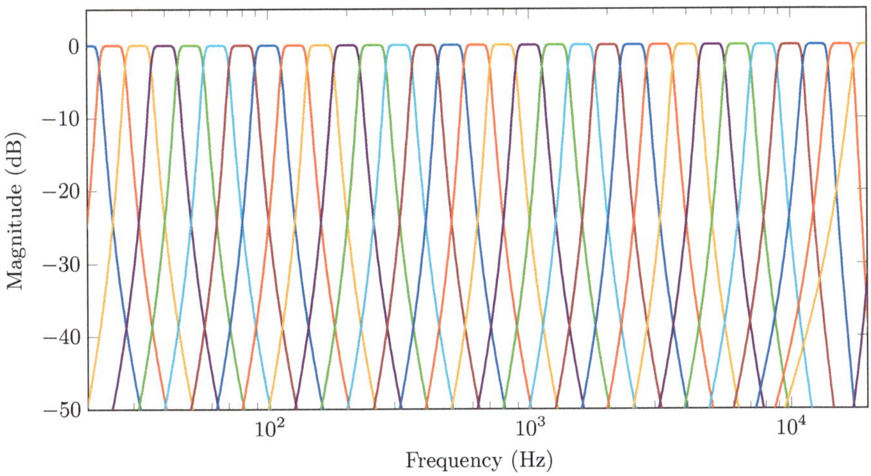

Figure 7.13 Third octave band filter shapes.

An octave is a factor of two in frequency. However, internationally agreed third octave band frequencies are an approximation, based on powers of 10 [10] rather than 2 as the numbers work out more conveniently. The third octave band *centre frequencies* are given by,

$$f_{n,centre} = 10^{n/10} \tag{7.105}$$

with $n \in \mathbb{Z}^+$. The lower (upper) limit is then given by the geometric mean of the centre frequency with that of the band below (above), so that,

$$f_{n,lower} = (f_{n-1,centre} \cdot f_{n,centre})^{\frac{1}{2}} = f_{n-1,upper} \tag{7.106}$$

which leads to,

$$\begin{aligned} f_{n,lower} &= 10^{(2n-1)/20} \\ f_{n,upper} &= 10^{(2n+1)/20}. \end{aligned} \tag{7.107}$$

For *octave*, rather than third octave bands, the indices are multiplied by 3 so that: $n \to 3n$ in Equation 7.105 and $(2n \pm 1)/20 \to [3 \times (2n \pm 1)]/20$ in Equation 7.107. These frequencies are tabulated in Table 7.1 (note that the International Standard preferred frequencies are slightly rounded for convenience so do not agree exactly with the calculated values).

Table 7.1

Centre frequencies for octave and third octave band spectra

Octave band			1/3 octave band		
Lower	Centre	Upper	Lower	Centre	Upper
Frequency (Hz)			Frequency (Hz)		
			22.4	25	28.2
22	31.5	44	28.2	31.5	35.5
			35.5	40	44.7
			44.7	50	56.2
44	63	88	56.2	63	70.8
			70.8	80	89.1
			89.1	100	112
88	125	177	112	125	141
			141	160	178
			178	200	224
177	250	355	224	250	282
			282	315	355
			355	400	447
355	500	710	447	500	562
			562	630	708
			708	800	891
710	1000	1420	891	1000	1122
			1122	1250	1413

Standard octave band filters do not have a sharp cutoff at the frequency limit, but have a tail which intersects that of the neighbouring band at the *half power* (−3 dB) points as shown in Figure 7.13. However, the most common approach to converting to and from narrow band results is to assume perfect a rectangular filter shape.

7.6.1 EXPORTING TO OCTAVE BANDS

The conversion to octave bands depends on whether we are dealing with *operational data* in the form of single value quantities (velocity, force, sound pressure, volume velocity, voltage, etc.), or *FRFs* and *transmissibilities* which are relative quantities.

Operational data

Considering first *operational data*, the important consideration is to preserve the *energy* in the frequency band. Thus, the conversion consists of summing the energy in the narrow band spectrum between the upper and lower limiting frequencies,

$$|v_{oct}| = \frac{1}{2}\left(\sum_{\omega_i=\omega_{lower}}^{\omega_{upper}} |v(\omega_i)|^2\right)^{\frac{1}{2}} \tag{7.108}$$

where the $|v(\omega_i)|$ are the magnitudes of the lines of the narrow band spectrum. In the unlikely event of a spectrum line falling precisely on the edge of a CPB band its energy should be shared between the two.

The factor of $\frac{1}{2}$ appears because band values are conventionally rms quantities whereas the narrow band spectrum is in terms of Fourier spectrum magnitude (a similar situation was encountered when dealing with power in 3.3).

It will be clear that any phase information is lost in the conversion and that the band values describe only magnitude.

Frequency response functions

When converting *relative quantities*, FRFs and transmissibilities, we require the *energetic average* value across the octave band,

$$|H_{oct}| = \frac{1}{N}\left(\sum_{\omega_i=\omega_{lower}}^{\omega_{upper}} |H(\omega_i)|^2\right)^{\frac{1}{2}} \tag{7.109}$$

where the $|H(\omega_i)|$ are the magnitudes of the lines of the narrow band spectrum and N is the number of lines in the frequency band.

7.6.2 IMPORTING OCTAVE BANDS AND PHASE SYNTHESIS

It is sometimes desirable to bring octave band data into a component-based simulation. For example, the reception plate method (3.3.7), SEA models (5.4) and sound insulation measurements will typically provide data in third octaves. However, conversion in this direction can only be achieved, at best, in an approximate sense because there is an absence of spectral detail in the CPB values.

More importantly, there is a lack of phase data which means that all information on time structure is lost. Thus, it is necessary to *synthesise* the phase if CPB band data is to be incorporated into a VAVP. Whether the synthesised phase provides satisfactory results is largely determined by perceptual criteria and this depends on whether it is the operational data, or the FRFs that are in CPB form as explained below.

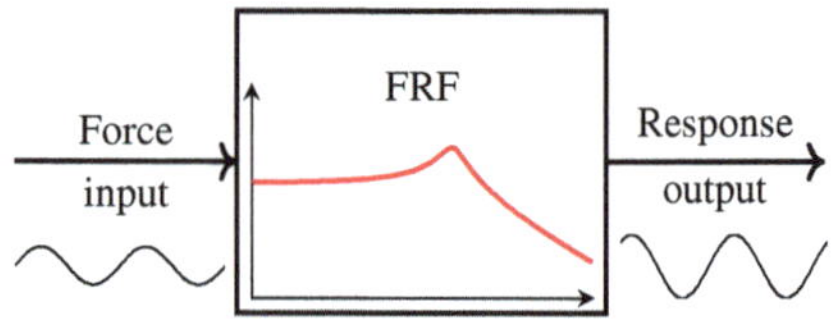

Figure 7.14 Input-output system as the simplest form of a VAVP.

Consider a VAVP in its simplest form as a single input-output system (Figure 7.14). The input, typically from blocked forces, is filtered by transmission through the assembly to give the audible output. In terms of perception, the most important features of the sound are usually found in the time structure of the operational input. The FRF on the other hand, will colour the sound by filtering, but rarely influences the time structure to an audible extent. There are some exceptions, such as an assembly that 'rings' strongly when excited by impulsive forces in which case the ringing may be audible on the output. However, for more steady state excitation the ringing of the filter will be buried in the output and may not be audible. Thus, while all information on time structure is missing from CPB values, for the above reasons this may matter less for FRFs than for operational data. In the following, we discuss the possibilities to synthesise FRFs, including phase, from CPB data. The more difficult problem of reconstructing operational data is then considered.

Frequency response functions

To construct a narrow band spectrum from a spectrum of CPB values we need to:

(a) convert the CPB values to a narrow band *magnitude* spectrum $|H(\omega_i)|$;
(b) add an appropriate synthesised *phase* spectrum, $\phi(\omega_i)$.

The complex narrow band spectrum is then given by,

$$H(\omega_i) = |H(\omega_i)|e^{i\phi(\omega_i)}. \tag{7.110}$$

Considering the *magnitude spectrum*, the simplest method is to assign the CPB value to every spectral line within the CPB band. For example, if we are aiming to represent a 125 Hz third octave band in narrow band with a resolution of 1 Hz we would assign the third octave band magnitude to a line every 1 Hz between 112 Hz and 141 Hz (see Table 7.1). This produces a flat narrow band spectrum within the frequency limits of each CPB band and Equation 7.109 is automatically satisfied. However, this may result in sudden jumps at the band edges. One would imagine that a continuous spectrum is less likely to generate audible artefacts. The spectrum shape can be smoothed if desired provided Equation 7.109 is satisfied, in other words provided the energy average of the FRF magnitudes across a CPB band equals the magnitude of the CPB band.

Regarding phase, if the filter phase is not audible in the output then almost any phase will do. Thus, there is a chance we can synthesise a phase spectrum for the filter and produce a realistic output sound, provided this choice does not itself generate audible artefacts! Broadly, the choices are a) constant (including zero) phase, b) random phase and c) minimum phase reconstruction.

For *constant* phase, one simply assigns a constant phase value for each frequency line, $\phi(\omega_i) = c$ where $0 \leq c \leq 2\pi$ is any phase value. For *random* phase, we assign $\phi(\omega_i) = \psi_i$ where the random phase $0 \leq \psi_i \leq 2\pi$ is uniformly distributed.

Minimum phase reconstruction exploits the fact that there is a known relationship between the magnitude and phase spectrum of a minimum phase FRF based on the Hilbert transform (as described in [16]) so that,

$$\phi(\omega_i) = -\mathscr{H}[\log|H(\omega_i)|] \tag{7.111}$$

where $\mathscr{H}[\square]$ is the Hilbert transform. In practice, this can be coded rather simply using the *Ceptstrum* function as described in [24]. Point FRFs are minimum phase functions and, if accurate magnitude data is available, the full FRF can be recovered. Transfer FRFs are not, in general, minimum phase but even here, use of a minimum phase spectrum will often produce a realistic auralisation. However, a magnitude spectrum that has been recovered from CPB data is, at best, an approximation and the minimum phase reconstruction may or may not produce a useful estimate of phase. Despite its limitations for CPB values however, minimum phase reconstruction is a useful technique that will often find application in practice.

The above steps will generate a complex, single-sided, narrow band spectrum that can be used in subsequent VAVP computations and to generate audible output by inverse DFT (see 7.2.2 and 7.5). However, it is stressed that the reconstruction of phase in this way should be considered *speculative* and the results should be subject to careful validation of audible realism before being used for any serious purpose.

Operational data

Since the time structure of the operational data is usually perceptually important, there is a limited amount one can do with CPB data due to lack of phase information. However, two possible applications are worth considering: (a) steady sounds and (b) combining recorded time histories with CPB magnitude data.

For compatibility with other data in the VAVP, the operational data is required in the form of a narrow band spectrum,

$$v(\omega_i) = |v(\omega_i)|e^{i\phi(\omega_i)} \tag{7.112}$$

where $\phi(\omega_i)$ is the phase spectrum.

If magnitude data is available in CPB bands then the magnitude of the DFT spectrum $|v(\omega_i)|$ can be generated using similar procedures to that described above for

FRFs. The only difference is that the spectrum *energy* (rather than the average value) must be consistent with the CPB, i.e. Equation 7.108 must be satisfied. Generally, there will be insufficient detail in a CPB magnitude spectrum to reproduce operational signals with tonal content. However, broadband spectra, such as that produced by aerodynamic noise, may be amenable to this analysis.

Having obtained the magnitude spectrum $|v(\omega_i)|$, the phase spectrum $\phi(\omega_i)$ must be synthesised.

Steady sounds

For *steady* broadband spectra, it may be possible to obtain a reasonable auralisation using a synthesised *random phase*: $\phi(\omega_i) = \psi_i$ where $0 \leq \psi_i \leq 2\pi$ is uniformly distributed.

Combined with the magnitude spectrum this will generate a complex, single-sided, narrow band spectrum that can be used in subsequent VAVP computations. If it is to be inverse transformed to the time domain then it will first need to be converted to a double-sided spectrum as described in 7.2.2.

Using measured phase

If the time structure of the operational signal is important then CPB is not sufficient to produce an auralisation. However, if even a single channel of recorded time history is available, then its time structure can be combined with the magnitude data from a CPB spectrum using a slightly adapted version of the overlap-add method described above (7.5).

The procedure is as follows: the recorded time history is split into blocks, transformed to the frequency-domain with the DFT and converted to a single-sided spectrum as per the usual overlap-add method. The only difference from the usual method is that, before solving the inverse problem, the magnitude spectrum is replaced by that generated from the CPB data. The analysis then continues according to the normal method, solving for the blocked forces and sound pressures before transforming back to the time-domain.

7.7 SUMMARY

Raw data is initially captured as analogue time signals and must first be *discretised.* This operation is usually done automatically in the data acquisition system but if done manually an anti-aliasing filter must be included to avoid aliasing. The discretised time sequence then needs to be split into blocks of length N samples prior to transformation to the frequency domain. Some form of *windowing* may be needed at this stage, its main purpose being to limit the effects of sudden truncation at the beginning and ends of the blocks.

Nearly all operations in a VAVP are conducted in the frequency domain. The primary tool for transformation from the time domain is the *Discrete Fourier Transform* (DFT). Raw DFTs are naturally double-sided where the conjugate of the spectrum at positive frequencies is reflected at negative frequencies. Since the *single-sided spectrum* contains all information, requires less computing and is more intuitive, it is normal to discard the negative frequency data and carry out subsequent processing with single-sided spectra. DFT phase is referenced to the beginning of the time window and so is not consistent from one block to the next. For averaging in the frequency domain therefore, the *auto-* and *cross spectra* are introduced which have a fixed *phase reference*.

In theory, an FRF is the ratio of the output to the input spectrum. In practice, averaging is needed to reduce the effect of noise so FRFs are typically constructed from auto and cross spectra which reduces bias. The so-called H_1 and H_2 estimates of FRFs are optimised for noise on respectively the output and input. Lesser used but potentially useful estimators include H_s and H_v. Most FRF measurements are of the *single-input multiple-output* (SIMO) type where a single input is applied at a time. Each excitation then generates a column of the FRF matrix. Multiple-input multiple-output (MIMO) measurement of FRF is also possible but requires more involved processing to extract individual FRFs. *Coherence* quantifies the linear dependence of the output of an FRF on its input and is a key indicator of the quality of the measurement. For multiple inputs, more complex metrics can be used such as *multiple* and *conditioned coherence*. Simple estimates of uncertainty can be made based on coherence.

Obtaining blocked forces requires *solving the inverse problem*. This problem is typically ill-posed meaning that the solution may be highly sensitive to errors on the input. In the *determined* problem, the unknown inputs are estimated from same number of responses using a square FRF matrix. However, it is recommended to *over-determine* the problem such that there are more responses than unknown inputs. This may then be solved with a pseudo-inverse matrix which provides a *least squares* estimate of blocked forces. An over-determined problem will often require some degree of *regularisation* to avoid undue influence of small, noisy singular values. Regularisation must be used with care to avoid 'tricking' an apparently good solution from poor data. In *Truncated Singular Value Discarding* (TSVD), singular values deemed unreliable are discarded, whereas in *Tikhonov* regularisation, they are de-emphasised with a weighting coefficient. For *partially correlated* sources, inversion has traditionally been based on cross spectral matrices.

In order to transform back to time domain an *overlap-add method* allows all time structure to be preserved so that it is applicable equally to fully, partially or uncorrelated blocked forces. This method is therefore recommended over traditional cross spectrum methods.

Octave and third octave bands are widely used in acoustic measurements but are not directly compatible with the narrow band spectra used for VAVPs. Narrow band outputs can easily be *exported* to (third) octave bands for display purposes. *Importing* octave band data is more problematic because it lacks phase information, although in limited situations it may be possible to *synthesise* the missing phase well enough to produce a realistic auralisation.

Bibliography

1. RJ Allemang, RS Patwardhan, MM Kolluri, and AW Phillips. Frequency response function estimation techniques and the corresponding coherence functions: A review and update. *Mechanical Systems and Signal Processing*, 162:108100, 2022.
2. JS Bendat. System identification from multiple input/output data. *Journal of Sound and Vibration*, 49(3):293–308, 1976.
3. JS Bendat. Statistical errors in measurement of coherence functions and input/output quantities. *Journal of Sound and Vibration*, 59(3):405–421, 1978.
4. JS Bendat and Allan G Piersol. *Random data: analysis and measurement procedures*. John Wiley & Sons, 2011.
5. D Calvetti, S Morigi, L Reichel, and F Sgallari. Tikhonov regularization and the l-curve for large discrete ill-posed problems. *Journal of Computational and Applied Mathematics*, 123(1-2):423–446, 2000.
6. JWR Meggitt and T Charity. Towards an indirect method for measurement of low frequency sound transmission. In *Proceedings of Acoustics 2024*, 2024.
7. HG Choi, AN Thite, and DJ Thompson. Comparison of methods for parameter selection in tikhonov regularization with application to inverse force determination. *Journal of Sound and Vibration*, 304(3-5):894–917, 2007.
8. AS Elliott, AT Moorhouse, T Huntley, and S Tate. In-situ source path contribution analysis of structure borne road noise. *Journal of Sound and Vibration*, 332(24):6276–6295, 2013.
9. GH Golub and CF Van Loan. An analysis of the total least squares problem. *SIAM Journal on Numerical Analysis*, 17(6):883–893, 1980.
10. ISO *266:1997 Acoustics Preferred Frequencies*, 1997.
11. Q Leclere, NB Roozen, and C Sandier. On the use of the hs estimator for the experimental assessment of transmissibility matrices. *Mechanical Systems and Signal Processing*, 43(1-2):237–245, 2014.
12. I Markovsky and S Van Huffel. Overview of total least-squares methods. *Signal Processing*, 87(10):2283–2302, 2007.
13. JWR Meggitt, AS Elliott, AT Moorhouse, G Banwell, H Hopper, and J Lamb. Broadband characterisation of in-duct acoustic sources using an equivalent source approach. *Journal of Sound and Vibration*, 442:800–816, 2019.
14. KL Napolitano. Using singular value decomposition to estimate frequency response functions. In *Topics in Modal Analysis & Testing, Volume 10: Proceedings of the 34th IMAC, A Conference and Exposition on Structural Dynamics 2016*, pages 27–43. Springer, 2016.
15. AV Oppenheim. *Applications of digital signal processing*. Prentice Hall, 1977.
16. AV Oppenheim and RW Schafer. Discrete-time signal processing. *Prentice Hall*, 1999.
17. A Pereira, Jérôme Antoni, and Quentin Leclere. Empirical bayesian regularization of the inverse acoustic problem. *Applied Acoustics*, 97:11–29, 2015.
18. MA Peres, C Kallmeyer, MC Witter, R Carneiro, Flávio Donizeti Marques, and LPR de Oliveira. Advantages of multiple-input multiple-output testing. *Sound and Vibration*, 49(8):8–12, 2015.
19. AW Phillips and RJ Allemang. An overview of mimo-frf excitation/averaging/processing techniques. *Journal of Sound and Vibration*, 262(3):651–675, 2003.
20. NG Polson and V Sokolov. Bayesian regularization: From tikhonov to horseshoe. *Wiley Interdisciplinary Reviews: Computational Statistics*, 11(4):e1463, 2019.
21. LR Rabiner and B Gold. *Theory and application of digital signal processing*. Englewood Cliffs: Prentice-Hall, 1975.
22. C Ramirez, V Kreinovich, and M Argaez. Why l_1 is a good approximation to l_0: A geometric explanation. *Journal of Uncertain Systems*, 7:203–207, 2013.
23. CE Shannon. Communication in the presence of noise. *Proceedings of the IRE*, 37(1):10–21, 1949.
24. JO Smith. *Introduction to Digital Filters with Audio Applications*. W3K Publishing, 2007.
25. MH Tan and JK Hammond. A non-parametric approach for linear system identification using principal component analysis. *Mechanical Systems and Signal Processing*, 21(4):1576–1600, 2007.
26. S Van Huffel and J Vandewalle. *The Total Least Squares Problem: Computational Aspects and Analysis*, volume 9. SIAM, 1991.

27. D Vandenbroeck and W Hendricx. Interior road noise analysis with principal components. In *Proceedings of the Acoustical Society of Korea Conference*, pages 854–859. The Acoustical Society of Korea, 1994.
28. GT Rocklin, JA Crowly, and H Vold. A comparison of h˙1, h˙2 and hv frequency response functions. In *Proceedings of the 3rd International Modal Analysis Conference, 1985*, pages 272–278, 1985.
29. PR White, MH Tan, and JK Hammond. Analysis of the maximum likelihood, total least squares and principal component approaches for frequency response function estimation. *Journal of Sound and Vibration*, 290(3-5):676–689, 2006.
30. A Wicks and W Shapton. Frequency response estimators for use in the presence of uncorrelated noise. *SAE Transactions*, pages 62–66, 1987.
31. E Zwicker and H Fastl. *Psychoacoustics: Facts and models*, volume 22. Springer Science & Business Media, 2013.
32. A Hjørungnes. Complex-valued matrix derivatives: with applications in signal processing and communications. Cambridge University Press, 2011.
33. DC Montgomery, and EA Peck, and GG Vining. Introduction to linear regression analysis. John Wiley and Sons, 2021.
34. FE Harrell. Regression modeling strategies: with applications to linear models, logistic regression, and survival analysis. Springer, 2001.
35. HW Engl and M Hanke, and A Neubauer. *Regularization of inverse problems*, volume 375. Springer Science & Business Media, 1996.
36. AN Tikhonov and V Arsenin. Solutions of ill-posed problems. John Wiley & Sons, 1977.
37. C Per Hansen. The truncated SVD as a method for regularization. BIT Numerical Mathematics, volume 27:534–553, Springer, 1987.

8 Uncertainties

8.1 INTRODUCTION

It might be considered that the presentation of the component-based approach is now complete since we have described everything needed to characterise components, to combine them and to 'switch on' the resulting assembly to produce a predicted output. However, one issue remains to be addressed: according to international consensus, this prediction can only really be considered complete once it is accompanied by some measure of its reliability [1]. To complete our VAVP therefore, we need to provide the predicted output with an accompanying statement of its inherent errors and uncertainties. The aim of this chapter is to provide the reader with the tools to achieve this. Although uncertainty in structural dynamics is a large topic, we focus on what is necessary for the component-based approach.

In 8.2 the main types of uncertainties in a VAVP are introduced and categorised. Generally, these can be classified as either *systematic* or *random*, which are discussed in detail in the two main sections of this chapter, 8.3 and 8.4 respectively.

The main sources of *systematic* error in a VAVP have been discussed throughout the book, the most important of which is lack of agreement between the true interface and its representation by an interface model. This will be categorised as *model uncertainty*. Since the systematic error due to model uncertainty cannot be reduced by averaging, it is good practice to identify its causes, quantify its effect and reduced it where possible. The main tools for this are tests of *completeness* and *consistency* as described in 8.3.

Random uncertainties (8.4), it will be seen, may be categorised as due to lack of *source* repeatability or *experimental* uncertainty, which can be further subdivided into *equipment* noise and *operator* error. Every measured quantity, whether an FRF or an operational response has an associated variance or (for matrix-vector quantities) co-variance, which can be estimated from repeated measurements. However, as should be clear by now, these (co-)variances may be amplified through inversion to produce a larger uncertainty on the output. Our main interest lies, therefore, not only in the uncertainties of directly measured quantities themselves, but in how they *propagate* onto the outputs of a VAVP. The *Law of Propagation of Uncertainty* (LPU) is well-known for this purpose and in 8.4 we will consider its adaption to the estimation of uncertainties in a VAVP. In general, to calculate the propagated uncertainty one needs: a) the *covariance* matrix for the measured variables and b) the so-called *Jacobian* for the underlying equation. In our case, the measured variables are the operational responses and the FRFs whose co-variance matrices are obtained in a standard way by repeated measurements. The main contribution of 8.4 then is to introduce the Jacobians for the VAVP equations, namely; the *inverse solution of blocked forces*, *substructuring*, and the *forward prediction of an operational response*. The derivations are quite involved but, once known, the Jacobians can be implemented in code

DOI: 10.1201/9781003399131-8

relatively easily to give uncertainty estimates on the output. This process of uncertainty propagation through a VAVP is illustrated through a numerical example.

Finally in this chapter, for large uncertainties we discuss the use of *Monte-Carlo methods*.

8.2 SOURCES OF UNCERTAINTY

In order to minimise the unwanted effects of uncertainty, it is necessary to understand its underlying cause. In the present context, i.e. the development of a VAVP (i.e. the application of blocked force and/or substructuring methods) the uncertainties encountered may be categorised as originating from the 'Model', the 'Source', or the 'Experiment'.

Table 8.1
Types of uncertainties encountered in vibro-acoustic problems.

Type	Description	Category
1) Model	Uncertainty resulting from an approximate model being used to describe the physical problem.	Systematic
2) Source	Uncertainty inherent to the operational behaviour of the source.	Random
3) Experimental:	Uncertainty due to experimentation.	
3a) Measurement	*Due to the cumulative effect of noise sources in the measurement signal path and post processing.*	*Random*
3b) Operator	*Due to human error in the measurement procedure.*	*Random*

Model uncertainty

Model uncertainties arise when an approximate model is used to describe the physical problem. In a VAVP, the most likely cause for model uncertainty is an inaccurate representation of the interface dynamics by the *interface model*. Common assumptions that may or may not be reliable are that the interface behaves as a number of point-like connections, or can be described by a limited number of DoFs. Another is that the passive properties of the assembly do not change under operational conditions. These examples will be considered as they relate to the notions of *completeness* and *consistency*. Strictly speaking , these forms of model uncertainty introduce a systematic or bias error as opposed to random uncertainty. Lack of linearity and time-invariance are also potential forms of model uncertainty but are not considered in detail.

Source uncertainty

Source uncertainty arises due to the inherent variation in the activity of any vibration source. This variation may be considered the temporal variation (i.e. fluctuating activity) exhibited by a single vibration source, or the variation in activity across an ensemble of nominally identical vibration sources. Source uncertainty is typically random in nature.

Experimental uncertainty

Experimental uncertainty can be further subdivided into '*Measurement*' and '*Operator*' uncertainty. Measurement uncertainty describes the cumulative effect of noise sources within the measurement signal path and beyond. Typical examples include, external disturbances, thermo-electrical noise, sampling error, etc. Measurement uncertainty is typically random in nature. Operator uncertainty describes the effect of human error in the measurement procedure. An important example is the inconsistent location and orientation of applied forces during the measurement of FRFs. Operator uncertainty is typically random in nature, although bias errors can be introduced if excitations are widely spread and not be applied at the desired location.

Having introduced the main types and categorisations of uncertainty (summarised in Table 8.1), the following sections will focus on their quantification and propagation.

8.3 MODEL UNCERTAINTY

As described above, model uncertainty emanates from the use of an approximate model to describe a physical problem. Its effect is to introduce a systematic error. Whilst its underlying cause may vary from application to application, there exist two notable sources worth discussing here. These are related to the notions of *completeness* and *consistency*, and are the main topics of this section.

- *Completeness* (see 4.5, 6.4.1) describes the degree to which the interface between two components has been correctly represented, for example, whether enough DoFs have been included (e.g. rotations/in-plane). The completeness of an interface description is essential if independent blocked forces are to be obtained. The completeness of an interface description is also an essential requirement in the coupling and decoupling of substructures.
- *Consistency* relates to the underlying dynamics shared by the active and passive properties of an assembly or component. For example, if the FRF of an assembly were to change whilst in operation, we would have an inconsistency between a measured operational response and the passive FRF. Inconsistency arises principally when active and passive quantities are obtained from separate experiments with different operating conditions.

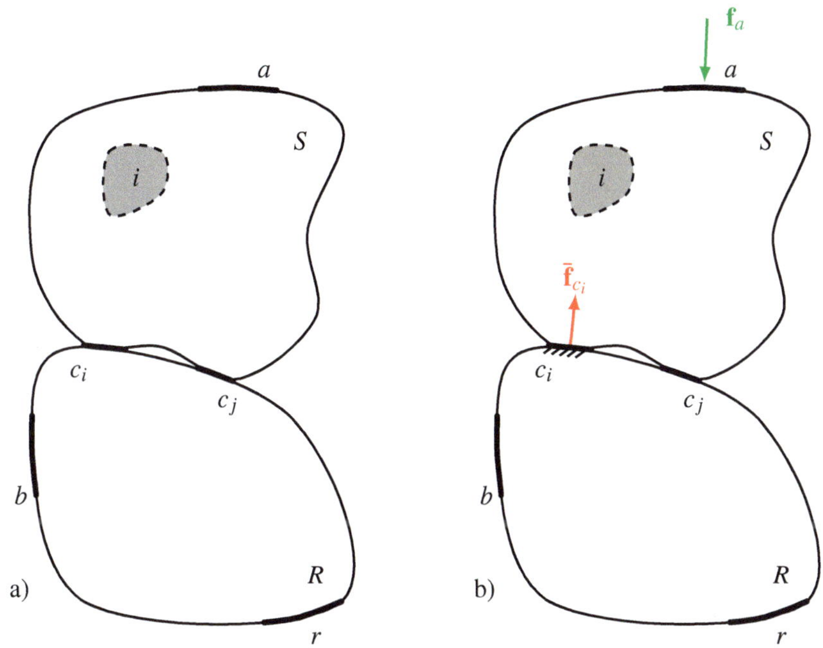

Figure 8.1 Unconstrained (a) and partially unconstrained (b) assemblies.

8.3.1 COMPLETENESS

Interface completeness is a general issue encountered whenever measurements are performed at the interface of one or more substructures. This includes not only the characterisation of blocked forces, but also the coupling and decoupling of substructures. To demonstrate the importance of completeness we will consider it from the perspective of a blocked force characterisation, though similar notions apply in other situations.

Shown in Figure 8.1a is a vibration source S coupled via two sets of DoFs c_i and c_j, to a receiver structure R. Suppose all of the interface DoFs (c_i and c_j) are constrained such that their responses (displacement, velocity and acceleration) are zero and an external force is applied remotely on the source. If the interface has been described *completely* (i.e. all DoFs are being constrained), then the resultant receiver response will be zero everywhere, since no vibration is permitted to cross the blocked interface. We call this a *complete* interface representation (or interface model). Had only a portion of the interface been constrained, say the c_i DoFs, then some vibration will propagate into the receiver and its response will no longer be zero. We call this an *incomplete* interface representation. This concept has previously been introduced in 4.5 and 6.4.1 and we will now introduce the so-called *Interface Completeness Criterion* as a convenient means of quantifying completeness.

Considering first a *complete* interface representation, where all interface DoFs are contained within the set c, the blocked force is obtained exactly as,

$$\bar{\mathbf{f}}_c = \mathbf{Y}_{Cbc}^{+}\mathbf{v}_c \tag{8.1}$$

where C denotes the coupled assembly, and b and c are sets of remote and interface DoFs, respectively. The response in some new assembly X at the target DoFs r can then be predicted exactly as,

$$\mathbf{v}_r = \mathbf{Y}_{Xrc}\bar{\mathbf{f}}_c. \tag{8.2}$$

However, what if the interface representation is not complete? Suppose only a subset of the interface DoFs, c_i, are measurable. It was already shown in 4.5 that neglecting a subset of interface DoFs (say c_j) when solving the inverse problems leads to an estimated blocked force of the form,

$$\tilde{\bar{\mathbf{f}}}_{c_i} = \bar{\mathbf{f}}_{c_i} + \mathbf{Y}_{Cbc_i}^{-1}\mathbf{Y}_{Cbc_j}\bar{\mathbf{f}}_{c_j} \tag{8.3}$$

where $\bar{\mathbf{f}}_{c_i}$ is the true blocked force at the measured DoFs c_i, and $\mathbf{Y}_{Cbc_i}^{-1}\mathbf{Y}_{Cbc_j}\bar{\mathbf{f}}_{c_j}$ is a *systematic error* arising due to an approximation in the underlying model, i.e. that the model did not include interface DoFs c_j. Had all DoFs been included in the interface description there would be no additional term and no systematic error. Importantly, this error is dependent on the dynamics of the assembly, and so renders the blocked force a property of the assembly rather than solely of the source, i.e. it is no longer independent or transferable.

As discussed in Chapter 3, one of the principle advantages of the blocked force over other source characterisation methods is its *transferability* from one assembly, say a test bench, to another, say a virtual prototype. This is possible because the blocked force is an independent source property. In the presence of an incomplete interface description though, the acquired blocked force becomes *dependent* on the assembly in which it is characterised. Let us now consider the prediction of a secondary assembly response, using a blocked force acquired with an incomplete interface description.

Using the subscript X to denote the secondary assembly, a remote receiver response prediction (indicated by the tilde accent $\tilde{\square}$) is given by,

$$\tilde{\mathbf{v}}_{Xr} = \mathbf{Y}_{Xrc_i}\tilde{\bar{\mathbf{f}}}_{c_i} = \mathbf{Y}_{Xrc_i}(\bar{\mathbf{f}}_{c_i} + \mathbf{Y}_{Cbc_i}^{-1}\mathbf{Y}_{Cbc_j}\bar{\mathbf{f}}_{c_j}). \tag{8.4}$$

Note that only the c_i DoFs are included, as the remaining c_j DoFs were considered unknown or inaccessible. Multiplying out the above then yields,

$$\tilde{\mathbf{v}}_{Xr} = \mathbf{Y}_{Xrc_i}\bar{\mathbf{f}}_{c_i} + \mathbf{Y}_{Xrc_i}\mathbf{Y}_{Cbc_i}^{-1}\mathbf{Y}_{Cbc_j}\bar{\mathbf{f}}_{c_j}. \tag{8.5}$$

This is the predicted response in the assembly X based on the (incomplete) blocked force $\tilde{\bar{\mathbf{f}}}_{c_i}$ from assembly C.

Based on the true blocked force, the remote receiver response in X is given by,

$$\mathbf{v}_{Xr} = \mathbf{Y}_{Xrc_i}\bar{\mathbf{f}}_{c_i} + \mathbf{Y}_{Xrc_j}\bar{\mathbf{f}}_{c_j}. \tag{8.6}$$

Notice that whilst the contribution of the available DoFs c_i is correct, the remaining c_j DoFs contribute through the FRF product $\mathbf{Y}_{Xrc_i}\mathbf{Y}_{Cbc_i}^{-1}\mathbf{Y}_{Cbc_j}$. In general, this product is not equal to the secondary assembly FRF, $\mathbf{Y}_{Xrc_j} \neq \mathbf{Y}_{Xrc_i}\mathbf{Y}_{Cbc_i}^{-1}\mathbf{Y}_{Cbc_j}$. Consequently, the bias error introduced into the blocked force, is propagated onto the response prediction.

The severity of this error will depend on a) the level of the blocked force $\bar{\mathbf{f}}_{c_j}$, b) its contribution to the measured response in the initial assembly $\mathbf{Y}_{Cbc_j}$, and c) the similarity of the two assemblies; if the secondary assembly is dynamically similar to the initial assembly $X \approx C$, it can be that the FRF product $\mathbf{Y}_{Xrc_j} \approx \mathbf{Y}_{Xrc_i}\mathbf{Y}_{Cbc_i}^{-1}\mathbf{Y}_{Cbc_j}$ and the bias error may be relatively small.

When characterising a component, it is of course beneficial to minimise the influence of bias errors due to incompleteness. Knowing this, it may be tempting to 'play safe' and ensure that the full 6 DoFs are included at every point. However, this may not be necessary, indeed if some DoFs do not contribute significantly then to include them would be undesirable a) because of the extra experimental effort involved and b) because the unnecessary measurements will add noise into the system without contributing to an accurate description of interface dynamics. For example, with resilient coupling it is often found that some DoFs are significantly softer than others and do not influence the interface dynamics to a great extent; rigid coupling at the interface on the other hand typically requires a greater number of DoFs to achieve a sufficient level of completeness. Therefore, there is great potential benefit in being able to *quantify* the completeness of the interface model prior to embarking on a full measurement campaign.

Quantification

To quantify completeness, we are interested in obtaining a measure of the bias error in Equation 8.3 relative to the true blocked force. However, this bias error, defined in terms of the unknown c_j DoFs, cannot be measured directly. Instead, we adopt the approach of comparing the *measured* response, which naturally includes transmission through *all* DoFs, with the response *predicted* from the blocked forces, which allows for transmission through the *known* DoFs, c_i, only. While this comparison can be made with operational excitation, there are several advantages in using artificial excitation, such as: (a) more repeatable excitation, (b) consistency between 'operational' and passive states and (c) ability to excite all DoFs at the interface through appropriate selection of artificial excitation locations and directions.

As shown in Chapter 3, the response produced by an artificial excitation at some remote location a on the source, can be reproduced identically by the corresponding (artificial) blocked force according to,

$$\mathbf{v}_{c_i} = \mathbf{Y}_{Cc_ic_i}\bar{\mathbf{f}}_{c_i} = \mathbf{Y}_{Cc_ia}\mathbf{f}_a. \tag{8.7}$$

From the above we obtain a transmissibility-based relation between the artificial excitation force and the resulting blocked force,

$$\bar{\mathbf{f}}_{c_i} = \mathbf{Y}_{Cc_ic_i}^{-1}\mathbf{Y}_{Cc_ia}\bar{\mathbf{f}}_a. \tag{8.8}$$

Using the above blocked force, we are able to predict the response at the remote receiver DoFs b,

$$\mathbf{v}_b = \overbrace{\mathbf{Y}_{Cbc_i}\mathbf{Y}_{Cc_ic_i}^{-1}\mathbf{Y}_{Cc_ia}}^{\mathbf{Y}_{Cba}^{(c_i)}}\mathbf{f}_a. \tag{8.9}$$

Noting that Equation 8.9 simply relates the external force at a to the response at b, the product on the right hand side is itself of the form of an FRF which we label $\mathbf{Y}_{Cba}^{(c_i)}$. The superscript $\square^{(ci)}$ is added to denote that it is a hypothetical, rather than physical, FRF reconstructed assuming transmission through the c_i DoFs only.

With a complete interface ($c = c_i$), the reconstructed FRF $\mathbf{Y}_{Cba}^{(c_i)} = \mathbf{Y}_{Cbc_i}\mathbf{Y}_{Cc_ic_i}^{-1}\mathbf{Y}_{Cc_ia}$ is identically equal to the directly measured FRF $\mathbf{Y}_{Cba}$ (in practice small discrepancies due to experimental error will be present). However, with an *incomplete* interface, the contribution of the neglected c_j DoFs will not be accounted for by $\mathbf{Y}_{Cba}^{(c_i)}$, though will be present in $\mathbf{Y}_{Cba}$. The level of disagreement between the reconstructed and directly measured FRFs will be proportional to the bias error in Equation 8.3. Therefore, to quantify completeness it is sufficient to assess the similarity of the direct and reconstructed FRFs, $\mathbf{Y}_{Cba}$ and $\mathbf{Y}_{Cba}^{(c_i)}$. To this end, a correlation-based measure of similarity was proposed in [7, 8] as the so-called *Interface Completeness Criterion* (ICC),

$$\text{ICC} = \text{Corr}\left(\underset{\rightarrow}{\mathbf{Y}}_{Cba}^{(c)}, \underset{\rightarrow}{\mathbf{Y}}_{Cba}^{(c_i)}\right) \tag{8.10}$$

$$\text{Corr}(\mathbf{a}, \mathbf{b}) = \frac{\left|\mathbf{a}^{\mathrm{H}}\mathbf{b}\right|^2}{\mathbf{a}^{\mathrm{H}}\mathbf{a} \cdot \mathbf{b}^{\mathrm{H}}\mathbf{b}} \tag{8.11}$$

where $\underset{\rightarrow}{\square}$ denotes matrix column-wise vectorisation (see Equation 8.28) and $\square^{\mathrm{H}}$ represents a conjugate transpose. Note that to form the necessary vectors, multiple excitation and/or response DoFs (b and a) should be considered when measuring the necessary FRFs. Ideally, the excitations a should be sufficient in number, and varied in direction, such that they are able to control all interface motions in the frequency range of interest. This requires the number of a points to be equal to or greater than the number of interface DoFs. Similarly, the responses positions b should be such that they are sufficient in number and orientation to observe all interface motions. This requires the number of b points to also be equal to or greater than the number of interface DoFs.

If the chosen interface representation is complete, that is, $|c_i| = |c|$, the ICC will be equal to one. If the interface description is incomplete, $|c_i| < |c|$ the ICC will be less than 1. In practice, it is important to note that the FRFs used in Equation 8.10 are obtained from measurement and are therefore subject to some degree of experimental uncertainty. Consequently, even in the case of a complete interface representation, an ICC equal to 1 is unlikely to be achieved. However, the authors have found that it is usually straightforward to distinguish between noise-induced discrepancies, which tend to be localised in frequency, and those from neglected DoFs which tend to cover a wide frequency range.

The ICC provides a measure of interface completeness irrespective of the operating nature of the connected components. For example, suppose a vibration source generates forces exclusively in the vertical z DoFs. Whilst a decent blocked force characterisation might be obtained using the z DoFs alone, this would not constitute a complete interface representation according to the ICC if other DoFs are important for the coupling. The ICC considers completeness more generally, specifically with respect to the applied forces at a and observed responses at b.

The ICC can be used also to assess the completeness of interface representations for structural coupling and decoupling problems. For decoupling, tests are required on the assembled structure (target component coupled to residual test bench), and so the ICC can be applied in the usual way. For coupling problems, one usually only measures the component in its free-interface configuration, from which the ICC cannot be computed. In this case, the component must be installed onto some test bench, or coupled to a dummy component, such that ICC can be computed its interface representation be assessed.

As a final point on completeness, it is worth noting that Equation 8.10 is not the only way to obtain a completeness metric. Any comparison between the reconstructed and directly measured FRFs can be used. For example, a conjugate-based formulation (see 6.4.3) was proposed in [3],

$$\text{ICC} = \langle \text{ICC}_{nm} \rangle \tag{8.12}$$

$$\text{ICC}_{nm} = \text{Conj}\left(\mathbf{Y}^{(c)}_{Cb_na_m}, \mathbf{Y}^{(c_i)}_{Cb_na_m}\right) \tag{8.13}$$

$$\text{Conj}\,(a,b) = \frac{(a+b)\,(a^*+b^*)}{2\,(aa^*+bb^*)} \tag{8.14}$$

where $\square^*$ denotes complex conjugate and $\langle\square\rangle$ an average over the indices nm.

8.3.2 CONSISTENCY

The incomplete representation of an interface has been shown above to introduce a systematic error on the acquired blocked force. A further source of systematic error can potentially arise if there is a change in the FRF characteristics of an assembly or component during the active and passive measurement phases. This might occur, for example due to a change in temperature or because of stiffening due to operational load. This effect is here termed as *inconsistency* [7].

As an example, consider that the operational response and blocked force are related by,

$$\mathbf{v}_c = \mathbf{Y}_{Ccc}\bar{\mathbf{f}}_c. \tag{8.15}$$

However, suppose that whilst the source is operating, the passive properties of the assembly change. In this case, Equation 8.15 becomes,

$$\mathbf{v}_c = \mathbf{Y}'_{Ccc}\bar{\mathbf{f}}_c. \tag{8.16}$$

where the $\square'$ is used to denote the *operational* FRF ($\mathbf{Y}'_{Ccc} \neq \mathbf{Y}_{Ccc}$).

Now, to characterise the blocked force we follow the standard procedure and measure the FRF whilst the source is turned off,

$$\tilde{\bar{\mathbf{f}}}_c = \mathbf{Y}_{Ccc}^{-1}\mathbf{v}_c \tag{8.17}$$

where $\tilde{\bar{\mathbf{f}}}_\mathbf{c}$ is the obtained blocked force ($\bar{\mathbf{f}}$ being the true blocked force). Substituting Equation 8.16 into the above we obtain,

$$\tilde{\bar{\mathbf{f}}}_c = \mathbf{Y}_{Ccc}^{-1}\mathbf{Y}'_{Ccc}\bar{\mathbf{f}}_c. \tag{8.18}$$

Note that if there is any inconsistency between the passive and the operational FRFs then $\mathbf{Y}_{Ccc}^{-1}\mathbf{Y}'_{Ccc} \neq \mathbf{I}$, and so we obtain a blocked force estimate with an inbuilt systematic error.

Tests for consistency are not fully developed at current state of the art. One possibility, mentioned in 6.4.3, is to compare the results of an on-board validation with real and artificial excitation. Any differences in the correlation obtained is a likely indication of inconsistency, although it would not give information on where or how the inconsistency occurred. Another similar approach might be to utilise the concept of generalised transmissibility which, as discussed in 2.4.2, can be obtained from both (passive) FRF-based testing and (operational) output-only measurements. A comparison of transmissibility obtained in these two different ways might provide some means of identifying changes in the passive properties whilst an assembly is operational.

8.4 RANDOM UNCERTAINTY

In 8.3 we considered role of *model uncertainty* and the introduction of systematic errors, focusing on the notions of completeness and consistency. In this section, attention is turned towards the role of *random uncertainty*.

Random uncertainties, or random errors, are those which, over a series of repeated measurements, *randomly* fluctuate about the true value of the measurand. Whilst all measurements are subject to some degree of random uncertainty, it is often not the measurements themselves that we are interested in. Rather, these measurements are used to infer or derive some other quantites. For example, we measure operational responses and FRFs, not for their own sake, but to calculate a blocked force or to perform substructuring. The question is then, how do the random errors in our measurements affect the estimates of derived quantities? That is, how does the uncertainty *propagate* through our VAVP equations?

The principle aim of this section is to introduce the reader to the fundamentals of uncertainty propagation, and provide the necessary formulae to both quantify and propagate random uncertainties in the context of a VAVP, including blocked force characterisation, substructuring, and response predictions. We will begin by introducing the necessary statistical quantities, before deriving the so-called *Law of Propagation of Uncertainty* (LPU, also termed the general law of error propagation) [1] for both single and multiple variables. The issue of complex variables and how best to treat their uncertainty will then be covered, after which the formulae for propagating uncertainty through inverse methods (e.g. blocked force characterisation), dynamic substructuring, and operational response predictions will be described.

8.4.1 STATISTICAL QUANTITIES

In this section, we briefly introduce some basic statistical concepts (mean, variance and covariance) for scalar, vector and matrix quantities that will be required later. There is no original material here and the familiar reader may skip this section.

When propagating uncertainty according to the LPU, three key statistical quantities are required for all measured variables: the mean, variance and covariance. These describe, respectively, the central tendency, spread of, and correlation between the acquired data.

Mean

The mean is a measure of central tendency. Given a discrete set of values, for example the repeated measurement of a quantity x, it is calculated as,

$$\mathbb{E}[x] = \frac{1}{N}\sum_{n}^{N} x_n = \mu_x \tag{8.19}$$

where x_n is the nth sample of the quantity x, and N is the total number of measured samples. For vector or matrix quantities we can similarly calculate the mean value as,

$$\mathbb{E}[\mathbf{x}] = \frac{1}{N}\sum_{n}^{N} \mathbf{x}_n = \mu_{\mathbf{x}} \tag{8.20}$$

where $\mathbf{x}_n$ is the nth sample of the vector (or matrix) quantity $\mathbf{x}$.

If the mean value is calculated from a number of repeated measurements, then it is called the sample mean. This is not to be confused with the true mean. In an experimental context, the true mean corresponds to the value of the measurand, in the absence of any random/and or bias errors. According to the Law of Large Numbers, in the absence of any systematic bias errors the sample mean obtained from a large number of repeated experiments should be close to the true mean, and tend towards it as the number of experiments increases.

The true mean, or the population mean, of a quantity is equal to the calculated mean only when every member of the 'population' is included. In an experimental setting the population of experiments is infinite, and so the true mean can only be approximated using the sample mean obtained from a sufficient number of repeated measurements. It is important to note that if an insufficient number of measurements are taken, the sample mean may differ considerably from the true mean.

Variance and covariance

Variance is a measure of how far a set of values are spread out from their mean. Given a discrete set of values, for example the repeated measurement of a quantity

x, the variance is calculated as,

$$\text{Var}(x) = \frac{1}{N-1}\sum_{n=1}^{N}(x_n - \mu_x)^2 = \sigma_x^2. \tag{8.21}$$

where σ_x is termed the standard deviation. The variance and standard deviation are strictly positive quantities, which get larger as the data becomes more widely spread.

Strictly speaking, Equation 8.21 is called the *sample* variance, as opposed to the population variance because it is based on the sample mean, which is not necessarily equal to the population mean. The sample *vs.* population mean can introduce a bias error on the estimate of σ_x^2. The use of $N-1$ as opposed to N ensures that the sample variance gives an unbiased estimate of the population variance. Clearly, this only has a noticeable effect when N is quite small.

Variance describes the variability of a measured quantity, say x. Covariance describes the *joint variability* between two measured quantities, say x and y. The sample covariance is calculated as,

$$\text{Cov}(x,y) = \frac{1}{N-1}\sum_{n=1}^{N}(x_n - \mu_x)(y_n - \mu_y) = \sigma_{xy}^2. \tag{8.22}$$

Unlike the variance, covariance can be positive or negative depending on how x and y vary relative to one another. For example, suppose we have a set of measurements for x and y. If the larger values of x correspond mostly with larger values of y, and the same holds for small values – x and y have similar behaviour – then their covariance is positive. Conversely, if the larger values of x correspond mostly with smaller values of y, and visa versa – x and y have opposite behaviour – then their covariance is negative. The sign of the covariance shows the linear tendency between x and y. However, unlike the variance, the magnitude of the covariance is not so easy to interpret. A popular normalisation of the covariance is Pearson's Correlation Coefficient (PCC), defined as,

$$\text{Corr}(x,y) = \frac{\sigma_{xy}^2}{\sigma_x \sigma_y} = \rho(x,y) \tag{8.23}$$

The PCC is a measure of the linear correlation between x and y. It has values between +1 and -1. A value of +1 indicates total positive correlation (large values of x tend to coincide with large values of y). A value of -1 indicates a total negative correlation. A value of 0 indicates no linear correlation. It is important to recognise that the PCC indicates *linear* correlation, and that it is possible for two quantities to be non-linearly correlated.

For vector quantities, the notion of covariance can be extended to a covariance matrix. Given a set of measurements for the vector quantity $\mathbf{x}$, its sample covariance matrix is calculated as,

$$\Sigma_{\mathbf{x}} = \frac{1}{N-1}\left[\left(\hat{\mathbf{x}} - \mu_{\mathbf{x}}\mathbf{1}^{\mathrm{T}}\right)\left(\hat{\mathbf{x}} - \mu_{\mathbf{x}}\mathbf{1}^{\mathrm{T}}\right)^{\mathrm{T}}\right] \tag{8.24}$$

where, to avoid summation notation, the repeated N measurements of the vector $\mathbf{x}$ are arranged into the columns of a measurement matrix $\hat{\mathbf{x}}$,

$$\hat{\mathbf{x}} = \begin{bmatrix} \mathbf{x}^{(1)} & \mathbf{x}^{(2)} & \cdots & \mathbf{x}^{(N)} \end{bmatrix} \tag{8.25}$$

from which the mean value can be subtracted by column stacking the mean vector $\mu_{\mathbf{x}}$ using,

$$\mu_{\mathbf{x}} \mathbf{1}^{\mathrm{T}} = \begin{bmatrix} \mu_{\mathbf{x}} & \cdots & \mu_{\mathbf{x}} \end{bmatrix} \tag{8.26}$$

where $\mathbf{1}$ is simply an N dimensional vector of 1s.

Note that the diagonal entries of $\Sigma_{\mathbf{x}}$ correspond to the variances of each element of $\mathbf{x}$, whilst the off-diagonal entries (covariances) satisfy $(\Sigma_{\mathbf{x}})_{ij} = (\Sigma_{\mathbf{x}})_{ji}$, i.e. the covariance matrix is symmetric.

We can similarly calculate a covariance matrix between two vector quantities $\mathbf{x}$ and $\mathbf{y}$,

$$\Sigma_{\mathbf{xy}} = \frac{1}{N-1} \left[\left(\hat{\mathbf{x}} - \mu_{\mathbf{x}} \mathbf{1}^{\mathrm{T}} \right) \left(\hat{\mathbf{y}} - \mu_{\mathbf{y}} \mathbf{1}^{\mathrm{T}} \right)^{\mathrm{T}} \right]. \tag{8.27}$$

Note that the diagonal elements of $\Sigma_{\mathbf{xy}}$ are not variances in this case, but covariances.

For matrix quantities $\mathbf{X}$ and $\mathbf{Y}$ we can calculate a covariance matrix $\Sigma_{\mathbf{XY}}$ in the same way as for the vectors $\mathbf{x}$ and $\mathbf{y}$. To do so we have to restructure each matrix into a (tall) vector. This is done by taking each column of the matrix $\mathbf{X} \in \mathbb{R}^{N \times M}$ (starting with the left most column) and stacking them vertically on top of one another such that $\mathrm{Vec}(\mathbf{X}) = \underset{\rightarrow}{\mathbf{X}} = \tilde{\mathbf{X}} \in \mathbb{R}^{NM}$. We call Vec( ) the vectorisation operator. Here is an example,

$$\mathrm{Vec}\left(\begin{bmatrix} A & B & C \\ D & E & F \\ H & I & J \end{bmatrix} \right) = \begin{pmatrix} A \\ D \\ H \\ B \\ E \\ I \\ C \\ F \\ J \end{pmatrix}. \tag{8.28}$$

Utilising the vectorisation operator, the covariance matrix of $\mathbf{X}$ from N measurements is given by,

$$\Sigma_{\mathbf{X}} = \frac{1}{N-1} \left[\left(\hat{\mathbf{X}} - \mu_{\underset{\rightarrow}{\mathbf{X}}} \mathbf{1}^{\mathrm{T}} \right) \left(\hat{\mathbf{X}} - \mu_{\underset{\rightarrow}{\mathbf{X}}} \mathbf{1}^{\mathrm{T}} \right)^{\mathrm{T}} \right] \tag{8.29}$$

where the measurement matrix $\hat{\mathbf{X}}$ is built from the vectorisation of each measured matrix,

$$\hat{\mathbf{X}} = \begin{bmatrix} \underset{\rightarrow}{\mathbf{X}}^{(1)} & \underset{\rightarrow}{\mathbf{X}}^{(2)} & \cdots & \underset{\rightarrow}{\mathbf{X}}^{(N)} \end{bmatrix} \tag{8.30}$$

and $\mu_{\underset{\rightarrow}{\mathbf{X}}}$ is the vectorisation of the matrix mean.

For the covariance between two matrices, say $\mathbf{X}$ and $\mathbf{Y}$ we have similarly,

$$\Sigma_{\mathbf{XY}} = \frac{1}{N-1} \left[\left(\hat{\mathbf{X}} - \mu_{\underset{\rightarrow}{\mathbf{X}}} \mathbf{1}^{\mathrm{T}} \right) \left(\hat{\mathbf{Y}} - \mu_{\underset{\rightarrow}{\mathbf{Y}}} \mathbf{1}^{\mathrm{T}} \right)^{\mathrm{T}} \right]. \tag{8.31}$$

We will use the above formulae for vector and matrix covariance matrices later to characterise the uncertainty in measured operational responses and FRFs.

Higher order statistics

The mean and variance describe the low order statistics of a measured quantity, i.e. its central tendency and spread. Higher order statistics can be computed in a similar way by generalising the variance equation above. For example, the next order statistic is *skewness*. It describes the asymmetry of the data relative to the mean and is obtained from,

$$\text{Skew}(x) = \frac{1}{N-1}\sum_{n=1}^{N}(x_n - \mu_x)^3 = \gamma_x. \tag{8.32}$$

Though uncertainty propagation is possible using these high order statistics, it is generally not done as it involves having to calculate ever high order derivatives which quickly becomes inconvenient for complex and multidimensional problems.

Relative (co)variance

Generally speaking, the variance of a quantity describes the amount of uncertainty in its measurement. The larger the variance, the greater the spread of measurements. Often we are not interested in the value of the variance specifically, but its value relative to the mean. Suppose given a set of measurements, we have that $\mu_x = 100$ and $\sigma_x^2 = 1$ – you might say the uncertainty is rather small, whilst if $\mu_x = 1$ and $\sigma_x^2 = 1$ you might say the uncertainty is rather large, yet they have the same variance. To avoid this issue of scaling we can define the *relative* variance,

$$\text{RelVar}(x) = \frac{\sigma_x^2}{\mu_x^2}. \tag{8.33}$$

For the covariance we can similarly define the relative covariance,

$$\text{RelCov}(x,y) = \frac{\sigma_{xy}}{\mu_x \mu_y}. \tag{8.34}$$

Similar definitions apply to vector and matrix quantities.

Normal distribution

Given some measured data, the mean and variance provide measures of central tendency and spread – they are single value representations of the data set. They do not provide information on how the individual measurements are distributed among the range of possible values. The probability of obtaining a particular value in a measurement can be characterised by a probability distribution which represents a functional form of the uncertainty present. Often this functional form is unknown and must be inferred from the data. However, in many practical situations, particularly when we consider mean values, it can be represented by the so-called normal (or Gaussian) distribution (the familiar bell shaped curve in Figure 8.2a.).

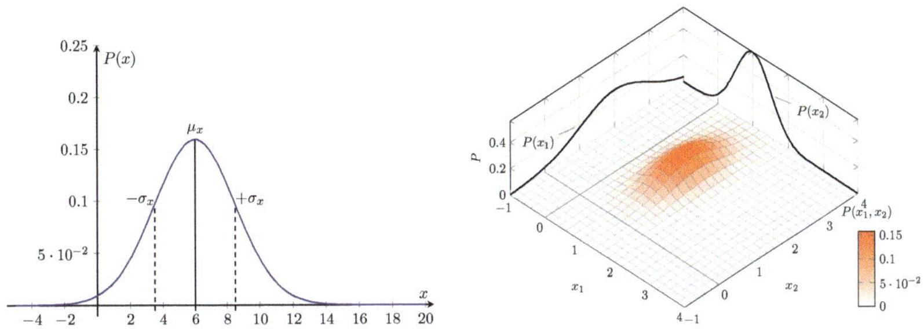

(a) Normal distribution. **(b)** Bivariate normal distribution.

Figure 8.2 Uni and bivariate normal distributions according to Equation 8.37 and 8.42.

The importance and widespread use of the normal distribution comes down to the so-called Central Limit Theorem. This states that, under appropriate conditions, the sum of a number of independent (uncorrelated) and identically distributed random variables with finite variances will tend to a normal distribution as the number of variables grows. This is especially useful when considering the sample mean, which is simply the sum of a series of random variables.

Consider the following; we do an experiment and obtain N measurements of x, from which we calculate the sample mean $\mu_x^{(1)}$. We repeat our experiment, obtain N new measurements, and calculate a new sample mean, say $\mu_x^{(2)}$. Being based on different samples, the two sample means will differ. There is some uncertainty in our mean value. It is natural to then ask, what is the variance of our sample mean? One option would be to repeat the experiment M times, and compute M sample means from which we could estimate the variance. Clearly, this could incur a significant experimental effort. However, thanks to the central limit theorem, we know that our sample mean μ_x will follow a normal distribution. In fact, it can be shown that the variance of the sample mean is related to the variance of the sample according to,

$$\sigma_{\mu_x}^2 = \frac{\sigma_x^2}{N}. \tag{8.35}$$

Similarly, in terms of standard deviation,

$$\sigma_{\mu_x} = \frac{\sigma_x}{\sqrt{N}}. \tag{8.36}$$

The quantity σ_{μ_x} is referred to as the standard error of the mean. It provides an estimate of standard deviation of the sample mean.

With the above in mind, the normal distribution is characterised by the function,

$$f(x) = \frac{1}{\sigma\sqrt{2\pi}} e^{-\frac{1}{2}\left(\frac{x-\mu}{\sigma}\right)^2} \tag{8.37}$$

where $f(x)$ is referred to as the probability density function (PDF) of the random variable x. The parameter μ is the mean value (central tendency) and σ is the standard deviation (spread of outcomes). An example of the normal distribution is shown in Figure 8.2a. Substituting μ for μ_x, and σ for σ_{μ_x}, $f(x)$ describes the probability density of the sample means. Note that $f(x)$ should not be interpreted as the probability of obtaining the value x. For a continuous variable, the probability of obtaining a specific value of x tends to 0. It is instead necessary to consider the probability of x falling within a particular range, say between a and b (rather than taking on a single value). This is done by integrating the probability density over the interval of interest,

$$\Pr(a \le x \le b) = \int_a^b f(x)dx. \tag{8.38}$$

From the above, if a sample mean and its standard deviation are known, the normal distribution can be used to estimate the probability of the sample mean falling within some specified range.

Given a normally distributed quantity (for example the sample mean), it is often useful to estimate the range which is likely to include an unknown population parameter, such as the population mean, a specified percentage of the time. We call this range a percentage Confidence Interval (CI). The confidence level defines the width of the interval, and so determines the percentage of the time that the confidence interval will contain the true parameter value. Typical choices for the confidence level are 0.90 (90%), 0.95 (95%), and 0.99 (99%). For the sample mean μ, the confidence interval is given by,

$$\mathrm{CI}_Z(\mu_x) = \mu_x \pm Z\frac{\sigma_x}{\sqrt{N}} \tag{8.39}$$

where Z is a value that dictates the confidence level, and N is the number of samples. The values of Z for various confidence levels are given in Table 8.2. For example, the 95% confidence interval

$$\mathrm{CI}_{95}(\mu_x) = \mu_x \pm 1.96\frac{\sigma_x}{\sqrt{N}} \tag{8.40}$$

defines a range that we would expect to contain the true mean 95% of the time.

Table 8.2
***Z* value for different confidence intervals**

Confidence level	Z
80%	1.282
85%	1.440
90%	1.645
95%	1.96
99%	2.576

If interest is not in the confidence interval of the mean value, but the data itself (i.e. what is the range of values we would expect a new observation to occur within, a specified percentage of the time), provided that the data is normally distributed, we can simply replace the standard error (standard deviation of the mean) in Equation 8.39 with the standard deviation of the sample,

$$\mathrm{CI}_Z(x) = \mu \pm Z\sigma_x. \tag{8.41}$$

The notion of sample mean and its variance can be extended to multiple variables $(x_1, x_2, \cdots, x_k)$, using the so-called multi-variate normal distribution. A special case of this, where only two variables are considered (bivariate normal distribution) is shown in Figure 8.2b. In general, the probability density function of a multi-variate normal distribution is given by,

$$f(x_1, x_2, \cdots, x_k) = \frac{1}{\sqrt{(2\pi)^k \det(\Sigma)}} e^{-\frac{1}{2}(\mathbf{x}-\mu_{\mathbf{x}})^{\mathrm{T}}\Sigma^{-1}(\mathbf{x}-\mu_{\mathbf{x}})} \tag{8.42}$$

where $f(\;)$ is the probability density of the random vector $\mathbf{x} = (x_1, x_2, \cdots, x_k)^{\mathrm{T}}$, Σ is a covariance matrix, and $\mu_{\mathbf{x}}$ is a vector of mean values. Hence, the multi-variate normal distribution is characterised by the two vector/matrix quantities, $\mu_{\mathbf{x}}$ and Σ.

With reference to the bivariate normal illustrated in Figure 8.2b, the mean $\mu_{\mathbf{x}}$ determines the position of maximum probability, the diagonal elements of Σ describe the width of the distribution in each direction (i.e. the variance of each element of $\mathbf{x}$), and the off-diagonal elements of Σ describe the orientation of the distribution (i.e. the correlation between elements of $\mathbf{x}$).

Like the univariate normal distribution $f(\;)$ should not be interpreted as a probability, as such. The probability of observing both $a_1 \leq x_1 \leq b_1$ and $a_2 \leq x_2 \leq b_2$, is given by,

$$\Pr(a_1 \leq x_1 \leq b_1, a_2 \leq x_2 \leq b_2) = \int_{a_1}^{b_1} \int_{a_2}^{b_2} f(x_1, x_2) dx_1 dx_2. \tag{8.43}$$

It is possible to similarly define a confidence region (a multi-dimensional generalisation of a confidence interval) for the multivariate normal. The reader is referred standard statistical textbooks for further details [5, 10].

8.4.2 LAW OF PROPAGATION OF UNCERTAINTY

Propagation of uncertainty (or error) is the procedure by which the uncertainties in one or more input variables (such as a set of measured quantities) are used to predict the uncertainty on the output of a function that depends on them. For example, in the context of a blocked force characterisation, we may wish to know the uncertainty in our estimate of the blocked force given the uncertainty in a measured FRF matrix.

Approaches towards uncertainty propagation can be categorised as probabilistic or non-probabilistic. Probabilistic approaches include sampling-based methods (Monte-Carlo simulations) and local expansion-based methods (e.g. Taylor expansions), among others. Non-probabilistic approaches include interval analysis, and fuzzy logic, among others.

In this chapter we will focus on probabilistic methods, in particular a local expansion-based method known as the Law of Propagation of Uncertainty (LPU). This method has advantages when being used with relatively small levels of uncertainty, or functions that do not express considerable non-linearity. In the presence of large uncertainty or very non-linear functions we will resort to sampling-based method using Monte-Carlo simulations (see 8.4.9).

The LPU is based on a first order (linear) Taylor series expansion of the function through which uncertainty is propagated. In this sense, we can think of LPU as also meaning Linear Propagation of Uncertainty. In what follows, we will derive the general LPU expressions, beginning with the simplified case of a single input single output function.

Single variable

For a single-input single-output function $G(\quad)$ the LPU can be derived as follows. Consider the variable x as the input to our function,

$$y = G(x). \tag{8.44}$$

Now suppose x is measured, and due to experimental error is accompanied by some small perturbation Δx. This small error on the input will yield a small error on the output,

$$y + \Delta y = G(x + \Delta x). \tag{8.45}$$

Equation 8.45 can now be expanded in the form of a Taylor series as,

$$y + \Delta y = G(x) + \frac{dG(x)}{dx}\Delta x + \frac{d^2G(x)}{dx^2}\frac{(\Delta x)^2}{2!} + O[(\Delta x)^3]. \tag{8.46}$$

where $O[(\Delta x)^3]$ denotes the contribution of higher order terms.

If the input error Δx is sufficiently small we can approximate the output error Δy using the first order term alone, giving

$$\Delta y \approx \frac{dG(x)}{dx}\Delta x. \tag{8.47}$$

This constitutes a linearisation of the propagation function $G(\quad)$.

Now suppose we perform N repeated measurements. For each we have,

$$\Delta y_n \approx \frac{dG(x)}{dx}\Delta x_n. \tag{8.48}$$

Squaring both sides of Equation 8.48, whilst summing over all measurements and dividing by $N-1$, yields,

$$\frac{1}{N-1}\sum_{n=1}^{N}\Delta y_n^2 \approx \left(\frac{dG(x)}{dx}\right)^2 \frac{1}{N-1}\sum_{n=1}^{N}\Delta x_n^2. \tag{8.49}$$

Recalling the definition of variance from Equation 8.21,

$$\sigma_x^2 = \frac{1}{N}\sum_{n=1}^{N}(x_n - \mu_x)^2 = \frac{1}{N}\sum_{n=1}^{N}\Delta x_n^2 \tag{8.50}$$

Equation 8.49 can be written as,

$$\sigma_y^2 \approx \left(\frac{\partial G(x)}{\partial x}\right)^2_{x=\mu_x} \sigma_x^2. \tag{8.51}$$

Equation 8.51 describes the propagation of uncertainty through the function $G(\ )$. The input uncertainty is characterised by the variance σ_x^2 and the output uncertainty by the variance σ_y^2. They are related by the squared derivative of the propagation function, with respect to the input variable. It is important to note that this derivative is evaluated at the true value of x which we approximate by its mean value μ_x.

In deriving Equation 8.51 we made two key assumptions. First, the level of input uncertainty is sufficiently small that only the first derivative is necessary in its Taylor expansion (also that the function is weakly non-linear such that higher order derivatives can be assumed small). Second, that the input uncertainty can be characterised by its variance alone. This is valid if input uncertainty is normally distributed.

An important consequence of Equation 8.51 is that for a linear propagation of uncertainty, the distribution of the output y has the same shape as the input x, albeit scaled by an amount proportional to the partial derivative of the function, and shifted according to the output of the function. Hence, a normally distributed input uncertainty yields a normal output uncertainty. We now consider an example.

Shown in Figure 8.3a, in green, is the output of the function,

$$y = G(x) = Mx + c \tag{8.52}$$

where M is the gradient of the straight line and c its intercept with the y axis. Along the x-axis in blue is the distribution of the input variable x. It is normal with mean μ_x and variance σ_x^2. Along the y-axis in red is the distribution of the output variable y. Its mean value is given by,

$$\mu_y = G(\mu_x) = M\mu_x + c. \tag{8.53}$$

Given that the propagation function $G(\ )$ is linear, its Taylor series has only a first order term. Consequently, the LPU provides an *exact* propagation of uncertainty. Since the input distribution is mapped linearly to the output distribution, its underlying shape is unchanged; the output remains normal. According to Equation 8.51 the variance of the output is given by,

$$\sigma_y^2 = M^2\sigma_x^2, \tag{8.54}$$

i.e. the output uncertainty is proportional to the gradient of the function; a larger gradient yields a larger output uncertainty.

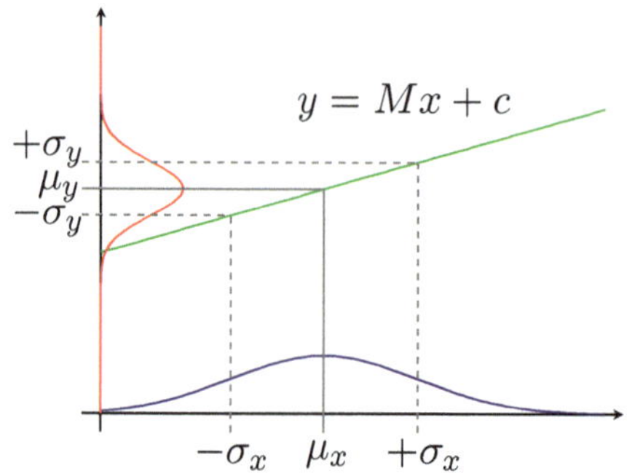

(a) Propagation of uncertainty through a linear function.

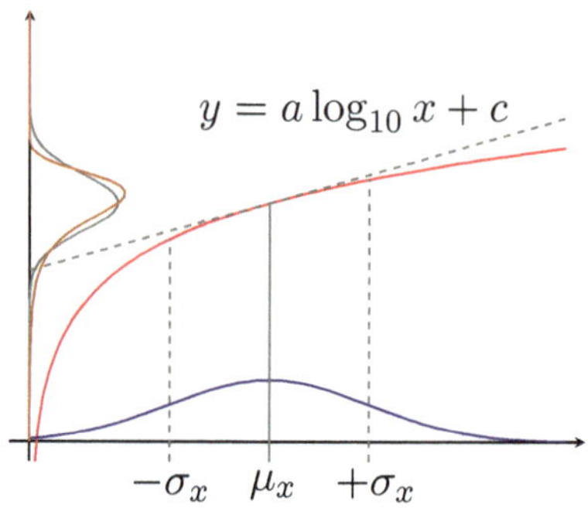

(b) Propagation of uncertainty through a non-linear function.

Figure 8.3 For small levels of input uncertainty (a) we can linearise the propagation function. The output uncertainty is then determined by the input uncertainty, and the propagating function's local gradient. For large levels of input uncertainty or highly non-linear functions (b) the output distribution is distorted, and we can no longer use the LPU or assume that the output has a normal distribution.

Figure 8.3b demonstrates the propagation of uncertainty through an illustrative non-linear function. In purple is the output of the function,

$$y = G(x) = a\log_{10} x + c. \tag{8.55}$$

Along the x-axis, in blue, is again the distribution of the input variable x. Along the y-axis two distributions are shown. In brown is the true output distribution, and in grey is the linearly propagated distribution. The true output distribution takes into account the non-linearity of the propagating function, causing a distortion in its shape; it is no longer normal. In contrast, the LPU approximates the non-linear function by its local gradient, shown by the dashed grey line. For small levels of input uncertainty this is a reasonable approximation. This is demonstrated by the normal output distribution in grey, whose mean value is given in this case by,

$$\mu_y = G(\mu_x) = a\log_{10} \mu_x + c \tag{8.56}$$

and variance by,

$$\sigma_y^2 = \left(\frac{a}{\mu_x \log(10)}\right)^2 \sigma_x^2. \tag{8.57}$$

Clearly, for large levels of input uncertainty the linear approximation will fail to accurately predict the output uncertainty. Even in the presence of small input uncertainty, if the propagation function is highly non-linear, a poor uncertainty estimate may be obtained. In these cases an alternative Monte-Carlo method, based on randomly sampling the available results and computing an output for each, may be used. This approach will be discussed in further detail in 8.4.9.

Multiple variables

In this section, we will extend the LPU to functions of multiple variables, and derive the appropriate equations. We begin by considering y_i and y_j, generally, as the outputs of some multi-variable functions,

$$y_i = G_i(x_1, x_2, \cdots, x_M) = G_i(\mathbf{x}) \tag{8.58}$$

$$y_j = G_j(x_1, x_2, \cdots, x_M) = G_j(\mathbf{x}), \tag{8.59}$$

where $\mathbf{x} = (x_1, x_2, \cdots, x_M)$ is a vector of input variables.

We follow the same steps as before. Suppose that we repeatedly measure each element of the input variable vector $\mathbf{x}$. Due to experimental error the result is perturbed by $\Delta\mathbf{x}$. This small change in the value of $\mathbf{x}$ will lead to a small change in the value of the output variables y_i and y_j. Assuming that the input error $\Delta\mathbf{x}$ remains small, the change in output Δy_i and Δy_j may be approximated by a first order Taylor series expansion,

$$\Delta y_i \approx \frac{\partial G_i(\mathbf{x})}{\partial x_1}\Delta x_1 + \frac{\partial G_i(\mathbf{x})}{\partial x_2}\Delta x_2 + \cdots + \frac{\partial G_i(\mathbf{x})}{\partial x_M}\Delta x_M \tag{8.60}$$

$$\Delta y_j \approx \frac{\partial G_j(\mathbf{x})}{\partial x_1}\Delta x_1 + \frac{\partial G_j(\mathbf{x})}{\partial x_2}\Delta x_2 + \cdots + \frac{\partial G_j(\mathbf{x})}{\partial x_M}\Delta x_M \tag{8.61}$$

where $\frac{\partial G_j(\mathbf{x})}{\partial x_M}$ is the partial derivative of the propagation function $G_j(\Box)$ with respect to the Mth input variable. Using subscript $\Box_n$ to denote the nth measurement, Equations 8.60 and 8.61 may be written more conveniently using summation notation,

$$\Delta y_{i_n} \approx \sum_m^M \frac{\partial G_i(\mathbf{x})}{\partial x_m}\Delta x_{m_n} \tag{8.62}$$

$$\Delta y_{j_n} \approx \sum_m^M \frac{\partial G_j(\mathbf{x})}{\partial x_m}\Delta x_{m_n}. \tag{8.63}$$

Now that we have two output variables, y_i and y_j, we are also interested in determining the covariance between them. To do so, we begin by multiplying together Equations 8.62 and 8.63,

$$\Delta y_{i_n}\Delta y_{j_n} \approx \left(\sum_m^M \frac{\partial G_i(\mathbf{x})}{\partial x_m}\Delta x_{m_n}\right)\left(\sum_m^M \frac{\partial G_j(\mathbf{x})}{\partial x_m}\Delta x_{m_n}\right). \tag{8.64}$$

Summing over N measurements and dividing both sides by $N-1$ yields,

$$\frac{1}{N-1}\sum_n^N \Delta y_{i_n}\Delta y_{j_n} \approx \frac{1}{N-1}\sum_n^N \left(\sum_m^M \frac{\partial G_i(\mathbf{x})}{\partial x_m}\Delta x_{m_n}\right)\left(\sum_m^M \frac{\partial G_j(\mathbf{x})}{\partial x_m}\Delta x_{m_n}\right). \tag{8.65}$$

By expanding the brackets,

$$\frac{1}{N-1}\sum_n^N \Delta y_{i_n}\Delta y_{j_n} \approx \frac{1}{N-1}\sum_n^N \left[\frac{\partial G_i(\mathbf{x})}{\partial x_1}\Delta x_{1_n}\left(\sum_m^M \frac{\partial G_j(\mathbf{x})}{\partial x_m}\Delta x_{m_n}\right) + \frac{\partial G_i(\mathbf{x})}{\partial x_2}\Delta x_{2_n}\right.$$
$$\left.\left(\sum_m^M \frac{\partial G_j(\mathbf{x})}{\partial x_m}\Delta x_{m_n}\right) + \cdots + \frac{\partial G_i(\mathbf{x})}{\partial x_M}\Delta x_{M_n}\left(\sum_m^M \frac{\partial G_j(\mathbf{x})}{\partial x_m}\Delta x_{m_n}\right)\right] \tag{8.66}$$

and regrouping terms we get,

$$\frac{1}{N-1}\sum_n^N \Delta y_{i_n}\Delta y_{j_n} \approx \sum_m^M \frac{\partial G_i(\mathbf{x})}{\partial x_m}\frac{\partial G_j(\mathbf{x})}{\partial x_m}\frac{1}{N-1}\sum_n^N \Delta x_{m_n}\Delta x_{m_n} +$$
$$\sum_{m\neq l}\frac{\partial G_i(\mathbf{x})}{\partial x_l}\frac{\partial G_j(\mathbf{x})}{\partial x_m}\frac{1}{N-1}\sum_n^N \Delta x_{l_n}\Delta x_{m_n} \tag{8.67}$$

Recalling the definitions of variance and covariance from Equations 8.21 and 8.22, Equation 8.67 can be rewritten as,

$$\sigma_{y_iy_j} \approx \sum_m^M \frac{\partial G_i(\mathbf{x})}{\partial x_m}\frac{\partial G_j(\mathbf{x})}{\partial x_m}\sigma_{x_m}^2 + \sum_{m\neq l}\frac{\partial G_i(\mathbf{x})}{\partial x_l}\frac{\partial G_j(\mathbf{x})}{\partial x_m}\sigma_{x_lx_m}^2 \tag{8.68}$$

where $\sigma_{y_iy_j}$ is the covariance between the output variables y_i and y_j, $\sigma_{x_m}^2$ is the variance of the input variable x_m, and $\sigma_{x_lx_m}^2$ is the covariance between the input variables x_l and x_m. For the case that $i = j$ Equation 8.68 reduces to,

$$\sigma_{y_i}^2 \approx \sum_m^M \left(\frac{\partial G_i(\mathbf{x})}{\partial x_m}\right)^2 \sigma_{x_m}^2 + \sum_{m\neq l}\frac{\partial G_i(\mathbf{x})}{\partial x_l}\frac{\partial G_i(\mathbf{x})}{\partial x_m}\sigma_{x_lx_m} \tag{8.69}$$

where $\sigma_{y_i}^2$ is the variance of the output variable y_i.

The first summation in Equations 8.68 and 8.69 describes the propagation of input variance onto the output uncertainty. The second summation describes the influence of input covariance (i.e. correlations between the input variables) on the output uncertainty. Often when dealing with uncertainty propagation, the simplifying assumption of uncorrelated input variables is made, that is $\sigma_{x_lx_{m\neq l}} = 0$. The output variance then reduces to,

$$\sigma_{y_i}^2 \approx \sum_m^M \left(\frac{\partial G_i(\mathbf{x})}{\partial x_m}\right)^2 \sigma_{x_m}^2. \tag{8.70}$$

Unfortunately, as will be discussed in 8.4.4, the experimental measurements associated with blocked forces and dynamic substructuring often yield highly correlated input variables (and consequently correlated output variables). For this reason, our uncertainty propagation must be conducted using Equation 8.68 rather than the simpler Equation 8.70.

Matrix form of LPU

When dealing with a large number of input and output variables it is far more convenient to express Equation 8.68 in matrix form. This is done by using the equation,

$$\sigma_{y_i y_j} \approx \mathbf{J}_i \Sigma_{\mathbf{x}} \mathbf{J}_j^{\mathrm{T}} \tag{8.71}$$

where,

$$\mathbf{J}_i = \begin{bmatrix} \frac{\partial G_i(\mathbf{x})}{\partial x_1} & \frac{\partial G_i(\mathbf{x})}{\partial x_2} & \cdots & \frac{\partial G_i(\mathbf{x})}{\partial x_M} \end{bmatrix} \tag{8.72}$$

and

$$\mathbf{J}_j = \begin{bmatrix} \frac{\partial G_j(\mathbf{x})}{\partial x_1} & \frac{\partial G_j(\mathbf{x})}{\partial x_2} & \cdots & \frac{\partial G_j(\mathbf{x})}{\partial x_M} \end{bmatrix} \tag{8.73}$$

are the so-called *Jacobian* matrices of the functions $y_i = G_i(\mathbf{x})$ and $y_j = G_j(\mathbf{x})$, respectively, and $\Sigma_{\mathbf{x}}$ is the covariance matrix of the input vector $\mathbf{x}$,

$$\Sigma_{\mathbf{x}} = \begin{bmatrix} \sigma_{x_1}^2 & \sigma_{x_1 x_2}^2 & \cdots & \sigma_{x_1 x_M}^2 \\ \sigma_{x_2 x_1}^2 & \sigma_{x_2}^2 & \cdots & \sigma_{x_2 x_M}^2 \\ \vdots & \vdots & \ddots & \vdots \\ \sigma_{x_M x_1}^2 & \sigma_{x_M x_2}^2 & \cdots & \sigma_{x_M}^2 \end{bmatrix}. \tag{8.74}$$

Now that we have a concise expression for $\sigma_{y_i y_j}$, we can go one step further by combing the Jacobian matrices for multiple output variables,

$$\mathbf{J} = \begin{bmatrix} \frac{\partial G_1(\mathbf{x})}{\partial x_1} & \frac{\partial G_1(\mathbf{x})}{\partial x_2} & \cdots & \frac{\partial G_1(\mathbf{x})}{\partial x_M} \\ \frac{\partial G_2(\mathbf{x})}{\partial x_1} & \frac{\partial G_2(\mathbf{x})}{\partial x_2} & \cdots & \frac{\partial G_2(\mathbf{x})}{\partial x_M} \\ \vdots & \vdots & \ddots & \vdots \\ \frac{\partial G_L(\mathbf{x})}{\partial x_1} & \frac{\partial G_L(\mathbf{x})}{\partial x_2} & \cdots & \frac{\partial G_L(\mathbf{x})}{\partial x_M} \end{bmatrix}. \tag{8.75}$$

The uncertainty propagation,

$$\Sigma_{\mathbf{y}} \approx \mathbf{J} \Sigma_{\mathbf{x}} \mathbf{J}^{\mathrm{T}} \tag{8.76}$$

then yields a covariance matrix for the output variable vector $\mathbf{y} = (y_1, y_2, \cdots, y_L)$,

$$\Sigma_{\mathbf{y}} = \begin{bmatrix} \sigma_{y_1}^2 & \sigma_{y_1 y_2}^2 & \cdots & \sigma_{y_1 y_L}^2 \\ \sigma_{y_2 y_1}^2 & \sigma_{y_2}^2 & \cdots & \sigma_{y_2 y_L}^2 \\ \vdots & \vdots & \ddots & \vdots \\ \sigma_{y_L y_1}^2 & \sigma_{y_L y_2}^2 & \cdots & \sigma_{y_L}^2 \end{bmatrix}. \tag{8.77}$$

Equations 8.74-8.77 describe, generally, the linear of propagation of uncertainty. As with the single variable case, it is important that the Jacobians are evaluated at the true value of $\mathbf{x}$, which we approximate by its mean value $\mu_{\mathbf{x}}$. These equations will form the basis of the VAVP uncertainty propagation to be described shortly.

It is important to note that like the single variable case considered before, Equations 8.74-8.77 are based on a first order Taylor series expansion of the now multi-variable function $G(\ )$. As such the same requirements are present; small input uncertainty and only weakly non-linear functions. If these requirements are satisfied then the LPU can provide a reasonable estimation of the output uncertainty. Like the single variable propagation, the multi-variable propagation performs a linear mapping of the input distribution to the output. This linear mapping has the effect of stretching or squeezing the input distribution (according to the Jacobian matrix), leaving its shape unchanged; a multi-variate normal input uncertainty yields a multi-variate normal output uncertainty.

8.4.3 PROPAGATING UNCERTAINTY OF COMPLEX VARIABLES

The LPU, as described above, has been formulated to deal with real-valued functions of real input variables. However, the VAVP equations namely blocked force characterisation, dynamic substructuring and response predictions, are all formulated in the frequency domain. Consequently, all our variables are complex, having both real and imaginary parts, or equivalently, magnitude and phase. To apply the LPU to a complex-valued function of a complex input variable some minor modification must be made. There are two approaches to do this, differing based on how the complex values are represented. One approach is to treat the complex value and its conjugate as independent variables, see for example the substructuring uncertainty propagation in [11]. Alternatively, the real and imaginary parts of the complex number can be treated as separate real variables. The authors find this approach the most intuitive and so the uncertainty framework presented here is done in this way.

Consider the complex variable $H = a + ib$. Separating its real and imaginary parts, we can rewrite H instead as the 2×1 column vector,

$$H \rightarrow \begin{pmatrix} \mathrm{Re}(H) \\ \mathrm{Im}(H) \end{pmatrix}. \tag{8.78}$$

Supposing H is a measured quantity, its mean value can similarly be given by,

$$\mu_H = \mu_{a+ib} = \mu_a + i\mu_b \rightarrow \begin{pmatrix} \mu_{\mathrm{Re}(H)} \\ \mu_{\mathrm{Im}(H)} \end{pmatrix}. \tag{8.79}$$

Having separated the real and imaginary parts of H, we can no longer define a single variance term to describe its uncertainty. Rather, we use the covariance matrix,

$$\sigma_H^2 \rightarrow \Sigma_H = \begin{bmatrix} \sigma^2_{\mathrm{Re}(H)} & \sigma^2_{\mathrm{Re}(H)\mathrm{Im}(H)} \\ \sigma^2_{\mathrm{Im}(H)\mathrm{Re}(H)} & \sigma^2_{\mathrm{Im}(H)} \end{bmatrix} \tag{8.80}$$

where $\sigma^2_{\mathrm{Re}(H)}$ is the variance of the real part of H, $\sigma^2_{\mathrm{Im}(H)}$ is the variance of the imaginary part of H, and $\sigma^2_{\mathrm{Re}(H)\mathrm{Im}(H)}$ is the covariance between the two. This *bivariate* covariance matrix describes the uncertainty in the complex variable H. The same idea can be used to express the covariance between two different complex variables H_1 and H_2,

$$\Sigma_{H_1H_2} = \begin{bmatrix} \sigma^2_{\mathrm{Re}(H_1)\mathrm{Re}(H_2)} & \sigma^2_{\mathrm{Re}(H_1)\mathrm{Im}(H_2)} \\ \sigma^2_{\mathrm{Im}(H_1)\mathrm{Re}(H_2)} & \sigma^2_{\mathrm{Im}(H_1)\mathrm{Im}(H_2)} \end{bmatrix}. \tag{8.81}$$

Hereafter, we will use the term '*bivariate*' more generally to describe any covariance or Jacobian matrix that has had its complex values separated into real and imaginary parts as described above.

Further extension to complex *vector and matrix* variables is straightforward. We can define the operator,

$$M_v(H) = \begin{pmatrix} \mathrm{Re}(H) \\ \mathrm{Im}(H) \end{pmatrix} \tag{8.82}$$

which separates and stacks vertically the real and imaginary parts of the complex variable H. Applying $M_v(\ \ )$ to each element of $\mathbf{x}$, $\mu_{\mathbf{x}}$, $\mathbf{y}$, and $\mu_{\mathbf{y}}$ Equations 8.24 and 8.27 will compute bivariate forms of the covariance matrices $\Sigma_{\mathbf{x}}$ and $\Sigma_{\mathbf{xy}}$. For a complex matrix variable, the vectorisation operator Vec( ) is applied first, before $M_v(\ \ )$ is applied element-wise. As an example, the bivariate form of Equation 8.74 is given by,

$$\Sigma_{\mathbf{x}} = \begin{bmatrix} \Sigma_{x_1x_1} & \Sigma_{x_1x_2} & \cdots & \Sigma_{x_1x_M} \\ \Sigma_{x_2x_1} & \Sigma_{x_2x_2} & \cdots & \Sigma_{x_2x_M} \\ \vdots & \vdots & \ddots & \vdots \\ \Sigma_{x_Mx_1} & \Sigma_{x_Mx_2} & \cdots & \Sigma_{x_Mx_M} \end{bmatrix} \tag{8.83}$$

where each univariate element has been replaced by a 2×2 bivariate covariance matrix,

$$\sigma^2_{x_ix_j} \rightarrow \Sigma_{x_ix_j} = \begin{bmatrix} \sigma^2_{\mathrm{Re}(x_i)\mathrm{Re}(x_j)} & \sigma^2_{\mathrm{Re}(x_i)\mathrm{Im}(x_j)} \\ \sigma^2_{\mathrm{Im}(x_i)\mathrm{Re}(x_j)} & \sigma^2_{\mathrm{Im}(x_i)\mathrm{Im}(x_j)} \end{bmatrix}. \tag{8.84}$$

To propagate a bivariate complex uncertainty it is necessary to redefine the Jacobian matrices accordingly. The bivariate form of the Jacobian is given by

$$\mathbf{J} = \begin{bmatrix} \frac{\partial \mathbf{G}_1(\mathbf{x})}{\partial x_1} & \frac{\partial \mathbf{G}_1(\mathbf{x})}{\partial x_2} & \cdots & \frac{\partial \mathbf{G}_1(\mathbf{x})}{\partial x_M} \\ \frac{\partial \mathbf{G}_2(\mathbf{x})}{\partial x_1} & \frac{\partial \mathbf{G}_2(\mathbf{x})}{\partial x_2} & \cdots & \frac{\partial \mathbf{G}_2(\mathbf{x})}{\partial x_M} \\ \vdots & \vdots & \ddots & \vdots \\ \frac{\partial \mathbf{G}_L(\mathbf{x})}{\partial x_1} & \frac{\partial \mathbf{G}_L(\mathbf{x})}{\partial x_2} & \cdots & \frac{\partial \mathbf{G}_L(\mathbf{x})}{\partial x_M} \end{bmatrix} \tag{8.85}$$

where we have simply replaced each partial derivative in Equation 8.75, with a 2×2 matrix containing the partial derivatives of the real and imaginary parts of the output, with respect to the real and imaginary parts of the input,

$$\frac{\partial G_i(\mathbf{x})}{\partial x_j} \rightarrow \frac{\partial \mathbf{G}_i(\mathbf{x})}{\partial x_j} = \begin{bmatrix} \frac{\partial \mathrm{Re}(G_i(\mathbf{x}))}{\partial \mathrm{Re}(x_j)} & \frac{\partial \mathrm{Re}(G_i(\mathbf{x}))}{\partial \mathrm{Im}(x_j)} \\ \frac{\partial \mathrm{Im}(G_i(\mathbf{x}))}{\partial \mathrm{Re}(x_j)} & \frac{\partial \mathrm{Im}(G_i(\mathbf{x}))}{\partial \mathrm{Im}(x_j)} \end{bmatrix}. \tag{8.86}$$

The bivariate covariance and Jacobian matrices described above are used in place of their univariate counterparts in Equation 8.76 to propagate complex uncertainty. The resulting output uncertainty, characterised by the covariance matrix $\Sigma_{\mathbf{y}}$, is also bivariate, with its elements describing the variance and covariance between the real and imaginary parts of its elements.

Before we can implement the uncertainty propagation described above, we need to evaluate the Jacobian matrices for each function of interest. These functions correspond to the inverse characterisation of blocked forces (Equation 5.100), the dynamic substructuring of assembly FRFs (Equations 5.61 or 5.69) and the forward prediction of an operational response (Equation 5.101); their Jacobians will be introduced in 8.4.5, 8.4.6 and 8.4.7, respectively.

8.4.4 COMPUTING THE FRF AND RESPONSE COVARIANCE

Now that we have introduced covariance formulae for both vector and matrix quantities, and discussed how to deal with complex uncertainty, we are ready to discuss their application to the FRF matrices and operational response vectors required to build a VAVP.

FRF covariance

It is standard practice when measuring FRFs to perform multiple repeated excitations (either by impact hammer or shaker). These repeated excitations are used to estimate the mean FRF using, e.g. the H_1 estimator (see 7.31). More often than not, these excitations are also used to compute the coherence, which provides a very simple measure of excitation repeatability. The FRFs due to individual excitations are typically not stored. They are however, necessary to perform a full uncertainty analysis. Whilst the coherence can be related to the FRF variance (see 7.3.2), this relation assumes uncorrelated uncertainty. It will be shown below in our discussion of inter- and intra-FRF uncertainty that this is not a valid assumption. To perform a full uncertainty analysis, the FRFs obtained from individual excitations must be computed, retained and used to estimate the FRF covariance matrix $\Sigma_{\mathbf{Y}}$. Equation 8.29 and 8.30 become,

$$\Sigma_{\mathbf{Y}} = \frac{1}{N-1}\left[\left(\hat{\mathbf{Y}} - \mu_{\underset{\rightarrow}{\mathbf{Y}}}\mathbf{1}^{\mathrm{T}}\right)\left(\hat{\mathbf{Y}} - \mu_{\underset{\rightarrow}{\mathbf{Y}}}\mathbf{1}^{\mathrm{T}}\right)^{\mathrm{T}}\right] \tag{8.87}$$

where the measurement matrix $\hat{\mathbf{Y}}$ is built from the FRFs obtained for N individual excitations,

$$\hat{\mathbf{Y}} = \begin{bmatrix} \underset{\rightarrow}{\mathbf{Y}}^{(1)} & \underset{\rightarrow}{\mathbf{Y}}^{(2)} & \cdots & \underset{\rightarrow}{\mathbf{Y}}^{(N)} \end{bmatrix} \tag{8.88}$$

and $\mu_{\underset{\rightarrow}{\mathbf{Y}}}$ is the H_1 estimate of the mean FRF. This H_1 estimate will also be used to compute the Jacobians, which are introduced in 8.4.5, 8.4.6, and 8.4.7.

Response covariance

The covariance of the operational response can be used to describe the time variation of a source's operation. Its calculation is quite similar to that of the FRF.

The operational response is recorded for a period of time, which is subsequently split into adjacent blocks. Each block is Fourier Transformed and converted into a single-sided complex spectrum, $\mathbf{v}^{(n)}$, where n denotes the nth block. Each block is phase referenced to a specific sensor before the measurement matrix $\hat{\mathbf{v}}$ is built,

$$\hat{\mathbf{v}} = \begin{bmatrix} \mathbf{v}^{(1)} & \mathbf{v}^{(2)} & \cdots & \mathbf{v}^{(N)} \end{bmatrix}. \tag{8.89}$$

The response covariance is then obtained from,

$$\Sigma_{\mathbf{v}} = \frac{1}{N-1}\left[\left(\hat{\mathbf{v}} - \mu_{\mathbf{v}}\mathbf{1}^{\mathrm{T}}\right)\left(\hat{\mathbf{v}} - \mu_{\mathbf{v}}\mathbf{1}^{\mathrm{T}}\right)^{\mathrm{T}}\right] \tag{8.90}$$

where $\mu_{\mathbf{v}}$ represents the time averaged response.

It is worth nothing that this approach to constructing the response covariance assumes that the responses are all coherent, so that the phase can be referenced to single sensor.

Inter- and intra-FRF Correlation

Often when propagating uncertainty it is convenient to assume that the input variables are uncorrelated, such that $\Sigma_{\mathbf{x}}$ is a diagonal matrix, as this greatly reduces the necessary calculations. In the presence of measurement uncertainty (e.g. due to thermal noise, etc.) this is a fair assumption as each sensor will be subject to its own independent and uncorrelated noise signal. This is however, an exception, and not the rule. It is important that if any correlations are present between the inputs, they are accounted for in the uncertainty propagation. Failure to account for such correlations can lead to large errors in the uncertainty estimates. An important example in the context of VAVP development is inter-FRF correlation, i.e. the correlation that occurs when measuring FRFs using single-input-multiple-output (SIMO) methods.

SIMO methods are based on the simultaneous measurement of multiple response variables due to a shared excitation. If there is any uncertainty in this shared excitation all of the response variables will be correlated with one another. This *inter-FRF* correlation is particularly relevant when impact-based FRF measurements are taken. These are of course subject to operator uncertainty, i.e. variation in excitation location and orientation. As an example, consider the 2×2 FRF matrix below,

$$\mathbf{Y} = \begin{bmatrix} Y_{11} & Y_{12} \\ Y_{21} & Y_{22} \end{bmatrix}. \tag{8.91}$$

Following a standard SIMO measurement, the columns of $\mathbf{Y}$ are measured one at a time. First, a series of impact excitations are made at position 1, the average of which is used to estimate Y_{11} and Y_{21}. Each excitation will be slightly different to the others, and so the FRFs Y_{11} and Y_{21} will be uncertain and correlated. Next, a series of impacts are made at position 2. Similarly, Y_{12} and Y_{22} will be uncertain and correlated. After vectorisation of the measured FRF, $\underset{\rightarrow}{\mathbf{Y}} = \begin{bmatrix} Y_{11} & Y_{21} & Y_{12} & Y_{22} \end{bmatrix}^{\mathrm{T}}$, and the element-wise application of $M_v(\square)$ to separate real and imaginary parts, we can compute a covariance matrix of the form,

$$\Sigma_{\mathbf{Y}} = \begin{bmatrix} \Sigma_{Y_{11}Y_{11}} & \Sigma_{Y_{11}Y_{21}} & \Sigma_{Y_{11}Y_{12}} & \Sigma_{Y_{11}Y_{22}} \\ \Sigma_{Y_{21}Y_{11}} & \Sigma_{Y_{21}Y_{21}} & \Sigma_{Y_{21}Y_{12}} & \Sigma_{Y_{21}Y_{22}} \\ \Sigma_{Y_{12}Y_{11}} & \Sigma_{Y_{12}Y_{21}} & \Sigma_{Y_{12}Y_{12}} & \Sigma_{Y_{12}Y_{22}} \\ \Sigma_{Y_{22}Y_{11}} & \Sigma_{Y_{22}Y_{21}} & \Sigma_{Y_{22}Y_{12}} & \Sigma_{Y_{22}Y_{22}} \end{bmatrix}. \tag{8.92}$$

The grey covariance terms, which relate uncorrelated FRFs (i.e. those obtained from different excitations), should tend to 0 as the number of measurements increases. In practice, the number of FRF measurements made will likely not be enough for these terms to diminish sufficiently. In this case, they can of course be set to zero by hand.

Equation 8.92 summarises the notion of inter-FRF correlation. Though we should not forget that each covariance term is in-fact a bivariate covariance between the real and imaginary parts of the respective FRFs. For a specific FRF Y_{ij} we have,

$$\Sigma_{Y_{ij}Y_{ij}} = \begin{bmatrix} \sigma^2_{\mathrm{Re}(Y_{ij})\mathrm{Re}(Y_{ij})} & \sigma^2_{\mathrm{Re}(Y_{ij})\mathrm{Im}(Y_{ij})} \\ \sigma^2_{\mathrm{Im}(Y_{ij})\mathrm{Re}(Y_{ij})} & \sigma^2_{\mathrm{Im}(Y_{ij})\mathrm{Im}(Y_{ij})} \end{bmatrix} \tag{8.93}$$

where we might term $\sigma^2_{\mathrm{Im}(Y_{ij})\mathrm{Re}(Y_{ij})} = \sigma^2_{\mathrm{Re}(Y_{ij})\mathrm{Im}(Y_{ij})}$, the uncertainty within an FRF, as intra-FRF uncertainty. Similarly we have for the covariance between two different FRFs,

$$\Sigma_{Y_{ij}Y_{kl}} = \begin{bmatrix} \sigma^2_{\mathrm{Re}(Y_{ij})\mathrm{Re}(Y_{kl})} & \sigma^2_{\mathrm{Re}(Y_{ij})\mathrm{Im}(Y_{kl})} \\ \sigma^2_{\mathrm{Im}(Y_{ij})\mathrm{Re}(Y_{kl})} & \sigma^2_{\mathrm{Im}(Y_{ij})\mathrm{Im}(Y_{kl})} \end{bmatrix}. \tag{8.94}$$

For two correlated FRFs, all terms in Equation 8.94 have the potential to be non-zero. Likewise, for the single FRF in Equation 8.93.

Example 8.1: Complex and correlated FRF uncertainty

In this example, we verify the presence of both inter- (between) and intra- (within) FRF uncertainty when obtained using impact hammer measurements. The presence of strong correlations justifies the use of a more rigorous uncertainty propagation framework, as described in 8.4.

To verify the presence of correlation, we revisit the results obtained for Example 6.1. A 2 × 2 FRF matrix is obtained using 20 repeated measurements on a steel beam. In Example 6.1 this measurement was repeated with ‘good’ and ‘poor’ excitation accuracy. Here we consider only the ‘good’ set of results.

The point and transfer FRFs Y_{11} and Y_{21} are shown in Figure 8.4, over an extended frequency range up to 5 kHz. To provide a clear visual indication of the correlation present, a sufficient number of frequency bins are required around the resonant frequency under study. For this reason we focus attention on the highlighted frequency range between 1900-2500 Hz. A frequency zoomed plot of the point and transfer FRFs is given in Figure 8.5.

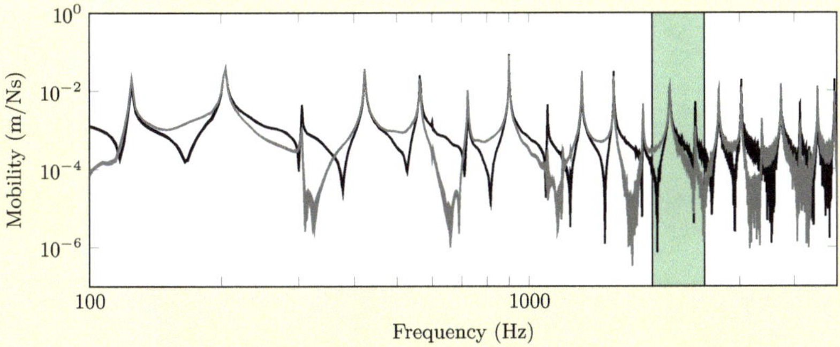

Figure 8.4 Repeated measurement of 20 point and transfer FRFs Y_{11} and Y_{21}.

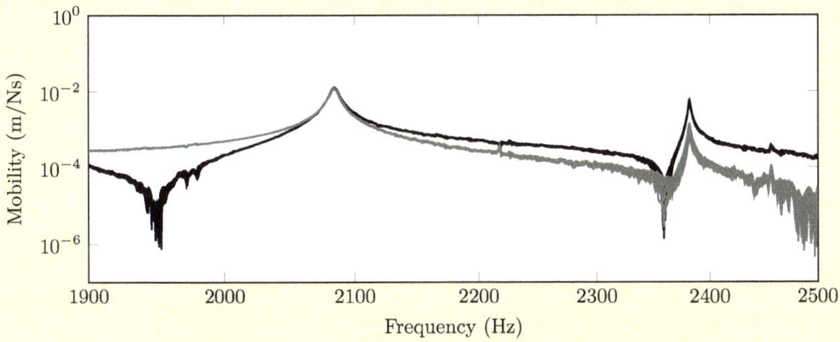

Figure 8.5 Frequency zoom of Figure 8.4.

To visualise intra-FRF uncertainty it is convenient to inspect the Nyquist plots of the point and transfer FRFs. These are shown, for the 20 repeated measurements, in Figure 8.6. Also shown in Figure 8.6 are the 95% confidence ellipses for each frequency bin. Note that in the presence of uncorrelated uncertainty, these ellipses would form circles around the mean of each frequency bin. From Figure 8.6 this is clearly not the case.

Not only are the confidence ellipses elliptical, demonstrating correlation, but also heavily frequency dependent, indicated by their rotation as they pass through resonance. This result justifies the representation of complex uncertainty using the bivariate form described in 8.4.3.

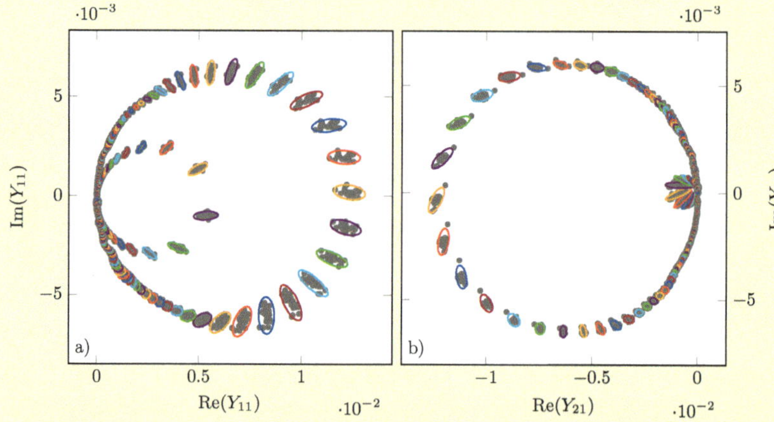

Figure 8.6 Nyquist plots for 20 repeated point (left) and transfer (right) mobility measurement shown in Figure 8.5, with 95% confidence ellipses.

To illustrate the inter-FRF correlation *between* the FRFs Y_{11} and Y_{21} we compute their bivariate covariances, and corresponding correlation coefficients. These are shown collectively in Figure 8.7. It can be seen that for the real-real and imaginary-imaginary covariances, strong correlations are present particularly around the resonant frequencies. Interestingly this does not appear to be the case for the cross real-imaginary covariance.

It is worth noting that the strong correlations in Figure 8.7 arise due to the shared excitation between the two FRFs Y_{11} and Y_{21}. The covariances between Y_{11} and Y_{12} do not present strong correlations owing to their independent excitations.

When propagating measurement uncertainty, it is important that these correlations are accounted for. Their neglect can lead to over or underestimated uncertainty estimates in any subsequent predictions.

To demonstrate the importance of correlation when propagating uncertainty, we consider the uncertainty in the magnitude of the point impedance $|Z_{11}|$. Using the bi-variate covariance matrix $\Sigma_{\mathbf{Y}}$ we are able to make a linear estimate of the impedance covariance using $\Sigma_{\mathbf{Z}} = \mathbf{J}\Sigma_{\mathbf{Y}}\mathbf{J}^{\mathrm{T}}$, where $\mathbf{J}$ is an appropriate Jacobian describing the derivative of a matrix inversion (see 8.4.5). Similarly, for the impedance magnitude we use $\sigma_{|Z_{11}|} = \mathbf{J}\Sigma_{Z_{11}}\mathbf{J}^{\mathrm{T}}$, where $\mathbf{J}$ here represents the Jacobian of the magnitude function (see 8.4.7).

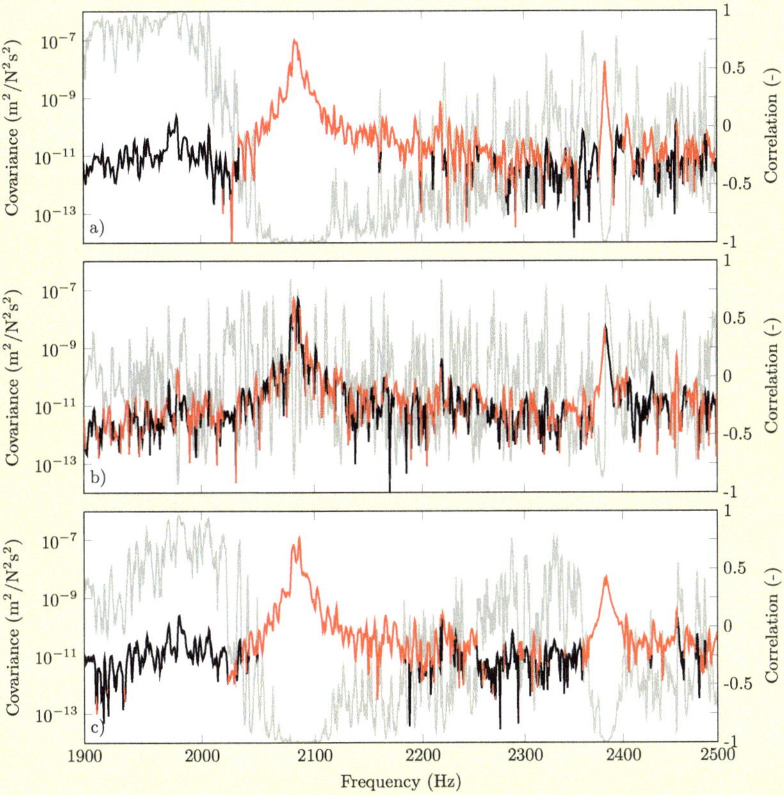

Figure 8.7 Covariance (black and red) and correlation coefficients (grey) between the real and imaginary parts of Y_{11} and Y_{21}: a) $\sigma^2_{\mathrm{Re}(Y_{11}),\mathrm{Re}(Y_{21})}$, b) $\sigma^2_{\mathrm{Re}(Y_{11})\mathrm{Im}(Y_{21})}$, c) $\sigma^2_{\mathrm{Im}(Y_{11})\mathrm{Im}(Y_{21})}$. The magnitude of the covariance is plotted to enable logarithmic scaling; red is used to indicate negative values.

To illustrate the effect of correlation, it is straight forward to perform the above uncertainty propagation using the full covariance $\Sigma_{\mathbf{Y}}$, which includes all correlations, and again using its diagonal values only $\mathrm{diag}(\Sigma_{\mathbf{Y}})$. It is clear from Figure 8.8 that the correlations present in an impact-based FRF measurement have a significant influence on the uncertainty of derived quantities. At some frequencies, correlations lead to an increase in the estimated uncertainty, for example around 2600 and 2360 Hz, though for the most part they have the effect of reducing the level of uncertainty. In any case, the inclusion of correlations provides a more accurate estimation of the true uncertainty, and this justifies the use of a more rigorous uncertainty propagation than standard (uncorrelated) variance formulas.

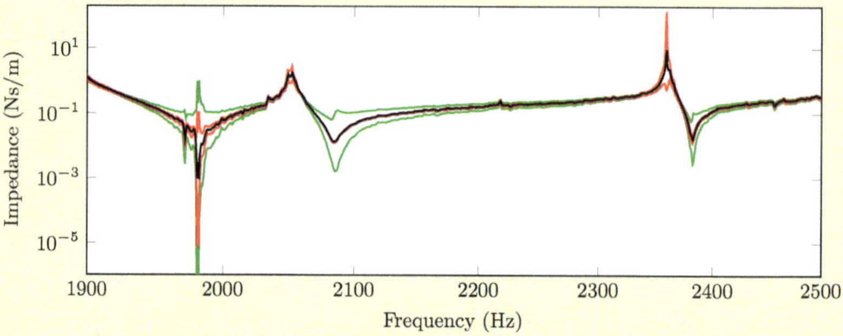

Figure 8.8 Point impedance magnitude (black) with 95% confidence bounds estimated based on linear propagation of uncertainty. Red bounds include the effects of correlation. Green bounds exclude the effects of correlation by using $\text{diag}(\Sigma_\mathbf{Y})$.

8.4.5 UNCERTAINTY PROPAGATION THROUGH INVERSE METHODS

Now that we have outlined the Law of Propagation of Uncertainty (LPU) and our strategy for dealing with complex valued quantities, we are ready to deal with the propagation of uncertainty through a VAVP. This will involve three distinct propagation steps: through the inverse characterisation of blocked forces (this section), through substructuring (8.4.6), and finally through a response prediction 8.4.7.

Whilst the focus of this section is on the propagation of uncertainty for *blocked forces*, it should be noted that the exact same set of equations will apply to other inverse problems of the same mathematical form, for example inverse force identification (3.1.2), pseudo-forces (3.1.4, and the characterisation of acoustic sources by equivalent volume velocities (3.4.4), etc.

The inverse blocked force relation is recalled from 3.2 as,

$$\bar{\mathbf{f}}_c = \begin{bmatrix} \mathbf{Y}_{Ccc} \\ \mathbf{Y}_{Cbc} \end{bmatrix}^+ \begin{pmatrix} \mathbf{v}_c \\ \mathbf{v}_b \end{pmatrix}. \tag{8.95}$$

where $\mathbf{Y}_{Ccc}$ and $\mathbf{Y}_{Cbc}$ are the interface and remote indicator FRF matrices, respectively, with superscript $\square^+$ denoting the matrix pseudo-inverse, and $\mathbf{v}_c$ and $\mathbf{v}_b$ are the corresponding vectors of operational responses. Writing Equation 8.95 in a DoF agnostic form we have simply,

$$\bar{\mathbf{f}} = \mathbf{Y}^+\mathbf{v} \tag{8.96}$$

with $\bar{\mathbf{f}} \in \mathbb{C}^N$, $\mathbf{Y} \in \mathbb{C}^{M\times N}$ and $\mathbf{v} \in \mathbb{C}^M$.

Both the FRF and operational response are measured quantities, and therefore subject to uncertainty. Under the assumption that the FRF and operational response are uncorrelated (this is a fair assumption as they will be measured as part of two separate tests) the LPU takes the form,

$$\Sigma_{\bar{\mathbf{f}}} = \mathbf{J}_{\mathbf{Y}}\Sigma_{\mathbf{Y}}\mathbf{J}_{\mathbf{Y}}^{\mathrm{T}} + \mathbf{J}_{\mathbf{v}}\Sigma_{\mathbf{v}}\mathbf{J}_{\mathbf{v}}^{\mathrm{T}} \tag{8.97}$$

where $\Sigma_{\mathbf{Y}} \in \mathbb{R}^{2NM\times 2NM}$, $\Sigma_{\mathbf{v}} \in \mathbb{R}^{2M\times 2M}$, and $\Sigma_{\bar{\mathbf{f}}} \in \mathbb{R}^{2N\times 2N}$ are the *bivariate* covariance matrices for the FRF matrix, operational response vector and blocked force, respectively. The Jacobians $\mathbf{J}_{\mathbf{Y}} \in \mathbb{R}^{2N\times 2NM}$ and $\mathbf{J}_{\mathbf{v}} \in \mathbb{R}^{2N\times 2M}$ contain the partial derivatives of each blocked force with respect to each FRF and operational response element, respectively. Note that the factor of 2 in the matrix dimensions account for their bivariate form (i.e. separating the real and imaginary parts).

To evaluate these Jacobians one must first separate Equation 8.96 into its real and imaginary parts, before differentiating each bivariate blocked force with respect to the each bivariate FRF and response element. To illustrate this process we will derive the Jacobian $\mathbf{J}_{\mathbf{v}}$ below.

We begin by taking the complex differential of the blocked force vector and using the product rule to separate the FRF and response differentials,

$$\mathrm{d}\bar{\mathbf{f}} = \mathrm{d}(\mathbf{Y}^{+}\mathbf{v}) = \mathrm{d}(\mathbf{Y}^{+})\mathbf{v} + \mathbf{Y}^{+}\mathrm{d}(\mathbf{v}). \tag{8.98}$$

With our aim being to derive $\mathbf{J}_{\mathbf{v}}$ (i.e. the Jacobian with respect to the operational response) we need only consider the second differential response term, $\mathbf{Y}^{+}\mathrm{d}(\mathbf{v})$. Now suppose the small change, represented by the differential $\mathrm{d}(\mathbf{v})$, is limited to a single element of the response vector. In this case the differential may be replaced by the scaled unit vector, $\mathbf{p}_i\mathrm{d}v_i$, where $\mathbf{p}_i = \begin{bmatrix} 0 & 0 & \cdots & 1 & \cdots & 0 \end{bmatrix}^{\mathrm{T}}$ is a zero vector bar the ith entry, whose value is one, and $\mathrm{d}v_i$ is a scalar differential term.

$$\mathrm{d}\bar{\mathbf{f}} = \mathbf{Y}^{+}\mathrm{d}(\mathbf{v}) \rightarrow \mathrm{d}\bar{\mathbf{f}} = \mathbf{Y}^{+}\mathbf{p}_i\mathrm{d}v_i \tag{8.99}$$

From Equation 8.99 the partial derivative of $\bar{\mathbf{f}}$, with respect to the single response variable v_i, is identified as,

$$\frac{\partial\bar{\mathbf{f}}}{\partial v_i} = \mathbf{Y}^{+}\mathbf{p}_i. \tag{8.100}$$

To construct the Jacobian matrix we must compute the derivative of $\bar{\mathbf{f}}$ with respect to *each* response variable. By horizontally aligning the unit vectors $\mathbf{p}_i$, the resultant matrix becomes the identity matrix,

$$\mathbf{P} = \begin{bmatrix} \mathbf{p}_1 & \mathbf{p}_2 & \cdots \end{bmatrix} = \begin{bmatrix} 1 & 0 & 0 \\ 0 & 1 & 0 \\ & & \\ 0 & 0 & \ddots \end{bmatrix} = \mathbf{I} \tag{8.101}$$

and the left-hand side becomes a matrix whose columns correspond to derivatives with respect to each element of the response vector. The partial derivative of $\bar{\mathbf{f}}$, with respect to the response vector $\mathbf{v}$ is then,

$$\frac{\partial\bar{\mathbf{f}}}{\partial\mathbf{v}} = \mathbf{Y}^{+}. \tag{8.102}$$

Note that in deriving Equation 8.102 we have not yet considered the complex nature of $\mathbf{Y}$ or $\mathbf{v}$. Indeed, Equation 8.102 is a *complex* derivative/Jacobian. To propagate uncertainty we need a *bivariate* Jacobian, one whose derivatives are with respect to the real and imaginary parts of the blocked force and operational response.

Taking the real part of the differential blocked force we have,

$$\mathrm{Re}(\mathrm{d}\bar{\mathbf{f}}) = \mathrm{Re}\left(\mathbf{Y}^+\mathbf{p}_i \mathrm{d}v_i\right). \tag{8.103}$$

Recalling that for the product of two complex numbers, $\mathrm{Re}\left([a+ib][c+id]\right) = ac - bd$, Equation 8.103 can be rewritten as,

$$\mathrm{Re}(\mathrm{d}\bar{\mathbf{f}}) = \mathrm{Re}(\mathbf{Y}^+\mathbf{p}_i)\mathrm{Re}(\mathrm{d}v_i) - \mathrm{Im}(\mathbf{Y}^+\mathbf{p}_i)\mathrm{Im}(\mathrm{d}v_i). \tag{8.104}$$

Similarly, recalling that $\mathrm{Im}\left([a+ib][c+id]\right) = ad + bc$, we can also write,

$$\mathrm{Im}(\mathrm{d}\bar{\mathbf{f}}) = \mathrm{Re}(\mathbf{Y}^+\mathbf{p}_i)\mathrm{Im}(\mathrm{d}v_i) + \mathrm{Im}(\mathbf{Y}^+\mathbf{p}_i)\mathrm{Re}(\mathrm{d}v_i). \tag{8.105}$$

From Equation 8.104 and 8.105 we can identify the bivariate derivatives,

$$\frac{\partial \mathrm{Re}(\bar{\mathbf{f}})}{\partial \mathrm{Re}(v_i)} = \mathrm{Re}(\mathbf{Y}^+\mathbf{p}_i) = \mathrm{Re}\left(\frac{\partial \bar{\mathbf{f}}}{\partial v_i}\right) \tag{8.106}$$

$$\frac{\partial \mathrm{Re}(\bar{\mathbf{f}})}{\partial \mathrm{Im}(v_i)} = -\mathrm{Im}(\mathbf{Y}^+\mathbf{p}_i) = -\mathrm{Im}\left(\frac{\partial \bar{\mathbf{f}}}{\partial v_i}\right) \tag{8.107}$$

$$\frac{\partial \mathrm{Im}(\bar{\mathbf{f}})}{\partial \mathrm{Re}(v_i)} = \mathrm{Im}(\mathbf{Y}^+\mathbf{p}_i) = \mathrm{Im}\left(\frac{\partial \bar{\mathbf{f}}}{\partial v_i}\right) \tag{8.108}$$

$$\frac{\partial \mathrm{Im}(\bar{\mathbf{f}})}{\partial \mathrm{Im}(v_i)} = \mathrm{Re}(\mathbf{Y}^+\mathbf{p}_i) = \mathrm{Re}\left(\frac{\partial \bar{\mathbf{f}}}{\partial v_i}\right). \tag{8.109}$$

Based on the above, a 2×2 bivariate element of the Jacobian matrix $\mathbf{J}_\mathbf{v}$ is given by,

$$\begin{bmatrix} \frac{\partial \mathrm{Re}(\bar{f}_j)}{\partial \mathrm{Re}(v_i)} & \frac{\partial \mathrm{Re}(\bar{f}_j)}{\partial \mathrm{Im}(v_i)} \\ \frac{\partial \mathrm{Im}(\bar{f}_j)}{\partial \mathrm{Re}(v_i)} & \frac{\partial \mathrm{Im}(\bar{f}_j)}{\partial \mathrm{Im}(v_i)} \end{bmatrix} = \begin{bmatrix} \mathrm{Re}\left(\frac{\partial \bar{f}_j}{\partial v_i}\right) & -\mathrm{Im}\left(\frac{\partial \bar{f}_j}{\partial v_i}\right) \\ \mathrm{Im}\left(\frac{\partial \bar{f}_j}{\partial v_i}\right) & \mathrm{Re}\left(\frac{\partial \bar{f}_j}{\partial v_i}\right) \end{bmatrix} = M\left(\frac{\partial \bar{f}_j}{\partial v_i}\right) \tag{8.110}$$

where $\frac{\partial \bar{f}_j}{\partial v_i}$ is the complex derivative, and we define the complex matrix mapping operator $M(\quad)$ as,

$$M(A) = \begin{bmatrix} \mathrm{Re}\,(A) & -\mathrm{Im}\,(A) \\ \mathrm{Im}\,(A) & \mathrm{Re}\,(A) \end{bmatrix}. \tag{8.111}$$

The entire bivariate Jacobian $\mathbf{J}_\mathbf{v}$ is obtained by applying the complex matrix mapping operator $M(\quad)$, element-wise,

$$\mathbf{J}_\mathbf{v} = M\left(\frac{\partial \bar{\mathbf{f}}}{\partial \mathbf{v}}\right) = M\left(\mathbf{Y}^+\right). \tag{8.112}$$

The resulting Jacobian $\mathbf{J_v}$ contains the partial derivatives of the real and imaginary blocked forces with respect to the real and imaginary operational responses. It can now be used to propagate uncertainty as part of the LPU.

A similar procedure, albeit more involved, can be used to evaluate the Jacobian $\mathbf{J_Y}$, i.e. the derivative with respect to each element of the FRF matrix. We will not go through the steps here, for the full derivation the reader is referred to [9]. The result is given below,

$$\mathbf{J_Y} = M_H(\mathbf{A}, \mathbf{B}) \tag{8.113}$$

with,

$$\mathbf{A} = -\left(\mathbf{Y}^{+}\mathbf{v}\right)^{\mathrm{T}} \otimes \mathbf{Y}^{+} \tag{8.114}$$

and,

$$\mathbf{B} = \left[\left(\left((\mathbf{I} - \mathbf{Y}\mathbf{Y}^{+})\mathbf{v}\right)^{\mathrm{T}} \otimes \mathbf{Y}^{+}\mathbf{Y}^{+\mathrm{H}}\right) + \left((\mathbf{Y}^{+\mathrm{H}}\mathbf{Y}^{+}\mathbf{v})^{\mathrm{T}} \otimes (\mathbf{I} - \mathbf{Y}^{+}\mathbf{Y})\right)\right]\mathbf{K} \tag{8.115}$$

where $\otimes$ is the Kronecker product and $M_H(\quad)$ is the modified complex matrix mapping operator defined as,

$$M_H(A,B) = \begin{bmatrix} \mathrm{Re}\,(A+B) & \mathrm{Im}\,(-A+B) \\ \mathrm{Im}\,(A+B) & -\mathrm{Re}\,(-A+B) \end{bmatrix}. \tag{8.116}$$

This operator is again applied element wise. The matrix $\mathbf{K}$ is the so-called *commutation* matrix. It is the matrix that transforms the vectorisation of a matrix to the vectorisation of its transpose [4].

The Jacobian given by Equation 8.113 represents the bivariate derivative of the blocked force with respect to each element of the FRF matrix. Its complexity comes from the fact that it is derived considering the generalised pseudo-inverse, and so is valid for non-square (over-determined) FRF matrices. If the FRF matrix is square, the second term in Equation 8.113 becomes zero, and the Jacobian reduces to a simpler form,

$$\mathbf{J_Y} = M\left(-(\mathbf{Y}^{-1}\mathbf{v})^{\mathrm{T}} \otimes \mathbf{Y}^{-1}\right). \tag{8.117}$$

Together, Equations 8.97, 8.112, and 8.113 (or 8.117) describe the linear propagation of uncertainty though an inverse problem, considered here to be the characterisation of blocked forces. Note that the Jacobians are computed using measured data (i.e. the mean FRF $\mathbf{Y}$ and response $\mathbf{v}$), and so no additional experimental effort is required beyond what would normally be undertaken for a blocked force characterisation.

8.4.6 UNCERTAINTY PROPAGATION THROUGH SUBSTRUCTURING

A key step in the development of a VAVP is to combine the FRFs of individual components, to form a model of the assembly. We call this process (dynamic) substructuring. In the context of uncertainty propagation, we are interested in propagating the uncertainty of each component's FRF matrix, onto that of the coupled assembly. The first consideration of uncertainty propagation in the context of substructuring was presented in [12]. The authors considered the dual formulation, and presented a linear propagation of uncertainty assuming uncorrelated errors. As discussed in 8.4.4, inter and intra-FRF correlation can have a large effect on the accuracy of an LPU-based uncertainty propagation. This was acknowledged in [11] and [6], the latter of which, based on the separation of real and imaginary parts, is presented below.

As discussed in 5.2 the equations for substructuring can be formulated in numerous ways. When considering the propagation of uncertainty we will focus attention on the primal and dual formulations, which are most commonly used in practice.

Both the primal and dual formulations take as an input a block diagonal matrix $\mathbf{Y}$ containing the free-interface (uncoupled) FRF matrices of each component, and output an FRF matrix of the coupled assembly $\mathbf{Y}_C$. In either case we can then interpret the substructuring equation as a matrix valued function,

$$\mathbf{Y}_C = G(\mathbf{Y}). \tag{8.118}$$

Hence we can apply the LPU generally as,

$$\Sigma_{\mathbf{Y}_C} = \mathbf{J}\Sigma_{[\mathbf{Y}]}\mathbf{J}^{\mathrm{T}} \tag{8.119}$$

where $\Sigma_{[\mathbf{Y}]}$ is a bivariate covariance matrix for the uncoupled FRF matrix $\mathbf{Y}$, and $\Sigma_{\mathbf{Y}_C}$ is the bivariate covariance matrix of the coupled FRF matrix $\mathbf{Y}_C$. The Jacobian matrix $\mathbf{J}$ contains the partial derivatives of each coupled FRF $Y_{C_{ij}}$ with respect to each uncoupled FRF $Y_{N_{ij}}$, where N denotes the Nth component. Its form will depend on whether a primal or dual assembly process is used. In what follows we present concise expressions for these Jacobians, their derivations are omitted, but can be found in [6].

Primal

The primal assemblage of component FRFs is given by the following equation,

$$\mathbf{Y}_C = \left(\mathbf{L}^{\mathrm{T}}\mathbf{Y}^{-1}\mathbf{L}\right)^{-1} \tag{8.120}$$

where $\mathbf{L}$ is a Boolean matrix that enforces compatibly and equilibrium constraints at the connecting DoFs. The corresponding Jacobian matrix is given by,

$$\mathbf{J}_{\mathbf{P}} = M\left(\left(\mathbf{Y}^{-1}\mathbf{L}\mathbf{Y}_C\right)^{\mathrm{T}} \otimes \left(\mathbf{Y}_C\mathbf{L}^{\mathrm{T}}\mathbf{Y}^{-1}\right)\right) \tag{8.121}$$

where the complex matrix mapping operator $M(\quad)$ is again applied element wise.

Dual

The dual assemblage of component FRFs is given by the following equation,

$$\mathbf{Y}_C = \mathbf{Y} - \mathbf{Y}\mathbf{B}^{\mathrm{T}}\left(\mathbf{B}\mathbf{Y}\mathbf{B}^{\mathrm{T}}\right)^{-1}\mathbf{B}\mathbf{Y} \tag{8.122}$$

where **B** is a signed Boolean matrix that enforces compatibly and equilibrium constraints at the connecting DoFs. The corresponding Jacobian matrix is given by,

$$\mathbf{J_D} = M\left(\mathbf{I} - \left[\left(\mathbf{B}^{\mathrm{T}}\left(\mathbf{BYB}^{\mathrm{T}}\right)^{-1}\mathbf{BY}\right)^{\mathrm{T}} \otimes \mathbf{I}\right] + \right.$$
$$\left[\left(\mathbf{B}^{\mathrm{T}}\left(\mathbf{BYB}^{\mathrm{T}}\right)^{-1}\mathbf{BY}\right)^{\mathrm{T}} \otimes \mathbf{YB}^{\mathrm{T}}\left(\mathbf{BYB}^{\mathrm{T}}\right)^{-1}\mathbf{B}\right] -$$
$$\left.\left[\mathbf{I} \otimes \mathbf{YB}^{\mathrm{T}}\left(\mathbf{BYB}^{\mathrm{T}}\right)^{-1}\mathbf{B}\right]\right). \tag{8.123}$$

where the complex matrix mapping operator $M(\quad)$ is again applied element wise. As for the blocked force uncertainty, the primal and dual Jacobians must be evaluated at the mean FRF. To do so, we simply compute them using, for example, the H_1 estimate of **Y**.

8.4.7 UNCERTAINTY PROPAGATION THROUGH RESPONSE PREDICTION

Having characterised the blocked force $\bar{\mathbf{f}}$, and measured or modelled the assembly FRF **H**, we are able to predict the operational response using the equation,

$$\mathbf{p} = \mathbf{H}\bar{\mathbf{f}}. \tag{8.124}$$

Note that we use **H** and **p** here to avoid confusion with **Y** and **v** which have already been used for the substructuring and blocked force uncertainty. Of course, the forward predicted response could be a sound pressure, or a structural response.

Equation 8.124 can be applied in a diagnostic or predictive context. In a diagnostic application the assembly FRF **H** is measured in the same assembly in which the blocked forces are characterised. In a predictive application the assembly FRF **H** is predicted using substructuring. The propagation of uncertainty will differ in these two cases.

Predictive application

In the predictive application we can assume that the coupled assembly FRF and blocked force are uncorrelated (since the component FRFs are measured as part of a separate test). According to the LPU the bivariate covariance matrix of the response vector **p** is given by,

$$\Sigma_{\mathbf{p}} = \mathbf{J_H}\Sigma_{\mathbf{H}}\mathbf{J_H}^{\mathrm{T}} + \mathbf{J_{\bar{f}}}\Sigma_{\bar{\mathbf{f}}}\mathbf{J_{\bar{f}}}^{\mathrm{T}} \tag{8.125}$$

where $\Sigma_{\mathbf{H}}$ and $\Sigma_{\bar{\mathbf{f}}}$ are the bivariate covariance matrices for the coupled FRF and blocked force respectively. These are given by Equation 8.119 and 8.97, respectively. The Jacobian matrices $\mathbf{J_H}$ and $\mathbf{J_{\bar{f}}}$ contain the partial derivatives of each response variable p_i with respect to each element of the FRF matrix and the blocked force vector, respectively. They are given by,

$$\mathbf{J_H} = M\left(\bar{\mathbf{f}}^{\mathrm{T}} \otimes \mathbf{I}\right) \tag{8.126}$$

and,

$$\mathbf{J}_{\bar{\mathbf{f}}} = M(\mathbf{H}). \tag{8.127}$$

In the case that the forward FRF matrix $\mathbf{H}$ is obtained by substructuring (i.e. in a VAVP application), Equation 8.125 is modified as so,

$$\Sigma_{\mathbf{p}} = \mathbf{J}_{\mathbf{H}}\left(\mathbf{J}_{P/D}\Sigma_{[\mathbf{Y}]}\mathbf{J}_{P/D}^{\mathrm{T}}\right)\mathbf{J}_{\mathbf{H}}^{\mathrm{T}} + \mathbf{J}_{\bar{\mathbf{f}}}\Sigma_{\bar{\mathbf{f}}}\mathbf{J}_{\bar{\mathbf{f}}}^{\mathrm{T}} \tag{8.128}$$

where $\mathbf{J}_{P/D}\Sigma_{[\mathbf{Y}]}\mathbf{J}_{P/D}^{\mathrm{T}}$ describes the propagation of FRF uncertainty through the substructuring process, with $\Sigma_{[\mathbf{Y}]}$ being the uncoupled FRF covariance matrix and $\mathbf{J}_{P/D}$ the Jacobian of the primal (Equation 8.121) or dual (Equation 8.123) method.

Diagnostic application

In a diagnostic application we can no longer assume that the coupled assembly FRF and blocked force are uncorrelated. Often the forward FRF matrix $\mathbf{H}$ is measured simultaneously to the FRF matrix $\mathbf{Y}_C$ that is used to characterise the blocked force. In this case, due to their shared excitation, the forward FRF and blocked force can exhibit strong correlation. It necessary to account for this correlation when propagating uncertainty. To do so, Equation 8.125 is modified as so,

$$\Sigma_{\mathbf{p}} = \mathbf{J}_{\mathbf{H}}\Sigma_{\mathbf{H}}\mathbf{J}_{\mathbf{H}}^{\mathrm{T}} + \mathbf{J}_{\bar{\mathbf{f}}}\Sigma_{\bar{\mathbf{f}}}\mathbf{J}_{\bar{\mathbf{f}}}^{\mathrm{T}} + \left[\mathbf{J}_{\mathbf{H}}\Sigma_{\mathbf{H},\mathbf{Y}}\mathbf{J}_{\mathbf{HY}}^{\mathrm{T}} + \mathbf{J}_{\mathbf{HY}}\Sigma_{\mathbf{Y},\mathbf{H}}\mathbf{J}_{\mathbf{H}}^{\mathrm{T}}\right] \tag{8.129}$$

where $\Sigma_{\mathbf{H},\mathbf{Y}}$ is the bivariate covariance matrix between the forward ($\mathbf{H}$) and inverse ($\mathbf{Y}_C$) FRF matrices. The Jacobian $\mathbf{J}_{\mathbf{H}}$ is given by Equation 8.126, and $\mathbf{J}_{\mathbf{HY}}$ by,

$$\begin{aligned}\mathbf{J}_{\mathbf{HY}} = M_H\big(-(\mathbf{Y}^+\mathbf{v})^{\mathrm{T}} \otimes \mathbf{HY}^+, \big[(((\mathbf{I}-\mathbf{YY}^+)\mathbf{v})^{\mathrm{T}} \otimes \mathbf{HY}^+\mathbf{Y}^{+\mathrm{H}}) + \\ ((\mathbf{Y}^{+\mathrm{H}}\mathbf{Y}^+\mathbf{v})^{\mathrm{T}} \otimes \mathbf{H}(\mathbf{I}-\mathbf{YY}^+))\big]\,\mathbf{K}\big).\end{aligned} \tag{8.130}$$

One of the useful benefits of the LPU is the ability to break down the uncertainty of an output into relative contributions from the various inputs. For example, identifying whether the FRF ($\mathbf{J}_{\mathbf{Y}}\Sigma_{\mathbf{Y}}\mathbf{J}_{\mathbf{Y}}^{\mathrm{T}}$) or operational response ($\mathbf{J}_{\mathbf{v}}\,\Sigma_{\mathbf{v}}\,\mathbf{J}_{\mathbf{v}}^{\mathrm{T}}$) contribute most to the uncertainty in the blocked force, or perhaps identifying which component FRF introduces the most uncertainty in the coupled assembly FRF. This sort of information can be used to identify troublesome measurements the retesting of which might reduce the output uncertainty most significantly. An example of this sort of application can be found in [13].

8.4.8 PRESENTING UNCERTAINTY ESTIMATES

The response prediction covariance matrix $\Sigma_{\mathbf{p}}$ output by Equation 8.125 provides the variance of, and covariance between, the real and imaginary parts of each predicted response. Assuming the input uncertainty follows a normal distribution, according to the LPU the real and imaginary parts of $\mathbf{p}$ will also be distributed normally. Their uncertainty can therefore be presented in the usual way as a mean value with a corresponding confidence interval. For a bivariate normal, this takes the form of a confidence ellipse, as illustrated in Figure 8.6. Alternatively, a relative variance can be presented, as described in 8.4.1.

More often however, we are not interested in the real and imaginary parts specifically, it is the magnitude that we would like to plot. How then do we determine the magnitude confidence interval of a complex quantity? Again, we can use the LPU.

Providing the output uncertainty is sufficiently small, we can use the LPU to propagate the bivariate uncertainty through the magnitude function,

$$|p_i| = \sqrt{\mathrm{Re}\,(p_i)^2 + \mathrm{Im}\,(p_i)^2}. \tag{8.131}$$

The uncertainty propagation takes the standard form,

$$\sigma^2_{|p_i|} = \mathbf{J}\Sigma_{p_i}\mathbf{J}^{\mathrm{T}} \tag{8.132}$$

where Σ_{p_i} is the bivariate covariance matrix of the response variable p_i,

$$\Sigma_{p_i} = \begin{bmatrix} \sigma^2_{\mathrm{Re}(p_i)\mathrm{Re}(p_i)} & \sigma^2_{\mathrm{Re}(p_i)\mathrm{Im}(p_i)} \\ \sigma^2_{\mathrm{Im}(p_i)\mathrm{Re}(p_i)} & \sigma_{\mathrm{Im}(p_i)\mathrm{Im}^2(p_i)} \end{bmatrix} \tag{8.133}$$

and $\mathbf{J}$ represents the Jacobian matrix of the magnitude function,

$$\mathbf{J} = \begin{bmatrix} \frac{\mathrm{Re}(p_i)}{\sqrt{\mathrm{Re}(p_i)^2+\mathrm{Im}(p_i)^2}} & \frac{\mathrm{Im}(p_i)}{\sqrt{\mathrm{Re}(p_i)^2+\mathrm{Im}(p_i)^2}} \end{bmatrix}. \tag{8.134}$$

The variance $\sigma^2_{|p_i|}$ can then be used to determine an appropriate confidence interval. Alternatively, it can be used to compute the relative variance,

$$\mathrm{RelVar}\,(|p_i|) = \frac{\sigma^2_{|p_i|}}{|p_i|^2}. \tag{8.135}$$

This LPU-based approach is clearly only appropriate for small levels of output uncertainty. The magnitude function is by definition strictly positive, the normal distribution however permits values from $-\infty$ to $+\infty$. An alternative sampling-based approach is described in 8.4.9.

8.4.9 MONTE-CARLO METHODS FOR PROPAGATING RANDOM UNCERTAINTY

In the presence of small levels of uncertainty, or when applied to weakly non-linear functions, the LPU is able to provide an accurate estimate of output uncertainty. If the above requirements are not satisfied however, the LPU can incur large errors and fail to produce an accurate estimate. In this scenario an alternative approach is required.

Monte-Carlo methods constitute a broad class of computational algorithms that rely on repeated random sampling to determine a numerical result. In the context of a VAVP, Monte-Carlo methods can be used to compute robust uncertainty estimates for blocked forces, substructured FRFs and response predictions.

Though computationally far more expensive, Monte-Carlo methods are advantageous for several reasons: 1) they can be applied in the presence of an arbitrarily large input uncertainty, 2) the function being evaluated can be highly non-linear, 3) there is no need to derive any complex Jacobian matrices, and 4) they can provide information on the precise distribution of the output.

With regards to 3) above, note that the LPU inverse uncertainty propagation (8.4.5) was presented for the least squares (over-determined) case. It cannot be used for a regularised solution, for example using a Tikhonov of tSVD scheme (7.4.3). This would require a new Jacobian to be derived. Such regularised solutions can easily be dealt with by the Monte-Calro approach.

Large input uncertainty

Whilst linearity is a fundamental assumption that underpins the component-based approach, it applies to the *components*, not necessarily the *equations* used to build a VAVP. Indeed, many of these equations, specifically those that contain *matrix inversions*, are non-linear. When we propagate uncertainty though such non-linear equations using the LPU, an approximation is introduced. For a blocked force characterisation, the non-linear propagation is between the coupled FRF $\mathbf{Y}$ and the blocked force $\bar{\mathbf{f}}$, i.e. between the covariances,

$$\Sigma_{\mathbf{Y}} \rightarrow \Sigma_{\bar{\mathbf{f}}}. \tag{8.136}$$

For the substructuring step, it is the propagation between the uncoulpled FRF matrix $[\mathbf{Y}]$ and the coupled FRF matrix $\mathbf{Y}_C$ that is non-linear,

$$\Sigma_{[\mathbf{Y}]} \rightarrow \Sigma_{\mathbf{Y}_C}. \tag{8.137}$$

For large levels of input uncertainty the LPU will fail to provide an accurate estimate of the above covariances. This propagation can however be performed using a Monte-Carlo approach, as will be described shortly.

It is worth noting that the propagation of operational response uncertainty,

$$\Sigma_{\mathbf{v}} \rightarrow \Sigma_{\bar{\mathbf{f}}} \tag{8.138}$$

forward FRF uncertainty,

$$\Sigma_{\mathbf{H}} \rightarrow \Sigma_{\mathbf{p}} \tag{8.139}$$

and blocked force uncertainty,

$$\Sigma_{\bar{\mathbf{f}}} \rightarrow \Sigma_{\mathbf{p}} \tag{8.140}$$

are all linear, as they do not involve any matrix inversions. As such, the LPU estimates for these contributions are exact. Though, in the case that the response prediction is made with a forward FRF that is correlated with the inverse FRF used to characterise the blocked force, a Monte-Carlo procedure may still be required.

Considering the FRF uncertainty propagation in Equation 8.136, a simple Monte-Carlo procedure may be outlined as follows:

1. Based on the N repeated FRF measurements, estimate the mean FRF and its covariance matrix.

2. Assuming the measured FRFs follow known distribution (e.g. normal), generate a random sample from the corresponding multivariate distribution with specified mean and covariance obtained from Step 1. Using the sampled FRF matrix compute a blocked force vector.
3. Repeat step 2 N times to obtain N blocked force vectors and estimate the bivariate covariance matrix $\Sigma^{MC}_{\bar{\mathbf{f}}_\mathbf{Y}}$ directly.
4. Substitute the Monte-Carlo covariance matrix $\Sigma^{MC}_{\bar{\mathbf{f}}_\mathbf{Y}}$ for the LPU propagation term $\mathbf{J}_\mathbf{Y}\Sigma_\mathbf{Y}\mathbf{J}_\mathbf{Y}^{\mathrm{T}}$ in Equation 8.97. The response uncertainty propagation term, being linear, remains the same as in Equation 8.97.

Note that, rather than sampling from a multivariate distribution, the individual FRF measurements themselves can be sampled. In this case it is important that the same excitation is used for all FRFs within a shared column, so as to capture the effects of inter-FRF correlation. Similar procedures can be followed to replace the substructuring LPU term $\mathbf{J}\Sigma_{[\mathbf{Y}]}\mathbf{J}^{\mathrm{T}}$ in Equation 8.119.

When using a Monte-Carlo propagation, it is important that the number of samples N used is sufficiently large such that the output covariance begins to converge. In [9] $N = 2000$ was shown to give good agreement with LPU estimates for low levels of uncertainty (with a 12 DoF numerical example). Best practice would be to include some convergence-based stopping criteria onto the covariance and use this to determine when sufficient N has been reached.

Note that the Monte-Carlo sampling described above must be repeated for all frequency bins of interest. This can drastically increase the computation effort required.

Magnitude Confidence Bounds

In 8.4.8 the LPU was applied to the magnitude function so to estimate the magnitude variance of a jointly correlated complex quantity. This can then be used to present uncertainty as a magnitude confidence interval. Whilst the LPU has been shown to provide excellent estimates of complex uncertainty, its application during this last step of obtaining a magnitude confidence interval usually provides suboptimum results. This is largely down to the assumption that the magnitude maintains a normal distribution which, being strictly positive, it does not. A more robust confidence interval estimation can be achieved quite easily using a Monte-Carlo-based approach as described below.

Assuming that the mean value μ_p and bivariate covariance matrix Σ_p of the response prediction p are available, and that we know the form of its underlying distribution (i.e. normal), a magnitude confidence interval can be determined as follows:

1. Based on μ_p and Σ_p, N random samples are drawn from a bivariate distribution (Figure 8.9a).
2. The magnitude of each sample pair (together representing the complex quantity p) is taken and used to estimate a probability distribution function (PDF) (Figure 8.9b).

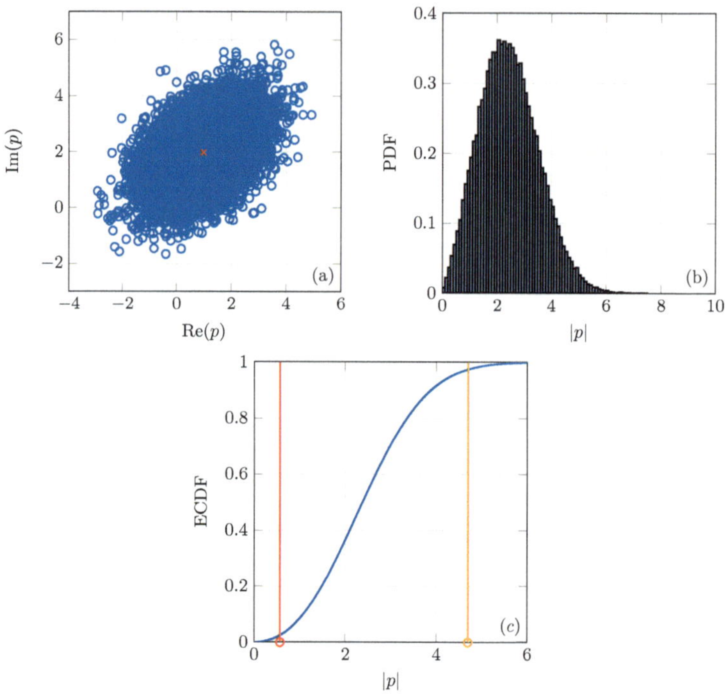

Figure 8.9 MC estimation of magnitude confidence interval using an Empirical Cumulative Distribution Function (ECDF).

3. An Empirical Cumulative Density Function (ECDF) is estimated by integration of the PDF (Figure 8.9c)).
4. The $\pm 95\%$ confidence interval of $|p|$ is then found by inspecting the values of $|p|$ for which the ECDF is equal to α and $1-\alpha$, where $\alpha = (1-0.95)/2$ (Figure 8.9c)).

Some examples of the confidence intervals obtained following this approach can be found in Figure 8.12 and 8.13. Of course, the exact same procedure can be applied to any complex quantity, e.g. blocked forces, substructured FRFs, etc. As before, the accuracy of the Monte-Carlo confidence interval will depend on the number of random samples drawn. Fortunately, estimation of the magnitude confidence interval is only necessary when plotting results, and so is only done for specific responses, making it quite quick to run.

Note that if the real and imaginary parts of the complex quantity are indeed normally distributed, analytical expressions are available for both the PDF and CDF of its magnitude and phase [2]. Given the mean and covariance terms, these can be used to derive an exact confidence interval, avoiding the MC sampling described above. The equations involved are, however, rather unpleasant (and so not shown here), and require the truncation of an infinite series so are not necessarily any more accurate that the sampling approach described above.

Example 8.2: VAVP uncertainty propagation

This numerical example illustrates the idea behind uncertainty propagation and shows how uncertainty estimates can be presented. The example includes a blocked force characterisation with an on-board validation, followed by a response prediction into a substructured assembly. For full details the reader is referred to [9].

Shown in Figure 8.10a) are two coupled plates, the smaller of which is considered our vibration source. The aim is to obtain the blocked force at the interface c, and use this to predict the response, first in the attached receiver, and then in a substructured assembly with a different receiver plate. To introduce representative uncertainties, all FRFs are simulated by randomly positioning 10 excitations around each DoF (as illustrated in Figure 8.10b). An example of the spread of FRFs is shown in Figure 8.11a. For the operational response, a unit force is applied within the source, and random noise added to simulate source uncertainty. An example is shown in Figure 8.11b.

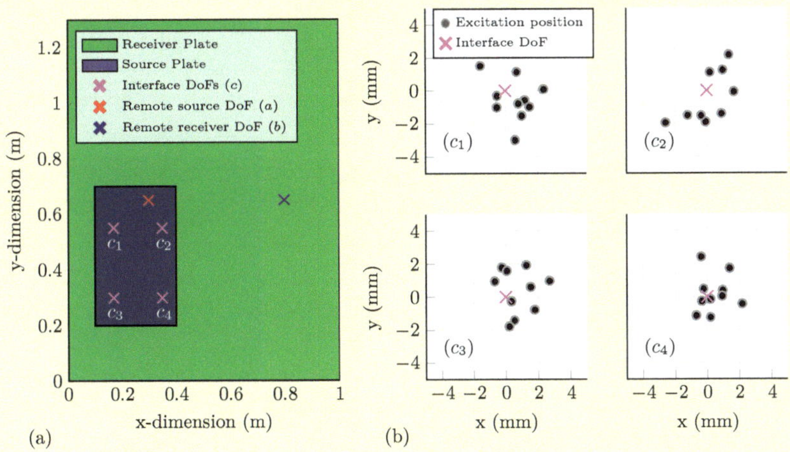

Figure 8.10 Illustration of numerical coupled plate example (a) with random FRF excitations (b).

The FRF and operational response covariance matrices $\Sigma_{\mathbf{Y}}$ and $\Sigma_{\mathbf{v}}$ are estimated from the simulated measurements. Together these are used to estimate the blocked force covariance $\Sigma_{\bar{\mathbf{f}}}$ according to Equation 8.97. The blocked force is then used to estimate the operational response at a remote point in the receiver. Its uncertainty is propagated, and combined with the forward

FRF uncertainty according to Equation 8.129. The result is shown in Figure 8.12.

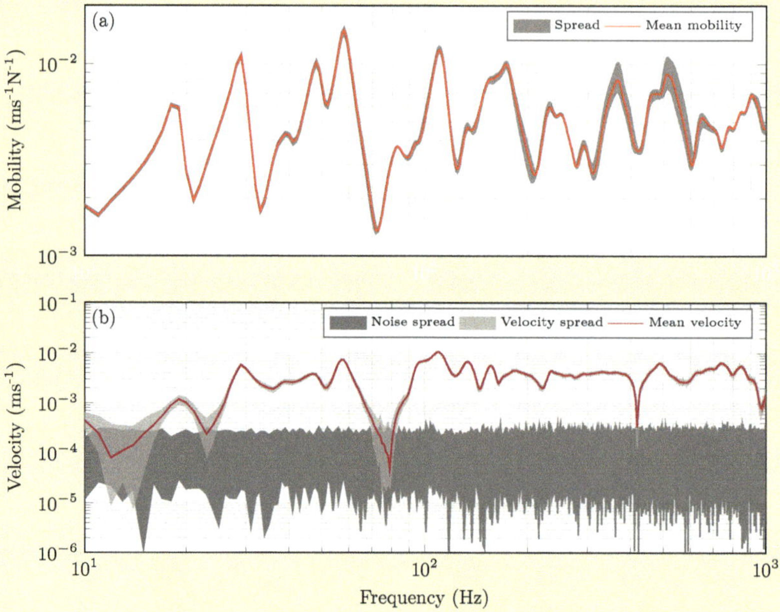

Figure 8.11 Example FRF (a) and operational response (b) measurements with simulated uncertainty.

Figure 8.12a shows the relative variance of the response prediction obtained using the LPU, compared against a Monte-Carlo (MC) estimation. In Figure 8.12b the same uncertainty result is presented instead as a 95% confidence interval surrounding the mean prediction. A MC-based bound is also shown. For the level of uncertainty considered, the LPU is in excellent agreement with the MC estimate.

Finally, the free-interface source and receiver FRFs are measured (again using randomised excitation positions). These are used to estimate the source and receiver covariance matrices $\Sigma_{\mathbf{Y}_S}$ and $\Sigma_{\mathbf{Y}_R}$. Using Equation 8.119 these are propagated through the substructuring equations to obtained the coupled assembly covariance $\Sigma_{\mathbf{Y}_C}$. Together with the blocked force covariance, Equation 8.125 is used to estimate the covariance of the VAVP response prediction. The result is shown in Figure 8.13.

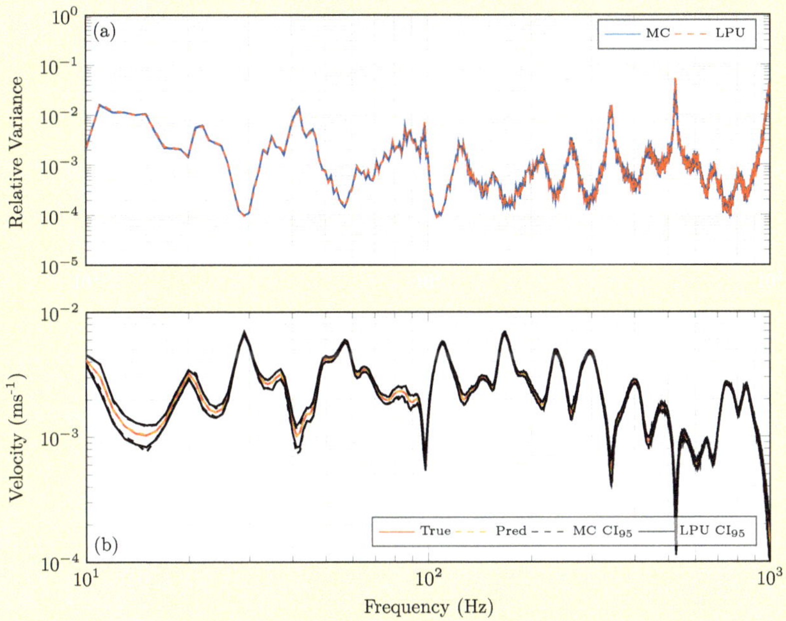

Figure 8.12 Uncertainty estimates for the blocked force on-board validation obtained using the LPU and Monte-Carlo sampling. a) Relative variance and b) 95% confidence interval.

As before, Figure 8.13 presents the uncertainty estimate in the form of a relative variance (a) and a confidence interval (b). The LPU estimation is in good agreement with MC approach across the entire frequency range.

Figure 8.13 provides an overview of the uncertainty in the VAVP response prediction. Further analysis could be done to separate the various contributions to this uncertainty, so to identify the greatest contributor [13].

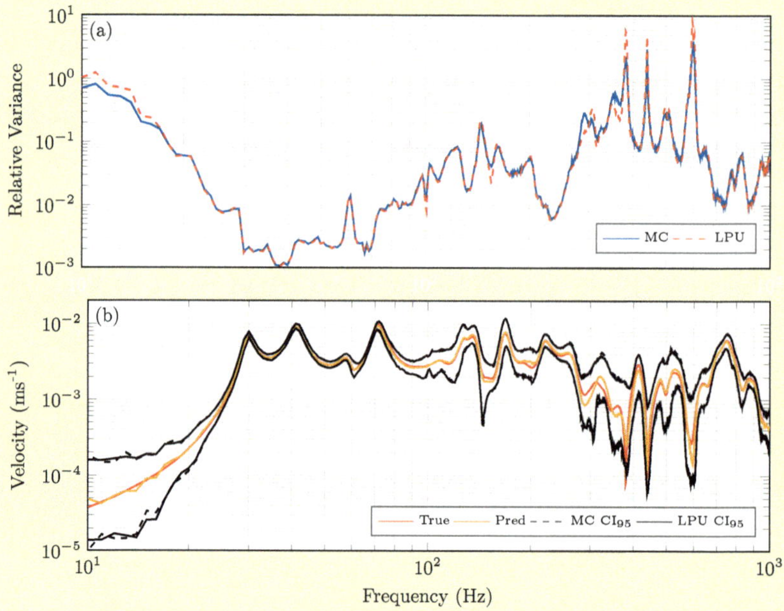

Figure 8.13 Uncertainty estimates for the VAVP response prediction (blocked force with substructuring) obtained using the LPU and Monte-Carlo sampling. a) Relative variance and b) 95% confidence interval.

8.5 SUMMARY

In general, all measured quantities should be accompanied by a statement of uncertainty. This requirement extends to any derived quantities, for example blocked forces, substructured FRFs and response predictions.

Within a VAVP, uncertainty can be categorised as originating from the *model*, the *source*, *experimental noise* or *human error*. Model error is *systematic*, the other types are *random*.

The main cause of *model* error is lack of agreement between the real assembly or component and its representation within the VAVP framework. The discrepancy can be due to lack of either *completeness* or *consistency*. Completeness describes the degree to which an interface is correctly represented, and can be quantified by the *Interface Completeness Criterion* (ICC). The ICC is essentially the correlation between a directly measured and reconstructed set of FRFs that traverse the interface. A value less than 1 indicates a lack of completeness. Lack of *consistency* may occur when the FRF characteristics of an assembly or component changes between operational and passive tests. It is a less well-developed concept but could potentially be quantified by comparing FRF and output-only-based transmissibilities.

In general, *random* variables are quantified by their *mean*, *variance* and *covariance*. For use in component-based simulation, these familiar concepts need to be extended to complex matrices and vectors. The mean, variance and covariance of measured variables (e.g. FRFs and responses) can be calculated based on repeated measurement. To do so, it is necessary to retain all measurements (i.e. individual impacts/time windows) when performing any component or assembly characterisation. Our main interest lies, however, not in the uncertainties of these directly measured quantities themselves but rather in how they *propagate* onto the outputs of the VAVP.

The propagation problem can be solved by applying the *Law of Propagation of Uncertainty* (LPU). The operations through which uncertainties are propagated in a VAVP comprise: *inverse solutions* (i.e. blocked force characterisation), *substructuring*, and *forward response predictions*. The main result from the LPU is that the covariance on the *output* is equal to the product of the covariance of the *input* (the directly measured operational responses and FRFs) multiplied by a *Jacobian* matrix. The Jacobian matrix depends on the propagating function considered (inversion, substructuring etc.). It is derived in terms of partial derivatives and is expressed in terms of combinations of the directly measured variables.

The *correlation* between the real and imaginary parts of frequency domain variables cannot, in general, be neglected. The LPU has therefore been extended to treat real and imaginary parts as separate variables. The uncertainties of these parts can be combined to obtain the uncertainty on the output *magnitude*, which can be presented as a relative variance or a confidence interval.

For large errors, the assumptions inherent in the LPU (small uncertainty and weak non-linearity) are not met and *Monte-Carlo* (MC) methods are required. Here, the mean and covariance of the input variables are used to generate a large number of random samples, which are individually passed through the function of interest and used to estimate the output uncertainty. MC methods are computationally expensive, but have several advantages, including handling of large uncertainty, non-linear functions and regularised inverse solutions.

Bibliography

1. IFCC BIPM, ILAC IFCC, IUPAC ISO, and OIML IUPAP. Evaluation of measurement data—guide to the expression of uncertainty in measurement, JCGM 100: 2008 GUM 1995 with minor corrections. *Joint Committee for Guides in Metrology*, 98, 2008.
2. P Dharmawansa, N Rajatheva, and C Tellambura. Envelope and phase distribution of two correlated gaussian variables. *IEEE Transactions on Communications*, 57(4):915–921, 2009.
3. M Haeussler, T Mueller, EA Pasma, J Freund, O Westphal, and T Voehringer. Component TPA: Benefit of including rotational degrees of freedom and over-determination. In *Proceedings of the international conference on noise and vibration engineering, Leuven, Belgium*, pages 7–9, 2020.
4. A Hjørungnes. *Complex-valued matrix derivatives: with applications in signal processing and communications*. Cambridge University Press, 2011.
5. G James et al. An introduction to statistical learning, New York, Springer, 2013.
6. JWR Meggitt and AT Moorhouse. A covariance based framework for the propagation of correlated uncertainty in frequency based dynamic sub-structuring. *Mechanical Systems and Signal Processing*, 136:106505, 2020.
7. JWR Meggitt and AT Moorhouse. On the completeness of interface descriptions and the consistency of blocked forces obtained in situ. *Mechanical Systems and Signal Processing*, 145:106850, 2020.

8. JWR Meggitt, AT Moorhouse, and AS Elliott. On the problem of describing the coupling interface between sub-structures: an experimental test for 'completeness'. In *Dynamics of Coupled Structures, Volume 4: Proceedings of the 36th IMAC, A Conference and Exposition on Structural Dynamics 2018*, pages 171–182. Springer, 2018.
9. JWR Meggitt, AT Moorhouse, K Wienen, and M Sturm. A framework for the propagation of uncertainty in Transfer Path Analysis. *Journal of Sound and Vibration*, 483:115425, 2020.
10. DC Montgomery and GC Runger. *Applied statistics and probability for engineers.* John Wiley & Sons, 2010.
11. F Trainotti, M Haeussler, and DJ Rixen. A practical handling of measurement uncertainties in frequency based substructuring. *Mechanical Systems and Signal Processing*, 144:106846, 2020.
12. SN Voormeeren, D De Klerk, and DJ Rixen. Uncertainty quantification in experimental frequency based substructuring. *Mechanical Systems and Signal Processing*, 24(1):106–118, 2010.
13. JWR Meggitt, AT Moorhouse, and AS Elliott. A covariance based framework for the propagation of uncertainty through inverse problems with an application to force identification. *Mechanical Systems and Signal Processing*, 124:275–297, Elsevier, 2019.

9 Building a Simple VAVP: A Coded Walk Through

9.1 INTRODUCTION

The purpose of Chapter 9 is to provide a coded 'walk-through' example of how to build a VAVP for a simple laboratory-style structure. The example will demonstrate the implementation of various techniques described through the book, including:

- Passive component characterisation, including free source/receiver FRF and (in situ) isolator transfer impedance (Chapter 2)
- Interface transformations using finite difference approximation (Chapter 4)
- Active source characterisation using in situ blocked force method (Chapter 3)
- Dynamic substructuring using primal formulation (Chapter 5)
- Operational response prediction and time synthesis (Chapter 5 and 7)

The chosen assembly has been designed to be simple enough to easily reproduce with limited resources, requiring low channel count and minimal instrumentation, whilst still demonstrating many of the key elements in the construction of a VAVP.

As a cautionary note, if you wish to replicate this study and would like to additionally compute the uncertainties, remember to store all the individual FRF impacts! Unfortunately, we forgot to...

9.2 DESCRIPTION OF ASSEMBLY AND MEASUREMENT PROCEDURE

The assembly considered is shown in Figures 9.1 and 9.2. It comprises a beam-like vibration source resiliently coupled to a plate-like receiver structure. Both source and receiver are made of perspex with dimensions $0.45 \times 0.04 \times 0.008$ m and $0.6 \times 0.45 \times 0.008$ m, respectively. Perspex was chosen here to avoid highly resonant behaviour, which would make for a more challenging case. Attached to both source and receiver are a pair of perspex blocks (dimensions $0.04 \times 0.04 \times 0.025$ m) which act as connections points. Internal forces are generated within the source through the operation of an attached shaker driven by a white noise signal. In addition, to simulate some time varying characteristics, external transient forces are applied to the source over the measurement period.

Our aim is to predict the operational response of this assembly at a target location on the receiver structure using the VAVP framework. Specifically, the following sections will detail the:

- characterisation of the source, including free-interface FRF and blocked force (obtained in situ);

DOI: 10.1201/9781003399131-9

Figure 9.1 Photo of test assembly for the coded VAP example.

- characterisation of the receiver structure's free-interface FRF;
- characterisation of the isolator's impedance matrix using the in situ method;
- substructuring of the component FRFs to predict the coupled FRFs of the assembly;
- prediction of the operational response, including time-domain response.

9.3 SOURCE CHARACTERISATION

The complete characterisation of a vibration source requires a characterisation of both its active and passive properties. As detailed though this book, we choose to characterise the active properties in terms of the blocked force obtained from in situ testing. For the passive properties we use the component's free-interface FRFs.

9.3.1 BLOCKED FORCE CHARACTERISATION AND VALIDATION

To characterise the blocked force of the source we use the coupled assembly illustrated in Figure 9.2. This is the same coupled assembly that we are attempting to build a VAVP of. The measurements made on this assembly, whilst providing a blocked force characterisation, will also provide validation data for the eventual VAVP. Note that this blocked force characterisation and its validation is entirely equivalent to a typical in situ TPA exercise.

The first step towards characterising the blocked force is to define a sign convention and decide which interface DoFs it is necessary to include. For the simple source considered here, we will include only the translational z and rotational x (α) DoFs at each connection point. This should be sufficient in the low frequency regime thanks to the resilient coupling. Had a rigid coupling been used, more DoFs would likely be required. To capture the z and α DoFs we use a finite difference approximation with a spaced sensor/force pair at each connection, as illustrated in Figure 9.2. In order to provide a validation of the blocked force an additional sensor is mounted to the receiver at DoF r; no excitation is required at this point.

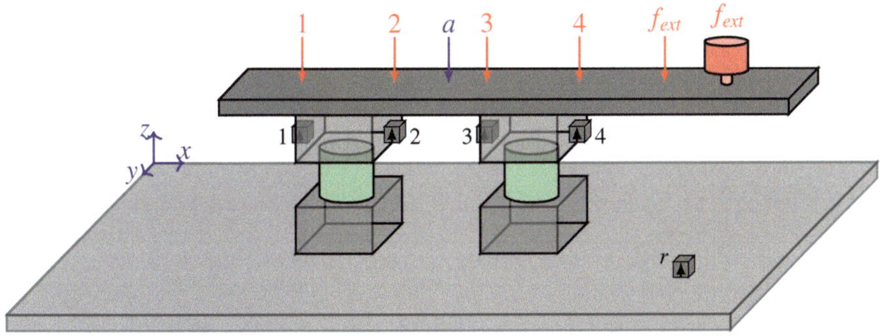

Figure 9.2 Diagram of test assembly. Two footed beam-like source, resiliently coupled to a plate-like receiver structure. Internal forces are generated by an operational shaker and external impact hammer. Artificial excitation is included to help assess the blocked force characterisation. Axes show the chosen sign convention.

Once the interface/validation points have been identified and instrumented, the next step towards implementing the in situ blocked force method is to measure the coupled assembly FRF matrix. To do this, excitations are applied one by one along the interface. In order to help assess the reliability of the blocked force, an artificial excitation is also considered; an additional force is applied to the remote source DoF a. The resulting measured accelerance matrix takes the form,

$$\hat{\mathbf{A}}_C = \begin{bmatrix} A_{11} & A_{12} & A_{13} & A_{14} & A_{1a} \\ A_{21} & A_{22} & A_{23} & A_{24} & A_{2a} \\ A_{31} & A_{32} & A_{33} & A_{34} & A_{3a} \\ A_{41} & A_{42} & A_{43} & A_{44} & A_{4a} \\ A_{r1} & A_{r2} & A_{r3} & A_{r4} & A_{ra} \end{bmatrix} \tag{9.1}$$

where $\hat{\square}$ is used to denote a measured matrix prior to any transformations being applied. The interface portion of $\hat{\mathbf{A}}_C$ (i.e. DoFs 1-4) will be used to obtain the blocked force. The transfer component between the interface and r will be used to make an on-board validation. The transfer component between the source DoF a and the interface will be used to investigate an artificial excitation.

For each excitation position three impacts were performed and the H_1 estimator used to compute the FRF. The sample rate was set at 20 kHz and the frequency resolution to 1 Hz, giving an FFT length of 20000 samples.

Once the passive measurement has been completed, the operational response of the assembly is recorded. The shaker is driven with a white noise signal and random impacts are applied to generate time varying behaviour. The operational response is measured for a period of 30 seconds. To ensure the same frequency resolution is used as in the passive test, an FFT length of 20000 samples was used. This gave 30 independent time windows across the 30 second recording. A rectangular window with no overlap was used. The acceleration vector for the nth time window takes the form,

$$\hat{\mathbf{a}}^{t_n} = \begin{pmatrix} a_1^{t_n} \\ a_2^{t_n} \\ a_3^{t_n} \\ a_4^{t_n} \\ a_r^{t_n} \end{pmatrix}. \tag{9.2}$$

In what follows we detail the main processing steps and give examples of their code implementation.

We begin by loading in the measurement data, including the accelerance, operational acceleration, and frequency vector. This particular step will vary depending on the acquisition system used, and may involve additional steps to build the required matrices from the exported data.

Code block 9.1: Load data

```
A_C = load('A_C.mat') % Coupled accelerance matrix
OppA_C = load('OppA_C.mat') % Operational acceleration matrix (
    time windowed)
f = load('f.mat') % Frequency vector
```

As we are performing all computations in the frequency domain, it is convenient to structure all measurement data as 3-dimensional arrays.

For measured FRF matrices, data is structured as in Figure 9.3. Dimensions 1 (rows) and 2 (columns) correspond, respectively, to the response and excitation positions of the accelerance matrix. Dimensions 3 (pages) represents the frequency axis. In this way, computations are performed page-by-page, with the key equations embedded within loops over frequency.

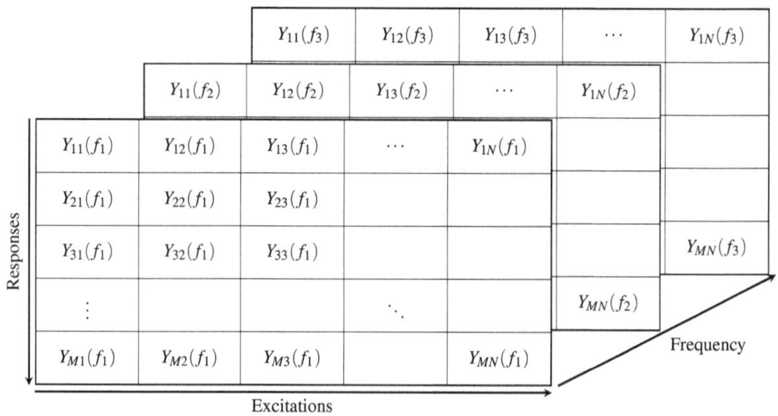

Figure 9.3 Structure of stored FRF matrix.

For the operational acceleration, the data is structured as in Figure 9.4. Dimension 1 similarly corresponds to the response position, and dimension 3 to the frequency axis. Dimension 2 however now corresponds to the time axis, i.e. each column represents the operational response over a particular time window. In this way, the standard matrix-vector multiplications used for blocked force calculations,

$$\bar{\mathbf{f}}_c = \mathbf{Y}_{Ccc}^{-1}\mathbf{v}_c \rightarrow \bar{\mathbf{f}}_c \tag{9.3}$$

can be replaced by matrix-matrix multiplications,

$$\left[\begin{array}{cccc} \bar{\mathbf{f}}_c^{t_1} & \bar{\mathbf{f}}_c^{t_2} & \cdots & \bar{\mathbf{f}}_c^{t_N} \end{array}\right] = \mathbf{Y}_{Ccc}^{-1}\left[\begin{array}{cccc} \mathbf{v}_c^{t_1} & \mathbf{v}_c^{t_2} & \cdots & \mathbf{v}_c^{t_N} \end{array}\right], \quad \bar{\mathbf{F}}_c = \mathbf{Y}_{Ccc}^{-1}\mathbf{V}_c. \tag{9.4}$$

where each column represents a different time window. This avoids the need for additional loops in the code.

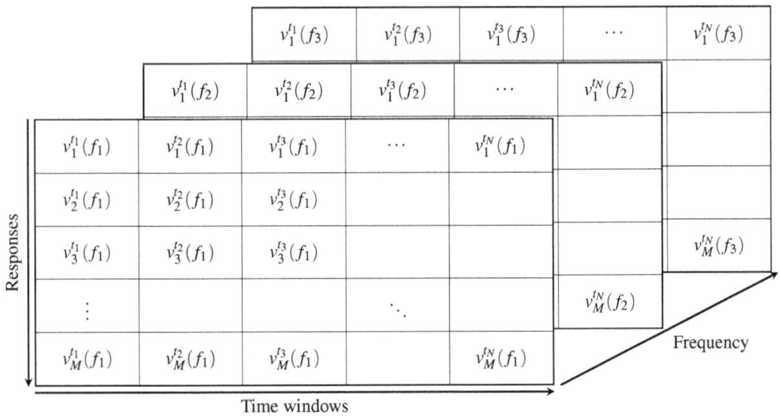

Figure 9.4 Structure of stored response matrix.

Once all data has been loaded in, it is time to perform any scaling, sign corrections and/or interface transformations.

Though not strictly necessary, as all calculations can be performed directly with accelerance and acceleration data, we will convert to mobility and velocity by frequency scaling,

$$\mathbf{Y} = \frac{1}{j\omega}\mathbf{A}, \quad \mathbf{v} = \frac{1}{j\omega}\mathbf{a}. \tag{9.5}$$

Because we have used the finite difference approximation, is also necessary to correct for force and response directions. As illustrated in Figure 9.2, all excitations were applied in the downward $-z$ direction, and all responses were orientated in the upwards $+z$ direction. The accelerance sign correction thus takes the form,

$$\hat{\mathbf{A}}_C^{\text{sc}} = \mathbf{S}_v \hat{\mathbf{A}}_C \mathbf{S}_f \tag{9.6}$$

$$\mathbf{S}_v = \begin{bmatrix} 1 & 0 & 0 & 0 & 0 \\ 0 & 1 & 0 & 0 & 0 \\ 0 & 0 & 1 & 0 & 0 \\ 0 & 0 & 0 & 1 & 0 \\ 0 & 0 & 0 & 0 & 1 \end{bmatrix}, \quad \mathbf{S}_f = \begin{bmatrix} -1 & 0 & 0 & 0 & 0 \\ 0 & -1 & 0 & 0 & 0 \\ 0 & 0 & -1 & 0 & 0 \\ 0 & 0 & 0 & -1 & 0 \\ 0 & 0 & 0 & 0 & -1 \end{bmatrix} \tag{9.7}$$

where $\square^{\text{sc}}$ denotes the sign corrected accelerance. Clearly in this case the sign corrected accelerance is simply the negative of the measured accelerance but things will not always be so simple. The operational acceleration correction is as follows,

$$\hat{\mathbf{a}}^{\text{sc}} = \mathbf{S}_v \hat{\mathbf{a}}. \tag{9.8}$$

Equations 9.5-9.8 are implemented in Code block 9.2.

Code block 9.2: Sign correction and frequency scaling

```
w = 2*pi*f; % Radian frequency vector
Sf = diag([-1 -1 -1 -1 -1]); % Force sign correction matrix
Sv = diag([1 1 1 1 1]); % Response sign correction matrix

for fi = length(f):-1:1 % Loop through frequency backwards to
    auto pre-allocate space
    Y_C(:,:,fi) = Sv*A_C(:,:,fi)*Sf./(1i*w(fi)); % Correct sign
        and convert to mobility
    opp_V(:,:,fi) = Sv*opp_A(:,:,fi)./(1i*w(fi));  % Correct sign
        and convert to velocity
end
```

Next we wish to extract the translational and rotational components by applying the finite difference approx.

$$\mathbf{Y}_C = \mathbf{T}_v \hat{\mathbf{Y}}_C \mathbf{T}_f^{\text{T}}, \quad \text{with} \quad \hat{\mathbf{Y}}_C = \begin{bmatrix} Y_{11} & Y_{12} & Y_{13} & Y_{14} & Y_{1a} \\ Y_{21} & Y_{22} & Y_{23} & Y_{24} & Y_{2a} \\ Y_{31} & Y_{32} & Y_{33} & Y_{34} & Y_{3a} \\ Y_{41} & Y_{42} & Y_{43} & Y_{44} & Y_{4a} \\ Y_{r1} & Y_{r2} & Y_{r3} & Y_{r4} & Y_{ra} \end{bmatrix} \tag{9.9}$$

In this walk through we will compare the blocked force on-board validation for the case of with (wr) and without (nr) rotations. So we build two different transformation matrices,

$$\mathbf{T}_{v,f}^{\text{wr}} = \begin{bmatrix} \frac{1}{2} & \frac{1}{2} & 0 & 0 & 0 \\ \frac{1}{2\Delta} & \frac{-1}{2\Delta} & 0 & 0 & 0 \\ 0 & 0 & \frac{1}{2} & \frac{1}{2} & 0 \\ 0 & 0 & \frac{1}{2\Delta} & \frac{-1}{2\Delta} & 0 \\ 0 & 0 & 0 & 0 & 1 \end{bmatrix}, \quad \mathbf{T}_{v,f}^{\text{nr}} = \begin{bmatrix} \frac{1}{2} & \frac{1}{2} & 0 & 0 & 0 \\ 0 & 0 & \frac{1}{2} & \frac{1}{2} & 0 \\ 0 & 0 & 0 & 0 & 1 \end{bmatrix} \tag{9.10}$$

where Δ is the sensor/force spacing (made the same in this example). The resulting FRF matrices take the form,

$$\mathbf{Y}_C^{\text{wr}} = \begin{bmatrix} Y_{Cc_{z1}c_{z1}} & Y_{Cc_{z1}c_{\alpha1}} & Y_{Cc_{z1}c_{z2}} & Y_{Cc_{z1}c_{\alpha2}} & Y_{Cc_{z1}a} \\ Y_{Cc_{\alpha1}c_{z1}} & Y_{Cc_{\alpha1}c_{\alpha1}} & Y_{Cc_{\alpha1}c_{z2}} & Y_{Cc_{\alpha1}c_{\alpha2}} & Y_{Cc_{\alpha1}a} \\ Y_{Cc_{z2}c_{z1}} & Y_{Cc_{z2}c_{\alpha1}} & Y_{Cc_{z2}c_{z2}} & Y_{Cc_{z2}c_{\alpha2}} & Y_{Cc_{z2}a} \\ Y_{Cc_{\alpha2}c_{z1}} & Y_{Cc_{\alpha2}c_{\alpha1}} & Y_{Cc_{\alpha2}c_{z2}} & Y_{Cc_{\alpha2}c_{\alpha2}} & Y_{Cc_{\alpha2}a} \\ Y_{rc_{z1}} & Y_{rc_{\alpha1}} & Y_{rc_{z2}} & Y_{rc_{\alpha2}} & Y_{ra} \end{bmatrix} \tag{9.11}$$

$$\mathbf{Y}_C^{\text{nr}} = \begin{bmatrix} Y_{Cc_{z1}c_{z1}} & Y_{Cc_{z1}c_{z2}} & Y_{Cc_{z1}a} \\ Y_{Cc_{z2}c_{z1}} & Y_{Cc_{z2}c_{z2}} & Y_{Cc_{z2}a} \\ Y_{rc_{z1}} & Y_{rc_{z2}} & Y_{ra} \end{bmatrix} \tag{9.12}$$

where z_n and α_n denote the translational and rotational DoFs at the nth connection. Note that it is also necessary to transform the operational responses so that they are consistent with the FRF matrices,

$$\mathbf{v}_C^{\text{wr}} = \mathbf{T}_{v,f}^{\text{wr}}\hat{\mathbf{v}}_C, \quad \mathbf{v}_C^{\text{nr}} = \mathbf{T}_{v,f}^{\text{nr}}\hat{\mathbf{v}}_C. \tag{9.13}$$

Equations 9.9-9.13 are implemented in Code block 9.3.

Code block 9.3: Apply finite difference transformation with and without rotations

```
D = 0.05; % Spacing between sensors/excitations
Twr = [0.5      0.5     0       0      0;
       0.5/D   -0.5/D   0       0      0;
       0        0       0.5     0.5    0;
       0        0       0.5/D  -0.5/D  0;
       0        0       0       0      1]; % FD transformation with rotations
Tnr = [0.5 0.5 0   0   0;
       0   0   0.5 0.5 0;
       0   0   0   0   1]; % FD transformation without rotations
for fi = length(f):-1:1
    Y_wr(:,:,fi) = Twr*Y_C(:,:,fi)*Twr.'; % Apply FD to mobility
    Y_nr(:,:,fi) = Tnr*Y_C(:,:,fi)*Tnr.'; % Apply FD to mobility

    oppV_wr(:,:,fi) = Twr*oppV(:,:,fi); % Apply FD to velocity
    oppV_nr(:,:,fi) = Tnr*oppV(:,:,fi); % Apply FD to velocity
end
```

Now that all measurement data has been scaled and transformed appropriately, the in situ blocked force method can be implemented. Before proceeding with the operational case, the artificial excitation is considered. For this, the operational response is replaced by a response normalised to its generating impact force, i.e. $\mathbf{v}_c \rightarrow \mathbf{Y}_{Cca}$, and the inverse blocked force equation becomes,

$$\bar{\mathbf{f}}_c^a = \mathbf{Y}_{Ccc}^{-1}\mathbf{Y}_{Cca} \tag{9.14}$$

where $\bar{\mathbf{f}}_c^a = \bar{\mathbf{f}}_c/1\text{N}$ represents the (dimensionless) blocking force due to the (1N) artificial excitation. Once computed, the blocked force is used as part of an on-board validation according to,

$$\mathbf{Y}_{Cra} = \mathbf{Y}_{Crc}\bar{\mathbf{f}}_c^a. \tag{9.15}$$

Equations 9.14-9.15 are implemented with and without rotations in Code block 9.4.

Code block 9.4: Calculate blocked force for artificial excitation and perform on-board validation

```
for fi = length(f):-1:1
    % With rotations:
    fBLK_a_wr(:,fi) = inv(Y_wr(1:4,1:4,fi))*Y_wr(1:4,5,fi); % Calculate blocked force
    obVal_a_wr(:,fi) = Y_wr(5,1:4,fi)*fBLK_a_wr(:,fi); % On-board validation prediction
    % Without rotations:
    fBLK_a_nr(:,fi) = inv(Y_nr(1:2,1:2,fi))*Y_nr(1:2,3,fi); % Calculate blocked force
    obVal_a_nr(:,fi) = Y_nr(3,1:2,fi)*fBLK_a_nr(:,fi); % On-board validation prediction
end

figure
loglog(f,squeeze(abs(Y_wr(5,5,:))))
hold on
loglog(f,squeeze(abs(obVal_a_wr)))
loglog(f,squeeze(abs(obVal_a_nr)))
xlabel('Frequency (Hz)')
ylabel('Mobility (m/Ns)')
xlim([10 2000])
legend('Meas', 'Pred - wr', 'Pred - nr')
```

Shown in Figure 9.5 are the results produced by Code block 9.4. It shows the directly measured response at the remote receiver DoF due to the artificial excitation (i.e. Y_{Cra}) compared against the response predictions made using the blocked force, with (red) and without (blue) rotations included.

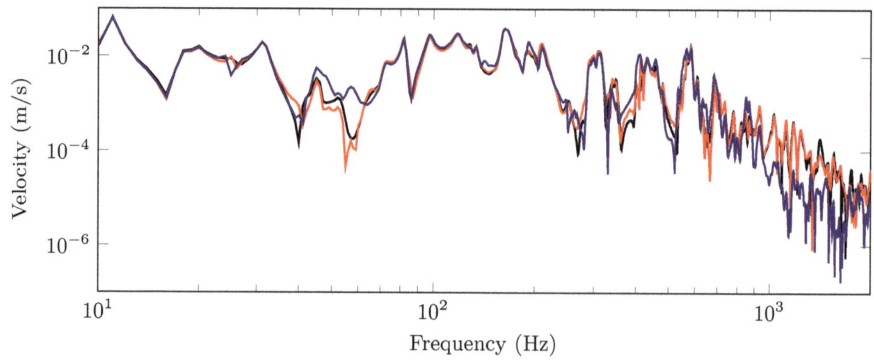

Figure 9.5 On-board validation results for blocked forces due to artificial excitation; measured response (black), prediction with rotations (red), prediction without rotations (blue).

Good agreement is obtained across the entire frequency range, particularly for the case including rotations. It can be seen that by neglecting rotations, the high frequency (above approx. 900 Hz) accuracy is reduced. Nevertheless, good agreement is retained up to 900 Hz for the translational only case.

Having completed the artificial excitation investigation, we move on to consider the operational blocked force. The key equations are recalled as,

$$\bar{\mathbf{f}}_c = \mathbf{Y}_{Ccc}^{-1}\mathbf{v}_c \tag{9.16}$$

and for the on-board validation,

$$\mathbf{v}_r = \mathbf{Y}_{Crc}\bar{\mathbf{f}}_c. \tag{9.17}$$

Equations 9.16-9.17 are implemented in Code block 9.5.

Code block 9.5: Calculate blocked force for operational excitation and perform on-board validation

```
for fi = length(f):-1:1
    % With rotations:
    fBLK_opp_wr(:,fi) = pinv(Y_wr(1:4,1:4,fi))*oppV_wr(1:4,:,fi); % Calculate blocked force
    obVal_opp_wr(:,:,fi) = Y_wr(5,1:4,fi)*fBLK_opp_wr(:,:,fi); % On-board validation
    % Without rotations:
    fBLK_opp_nr(:,:,fi) = pinv(Y_nr(1:2,1:2,fi))*oppV_nr(1:2,:,fi); % Calculate blocked force
    obVal_opp_nr(:,:,fi) = Y_nr(3,1:2,fi)*fBLK_opp_nr(:,:,fi);  % On-board validation
end

figure
loglog(f,squeeze(sqrt((mean(oppV_wr(5,:,:).*conj(oppV_wr(5,:,:)),2)))))
hold on
loglog(f,squeeze(sqrt((mean(obVal_opp_wr.*conj(obVal_opp_wr),2)))))
loglog(f,squeeze(sqrt((mean(obVal_opp_nr.*conj(obVal_opp_nr),2)))))
xlabel('Frequency (Hz)')
ylabel('Velocity (m/s)')
xlim([10 2000])
legend('Meas', 'Pred - wr', 'Pred - nr')
```

Shown in Figure 9.6 are the results produced by Code block 9.5. To plot the results of the on-board validation, we compute the square rooted time averaged auto-spectrum for the directly measured velocity, and those predicted using the blocked force with and without rotations. It shows the directly measured response at the remote receiver DoF due to an operational excitation (i.e. shaker driven by pink noise with additional external excitations) compared with the response predictions made using the blocked force, with (red) and without (blue) rotations included.

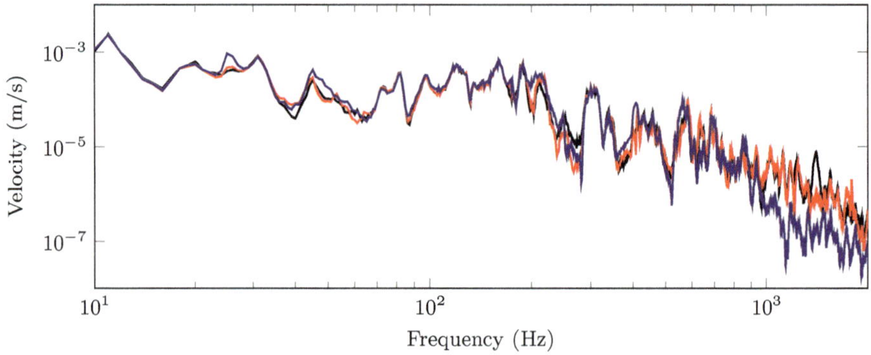

Figure 9.6 On-board validation results for blocked forces due to operational excitation; measured response (black), prediction with rotations (red), prediction without rotations (blue).

Like the artificial excitation results, good agreement is obtained across the whole frequency range, with rotations becoming increasingly important at high frequencies.

Note that the on-board validation result of Figure 9.6 is essentially an in situ TPA. As discussed in Chapter 3, it is possible to split the blocked force up and investigate the contribution of individual forces or groups thereof. To illustrate this, Code block 9.6 predicts the contribution of each foot (inc. translation and rotation) separately.

Code block 9.6: Calculate blocked force for operational excitation and perform on-board validation

```
for fi = length(f):-1:1
    obVal_tran(:,:,fi) = Y_wr(5,[1 2],fi)*fBLK_opp_wr([1 2],:,fi)
        ; % Foot 1
    obVal_rot(:,:,fi) = Y_wr(5,[3 4],fi)*fBLK_opp_wr([3 4],:,fi);
        % Foot 2
end

figure
loglog(f,squeeze(sqrt((mean(oppV_wr(5,:,:).*conj(oppV_wr(5,:,:))
    ,2)))))
hold on
loglog(f,squeeze(sqrt((mean(obVal_tran.*conj(obVal_tran),2)))))
loglog(f,squeeze(sqrt((mean(obVal_rot.*conj(obVal_rot),2)))))
loglog(f,squeeze(sqrt((mean(obVal_a_wr.*conj(obVal_a_wr),2)))))
xlabel('Frequency (Hz)')
ylabel('Velocity (m/s)')
xlim([10 2000])
legend('Meas', 'Trans', 'Rot', 'Tot')
```

Shown in Figure 9.7 are the results produced by Code block 9.6. It shows the individual contribution of each foot towards the total response. From it we can see that over certain frequency ranges, one foot dominates over the other. In a practical context, this information can be used to help design mitigation strategies for troublesome contributions.

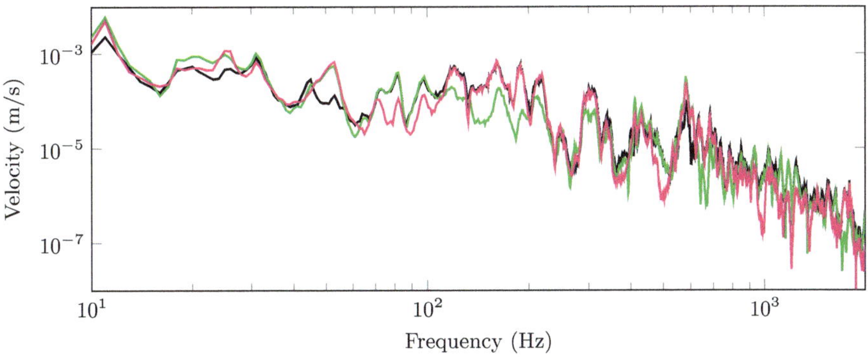

Figure 9.7 Contribution analysis using blocked forces; total response prediction (black), contribution of foot 1 inc. rotations (green), contribution of foot 2 inc. rotations (magenta).

Based on the results of the blocked force validation above, moving forward we will neglect the rotational DoFs. This will simplify the construction of the VAVP, and should provide reasonable results up to approx 900 Hz.

9.3.2 FREE SOURCE MOBILITY

Having obtained the blocked force, it is necessary to measure the free-interface FRF of the source so as to complete the source characterisation. To do so, the source was placed on two foam pads, located at either end of the beam so as to limit their effect on the interface dynamics. The source instrumentation from the in situ blocked force characterisation was retained, and forces were similarly applied along the interface, as illustrated in Figure 9.8.

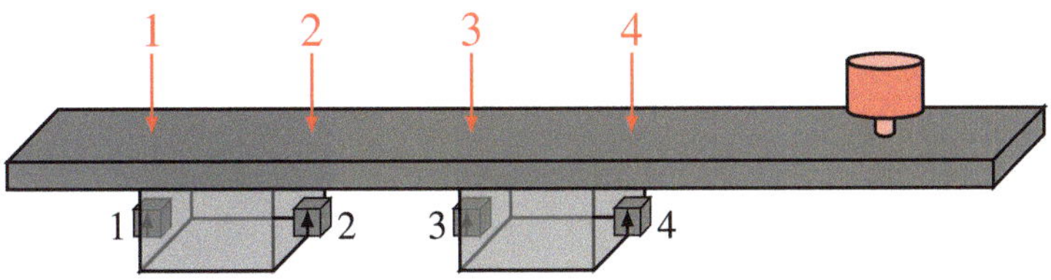

Figure 9.8 Diagram of free source.

Supported by the results of the previous on-board validation, only the translational z DoFs will be retained. The finite difference transformation matrix thus takes the form,

$$\mathbf{T}^{\mathrm{nr}}_{v,f} = \begin{bmatrix} \frac{1}{2} & \frac{1}{2} & 0 & 0 \\ 0 & 0 & \frac{1}{2} & \frac{1}{2} \end{bmatrix} \tag{9.18}$$

which yields the free-interface source FRF,

$$\mathbf{Y}_S^{\mathrm{nr}} = \begin{bmatrix} Y_{Sc_{z1}c_{z1}} & Y_{Sc_{z1}c_{z2}} \\ Y_{Sc_{z2}c_{z1}} & Y_{Sc_{z2}c_{z2}} \end{bmatrix}. \tag{9.19}$$

The necessary frequency scaling, sign corrections and transformations are implemented in Code block 9.7.

Code block 9.7: Apply sign correction, frequency scaling and finite difference approx.

```
A_S = load('A_S.mat') % Load Source accelerance data
Sf = diag([-1 -1 -1 -1]); % Force sign correction matrix
Sv = diag([1 1 1 1]); % Response sign correction matrix
Tnr = [0.5 0.5 0    0   ;
       0   0   0.5  0.5]; % FD transformation without rotations

for fi = length(f):-1:1
    Y_S(:,:,fi) = Sv*A_S(:,:,fi)*Sf./(1i*w(fi)); % Correct sign
        and convert to mobility
    Y_S_nr(:,:,fi) = Tnr*Y_S(:,:,fi)*Tnr.'; % Apply FD
end
```

9.4 RECEIVER CHARACTERISATION

Having completed the source characterisation, we must now characterise the remaining components, starting with the receiver structure. The accelerometers mounted to the source interface are detached and adhered to the receiver interface, as illustrated in Figure 9.9.

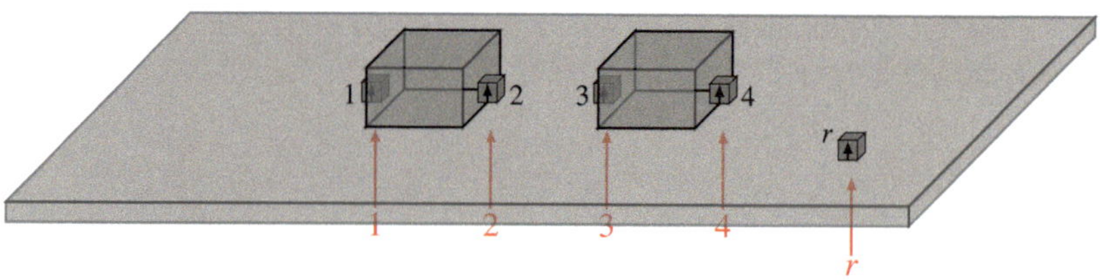

Figure 9.9 Diagram of free source.

Note that the remote target sensor remains attached from the blocked force characterisation. The receiver structure is mounted in the same way as in the initial coupled assembly measurement, with a foam pad resting under each corner of the plate. The free accelerance was then measured by applying forces along the interface, as well as at the target sensor location. As with the free source mobility, only the translational z DoFs are retained during the finite difference approximation, with

$$\mathbf{T}_{v,f}^{\mathrm{nr}} = \begin{bmatrix} \frac{1}{2} & \frac{1}{2} & 0 & 0 & 0 \\ 0 & 0 & \frac{1}{2} & \frac{1}{2} & 0 \\ 0 & 0 & 0 & 0 & 1 \end{bmatrix} \tag{9.20}$$

which yields the free-interface receiver FRF,

$$\mathbf{Y}_R^{\mathrm{nr}} = \begin{bmatrix} Y_{Rc_{z1}c_{z1}} & Y_{Rc_{z1}c_{z2}} & Y_{Rc_{z1}c_r} \\ Y_{Rc_{z2}c_{z1}} & Y_{Rc_{z2}c_{z2}} & Y_{Rc_{z2}c_r} \\ Y_{Rc_rc_{z1}} & Y_{Rc_rc_{z2}} & Y_{Rc_rc_r} \end{bmatrix}. \tag{9.21}$$

The necessary frequency scaling, sign corrections and transformations are implemented in Code block 9.8.

Code block 9.8: Apply sign correction, frequency scaling and finite difference approx.

```
A_R = load('A_R.mat') % Load receiver accelerance data
w = 2*pi*f; % Radian frequency vector
Sf = diag([1 1 1 1]); % Force sign correction matrix
Sv = diag([1 1 1 1]); % Response sign correction matrix
Tnr = [0.5 0.5 0    0   0;
       0   0   0.5  0.5 0;
       0   0   0    0   1]; % FD transformation without rotations

for fi = length(f):-1:1
    Y_R(:,:,fi) = Sv*A_R(:,:,fi)*Sf./(1i*w(fi)); % Correct sign
        and convert to mobility
    Y_R_nr(:,:,fi) = Tnr*Y_R(:,:,fi)*Tnr.'; % Apply FD
end
```

9.5 ISOLATOR CHARACTERISATION

Unlike the source and receiver components, which can easily be characterised by their free-interface FRFs, the vibration isolators cannot be measured freely suspended. They must be installed within an assembly to provide some means of instrumentation. Independent isolator properties (i.e. transfer impedance) must then be extracted from the assembly measurements. This is done using the in situ method, as detailed in Chapter 2 (2.3.2).

A single isolator is installed between two small masses and instrumented as illustrated in Figure 9.10. The assembly accelerance matrix is measured by applying forces above/below each spaced sensor pair. In keeping with the source and receiver characterisation, the finite difference transformation,

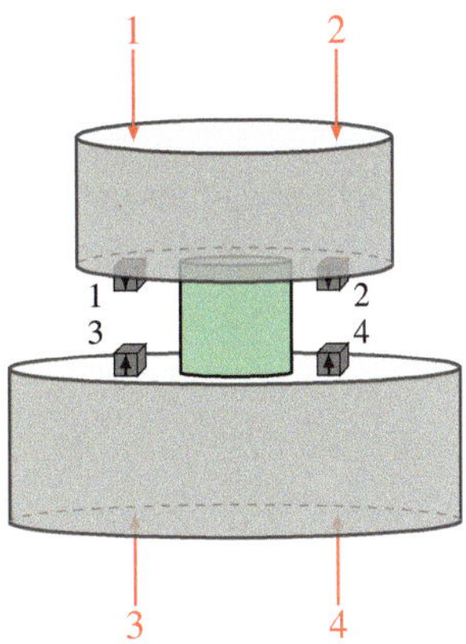

Figure 9.10 Diagram of mass-isolator-mass assembly used for isolator characterisation.

$$\mathbf{T}_{v,f}^{\mathrm{nr}} = \begin{bmatrix} \frac{1}{2} & \frac{1}{2} & 0 & 0 \\ 0 & 0 & \frac{1}{2} & \frac{1}{2} \end{bmatrix} \tag{9.22}$$

retains only the translational DoFs above and below the isolator. In combination with frequency scaling and sign correction, we obtain the coupled interface FRF,

$$\mathbf{Y}_C^{\mathrm{nr}} = \begin{bmatrix} Y_{Cc_{z1}c_{z1}} & Y_{Cc_{z1}c_{z2}} \\ Y_{Cc_{z2}c_{z1}} & Y_{Cc_{z2}c_{z2}} \end{bmatrix}. \tag{9.23}$$

Frequency scaling, sign correction and finite difference transformations are implemented in Code block 9.9.

Code block 9.9: Apply sign correction, frequency scaling and finite difference approx.

```
A_MIM = load('A_MIM.mat') % Load Mass-Isolator-Mass accelerance data
Sf = diag([-1 -1 1 1]); % Force sign correction matrix
Sv = diag([-1 -1 1 1]); % Response sign correction matrix
Tnr = [0.5 0.5 0    0   ;
       0   0   0.5  0.5]; % FD transformation without rotations

for fi = length(f):-1:1
    Y_MIM(:,:,fi) = Sv*A_MIM(:,:,fi)*Sf./(1i*w(fi)); % Correct sign and convert to mobility
    Y_MIM_nr(:,:,fi) = Tnr*Y_MIM(:,:,fi)*Tnr.'; % Apply FD
end
```

Inversion of the coupled FRF yields the impedance matrix,

$$\mathbf{Z}_C^{\mathrm{nr}} = \begin{bmatrix} Z_{AIc_{z1}c_{z1}} & Z_{Ic_{z1}c_{z2}} \\ Z_{Ic_{z2}c_{z1}} & Z_{IBc_{z2}c_{z2}} \end{bmatrix} \tag{9.24}$$

of which the off-diagonal elements represent the independent transfer impedances of the isolator.

Extracting one of the off-diagonal transfer impedances, and making the assumption of force conservation across the element (this is a fair assumption given approx 800 Hz frequency limit considered here), a complete isolator impedance matrix can be built as,

$$\mathbf{Z}_I^{\mathrm{nr}} \approx \begin{bmatrix} -Z_{Ic_{z1}c_{z2}} & Z_{Ic_{z1}c_{z2}} \\ Z_{Ic_{z1}c_{z2}} & -Z_{Ic_{z1}c_{z2}} \end{bmatrix}. \tag{9.25}$$

While expecting the measured data, a poor result was noted at low frequency, due to small errors in reciprocity. With vibration isolators exhibiting simple dynamics in the low frequency range, a simple linear interpolation was therefore implemented to provide a more reliable low frequency transfer impedance.

The transfer impedance extraction, interpolation, and construction of $\mathbf{Z}_I^{\mathrm{nr}}$ is implemented in Code block 9.10.

Code block 9.10: Implement in situ characterisation and build isolator impedance matrix

```
for fi = length(f):-1:1
    Z_MIM(:,:,fi) = inv(Y_MIM(:,:,fi)); % Invert to obtained coupled impedance matrix
end
% Fit linear models for LF extrapolation
Zintep_R = fit(f(200:1000), real(squeeze(Z_MIM(2,1,200:1000))),'poly1');
Zintep_I = fit(f(200:1000), imag(squeeze(Z_MIM(2,1,200:1000))),'poly1');
Zintep(1,1,:) = Zintep_R(f(1:300))+1i.*Zintep_I(f(1:300));

% Extract transfer impedance and build full impedance matrix assuming force conservation:
Z_I_high = [-Z_MIM(2,1,300:end),  Z_MIM(2,1,300:end);
             Z_MIM(2,1,300:end), -Z_MIM(2,1,300:end)]; % High freq. impedance.
Z_I_low = [-Zintep(1,1,1:299),  Zintep(1,1,1:299);
            Zintep(1,1,1:299), -Zintep(1,1,1:299)]; % Low freq. impedance
Z_I = cat(3,Z_I_low,Z_I_high); % Combine low and high.

figure
loglog(f,abs(squeeze(Z_MIM(2,1,:))))
hold on
loglog(f(1:300),abs(squeeze(Zintep(1:300))))
xlabel('Frequency (Hz)')
ylabel('Impedance (Ns/m)')
xlim([10 2000])
```

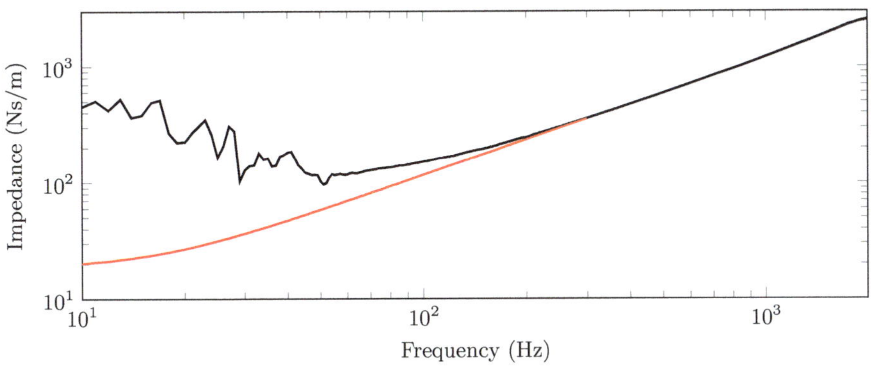

Figure 9.11 Isolator dynamic transfer impedance obtained using in situ method; measured transfer impedance (black) extrapolated transfer impedance (red).

Shown in Figure 9.11 is the isolator transfer impedance obtained by the in situ method, as well as the low frequency linear interpolation, used to take over from the experimental result below 300 Hz.

9.6 SUBSTRUCTURING

Having characterised all of the components within our assembly, we are ready to build a passive model of the assembled structure. In this example we will use the primal substructuring formulation.

The first step in implementing the primal substructuring formulation is to build the block diagonal matrix that contains the impedance matrix of each component. For the example considered here this takes the form,

$$\mathbf{Z}=\begin{bmatrix} \begin{bmatrix} Y_{Sc_1c_1} & Y_{Sc_1c_2} \\ Y_{Sc_2c_1} & Y_{Sc_2c_2} \end{bmatrix}^{-1} & & & \\ & \begin{bmatrix} -Z_{Ic_1c_2} & Z_{Ic_1c_2} \\ Z_{Ic_1c_2} & -Z_{Ic_1c_2} \end{bmatrix} & & \\ & & \begin{bmatrix} -Z_{Ic_1c_2} & Z_{Ic_1c_2} \\ Z_{Ic_1c_2} & -Z_{Ic_1c_2} \end{bmatrix} & \\ & & & \begin{bmatrix} Y_{Rc_1c_1} & Y_{Rc_1c_2} & Y_{Rc_1c_r} \\ Y_{Rc_2c_1} & Y_{Rc_2c_2} & Y_{Rc_2c_r} \\ Y_{Rc_rc_1} & Y_{Rc_rc_2} & Y_{Rc_rc_r} \end{bmatrix}^{-1} \end{bmatrix} \tag{9.26}$$

where, for convenience, we have omitted z from the subscripts.

From this uncoupled impedance matrix, we can construct the Boolean matrix $\mathbf{L}$ that satisfies continuity between the connecting DoFs,

$$\mathbf{L}^{\mathrm{T}}=\begin{bmatrix} 1 & 0 & 1 & 0 & 0 & 0 & 0 & 0 & 0 \\ 0 & 1 & 0 & 0 & 1 & 0 & 0 & 0 & 0 \\ 0 & 0 & 0 & 1 & 0 & 0 & 1 & 0 & 0 \\ 0 & 0 & 0 & 0 & 0 & 1 & 0 & 1 & 0 \\ 0 & 0 & 0 & 0 & 0 & 0 & 0 & 0 & 1 \end{bmatrix}. \tag{9.27}$$

The primal substructuring is then completed using the equation,

$$\mathbf{Y}_C^{\mathrm{ss}} = \left(\mathbf{L}^{\mathrm{T}}\mathbf{Z}\mathbf{L}\right)^{-1} \tag{9.28}$$

where $\square^{\mathrm{ss}}$ denotes the substructured prediction. The resulting FRF matrix takes the form,

$$\mathbf{Y}_C^{\mathrm{ss}}=\begin{bmatrix} Y_{Cc_1c_1} & Y_{Cc_1c_2} & Y_{Cc_1c_3} & Y_{Cc_1c_4} & Y_{Cc_1r} \\ Y_{Cc_2c_1} & Y_{Cc_2c_2} & Y_{Cc_2c_3} & Y_{Cc_2c_4} & Y_{Cc_2r} \\ Y_{Cc_3c_1} & Y_{Cc_3c_2} & Y_{Cc_3c_3} & Y_{Cc_3c_4} & Y_{Cc_3r} \\ Y_{Cc_4c_1} & Y_{Cc_4c_2} & Y_{Cc_4c_3} & Y_{Cc_4c_4} & Y_{Cc_4r} \\ Y_{Crc_1} & Y_{Crc_2} & Y_{Crc_3} & Y_{Crc_4} & Y_{Crr} \end{bmatrix} \tag{9.29}$$

where c_1 and c_2 represent the source side interface DoFs, c_3 and c_4 the receiver side interface DoFs, and r the remote target receiver DoF.

Equations 9.26-9.28 are implemented in Code block 9.11.

Code block 9.11: Implement primal substructuring

```
L = [1 0 1 0 0 0 0 0 0;
     0 1 0 0 1 0 0 0 0;
     0 0 0 1 0 0 1 0 0;
     0 0 0 0 0 1 0 1 0;
     0 0 0 0 0 0 0 0 1]; % Boolean coupling matrix

for fi = length(f):-1:1
    % Build block diagonal impedance matrix:
    Z(:,:,fi) = blkdiag(inv(Y_S_nr(:,:,fi)), Z_I(:,:,fi), Z_I
        (:,:,fi), inv(Y_R_nr(:,:,fi)));
    Y_ss(:,:,fi) = pinv(L*Z(:,:,fi)*L.'); % Implement primal
        substruturing
end

figure
subplot(2,1,1)
loglog(f,squeeze(abs(Y_nr(3,1,:))))
hold on
loglog(f,squeeze(abs(Y_ss(5,1,:))))
xlim([10 2000])
xlabel('Frequency (Hz)')
ylabel('Mobility (m/Ns)')
subplot(2,1,2)
loglog(f,squeeze(abs(Y_nr(3,2,:))))
hold on
loglog(f,squeeze(abs(Y_ss(5,2,:))))
xlim([10 2000])
xlabel('Frequency (Hz)')
```

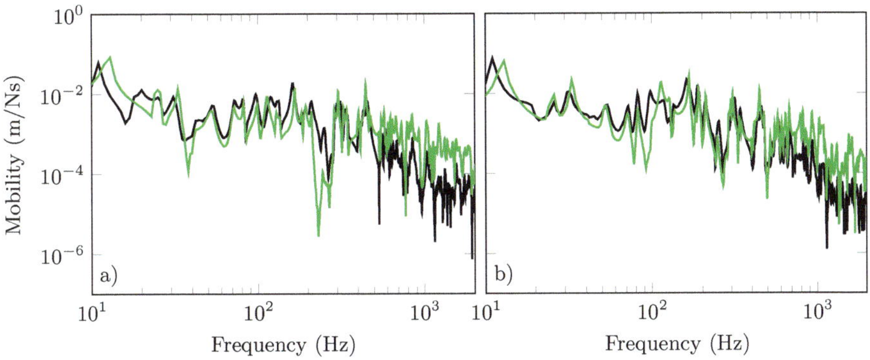

Figure 9.12 Coupled transfer FRFs between interface (a – Y_{Crc_1}, b – Y_{Crc_2}) and target DoF; direct measurement (black), substructured prediction (green).

Shown in Figure 9.12 are the directly measured and substructured predictions for the coupled transfer mobility between the two interface DoFs, and the remote target

DoF. Note that these predictions are made using the characterisation of a single nominally identical isolator (not the specific ones used in the physical assembly). Despite using a nominal isolator, good agreement is obtained across the low-mid frequency frequency range, with an over estimation occurring above approx 800 Hz. This is likely due to the neglect of rotational DoFs and the breakdown of force conservation assumption. Given that our blocked force on-board validation suggested reliable estimates only below 800 Hz, this is not considered a problem.

9.7 VAVP PREDICTION

Having built a model for the passive assembly, the final step in constructing a VAVP is to apply the obtained blocked force so to predict an operational response. To do this, we first embed the blocked force within a zero vector such that its dimensions match those of the coupled FRF,

$$\bar{\mathbf{f}} = \begin{pmatrix} \bar{\mathbf{f}}_c \\ 0 \\ 0 \\ 0 \end{pmatrix} \tag{9.30}$$

before multiplying with the substructured FRF,

$$\mathbf{v}^{SS} = \mathbf{Y}_C^{SS}\bar{\mathbf{f}}. \tag{9.31}$$

The resulting $\mathbf{v}^{SS}$ is the predicted operational response across all the structure's DoFs. Recalling that the initial operational response measured during the blocked force characterisation was structured as in Figure 9.4, the predicted response $\mathbf{v}^{SS}$ similarly contains predictions for each time window across its columns. Equations 9.30–9.31 are implemented in Code block 9.13.

Code block 9.12: Combine blocked force and substructured FRFs.

```
fBLK_opp_nr_zeros = [fBLK_opp_nr; zeros(3,1,length(f)]
for fi = length(f):-1:1
    v_ss(:,:,fi) = Y_ss(:,:,fi)*fBLK_opp_nr_zeros(:,:,fi); % Apply blocked force to substructured mobility
end

figure
loglog(f,squeeze(sqrt((mean(oppV_wr(5,:,:).*conj(oppV_wr(5,:,:)),2)))))
hold on
loglog(f,squeeze(sqrt((mean(obVal_opp_nr.*conj(obVal_opp_nr),2)))))
loglog(f,squeeze(sqrt((mean(v_ss.*conj(v_ss),2)))))
xlabel('Frequency (Hz)')
ylabel('Velocity (m/s)')
```

```
xlim([10 2000])

figure
subplot(2,1,1)
pcolor(time,freq,squeeze(20*log10(abs(oppV_wr(5,:,:)))).');
ylim([10 2000])
set(gca,'YScale','log')
ylabel('Frequency (Hz)')
subplot(2,1,2)
pcolor(time,freq,squeeze(20*log10(abs(VAP_pred(1,:,:)))).');
set(gca,'YScale','log')
ylim([10 2000])
xlabel('Time (s)')
ylabel('Frequency (Hz)')
```

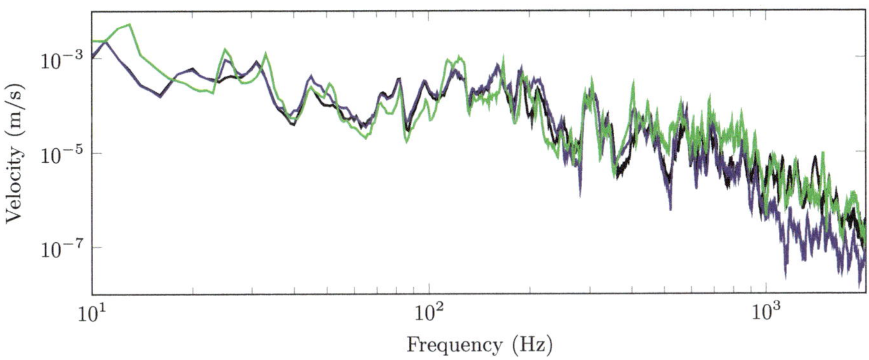

Figure 9.13 Time averaged VAVP response prediction; direct measurement (black), prediction with measured (blue) and substructured (green) FRF.

Shown in Figure 9.13 are the time averaged operationally response predictions for the assembly using both directly measured (blue) and substructured (green) FRFs. Note that the prediction using measured FRFs is identical to the on-board validation result (without rotations) presented in Figure 9.6. The response prediction with substructured FRFs represents our final VAVP prediction. We see good agreement across the entire frequency range. However, the agreement above 800 Hz is likely a coincidence – the under estimation of the blocked force is being compensated by the over estimation of the substructured transfer FRFs in this case.

Shown in Figure 9.14 are the target DoF time-frequency spectrograms of the directly measured operational assembly (a) and the VAVP prediction (b). It can be seen that by performing blocked force and response prediction calculations repeatedly for each time, we are able to retain the non-stationary time structure of the operational source.

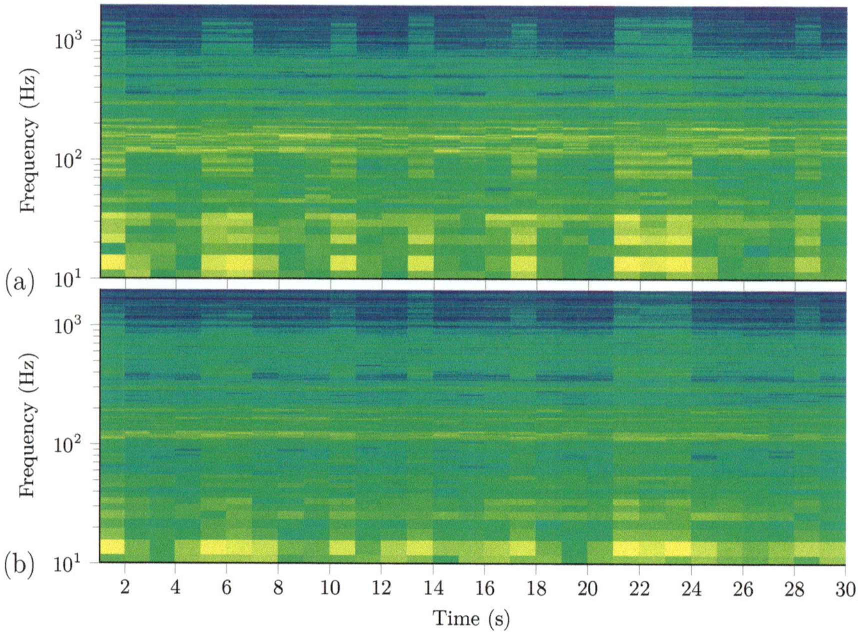

Figure 9.14 Spectrogram prediction of operational response at the target DoF; a) directly measured, b) VAVP prediction with substructured FRF.

Code block 9.13: Convert predicted response back to time domain.

```
fs = 20000 % Sample rate
N = 20000 % Window length
for fi = length(f):-1:1
    a_meas(:,:,fi) = oppV_wr(:,:,fi).*1i*w(fi); % Convert measured V to acceleration
    a_ss(:,:,fi) = v_ss(:,:,fi).*1i*w(fi); % Convert predicted V to acceleration
end

for t = 1:30 % Loop over each time window
    % IFFT freq domain signals and scale accordingly:
    OppVt(:,t) = N.*ifft([0;squeeze(oppA_wr(5,t,2:end));
                                            conj(flip(squeeze(oppV_wr(5,t,2:end))))])/2;
    OppVpt(:,t) = N.*ifft([0;squeeze(a_ss(5,t,2:end));
                                            conj(flip(squeeze(v_ss(5,t,2:end))))])/2;
end

[b,a] = butter(8,800/fs/2,'low'); % Create low pass filter with 800 cut off
OppVt_filt = filter(b,a,OppVt(:)); % Apply filter to (vectorised) measured signal
```

```
OppVpt_filt = filter(b,a,OppVpt(:)); % Apply filter to (
    vectorised) predicted signal

time = 0:1/fs:1/fs*(length(OppVt_filt)-1); % Create time vector

figure
plot(time,OppVpt_filt)
hold on
plot(time,OppVt_filt)
legend('Pred','Meas')
xlabel('Time (s)')
ylabel('Acceleration (m/s)')
xlim([0 30])
```

As a final step, the individual time windows are scaled back to acceleration, inverse Fourier transformed and recombined to yield the time domain response (see 7.5). To enable fair comparison against the directly measured acceleration signal, both measured and predicted signals are filtered with an 8th order low pass Butterworth filter with a cut off frequency of 800 Hz. Note that the DC (0 Hz) component of the two-sided spectrum (required for the inverse FFT), is replaced by a zero to avoid 'divide by zero' errors when scaling to mobility.

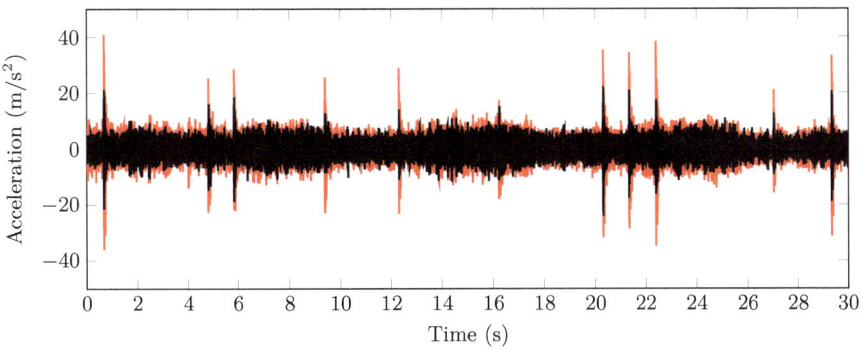

Figure 9.15 Reconstructed time domain response at the remote target location: red – predicted, black – directly measured.

Shown in Figure 9.15 is the time domain operational response prediction compared with the directly measured response. It is clear that the non-stationary characteristics of the vibration source are captured and correctly predicted.

In summary, Figures 9.13- 9.15 are predictions of the operational response of the structure illustrated in Figures 9.1 and 9.2, due to complex, time-varying forces acting internally to its vibration source. This was achieved by characterising independently the assembly's constituent components and recombining them mathematically to yield a model of the assembled structure in operation. Though a simple assembly, and an even more simplified model, the above example demonstrates many of the important steps required to build a VAVP.

9.8 SUMMARY

The chapter has presented a walk-through example of the construction of a VAVP drawing on the principles and techniques described throughout the book

Part 3: Case Studies

10 Invited Industrial Case Studies

10.1 INTRODUCTION

Having introduced theory and concepts in Part 1 of the book and considered their practical implementation in Part 2, we now move on to Part 3 in which we aim to offer a view of the current and potential future state-of-the-art. In this chapter, we present real-life examples of the principles in operation through a series of industrial case studies. In Chapter 11, we will summarise the main findings of the book and, by way of conclusion, attempt to establish their place within a broader context of industrial and academic practice.

The case studies presented in this chapter are drawn from a range of industries and involve sources that vary significantly in every respect, in size, power, material and their source mechanisms which include a range mechanical excitation types as well as electrodynamic and fluid-dynamic forces. These sources are both rigidly and resiliently mounted to a range of light and heavyweight support structures with significantly different dynamic properties. Despite all this diversity, the component-based approach is applied in essentially the same way in every case: the common elements of the analysis are (a) *in situ* characterisation of blocked forces (b) characterisation of components or an assembly with FRFs and (c) prediction of response in an assembly by combining blocked forces with FRFs. One aim of presenting this selection therefore is to demonstrate the *flexibility* of the component-based framework across a broad range of applications.

There *are* differences in the case studies of course, particularly in the manner of obtaining the FRFs used for prediction of the output: in some cases the assembly FRFs are measured directly (as in *in situ* TPA), in another they are computed from the separate component properties by *substructuring* and in another from a *hybrid* of measured and numerically modelled FRFs. The examples therefore span a range of uses from diagnostics within a single assembly, to prediction in new assemblies, to prediction in not-yet-existing assemblies which takes us firmly into the realm of virtual prototyping. Again, the fact that the same method is employed across this diversity of applications shows the power and flexibility of the component-based paradigm. Moreover, the challenging industrial setting of some cases shows the *maturity* of the methods and proves that they have now progressed well beyond the laboratory.

As well as conforming to a common framework, the case studies have each been selected to illustrate some unique feature:

- in 10.2, *In situ TPA of a quarter vehicle*, we present the first application of *in situ* TPA and a detailed comparison with classical TPA;

DOI: 10.1201/9781003399131-10

- in 10.3, *Blocked forces from a car door woofer*, we present an unusual application to loudspeaker squeak and rattle encompassing hybrid analysis in which measurements and FEM models were combined;
- in 10.4, *Wheel-hub component-based TPA*, we present the highly challenging case of a rotating hub as a source, a problem which would have been intractable until relatively recently but can now be convincingly solved with *in situ* methods. This case also provides an example of *inconsistency* of passive and operational FRFs which was solved by *in situ* measurement;
- in 10.5, *Structure-borne noise due to gyms*, we present one of few examples of methods being applied to large scale and heavy weight structures i.e. buildings;
- in 10.6, *Rail rolling-stock virtual prototype*, we present a full VAVP bringing together measurements made across manufacturer's sites in different countries and in challenging industrial environments. This case also demonstrates OEMs and suppliers collaborating around a commonly understood framework provided by the component-based approach.

The reporting of these cases is kept brief, omitting technical details which are covered elsewhere in the book. Readers are referred to original publications for more detail on each.

10.2 *IN SITU* TPA OF A QUARTER VEHICLE

Contributed by Simon Tate of Ricardo UK, with support from Dr Andy Elliott of the University of Salford Acoustic Laboratories

One of the main noise sources in road vehicles of concern to NVH engineers is *road noise*. As tyres run over an uneven road surface, fluctuating forces are generated which are transmitted through the vehicle suspension and into the body from where they are radiated as structure-borne noise. This mechanism provides one of the dominant noise source for vehicle passengers. Usually, the NVH engineer cannot do much about the tyre-road forces themselves so attention is typically placed on limiting the *transmission* into the passenger compartment by redesign of the vehicle suspension. Transfer Path Analysis (TPA) was developed to assist redesign by quantifying the contribution of the various transmission paths to the overall transmission. By identifying the dominant path, design modifications could be correctly targeted. Thus, the primary objective of TPA is to quantify the contributions of the various paths between the source and the receiver as an aid to possible redesign.

Classical TPA is however limited to a diagnostic role in a single given assembly since it is based on contact forces which change following design modifications. In this paper we present an equivalent method, called variously *in situ TPA*, *iTPA* or *blocked force TPA* which follows all the same steps as classical TPA but is based on *blocked* forces rather than *contact* forces. The new iTPA approach allows path contributions to be quantified in a similar way to classical TPA but has an important advantage in that the blocked forces *do not* change when the assembly is modified. In addition to providing similar diagnostic information to classical TPA then, iTPA potentially opens up a whole realm of *predictive* possibilities in new assemblies. The aim of this case study is to compare iTPA with (classical) TPA which, at the time, was the most established method. The comparison is made both in terms of the results and also of the practicalities involved in achieving these results.

The study was the first example of *in situ* TPA, which has now mostly become the standard method and the forerunner of component-TPA. It was first reported in [1] and was funded in the UK by EPSRC whose support is gratefully acknowledged (Grant ref: EP/G066582/1).

10.2.1 METHODOLOGY

The study comprises back-to-back measurements on a quarter vehicle using both classical TPA and iTPA. For both methods, a two phase test is conducted comprising a passive and an operational phase. In the passive phase, FRFs were measured at connection points on the vehicle suspension (described in more detail later). FRFs were measured twice, once (for TPA) with the vehicle's OSF suspension components removed and secondly (for iTPA) with the vehicle intact. Microphones were also included at the target locations at the ear positions of the driver and three passengers inside the vehicle cabin.

In the operational phase, the vehicle, a 2007 VW Polo, was motored by a rolling road (engine off) with a coarse chip roller on the off side and a smooth roller on the near side. Three road speeds were tested, 30, 50 and 70 kph, although the roller speed was fluctuated in each case by about 10% to avoid periodic patterns in the excitation. The same operational test was used for both TPA and iTPA. Once the FRFs and operational data was obtained, the cabin sound pressure was predicted and compared with direct measurement. In the original study, the processing made use of principal component analysis (PCA), details of which are not reported here but can be found in [1, 3]. Although the path contributions are of primary interest in TPA, the comparison between methods was made on the basis of their ability to reconstruct the output.

Shown in Figure 10.1 is an illustration of a typical passenger car's front suspension system which is representative of the case study presented here (see Figure 10.2). By performing a quarter vehicle analysis the problem is reduced to include only one source of vibration, the off-side front wheel in our case, but the routes by which vibration enters the vehicle are of primary interest. For this reason the vibration source is considered to include the suspension system because there are clearly identifiable junctions which can be used to define the possible vibration transmission paths.

The interface between source and receiver was chosen to comprise three connection points on the suspension: the wishbone front (WBF) and rear (WBR) and the strut (S). (The steering rack to subframe connections were also initially included but were found not to be of importance.) The *source* substructure is therefore defined as the suspension system below these connections including the tyre and road, and the

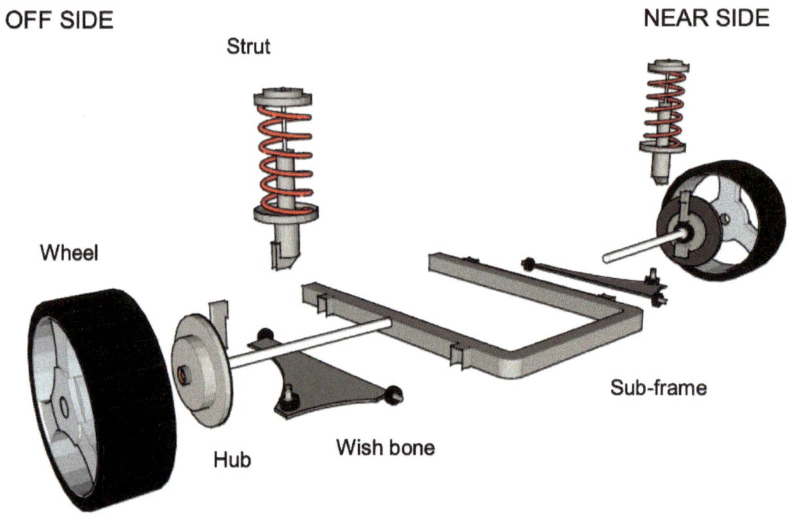

Figure 10.1 Exploded view of suspension system showing the connection points on the wishbone and strut.

receiver substructure as the rest of the vehicle above the interface. Each connection was considered to have 5 degrees of freedom (3 translational and two rotational). The x, y and z directions were defined as x: front to rear of the vehicle, y: near side to off side and z: the positive vertical direction. For the wishbone to subframe mountings, rotations about the x-axis were neglected because the wishbone is designed to pivot naturally about this axis. For the strut, rotations about the vertical z-axis were neglected for similar reasons. In total then, 15 motions (3 forces and 2 moments for each transfer path) were included. The aim is to relate the forces for these degrees of freedom to the four target sound pressures inside the vehicle. In addition to the 5 DoFs included for each interface an additional 8 indicator points were used to over-determine the problem.

To ensure a fair comparison, the same accelerometers were used for TPA and iTPA. For the same reason, it was decided not use regularisation to obtain the blocked forces since it might affect the TPA and iTPA results differently.

10.2.2 RESULTS

Shown in Figure 10.3 are the measured and predicted spatially averaged sound pressures inside the vehicle including the transfer paths wish bone front (WBF), wishbone rear (WBR) and the strut (S). The top Figure shows the narrow band response, and the middle plot the A-weighted octave band response. The bottom plot shows the overall A-weighted contribution to the total response from the three transfer paths (recall that each transfer path is the combination of 5 DoFs, three translations and

Figure 10.2 Some photos of interface instrumentation.

two rotations). Also presented is the overall A-weighted SPL for each TPA approach and that of the directly measured response (in black). Figure 10.3 was obtained at a speed of 50 kph and broadly similar results were also obtained at 30 and 70 kph as shown in the original paper [1].

In practice, a TPA would generally be considered trustworthy if the sum of partial pressure contributions from individual transfer paths were in agreement with the target sound pressure(s) measured during operational measurements. Here we see similar agreement for TPA and iTPA, if anything slightly better for the new method.

Regarding the rank ordering of transfer paths, classical TPA may be considered as a benchmark as it has been so widely used in the past. Note that the path contributions from the contact forces (as used in TPA) and the blocked forces (in iTPA) are not theoretically the same. Nevertheless, one would not expect to obtain widely differing conclusions from the two methods.

It is demonstrated in Figure 10.3 that the main contributor to the A-weighted interior noise was the WBF connection to the subframe followed by the strut and wishbone rear connection, both methods were in agreement on this. However this was only shown in the form of the overall A-weighted levels which does not allow a detailed understanding of the individual paths. Thus, shown in Figure 10.4 are the individual sound pressure contributions calculated using both methods for the strut, wishbone front and wishbone rear paths.

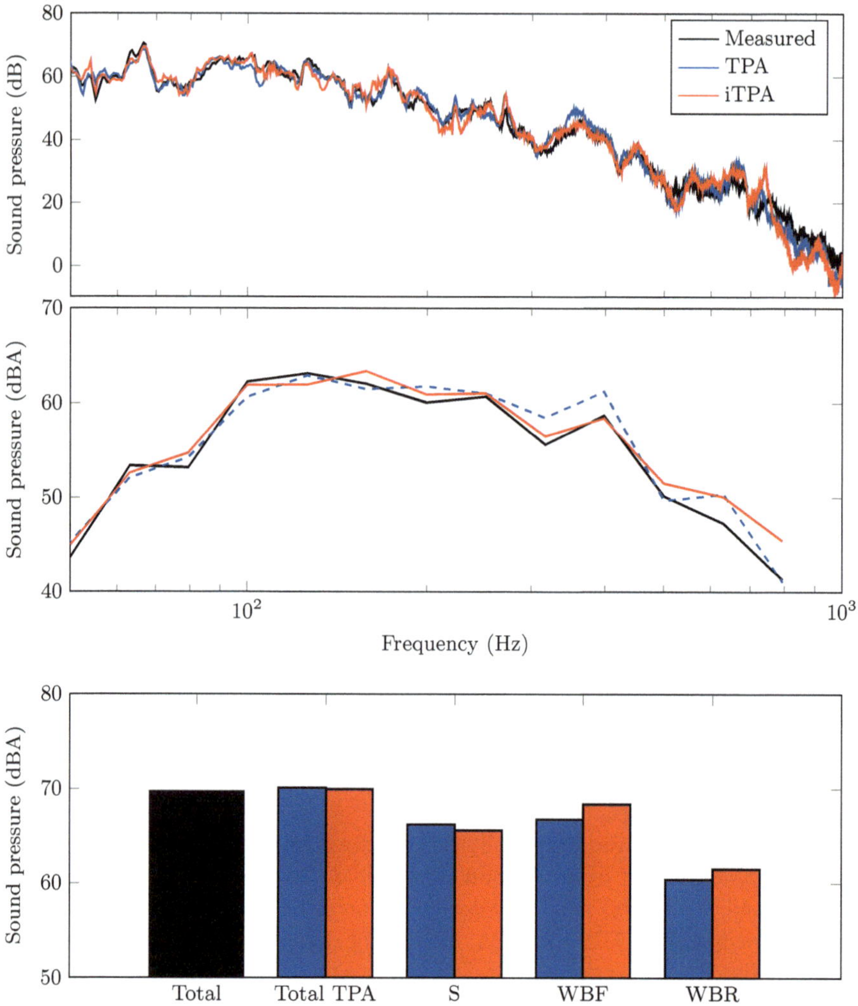

Figure 10.3 Comparison of spatially averaged sound pressure level measured at a speed of 50 kph, against those obtained using classical TPA (contact force) and in situ (blocked force) iTPA. Top – narrow band response. Middle – octave band response. Bottom – total A-weighted contribution of various sources.

It can be seen that at around 100Hz the iTPA result gives slightly higher predictions of the interior noise for the WBF source path. This appears to be the result of an underestimation of the WBF contribution by conventional TPA because Figure 10.3 shows clearly that iTPA predicts better the sound pressure at 100 Hz than does conventional TPA.

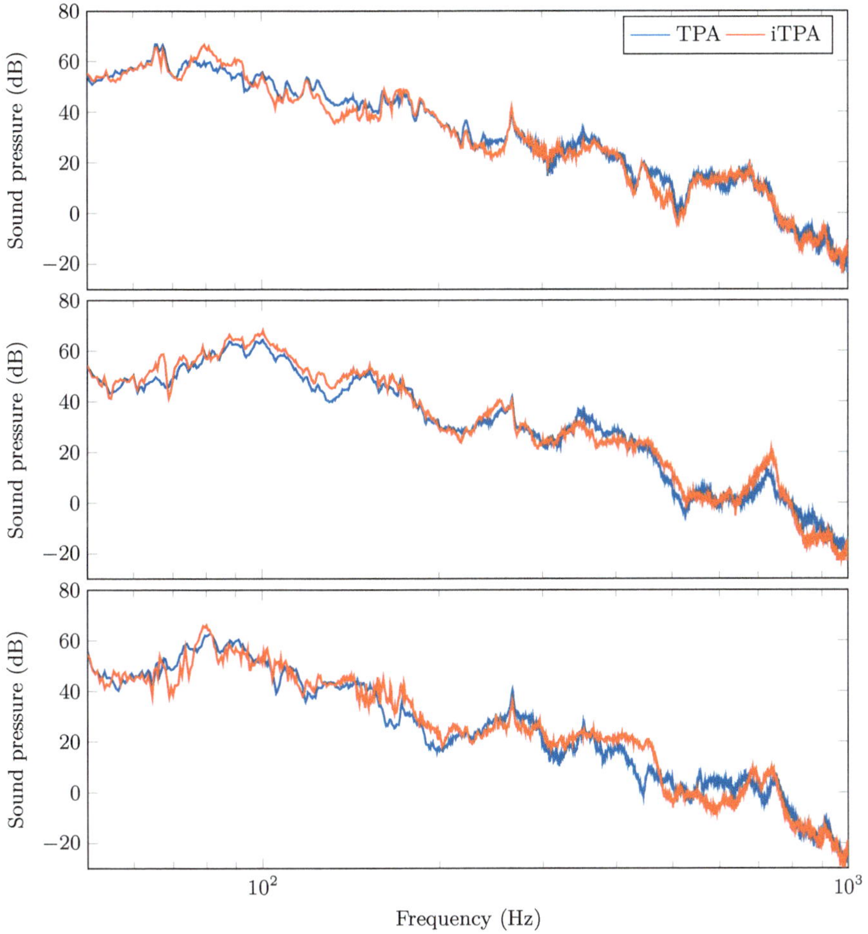

Figure 10.4 Comparison of path contributions predicted by TPA and iTPA. Top – strut (S), middle – wish bone front (WBF) and bottom – wishbone rear (WBR).

Based on these results, it can be concluded that both TPA methods were successful in reconstructing the interior noise and delivered the same rank ordering of paths. It could also be argued that, of the two methods, the in situ TPA approach was found to give slightly better results whilst also taking less than half the time to perform.

It is often suggested that both translational and rotational degrees of freedom should be included when performing a TPA analysis but this is rarely carried out in practice because of the difficulties associated with moment excitation and the measurement of angular velocity. In the results presented above, rotational degrees of freedom were included by means of the finite difference approximation [2]. It is interesting to investigate their significance as an aside to the main aim.

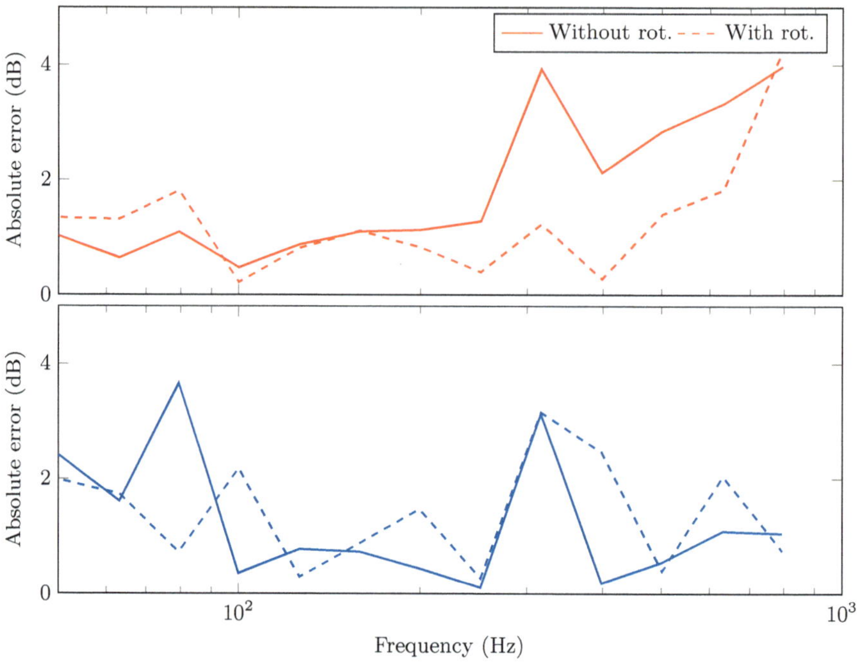

Figure 10.5 One third octave-band TPA sound pressure reconstruction errors averaged over three operating conditions (30, 50 and 70 kph). Solid line – reconstruction error with rotations neglected. Dashed line – reconstruction error when rotations are included. Top – iTPA. Bottom – TPA.

Having included rotational DoFs in the preceding results, it is straightforward to remove rotations from the analysis by eliminating data from the FRF and operational matrices so that every connection was described only by translation in x, y and z DoFs. Shown in Figure 10.5 are the average one third octave band sound pressure reconstruction errors for in situ and conventional TPA. In Figure 10.5 there are two plots (one for each method) and on each plot there are two lines; solid lines show the error when rotations were neglected, whilst dashed lines show the error when rotations are included in the analysis. The errors in both plots were calculated in the same way by taking the dB reconstruction error for each running speed (30, 50 and 70 kph), converting these errors to the equivalent in Pascals, averaging over the three running speeds and finally converting back to dB to give an average dB error.

For iTPA, it is clear that the sound pressure reconstruction errors in the one third octave bands (50–400 Hz) do not exceed 2 dB on average when rotations are included in the analysis. However the same reconstruction error, observed when rotations were neglected, was found to approach 4 dB. Thus, it appears that including rotations in the analysis of the in situ TPA data resulted in a generally lower sound pressure reconstruction error. However, the same argument cannot easily be made for the conventional TPA approach because the sound pressure reconstruction errors were found to be similar regardless of whether rotations were included in the analysis.

10.2.3 DISCUSSION AND CONCLUSIONS

We conclude on the basis of these results that iTPA is as reliable as TPA, possibly slightly more so. Naturally, the results only apply to this case and more cases are needed to form a general conclusion, nevertheless this assembly is fairly representative of those used in TPA generally. Both methods rely on effectively the same assumptions and the same, or very similar measurements, so the only significant difference is the need to remove the suspension for the TPA test. Intuitively, one would expect some uncertainty to be introduced by this action, particularly if sensors are removed and replaced between the FRF and operational test phases. Therefore, it is logical, even expected, that iTPA would, if anything, give slightly more reliable results. On the negative side, leaving the suspension in place may in some cases make access more difficult.

Avoiding the need to remove parts of the structure is a major practical advantage of the new methodology. Indeed, the iTPA test took around half the time of the conventional TPA test and was logistically much simpler. Importantly though, although with faster testing a compromise on reliability is usually expected ([]), in the case of iTPA there was no such compromise, in fact reliability was if anything slightly improved.

Lastly, it is highly significant that the blocked forces obtained in the iTPA test would remain valid irrespective of any modifications on the receiver side of the interface. Although not tested in this study, the theoretical possibility is open to use the same blocked forces to excite a modified vehicle body, a completely new vehicle, or even a non-existent vehicle represented by a model. Thus, iTPA can be seen to provide a seamless link between what was previously a purely diagnostic approach and the powerful virtual prototyping methodology that has since become known as component-based TPA.

Bibliography

1. AS Elliott, AT Moorhouse, T Huntley, and S Tate. In-situ source path contribution analysis of structure borne road noise. *Journal of Sound and Vibration*, 332(24):6276–6295, 2013.
2. AS Elliott, AT Moorhouse, and G Pavić. Moment excitation and the measurement of moment mobilities. *Journal of Sound and Vibration*, 331(11):2499–2519, 2012.
3. D Vandenbroeck and W Hendricx. Interior road noise analysis with principal components. In *Proceedings of the Acoustical Society of Korea Conference*, pages 854–859. The Acoustical Society of Korea, 1994.

10.3 BLOCKED FORCES FROM A CAR DOOR WOOFER

Contributed by Dr David Lennström of the Volvo Car Group

In the automotive industry, there is a constant need for being more efficient and confident in the prediction capability for various attributes. A single part or system is often shared by many car models that have a common vehicle design. Many of these components are needed to be considered for potential noise, vibration and squeak and rattle. The method presented here can help the automotive industry to improve the prediction accuracy and efficiency for various vibration-related attributes. This case study presents a method for estimating the vibration effects of a woofer on a door panel, which may cause unwanted noise and rattles. The method uses blocked forces, which are measured indirectly from the woofer's frequency response and acceleration, to characterise the woofer as a vibration source. Examples of published case studies of indirect measurements of blocked forces include power steering systems [7, 8], electric vehicles and motors [1–3, 6], wind turbines [4] and road noise [5]. The blocked forces are then applied to a finite element model of the door panel to simulate the vibration response when the woofer is placed at a different location in a new car model. The method is validated by comparing the simulated and measured responses for the same car model, showing good agreement. The woofer was available in an already existing car, but the upcoming car was only available as a finite element (FE) model. The source characterisation was based on the blocked forces at the points where the woofer was attached to the door panel. This seemed to be the most suitable method for an independent source characterisation in this case, because the woofer's loading depends on the 'woofer box' around it, which is the door panel in this situation.

The contribution of the work presented here is to provide an industrial case study where blocked forces and numerical data were combined in order to achieve a hybrid prediction model. The novelty of this work is the application case, which is a rare but clear illustration of how the source activity is affected by the receiver structure.

10.3.1 METHODOLOGY

To measure the blocked forces, a Volvo XC90 from 2015 was equipped with ten uniaxial accelerometers of 100 mV/g sensitivity, as shown in Figure 10.6. They were attached to the door's inner panel with cyanoacrylate glue. The accelerometers were placed close to the points where the woofer was fixed to the panel. Areas where the panel was very flexible was avoided. The blocked forces were calculated at the four screw heads that held the woofer to the panel.

The FRF testing involved using an impact hammer and a shaker. The shaker produced chirp signals (sinusoidal waves that change frequency). The transfer accelerances from the top right screw to the 9 accelerometers are shown in Figure 10.7.

The accelerances are generally similar between the impact hammer and shaker case, although some discrepancies can be noted. This is expected since a stinger

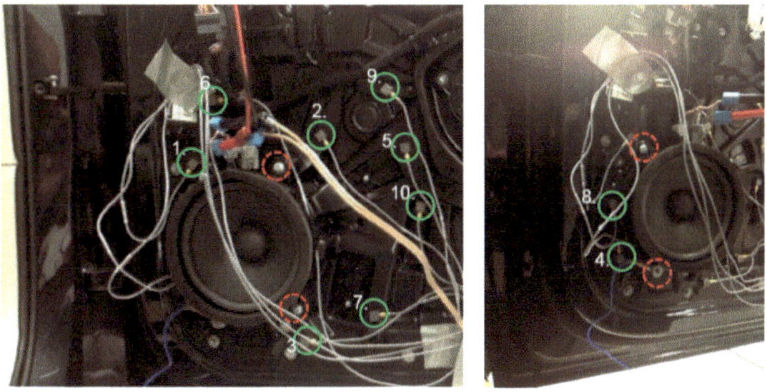

Figure 10.6 Instrumentation for blocked force measurements. The ten green solid circles denote uni-axial accelerometer positions and the four red dashed circles denote the blocked force positions.

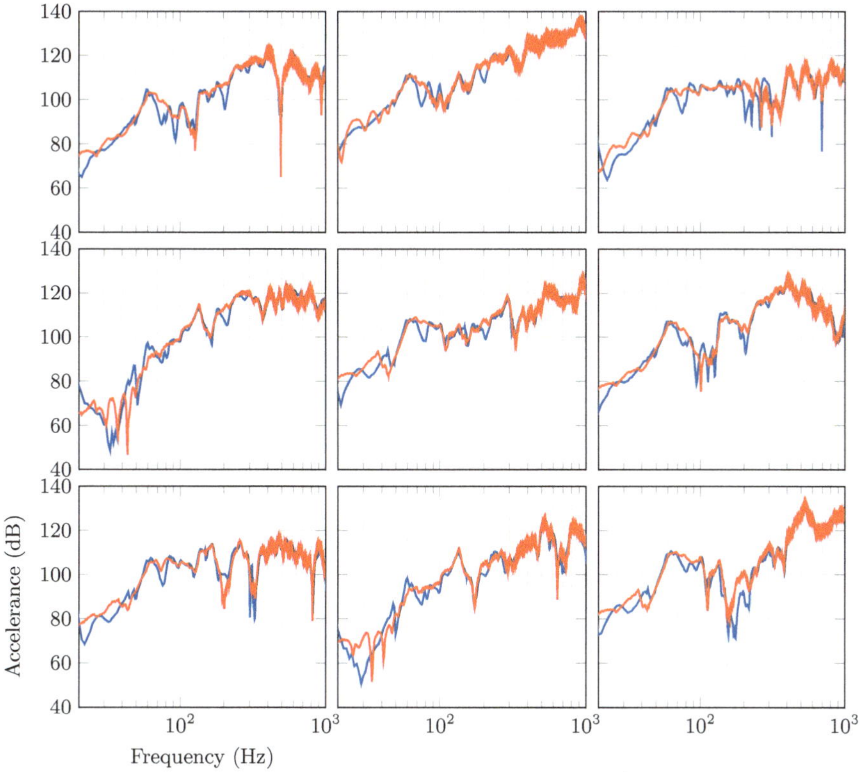

Figure 10.7 Transfer accelerances from top right screw to accelerometer points 1-9. Red – FRFs from shaker excitation, blue and dashed – FRFs from impact hammer excitation.

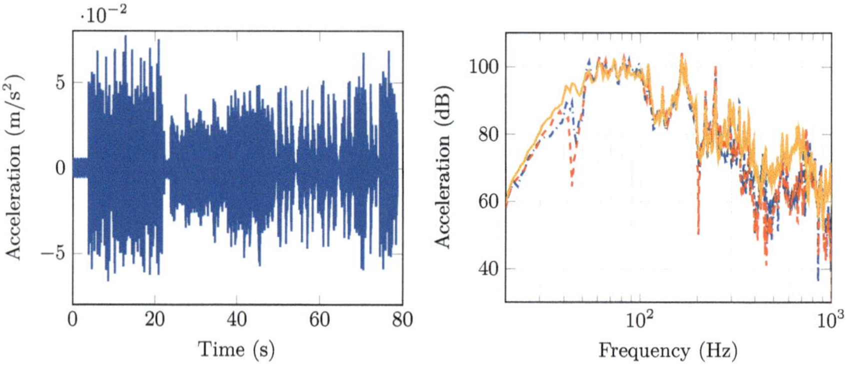

Figure 10.8 Left) Time trace of the validation audio test tracks, measured acceleration at accelerometer 10. Right) Frequency spectrum of the audio test track. Red – measured acceleration at accelerometer 10. Blue and dash dotted – On board validation of accelerometer 10 using shaker excitation during FRF tests. Brown and dashed – On board validation of accelerometer 10 using impact hammer excitation during FRF tests.

from a shaker often adds some extra damping to the structure for instance. These 9 points were used as indicators to calculate the blocked forces at the four places where the structure was attached. Point 10 was a reference point to check the accuracy of the on-board validation.

The TPA process used Siemens LMS Test Lab TPA software to compute the blocked forces and the expected response. The transfer accelerance matrix, $\mathbf{Y}_{Cbc}$, was well conditioned with condition numbers below 80 for all frequencies for both the impact hammer and the shaker excitation cases. A low condition number of the matrix was in line with the expectations since only the vertical degree of freedom was included in the measurements. However, it is difficult to set a limit for an acceptable condition number in this case.

The operational measurements used one audio tracks to find squeak and rattle noises in vehicles. This track was played back in the car's stereo at an equivalent sound pressure level as during generic squeak and rattle subjective assessments. Figure 10.8 (left) shows the dynamics of the time trace of the sound track that was being used for the validation case. Note however that woofer itself is not expected to induce squeak and rattle, it is rather other parts inside the car structure.

10.3.2 RESULTS

On-board validation

The operational cases had two validation sets each because the shaker and the impact hammer measured the FRFs. The accelerance matrix was not regularised when computing the blocked forces. The Siemens LMS Test.Lab Time TPA module used the

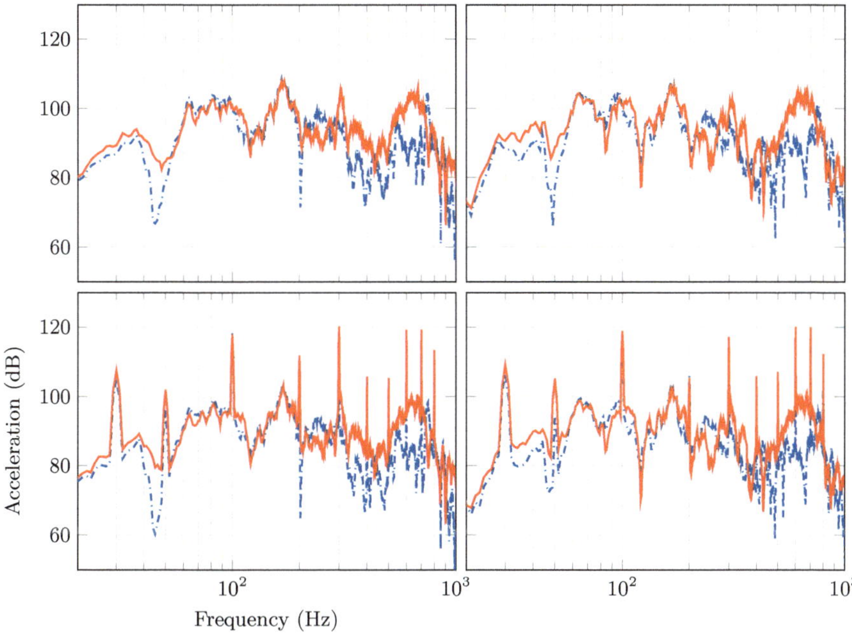

Figure 10.9 Frequency spectrum of the white noise and the white noise including tones operational case. Red – measured acceleration at accelerometer 10. Blue and dash dotted – on board validation of accelerometer 10.

FFT convolution method to get the blocked forces and the expected response in the time domain. Figure 10.9 (right) displays the spectrum of the acceleration that was measured and validated on board. It shows that the measured and predicted signals match well up to at least 200 Hz, except in the 30 to 50 Hz frequency range. In this range, the hammer excited FRF gives a better prediction than the shaker excited FRF. For other frequencies, the prediction is almost the same for both the hammer and the shaker excited FRFs. The lower end of the frequency range is the most important one for squeak and rattle excitation and therefore logarithmic x-axis were chosen.

The woofer also played white noise and white noise including a number of tones to validate the response for operational load cases that were not as dynamic as the audio test track. Both the impact hammer and the shaker excitation were used for the on board validation for this case.

Figure 10.9 shows how the measured and the validated acceleration compare for those cases. The measured and predicted response agree well both cases, except in the 30-50 Hz frequency band. The prediction is satisfying between 50 and 200 Hz. In essence, the prediction accuracy is very similar as for the case with the audio test track, indicating that the linear model used in both cases works well.

Hybrid prediction combining blocked forces and an FE model

To predict the response in a different car model, an FE model of the door panel with the woofer, shown in Figure 10.10, was employed. It was a Nastran linear model with 5 mm mesh size. The woofer was a solid element without any membrane and the air cavities were not modelled. Since the door model also contained the woofer, yet simplified, the obtained blocked forces could be utilised directly without the need of recalculations to contact forces in order to predict the vibrations. In essence, the transfer functions from the screw points to the target point was extracted from the FE model and this matrix was multiplied with the blocked forces obtained from the measurements.

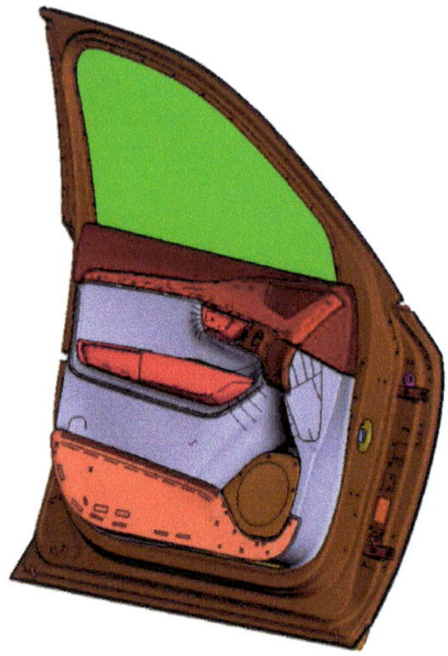

Figure 10.10 FE model of the door for squeak & rattle prediction.

The hybrid prediction case could then be compared with the operational measurement response in the same point on the door panel. Figure 10.11 (left) shows the time traces and frequency spectrum of these two cases respectively.

10.3.3 DISCUSSION AND CONCLUSION

This study aimed to get vibration data from a woofer to do squeak and rattle prediction using finite element models of the woofer and the door, as part of the full car structure. Since the aim was to predict the vibrational response from the woofer when installed in various car doors, independent source data was needed. Initially in the selection of possible methods, measuring blocked forces directly with force transducers between the woofer's attachment points and a very rigid support was considered. However, since the acoustic impedance that the woofer membrane faces in operation would likely affect the forces at the attachment points, it was chosen to measure the blocked forces in situ instead to have a realistic loading.

The blocked forces were estimated at the four screw heads of the woofer attachment to the door panel. The study only considered vertical forces because it only needed low frequency content (up to 200 Hz) for squeak and rattle prediction. Actually, the interface between the woofer and the door panel is a continuous circle in reality but the four-point description was good enough to describe the behaviour up to 200 Hz (see Figure 10.9). However, there was an exception for a frequency band between 40 and 50 Hz, where the acceleration prediction was too low. This was clear for all operational load cases (the audio test track and the white noise cases) and for both impact hammer and shaker excitation, but the impact hammer performed a bit better. For frequencies above 200 Hz, clear deviations were observed which were likely due to an inadequate source characterisation, i.e. four points and only vertical forces was probably not enough for those higher frequencies.

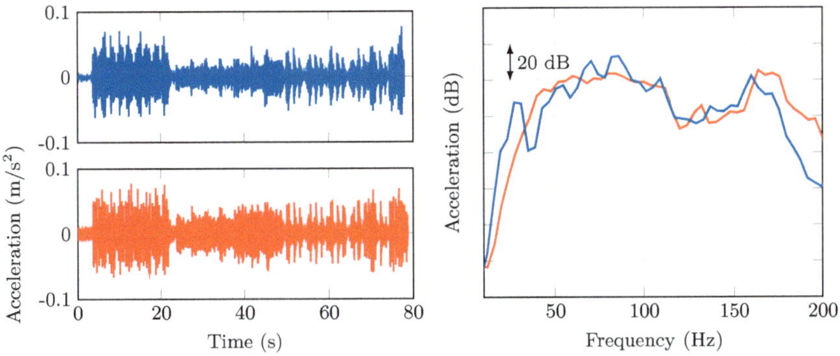

Figure 10.11 Left) Time series of response on measurement point 10. Right) Frequency spectrum of the audio test track. Red – measured acceleration at accelerometer 10. Blue and dash dotted – Predicted by FE model data of the door panel in combination with measured blocked forces.

Finally, the study compared the measured response and the response from loading the FE model of the door with the woofer with the blocked forces time data. The FE model was not calibrated, so it was not clear how well it represented the real door and the real woofer. Still, the magnitudes from the two cases had similar patterns over frequency (see Figure 10.11). Therefore, this hybrid modelling method looks promising and can be useful in cases where the vibration source's internal forces are unknown.

Bibliography

1. G Banwell and R Faventi. Assessment of experimental techniques to characterise the vibration source strength of a motor radially mounted with resilient elements. *In Proceedings of ISMA, Leuven*, 2016.
2. PO Sturesson, J Andersson, D Lennström, and H Olsson. Nvh development of the all new volvo xc90 twin engine plug-in hybrid. In *Proceedings of Aachen Acoustics Colloquium 2015*, pages 67–77, 2015.
3. A Diez-Ibarbia, M Battarra, J Palenzuela, G Cervantes, S Walsh, M De-la Cruz, S Theodossiades, and L Gagliardini. Comparison between Transfer Path Analysis methods on an electric vehicle. *Applied Acoustics*, 118:83–101, 2017.
4. AS Elliott, AT Moorhouse, et al. In-situ characterisation of structure borne noise from a building mounted wind turbine. In *Proceedings of ISMA*, pages 2055–2068. Citeseer, 2010.
5. AS Elliott, AT Moorhouse, T Huntley, and S Tate. In-situ source path contribution analysis of structure borne road noise. *Journal of Sound and Vibration*, 332(24):6276–6295, 2013.
6. D Lennström, M Olsson, F Wullens, and A Nykänen. Validation of the blocked force method for various boundary conditions for automotive source characterization. *Applied Acoustics*, 102:108–119, 2016.
7. M Sturm, T Alber, AT Moorhouse, D Zabel, and Z Wang. The in-situ blocked force method for characterization of complex automotive structure-borne sound sources and its use for virtual acoustic prototyping. *Proceedings of ISMA, Leuven, Belgium*, 2016.

8. M van der Seijs, EA Pasma, D de Klerk, and DJ Rixen. A comparison of two component tpa approaches for steering gear noise prediction. In *Dynamics of Coupled Structures, Volume 4: Proceedings of the 33rd IMAC, A Conference and Exposition on Structural Dynamics, 2015*, pages 71–79. Springer, 2015.

10.4 WHEEL-HUB COMPONENT-BASED TPA

Contributed by Fabio Bianciardi of Siemens Industry Software NV, with support from Nicola Pieroni and Peter Kindt from Goodyear Dunlop Tires Operations

The shift towards electric vehicles in the automotive industry has increased the importance of tyre-road interaction as a significant source of noise in vehicle cabins. This is mostly due to the decreased masking effect generated from traditional powertrain noise. In addition, the shortening of the development cycle, and the increased vehicle variants, have pushed the automotive industry to seek technologies that enable the prediction of road noise contributions in the early stage of the vehicle development process. Thus, avoiding costly and time-consuming design iterations. Various methods to characterise excitation for vehicle interior noise prediction have been explored [5]. While spindle forces derived from full vehicle road measurements offer accurate predictions for specific vehicle configurations, they are not easily transferable to other vehicles due to dependencies on the full vehicle assembly.

In recent years, component-based (CB-)TPA has become a powerful methodology to predict the NVH performance of a virtual prototype assembly starting from independent sources and component models. In this invited contribution we show the application of component-based TPA to road noise prediction and analysis. A key step in the methodology is to identify the invariant input loads (i.e. blocked forces) at the spindle location. These forces, characteristic of the tyre-wheel assembly (i.e. the source component) when the spindle is 'blocked', are independent of suspension and vehicle dynamics. They enable vehicle interior noise prediction without the need of a physical vehicle, enabling predictive engineering for various vehicle and suspension configurations.

In this context, the CB-TPA methodology allows, on one hand, the manufacturers to measure the blocked forces of their s on a test-rig and, on the other hand, it enables the automotive OEMs to perform the prediction of the full vehicle behaviour starting from the test-rig measurements. The vehicle interior road noise is obtained by propagating the test-rig blocked forces through the coupled full-vehicle FRFs. Those can be either obtained from experimental substructuring of the uncoupled components, or experimentally measured if the vehicle prototype is available.

The use case presented here focuses on structure-borne road noise. A frequency range of 50 Hz to 600 Hz is investigated. A slick tyre, selected for this analysis, is characterised on a test-rig by a set of blocked forces. Vehicle noise prediction is obtained using substructured FRF with these forces. The method is validated by comparison with direct vehicle measurements.

10.4.1 BLOCKED FORCES IDENTIFICATION ON A TEST-RIG

The source under investigation consists of a slick tyre-wheel on a drum (here termed component *A*). The tyre-wheel was mounted on a rigid test-rig as the passive receiver (component *B*), as presented in Figure 10.12. To represent the same conditions as on the vehicle, a static load of 2864 N was applied to the on the rig. The tyre-wheel was rigidly connected to the rig using five bolted joints. A bearing allowed for the free rotation of the tyre, driven only by the drum.

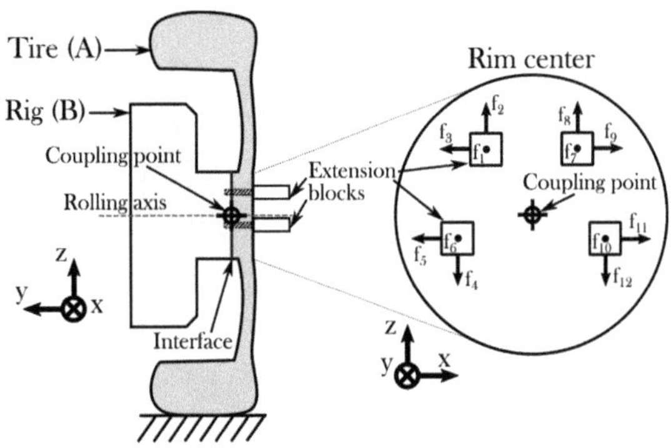

Figure 10.12 Schematic depiction of the test-rig setup.

A set of eight tri-axial accelerometers were placed on the rig (component B) to be used as indicators for the in situ estimation of the blocked forces. To account for translational and rotational DoFs, 12 excitations in different directions were applied on four lightweight aluminum extension blocks placed on the bolted joints, as presented in Figure 10.12 (see also Figure 10.16 for photo of the extension blocks). The FRFs of the coupled system were measured using an instrumented shaker (Simcenter-Qsource, Q-ISH). The excitations and measured responses were then geometrically reduced to the centre of the connection using the Virtual Point Transformation, excluding the rotational DoF aligned with the rolling direction, assuming that the loads around the rotation axis would be negligible.

For the operating conditions, a rough coating was used on the drum to act as the road profile. The accelerations at indicator locations were measured while the drum was rolling at a constant speed of 20 kph over a period of 200 seconds. The blocked forces corresponding to these operating conditions were estimated using the indicator accelerations, according to the in situ method. The time averaged auto-power spectrum for each blocked force DoF is presented in Figure 10.13. For road noise applications at low frequencies, the direct measurement of blocked forces using a load sensor can be an effective option, according to its definition on a highly rigid test-rig. Nonetheless, as frequency approaches the first resonance mode of the test-rig, the measured interface forces will diverge from the actual blocked forces. It is then recommended to estimate blocked forces in situ, especially for higher frequencies, as described in above.

10.4.2 THE VEHICLE PREDICTION FROM TEST-RIG MEASUREMENTS

In this section, the prediction of the structure-borne road noise transmitted toward the interior of a vehicle (termed component *C*) is described and assessed. First, the

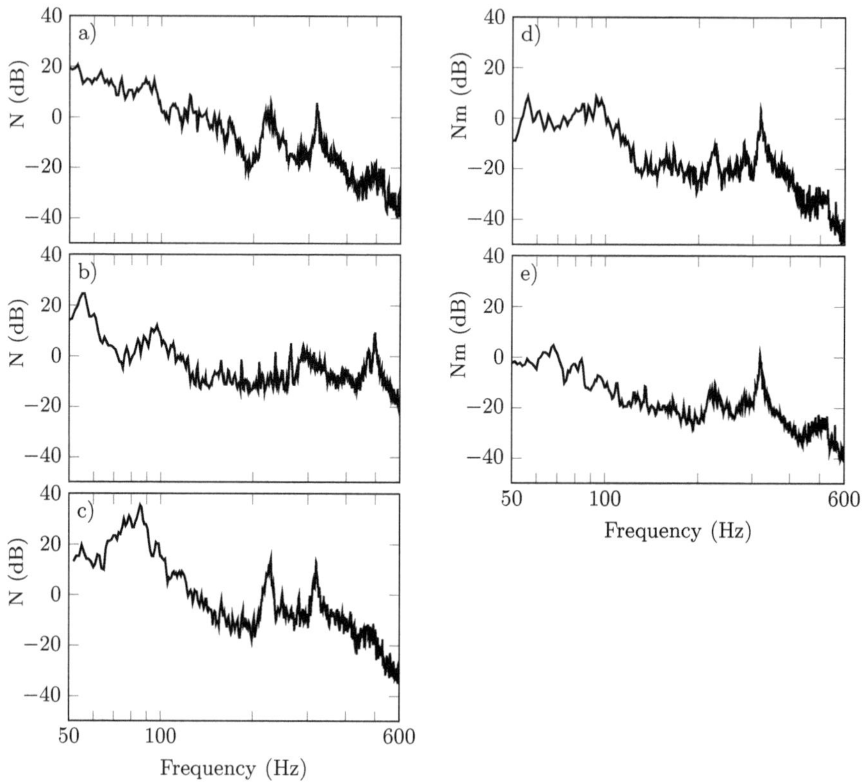

Figure 10.13 blocked forces estimated in situ from test-rig measurements; translations in the a) x, b) y, c) z directions and moments about the d) x and e) z axes.

full-vehicle coupled FRFs are obtained from measurements of the uncoupled vehicle and -wheel components using dynamic substructuring (FBS). The full-vehicle coupled FRFs are also measured directly, and used to validate the FRFs obtained from sub structuring. Once the coupled FRFs are obtained, the blocked forces estimated in situ on the test-rig are used as input to predict the operational response of the assembly. To validate the interior predicted noise, operational measurements with the full vehicle running on a chassis dynamometer on its rear wheels were performed.

Experimental dynamic substructuring for tyre-vehicle integration

Experimental dynamic substructuring aims to generate the FRFs of the coupled system starting from the uncoupled component FRFs (i.e. tyre-wheel and vehicle). For the uncoupled tyre-wheel FRF measurements (component A), a dedicated setup, available at the engineering facility of Siemens in Leuven, was used.

The FRF measurements consisted of twelve excitations as inputs, located on four aluminium blocks glued to the inner side of the wheel rim, and four tri-axial accelerometers as response outputs, as shown in Figure 10.14. Both the excitations and the responses of the acquired FRFs were geometrically reduced, using again the Virtual Point Transformation, at the coupling point located at the centre of the wheel, thus obtaining the translational and rotational FRFs. The uncoupled -wheel FRFs were identified under static vertical loading conditions while the was not rolling.

For the measurements of the uncoupled vehicle FRFs (component *C*), both rear tyres were removed and, while supported only on the front tyres, the vehicle was suspended on its rear side using elastic bungees. In this way, the vehicle suspension was still compressed under the actual vehicle weight while achieving a free-interface condition. For the FRFs, 12 shaker excitations and four tri-axial accelerometers were employed on each side. The uncoupled vehicle setup is shown in Figure 10.15. During the FRF measurements, an additional vertical low frequency excitation was applied to enforce on the suspension the same displacement the vehicle suspension is subjected to during operational conditions [2]. A microphone, placed in the vehicle interior cabin, was used as a validation target. The FRFs were geometrically reduced to the centre of the hub. Two sets of FRFs were obtained: the FRFs at the coupling interface; and the FRFs between the coupling interface excitations and the microphone target response. To virtually assemble the same model to both rear sides of the vehicle, the FRF matrix was duplicated and mirrored. In this way, both rear tyres were properly oriented to match the vehicle coordinate system.

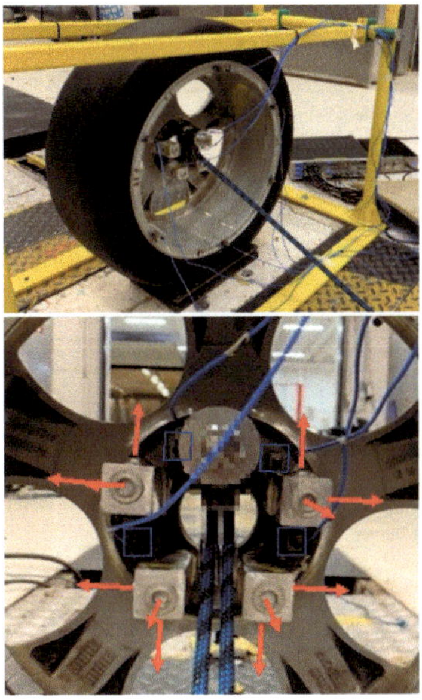

Figure 10.14 Uncoupled -wheel setup for static FRF measurement.

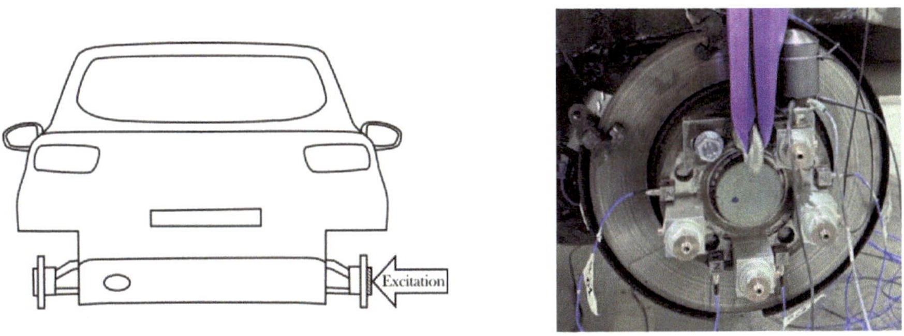

Figure 10.15 Uncoupled vehicle setup for FRFs measurement.

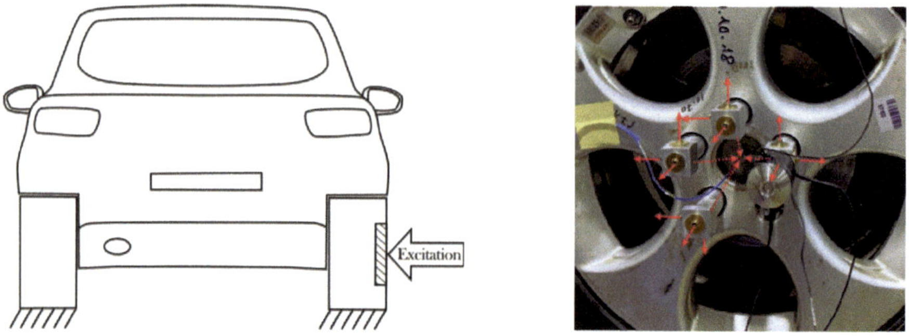

Figure 10.16 Setup for coupled FRFs measurements on the rear-right side.

To validate the coupled FRFs of the coupled vehicle/-wheel obtained from substructuring, coupled FRFs of the full assembly were also experimentally measured. This validation step was performed to verify the predictive capabilities of the substructuring. The acquisitions of the coupled FRFs of the fully assembled vehicle were performed following the same procedure described above for the uncoupled vehicle: 12 excitations in different directions were applied on blocks on the wheel bolted joints, and afterward geometrically reduced to the centre of the interface. The coupled vehicle setup is shown in Figure 10.16. After reducing the FRFs, the translational and rotational FRFs between the coupling interface excitations and the microphone target were obtained. Additionally, a low frequency excitation was applied to the vehicle to enforce the same displacement the vehicle suspension is subjected to during operational conditions. The FRFs were measured in static condition while the was not rolling.

In Figure 10.17, a comparison between the coupled FRFs experimentally measured and obtained from substructuring are shown. The comparison of the FRFs between excitations at the rear wheel on the right side and the target microphone show a very good agreement, although some errors still appear due to the sensitivity of substructuring to error propagation especially in the higher frequency range above 500Hz [6].

Vehicle noise prediction from blocked forces

This section describes the prediction of vehicle interior noise based on the blocked forces identified on the test-rig, and the substructured FRFs described above. The procedure involves the propagation of these identified forces through the coupled FRFs of the vehicle derived from substructuring, specifically focusing on the road noise excitation generated by only one wheel rolling, in this case the rear right wheel. The approach presented in the previous section for deriving coupled vehicle FRFs from FBS was used, therefore the effect that the rolling might have on the coupled vehicle FRFs was excluded. A comparison between the predicted noise levels and the actual measurements at the driver seat microphone is presented in Figure 10.18.

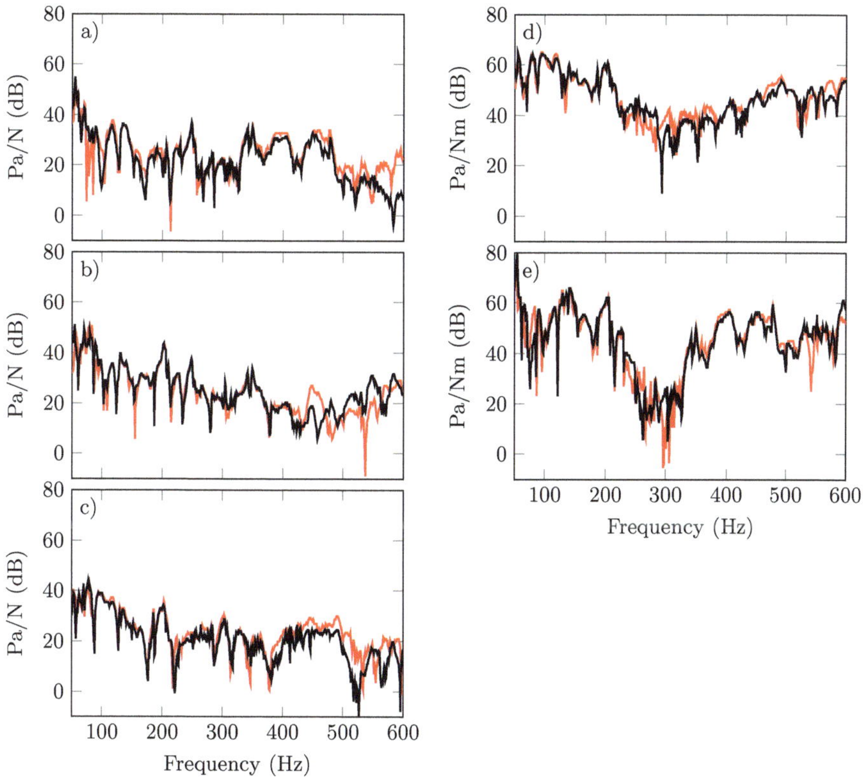

Figure 10.17 Comparison between measured (red curve) and substructured (black) coupled vibro-acoustic FRFs, due to forces in a) *x*, b) *y*, c) *z* directions and moments about the d) *x* and e) *z* axes.

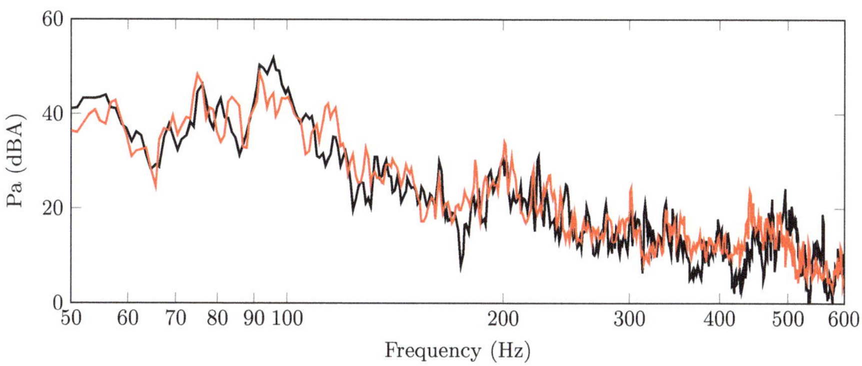

Figure 10.18 Driver seat microphone response: measured (red curve) vs predicted from test-rig blocked forces and substructuring (black curve).

The comparison shows a similar agreement with slight overestimation around 95 Hz. This phenomenon has been extensively investigated and it is related to a limitation of the tyre-wheel FRF used in the substructuring equations. It was observed that the static FRF was not representative of the tyre behaviour under rolling conditions in specific frequency ranges when the was coupled to a vehicle. The authors believe that having a set of tyre-wheel FRFs which include the rolling effect of the would improve the prediction. However, as known in literature, the FRFs of a tyre-wheel depend not only on the static structure, but also on the rolling condition [1, 4]. Considering that the is rigidly coupled to the vehicle, the behaviour under rolling conditions affects the full vehicle FRFs as well.

To measure the vehicle coupled FRFs under rolling conditions, an alternative experimental method was investigated by positioning the exciters on the inner side of the knuckle. The setup is depicted in Figure 10.19. The right rear knuckle was instrumented with nine excitation locations on different parts of the structure. Aluminium blocks were glued to provide excitations on different directions. The measurement of the operational FRFs was performed while the tyre was rolling at 20km/h. The smooth coated surface of the drum was used to minimise the additional excitation generated by the road. The measured FRFs were afterward geometrically reduced to the centre of the wheel interface.

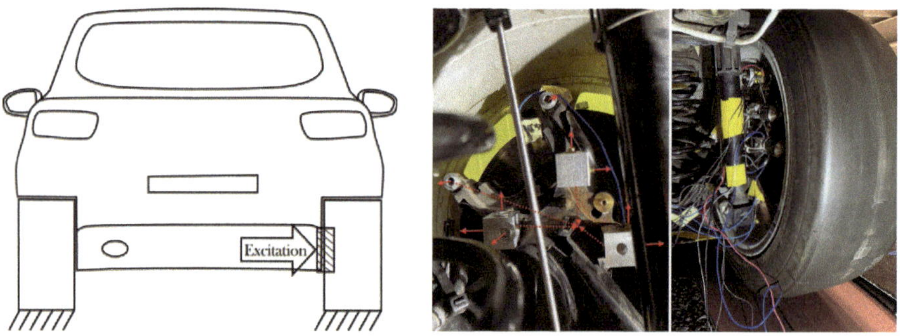

Figure 10.19 Setup for operational FRFs measurements on the rear-right side.

It was observed that wheel centre and knuckle behave as rigid regions only for low frequencies below 120Hz. Above 120 Hz, the components between the wheel and the knuckle present structural flexibilities limiting the applicability of the Virtual Point Transformation. The effect of the rolling condition was mainly noticeable at around 95 Hz.

In Figure 10.20, the prediction of interior noise level is shown by propagating the blocked forces obtained from the test-rig, through the static and operational coupled vehicle FRFs obtained directly from measurement (not by substructuring). It is worth noting that the predicted noise obtained using the operational FRF is plot only up to 120 Hz, above which it was deemed unreliable due to the knuckle flexibility. These results validate the previous hypotheses supporting the need for a FRF model under rolling condition in order to improve the VAVP results [3, 4].

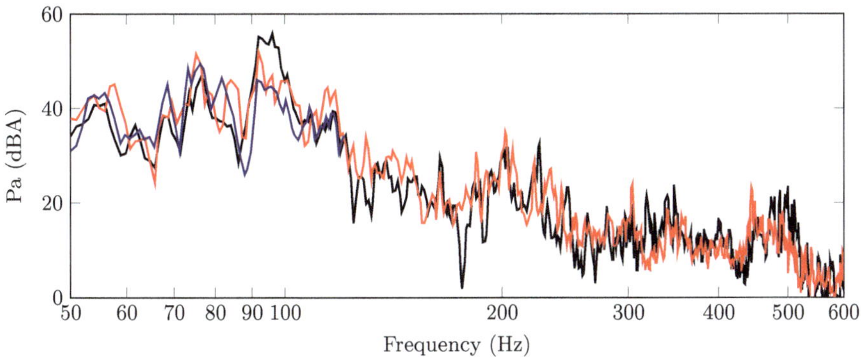

Figure 10.20 Interior target microphone, measured (red curve) vs predicted from test-rig blocked forces using measured static (black curve) and operational FRFs (blue curve).

10.4.3 CONCLUSIONS

This industrial contribution validates the component-based TPA methodology for vehicle road noise prediction. The effect of rolling on predicted noise was investigated by characterising the behaviour on a test-rig and using the identified data to predict interior vehicle noise. The results obtained from the prediction were in good agreement with the measurements for the frequency range of interest. This methodology offers significant value to the industry, streamlining component development, enabling manufacturers to set targets for blocked forces, and facilitating rankings up to 600 Hz. The research demonstrates experimental feasibility and aims to build confidence in the methodology for industry adoption.

A key methodological innovation resides in the assessment of the impact of the rolling on the full-vehicle FRFs and providing guidelines for experimental evaluation. Notably, the rolling effect on test-rig FRFs was found to be negligible due to the higher rigidity of the rig, thus not considered in experiments where only stationary FRFs were used. Substructuring was validated for this road noise analysis, showing consistent noise predictions when transferring blocked forces from the test-rig to the vehicle. However, a limitation arises in accounting for rolling effects within substructuring due to uncoupled FRFs being measured in a static condition. On-going research is addressing this limitation through numerical modelling or experimental substructuring techniques to incorporate the rolling effects into the substructuring prediction. This framework could be applied during the virtual vehicle development process in several applications, such as, road noise synthesis, what-if analysis and road noise auralisation.

Bibliography

1. JO Almirón, F Bianciardi, P Corbeels, N Pieroni, P Kindt, and W Desmet. Vehicle road noise prediction using component-based Transfer Path Analysis from tire test-rig measurements on a rolling tire. *Journal of Sound and Vibration*, 523:116694, 2022.

2. CTT Geluk. Vehicle vibration comfort: the influence of dry friction in the suspension. Master's thesis, Technische Universiteit Eindhoven Department Mechanical Engineering Dynamics and Control Group, Eindhoven, 2005.
3. D Minervini and A Stella. Experimental frequency-based decoupling for tire characterization. In *Proceedings of ISMA*, 2022.
4. D Minervini, SY Park, T Dirickx, and T Geluk. FBS decoupling at suspension level for road noise applications. *SAE International Journal of Advances and Current Practices in Mobility*, 5(2022-01-0978):937–951, 2022.
5. J Park, P Gu, J Juan, A Ni, and J Van Loon. Operational spindle load estimation methodology for road nvh applications. *SAE Transactions*, pages 2055–2063, 2001.
6. DJ Rixen. How measurement inaccuracies induce spurious peaks in frequency based substructuring. In *International modal analysis conference IMAC-XXVI*, pages 1–15, 2008.

10.5 STRUCTURE-BORNE NOISE DUE TO GYMS

Contributed by Martin McNulty of Ramboll

Historically, gym and fitness facilities are often situated in out-of-town locations so as to reduce the potential for adverse noise and vibration effects to sensitive neighbours. In recent times however, the growth of the fitness industry has led to commercial gym operators and private developers targeting inner-city locations, often within mixed-use buildings. Despite structural serviceability criteria [2] to limit excessive floor vibration during repeated rhythmic activity, established good practise and guidance has only recently been introduced within the UK [1]. This has partly been in response to complaints from those affected in neighbouring areas and partly due to a general lack of consistency in assessment and mitigation approaches. The UK-guidance favours a measurement-based approach however such methods can be impractical to implement for all source types. An engineering-grade methodology for prediction is provided in the guidance document, aimed to be computed within a spreadsheet-executable package so as to permit widespread use in the industry. A precision-grade assessment is considered beyond the scope of the guidance document.

This case study aims to predict noise and vibration from treadmill use at a proposed gym location. The techniques utilised are based upon those which are well developed in the automotive industry but are not widely used in the specialism of building acoustics, thus the case presents a novel application. An existing gym facility is used for the purposes of source characterisation, with blocked forces derived for a single treadmill. At the proposed facility, vibro-acoustic transfer paths were measured with excitation positions at the potential treadmill locations and response locations in areas likely to be affected. The resulting transfer paths were used to predict the noise and vibration impact and inform future gym layouts by avoiding regions where increased noise and vibration occur.

Figure 10.21 Experimental set-up and testing.

10.5.1 METHODOLOGY

In providing an estimate of noise and vibration exposure within a noise and/or vibration-sensitive receiver, in this case an office, a two-stage approach is required. First, it is necessary to determine the source properties of treadmills in a manner which is transferable to that of the candidate receiver. Once source properties are determined, tests at the candidate location are required to determine the efficiency of vibro-acoustic paths between proposed treadmill locations and general office spaces. By combination of these elements, predictions for multiple use-cases can be generated and compared with target criteria. A particular benefit of the approach is that source-receiver transmission can reveal dominant transmission paths which might be problematic under commercial operation.

The activities undertaken are summarised in Table 10.1. Assessment with regards to disturbance is not presented in this case study, the focus instead being on the methods used to predict the response.

Source activity

Due to the impracticality of installing multiple treadmills at the proposed gym location and testing them under various operational scenarios, a single treadmill, the type of which was representative of those intended for the gym facility, was selected for evaluation at an existing gym. The source strength and operational characteristics of the treadmill are not constant, and vary with running speed, user running style, and body weight. Given the difficulty of capturing all possible variations experimentally, the treadmill was operated within a limited range of speeds using a person of average weight. This approach aligns with common assumptions made when assessing floor serviceability. The resultant source characterisation data for the single treadmill can be applied to multiple proposed locations in the candidate gym facility using Transfer

Table 10.1
Activities undertaken.

Item	Description	Quantities determined	Location
1	Source activity measurements of an operational treadmill	Blocked forces at each treadmill mount in contact with the floor for a range of operating speeds.	Existing gym location.
2	Transfer Path Analysis at proposed treadmill locations	Vibration and Vibro-acoustic Frequency Response Functions describing the relationship between force (applied on the proposed gymnasium floor) and sound pressure/vibration (measured in the office space)	Proposed gym location.

Path Analysis (TPA) approaches. In this way both the source mechanism and transmission path are accounted for, including complexities associated with phase and unique transmission routes. To characterise the blocked force of the treadmill, the coupled frequency response functions at its contact points to the floor must be measured, before an operational measurement is taken. These steps are described in more detailed below.

Passive measurement – Accelerance Frequency Response Functions (FRFs) were measured as close as practicably possible to each of the four treadmill contact points with the floor, comprising two wheels at the front of the treadmill and two resilient mounts to the rear. The resulting FRF matrix contains all passive transfer function information required for characterisation. Two additional sensors were positioned on the same floor but at remote indicator locations for the purposes of on-board validation (discussed later). Excitations were provided by a force hammer, and responses measured using general-purpose accelerometers sufficient to capture responses over a broad frequency range. Measurements captured vertical (z) and horizontal components of vibration (x). The quality of the FRF measurements was confirmed via inspection of the coherence between excitation and response location. In the case of the remote indicator locations, FRFs were measured using these positions as both as excitation and response locations to confirm system linearity via reciprocity. Such checks are advantageous in field conditions where ideal testing conditions may be limited by background noise (noise and/or vibration) or other site constraints.

Active elements – Since FRF information relates only passive input force and output accelerations, it is necessary to capture operational vibration and use both measurement sets to determine the blocked force of the source. Accordingly, operational measurements of the treadmill were performed for a range of speed settings, loosely categorised as slow, medium and fast paced (treadmill display speed readouts 10, 15 and 20). The mass of the participant running on the treadmill was approximately that of the 'average' mass stated in serviceability assessment guidance [2] (76 kg). Sensor placement locations were identical to those described for the purpose of FRF measurements.

By combination of passive and operational measurements, the blocked forces for the user under each treadmill speed setting can be determined according to the methods outlined in Chapter 6.

Analysis of transfer paths

Vibration and vibro-acoustic transfer paths were assessed at the candidate gym location to relate the force input from proposed treadmills to measurable components of noise and vibration within sensitive spaces in the office below. Excitation points across the floor plate were chosen to match the preferred layout provided by the gym operator. This indicated a preference that the perimeter of the floor plate be utilised.

At each treadmill location, four force excitations were applied at points corresponding to the treadmill contact points at the proposed location. Importantly, it was necessary to ensure that the distances between each set of contact points, representing the treadmill footprint, matched those of the candidate treadmill during the source characterisation element.

FRF responses were measured in the noise and vibration-sensitive space immediately below at various locations using microphones and accelerometers. The accelerometers were arranged triaxially to capture horizontal and vertical components arising from the vertical forces applied above.

It is important to note that given the heavy weight nature of the receiver, it was assumed that the coupled assembly FRF of the target location was approximately equal to that of the receiver alone. That is, the treadmill does not significantly affect the coupled assembly FRF. This avoids the need to perform a substructuring prediction, itself requiring free-interface FRF characterisation of the treadmill.

10.5.2 RESULTS

In this section, we present a selection of results from the different steps in the measurement campaign and subsequent analysis.

Operational data

Operational data (acceleration) for each speed setting indicate that vibration amplitudes are speed-dependent, with observable components of energy across a wide range frequency range. This observation suggests that the model sometimes used of the treadmill as a single degree of freedom (SDoF) system, comprising combined mass of runner and treadmill with floor stiffness, is somewhat simplistic . Rather, it

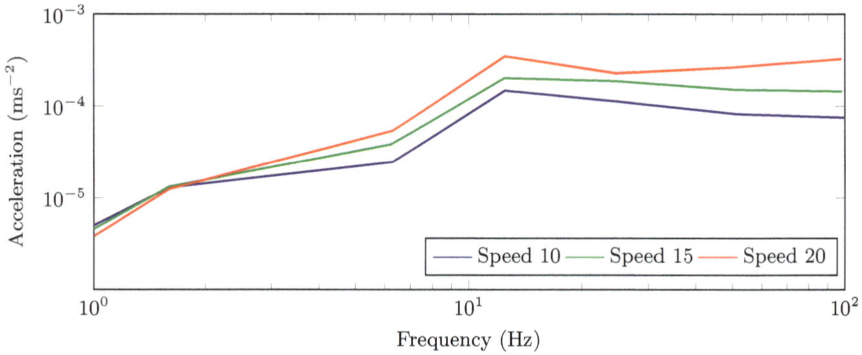

Figure 10.22 One-third octave band response showing variable acceleration spectra. Higher magnitudes are generally seen to occur at higher speed settings.

is evident that there is a degree of impulsivity and complexity to the forcing of the source mechanism, resulting in the wide band of excitation frequencies observed.

In general, it was observed that faster speed settings resulted in higher levels of vibration, as shown in Figure 10.22. It should be noted however that response spectra captured during operational events may also be sensitive to running style and user mass, though it was not practicable to investigate such sensitivities in the site-based appraisal described here. Operational measurements further confirmed that vertical components of vibration were dominant.

Blocked forces

By combination of FRF and operational data, blocked forces were calculated for each of the three speed settings used during operational measurements. Figure 10.23 provides an example of the narrow-band blocked forces obtained for a single speed setting at each of the four treadmill mounts. The initial peak in the response occurs at circa 3 Hz and is likely related to the pace frequency of the treadmill. The dominant peak generally occurs at frequencies close to 6 Hz, (possibly the 2nd harmonic) for the front right and rear left/right mounts. At the front left mount the maximum occurs at approximately 10 Hz and possibly coincides with higher harmonics of the forcing component.

Force magnitudes remain relatively high up to around 50 Hz. Excitation at the higher end of this range could elicit a response in the audible frequency range which would not be accurately modelled in a single DoF model and might therefore lead to ill-suited mitigation measures.

On-board validation

A benefit of experimentally deriving the blocked forces of a source is the ability to undertake quality and validation checks of the characterisation data. The process of 'on-board validation' took place at the existing gym facility. Derived blocked forces,

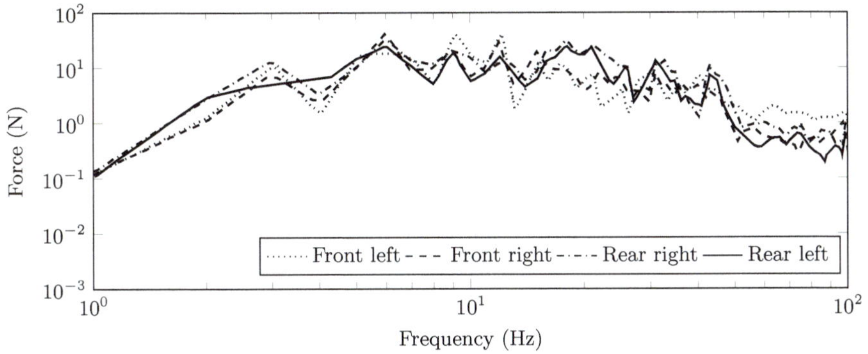

Figure 10.23 Derived blocked forces at each treadmill mount.

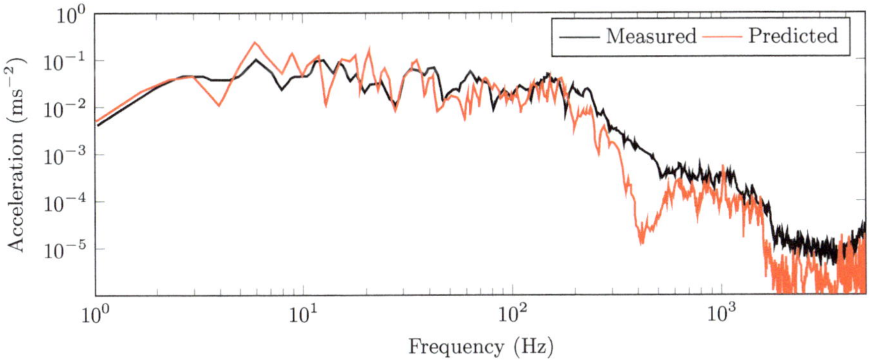

Figure 10.24 On-board validation of the blocked forces.

along with transfer FRFs to the remote sensor locations, described in the methodology section, were combined to reproduce the vibration response at the remote locations for which the true response is also known. The results of the validation for a remote sensor in the z-orientation normal to the floor are provided in Figure 10.24.

The plot shows measured and predicted narrow-band acceleration levels at a single operational speed setting. Agreement between the two curves is good up to approximately 100 Hz, beyond which signal to noise issues begin to influence the quality of the blocked forces and the corresponding transfer FRFs. It should be noted that the gym was occupied during testing and was subject to background noise from music sound systems, etc. Given such limitations, data above 100 Hz (relevant to structure-borne noise only) was regarded as unverified.

More detailed analysis indicated that the z-axis (normal to the floor) was the dominant contribution, and thus remained the focus of the study since this best captures bending waves in the structure which are expected to contribute most to structure-borne sound radiation.

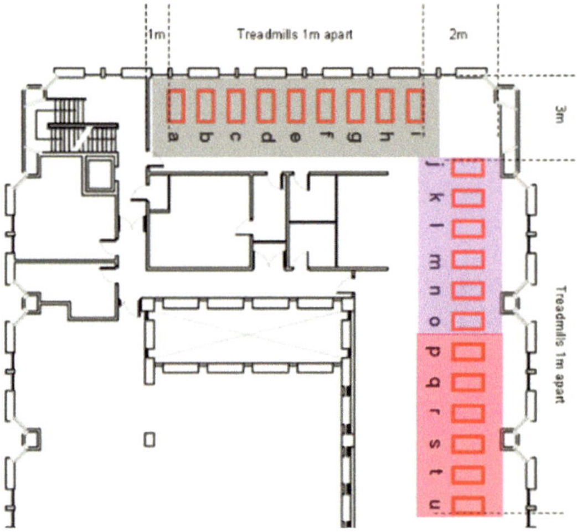

Figure 10.25 Potential locations for treadmills in the target environment.

Analysis of transfer paths

To predict the vibro-acoustic response due to (hypothetical) treadmills at the proposed gym location, FRF measurements are required with excitations at the proposed treadmill locations and responses measured within noise/vibration sensitive areas below. Despite this necessity for prediction, the procedure has an additional benefit in that prominent vibration transmission paths can be quantified and compared. Figure 10.26 summarises the one-third octave band vibration predictions for each treadmill location considered in the assessment within the office space for one particular speed setting. In making the prediction, it was assumed that the added mass of the treadmills at each location presented negligible impedance such that the FRFs measured at each proposed treadmill contact point were largely representative of the FRF with the treadmill in place.

Since the same blocked force is applied at each location, the difference between the respective curves is attributable to the variation in the transfer FRFs at each treadmill location. Most one-third octave band responses are similar, however significant variations, close to an order of magnitude, are observed in the region of the 20 Hz band. These variations, it was found, were attributable to a section of the floor that was previously a staircase opening, subsequently infilled with a lighter material. This lighter material, with higher accelerance, presents a more efficient pathway to sensitive spaces compared with that of other regions comprising heavier and stiffer constructions. The approximate area corresponding to this opening is highlighted as the red-shaded region of Figure 10.25 encompassing treadmill locations ‘p’ through to ‘s’.

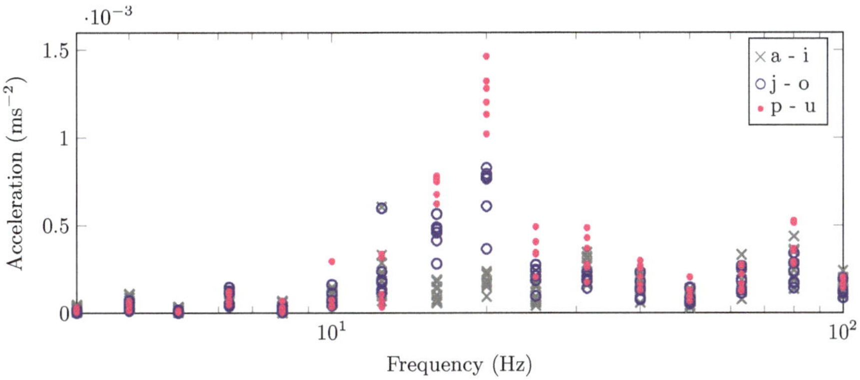

Figure 10.26 Operational response prediction in target environment due to hypothetical treadmills at various locations (see Figure 10.25).

10.5.3 DISCUSSION AND CONCLUSIONS

The study illustrates techniques that can be employed to facilitate experimentally-based predictions where they would normally be challenging or inherently impractical using 'mock-up' methods traditionally used in building acoustics investigations. With regards to source characterisation, on-board validation results provided good agreement across a wide range of frequencies despite characterisation measurements being conducted in a noisy environment with signal to noise issues above 100 Hz. Further work on increasing the range of frequencies obtained via the method, in addition to including additional degrees of freedom, such as rotations, would be a welcome next step to facilitate a more complete characterisation of the treadmill in question.

On acquiring a satisfactory representation of the source, the prediction of sound transmission into a receiving area is limited only by the acquisition of receiver FRFs and their quality. It was shown that paying due care to the positioning and measurement of transfer FRFs not only facilitated the prediction process but also highlighted areas where increased transmission occurred, owing to a critical, but at the time unknown, change in construction. This information is highly valuable to clients seeking to minimise the adverse effects that their operation may have upon their neighbours. Consequently, the rigour applied to capturing high-quality transfer FRFs between potential source and receiver locations cannot be understated.

With regards to prediction, the added mass of the treadmill was assumed to present negligible impedance to that of the baseline floor in the candidate location. This greatly simplified the prediction methodology, avoiding the need for substructuring. A validation exercise to confirm such assumptions would therefore be of interest. Additionally, it is acknowledged that there are certain limitations which may arising from a prediction based upon characterising a single treadmill with a single user operating the machine at a limited range of frequencies on a specific treadmill model and type. This limitation presents opportunities for suppliers to undertake product testing of their own treadmill variants for a range of users and speed settings which, in turn, could be used by practitioners to avoid the characterisation process, leaving only FRFs to be measured. Additionally, the use of off-the-shelf blocked forces for a range of treadmills, user masses, running speeds and/or styles could be utilised to create uncertainty bounds on the prediction. This would undoubtedly result in more reliable predictions when considering impact in noise and vibration sensitive settings.

Bibliography

1. *ProPG: Gym Acoustics Guidance (GAG).* Association of Noise consultants, 2023.
2. AL Smith, SJ Hicks, and PJ Devine. *Design of floors for vibration: A new approach.* Steel Construction Institute Ascot, Berkshire, UK, 2007.

10.6 RAIL ROLLING-STOCK VIRTUAL PROTOTYPE

Contributed by Nezam Sanei of Alstom

Accurate prediction of train interior structure-borne noise is extremely challenging but is an essential requirement for rolling stock manufacturers seeking to provide the highest levels of passenger comfort. This case study presents the construction and testing of a VAVP of an air compressor installed in a train and summarises work reported in [1–3]. The source of vibration, a large AGTU (Air Generation and Treatment Unit), was characterised (blocked force and free mobility) by test bench measurements at the manufacturers' site in Italy. The receiver structure, i.e. the train, was characterised by its free-interface FRF, measured at a plant in Poland. The blocked forces were combined with the source and receiver FRFs and used to predict the sound pressure in the passenger compartment of the train. The FRFs for the coupled AGTU-train assembly were obtained by substructuring. Validation was achieved by comparing predicted sound pressure levels with values measured directly on the train.

10.6.1 MEASUREMENT AND VALIDATION OF BLOCKED FORCES

The source was an air generation and treatment unit (AGTU) equipped with Buran 5 oil free piston compressor designed and manufactured by the rolling stock supplier Faiveley Transport. Figure 10.28 shows the unit alone and Figures 10.27 and 10.32 illustrate respectively the unit installed on the test bench and in the train. An array of seven accelerometers was employed at each connection point to capture six degrees of freedom (see Figure 10.29) using an equivalent multi-point scheme.

In order to provide a cross check on the blocked forces, the source was characterised in two very different test bench configurations, rigidly and resiliently mounted as shown in Figure 10.27. Blocked forces were obtained using the *in situ* method

Figure 10.27 AGTU resiliently mounted (left) and rigidly mounted (right). The accelerometer on the test rig support in the rigid mounted case was used as a reference for on board validation.

separately for each configuration. Figure 10.30 compares the blocked forces for *artificial* excitation. It can be seen that the blocked forces obtained from the two extreme boundary conditions of the source are in good agreement with some errors at low frequency due to noise. In order to cover a wide frequency range a plastic hammer tip was used with 100mV/g accelerometers but the results could be optimised in specific frequency ranges by using softer or harder hammer tips together with different sensitivities of accelerometers.

Figure 10.28 The AGTU unit – the vibration source.

The same exercise was also performed with *real* excitation of the AGTU operating under load as it would do on a train after a 10 minute warm up period. Results are shown in Figure 10.31. The blocked forces from the two configurations are again in good agreement and appear improved at low frequency, which may be due to stronger excitation from the source in this frequency range. The results in Figures 10.30 and 10.31 therefore partially validate the measured blocked forces from the two different test configurations and suggest that the source operation is not affected by the mount condition. It is worth noting that such good agreement would not be obtained had only the vertical DoF been included. Such independence requires a complete interface representation.

Predicted responses at a reference point on the mounting were also compared, as part of an on-board validation. Results were in satisfactory agreement but are not shown here for brevity, although they can be seen in [1, 3].

Figure 10.29 Sensor array at one of the interface points on the AGTU.

Note that the resiliently mounted test bench configuration also made it possible to measure the free-interface FRFs of the uncoupled source (AGTU). While these source FRFs were not used to obtain the blocked forces, they *are* required for a full characterisation of the source and were subsequently used for substructuring calculations of the AGTU installed in the train as described in the following section.

10.6.2 VALIDATION IN THE TRAIN

Having validated the blocked forces measured on the test bench assembly and satisfied ourselves that they were transferable it was then possible to use them to make predictions of structure borne noise and vibration for the unit installed in a train as shown in Figure 10.32. The AGTU was a nominally identical unit to the one used for source characterisation described in the previous section.

Although the FRFs of the coupled AGTU-train could be measured directly, it was decided to obtain them by substructuring calculations so as to provide a validation of the full VAVP methodology. To this end, the AGTU was removed and free-interface FRFs of the train alone were measured in six degrees of freedom at the interface connections points whilst also capturing the acceleration at a point on the train floor and the sound pressure at three interior microphone positions. FRFs for the source had already been obtained from the resiliently mounted AGTU during the source characterisation tests reported in the previous section. The source and receiver FRFs were then combined using substructuring calculations. An example of the results for a transfer FRF from the interface to the reference accelerometer are compared with directly measured values in Figure 10.33.

Shown in Figure 10.34 is a prediction of the floor acceleration (orange) by combining the substructured FRFs and blocked forces. This is compared with the

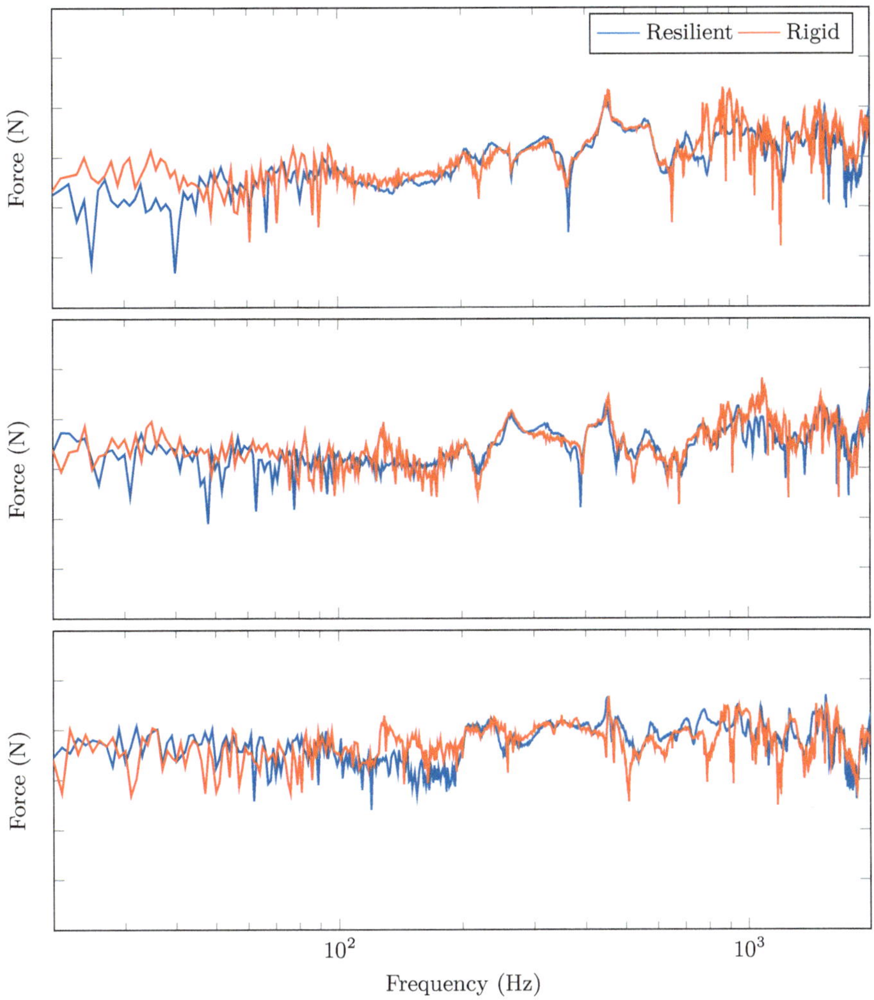

Figure 10.30 Comparison of blocked forces due to artificial excitation from rigid (orange) and resilient (blue) mounting conditions. Top, middle and bottom plots correspond to responses in the z, y and x directions at attachment point 1 respectively .

acceleration measured with the air compressor installed and operational (blue). Good agreement is obtained across the entire frequency range of interest, with a small over estimation of the tonal components around 200 Hz.

Shown in Figure 10.35 is a prediction of the train interior noise (orange), again obtained by combining the substructured *vibro-acoustic* FRFs with the same blocked forces as before. The directly measured pressure response is also shown (blue). Unlike the structural FRFs and directly measured acceleration response prediction, the vibro-acoustic FRF and operational sound pressure levels were subject to

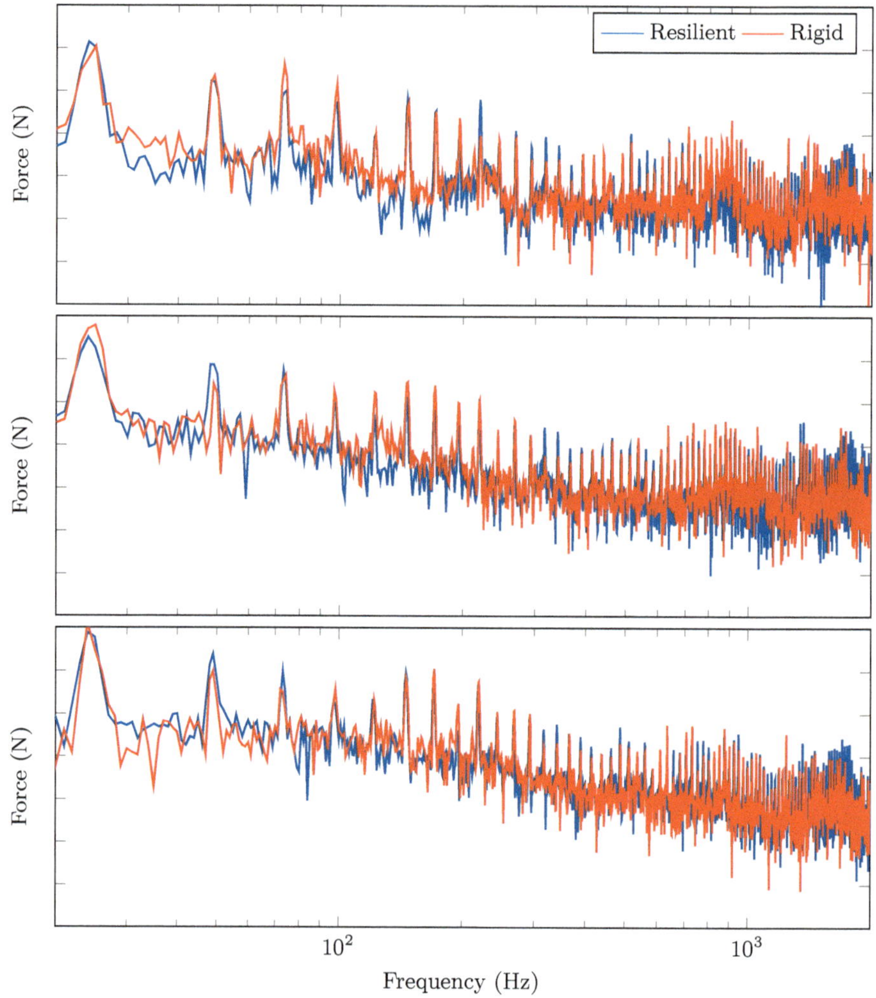

Figure 10.31 Comparison of *operational* blocked forces from rigid (orange) and resilient (blue) mounting conditions. Top, middle and bottom plots correspond to responses in the z, y and x directions at attachment point 1 respectively.

considerable background and flanking noise. Indeed, the compressor is also a source of airborne sound and that some of the carriage doors had not yet been installed. Nevertheless, we still obtain good agreement between the predicted and directly measured response, particularly between 100Hz and 1kHz, where the overall response is within 3dBA.

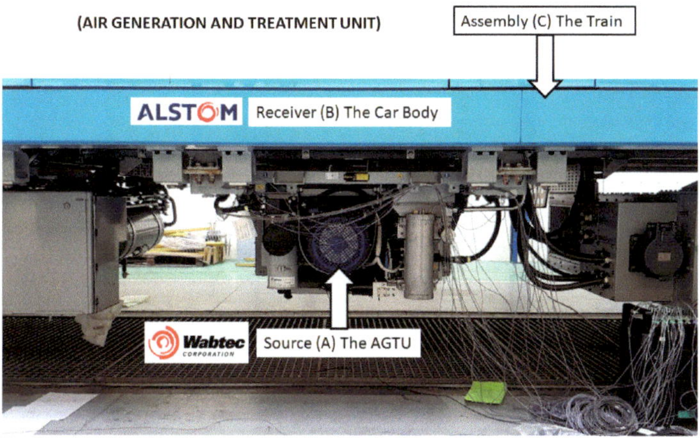

Figure 10.32 Photograph of the source-receiver assembly.

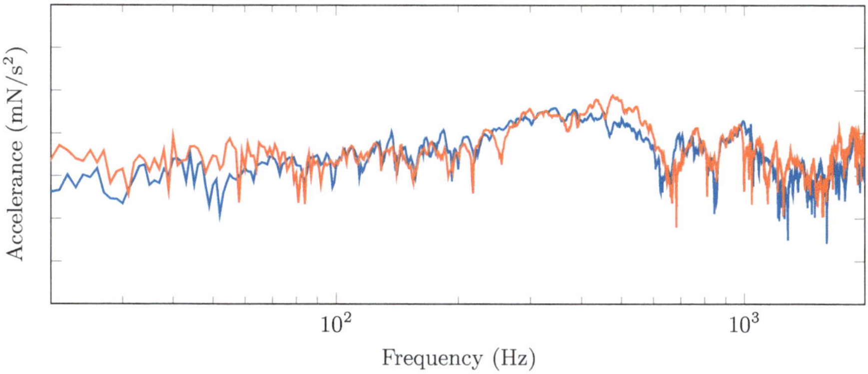

Figure 10.33 Example transfer FRF between the interface *c* and remote receiver point *r* on the coupled AGTU-train assembly. Predicted by substructuring (orange) and directly measured (blue).

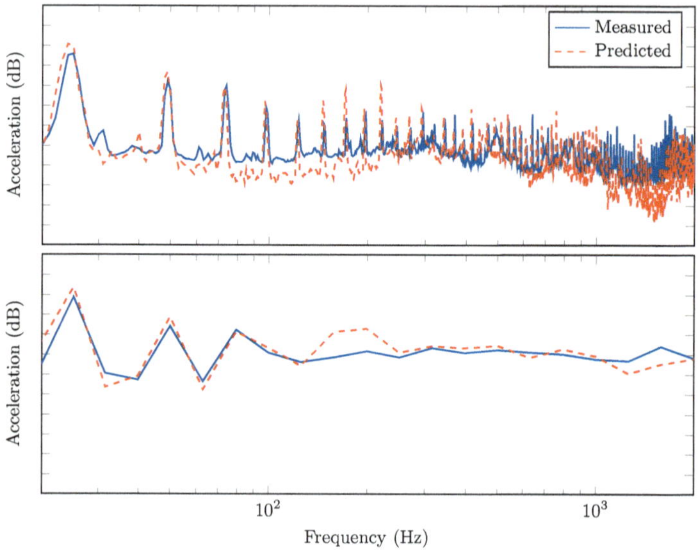

Figure 10.34 Operational acceleration response on the floor of the AGTU-train assembly in narrow band (top) and third octaves (bottom). Predicted (orange) and directly measured (blue).

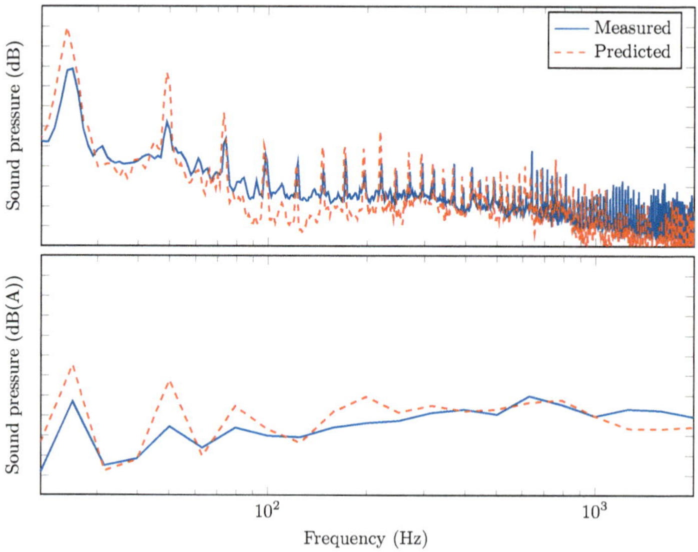

Figure 10.35 Operational sound pressure response in the carriage of the AGTU-train assembly in narrow band (top) and third octaves (bottom). Predicted (orange) and directly measured (blue).

10.6.3 CONCLUSIONS

A novel feature of the case study is the source characterisation of a real industrial source using six DoF at each mounting point. Secondly, this is believed to be the first example of a full validated VAVP prediction for a large industrial source, i.e. combining in situ measured blocked forces with FRFs obtained by substructuring (including vibro-acoustic FRFs). All measurements had to be conducted in an industrial environment with all the challenges that brings, for example there was significant flanking sound transmission through the open doors of the train. Satisfactory agreement has been obtained despite these difficulties which demonstrates that the VAVP methodology is practical even in the non-ideal measurement conditions often found in industry.

Bibliography

1. AS Elliott, AT Moorhouse, N Sanei, M Glesser, J Sapena, and S Ferella. Structure borne noise characterization of an air generation and treatment unit (agtu) for a train using the mobility method and sub-structuring. In *INTER-NOISE and NOISE-CON Congress and Conference Proceedings*, volume 259, pages 6707–6715. Institute of Noise Control Engineering, 2019.
2. AT Moorhouse, AT Elliott, and JWR Meggitt. Virtual acoustic prototyping-a story of four decades. In *INTER-NOISE and NOISE-CON Congress and Conference Proceedings*. Institute of Noise Control Engineering, 2022.
3. N Sanei, M Glesser, J Sapena, S Ferella, and AS Elliott. Structure borne noise characterization of an air generation and treatment unit (agtu) for a train by using blocked forces method. In *INTER-NOISE and NOISE-CON Congress and Conference Proceedings*, volume 259, pages 6624–6636. Institute of Noise Control Engineering, 2019.

11 Conclusions

There is no doubt that we are entering the age of *zero physical prototyping*. In the Introduction, the reasons given for this movement were that it is cheaper, faster, safer and less resource intensive to work with a virtual prototype than with a physical one. There is a perception that computer simulation allows engineers to do in minutes what would have taken days, or even weeks to achieve by conventional (physical) means. However, it has become clear throughout the book that, in the case of vibro-acoustics, building a virtual prototype is neither quick nor simple.

To be useful, a virtual prototype must accurately mimic every aspect of a mechanical product's performance and for vibro-acoustic performance this means it must sound and feel exactly like the real thing. Perhaps unsurprisingly, given the complexity of the various internal mechanisms and the sophistication of human hearing, this has turned out to be an exceptionally difficult challenge. In the 1980s, this aim might have seemed like a pipe dream. Even in the early 2000s, lack of sufficiently reliable prediction methods for vibro-acoustics appeared to be in danger of holding up the whole virtual prototyping agenda.

Whereas in some fields, the demands of virtual prototyping could be met solely by numerical modelling, vibro-acoustic engineers have not been so fortunate. The complexities that underlie vibro-acoustics, even of the most simple vibration sources and common structures, have proven beyond the capabilities of today's numerical simulation, especially when subjective realism is required. We have seen that things could only be moved forward by integrating *measured* data within an overarching *Vibro-Acoustic Virtual Prototyping (VAVP) framework*. Whilst the theory underlying the VAVP framework is now well developed, considerable ingenuity remains necessary to obtain experimental data compatible with this scheme. This book is about what data is required to build a VAVP, how this data can be obtained from measurement, and how to evaluate and minimise errors and uncertainties.

11.1 THE STORY SO FAR

Through this book we have aimed to provide the reader with all of the ingredients required to build a complete VAVP. In Part 1 we discussed the more theoretical elements of this scheme – *passive* and *active component characterisation*, the *interface* representations, and modelling of *assemblies*. In Part 2 we considered the implementation and validation of the *measurements* needed, *processing* of the data and evaluation of *uncertainties*. Part 3 featured various *applications* where the VAVP approach has been applied to real world problems.

VAVPs, just like physical prototypes, are built from *components*. Within the VAVP framework these components are treated as *black boxes*; they are characterised entirely by the dynamic properties at interfaces. All components have passive properties and some (sources), in addition, also have active properties to represent internal

DOI: 10.1201/9781003399131-11

operational forces. The *passive vibro-acoustic properties* of a black box component are defined by matrices of FRFs describing the force-response relation at and between all interface DoFs, plus any additional target locations. Regarding *source characterisation*, active properties are represented by blocked forces, or sometimes free velocities, again defined only at the interface DoFs – the actual location of the internal operating forces is not considered important, only their net effect at interfaces.

Of great practical significance in the component-based approach is that each component must be represented in an *interchangeable* (or transferable) manner. This requires that their properties are *independent* which, in turn, implies they are defined with the interfaces either *free* or *fixed*. Sometimes, these conditions can be realised physically for the test, but when this is not the case, measurements must be done *in situ*, i.e. with the component installed in an assembly. One of the main areas of development then has been in methods able to extract independent properties from measurements made *in situ*.

Of central importance to the VAVP scheme is the *in situ* blocked force method, described at length in Chapters 3, 5 and 6. This method, remarkably, allows independent blocked forces to be obtained entirely from *in situ* measurements. To achieve independence, fixed boundary conditions are, in effect, imposed mathematically rather than physically. It is the blocked force that, crucially, allows us to *switch on* VAVPs. Regarding *passive* properties, some, but not all, can be extracted from *in situ* measurements using the same trick of mathematically blocking the boundary, for example as used for the *in situ* measurement of isolator transfer impedance in Chapter 2. Other in-direct approaches include *decoupling* procedures to remove the effect of attached components (Chapter 2) and use of model updating schemes.

The importance of a reliable *interface model* is stressed throughout the book. Indeed, the success of any experimental characterisation is reliant on the definition of an appropriate interface representation. To this end, Chapter 4 describes how various types of interface geometries may be represented in discrete form. The most common assumption is that structural connections are *point-like* with up to six DoF at each. Vibro-acoustic interfaces on the other hand are continuous and may be discretised into patches.

Having explored component characterisation in Chapters 2 and 3, and interfaces in Chapter 4, Chapter 5 discusses the key steps required to model an *assembly*. First, one must assemble its constituent components to form a passive system (represented by its FRF matrix). This is achieved by means of substructuring. The substructuring methods presented (direct, primal and dual) differ in detail but are mathematically identical, in that they each enforce the same conditions of continuity and equilibrium.

Following the assembly of components, we discussed the various ways one can predict the operational response of a system, which led to discussion of various formulations of Transfer Path Analysis (TPA). Whilst most variants of TPA do not satisfy the requirements of a VAVP (i.e. interchangeability of components), attention was drawn to *in situ TPA*. This uses the *in situ* blocked force method to obtain blocked forces which, unlike contact forces, are transferable to other assemblies. *In situ* TPA appears to have become the leading diagnostic method for NVH and is used extensively across the automotive industry, among others. It does not however, constitute a full VAVP; a natural extension of in situ TPA is to introduce substructuring to replace what was previously a measured FRF. The result, *component-based TPA*, provides a seamless link between TPA, which was previously restricted to diagnostic applications on a single assembly, and the entire predictive capabilities of VAVPs.

In *Part 2*, we moved on to the practical implementation of the now established VAVP framework. The main themes here are (a) the need for systematic evaluation and minimisation of uncertainties and (b) the need to solve the inverse problem. We have seen that matrix inversions are required for both blocked force and substructuring computations, and hence all measured data is at some stage inverted when building a VAVP. Owing to the sensitivity of matrix inversions, it is essential that measured data is both *accurate* and *fully representative*.

Accuracy is required because errors in any single element will typically translate to the entire solution and may also be amplified on inversion. To minimise this, careful checking of every measurement is recommended (coherence, reciprocity, passivity, linearity and condition number). In line with good practice, an evaluation and propagation of experimental uncertainties should accompany any estimates, a framework for which is discussed in Chapter 8.

Even with perfect measurements though, things can still go wrong. Indeed, the most likely place for an error to occur when building a VAVP is in the *interface model*: *fully* representing all dynamic interactions at the interface by correct identification of important DoFs is of crucial importance. The *completeness criterion* provides an important check on the reliability of the interface model. For a blocked force characterisation, a staged programme of validation checks is also recommended, starting with on-board validation (essentially the same as *in situ* TPA) and progressing to transferability validation and full VAVP validation. The need to ensure the quality of the data prior to regularisation of the inverse solution is emphasised.

The case studies in Part 3 illustrate, through their great diversity, the flexibility of VAVPs and the component-based approach. Furthermore, they demonstrate that the techniques have matured sufficiently to allow application to complex products in challenging industrial environments. Above all though, they prove that, with appropriate care, it is now possible to produce VAVPs that are sufficiently reliable to fit the virtual prototyping agenda.

11.2 PERSPECTIVES

While the case studies illustrate that the component-based approach has reached a level of maturity appropriate for summary in a book, there are undoubtedly aspects with potential for expansion and development. Here we consider some trends that are already underway and also allow ourselves to speculate on some possible future developments.

First, the component-based approach provides a framework that clarifies the communication between suppliers, who provide the components, and OEMs, who build them into assemblies. The Alstom case study in 10.6 (see also comments in the Foreword) illustrates OEMs and suppliers working together within the component-based framework. The commonly understood structure allows contractual responsibilities to be clearly delineated in a way that was not previously possible.

A second trend that is already underway but looks set to continue is that of *mass testing*. It is evident from the case studies that a high level of skill is required for blocked force measurements which means that only small numbers of a product are typically tested. However, this could be revolutionised by an automated system including a sensor array that, once designed and tested by a highly skilled engineer, can be rolled out quickly and reliably across nominally identical products. Much larger samples of a production run could then be tested, potentially even extending to end-of-line blocked force tests. Tests at scale would allow meaningful ensemble statistics to be gathered that would be valid for any subsequent installation. Development of robust criteria for acceptance/ rejection would then be possible whilst avoiding uncertainties associated with different installations. These could potentially even be developed into fault/ no-fault criteria for condition monitoring applications.

The ability to extract passive component properties from *in situ* measurements, as described in Chapter 2, may also have application in *structural health monitoring*.

On the modelling side, there is much scope for improved modelling of *source mechanisms*. The component-based framework provides an important platform here both to clarify what form the outputs of the modelling should take (blocked forces or free velocities at the interfaces) as well as how to couple the source models to those for passive components.

Also on the modelling side we can expect more *hybridisation*, taken in a broad sense to include measurements combined with models as well as different modelling approaches within the same framework. This seems essential since no one approach is able to handle the whole vibro-acoustics frequency range. Combining measured and modelled data is in theory straightforward, yet to date there are relatively few published examples where this has been convincingly achieved. It seems that development here will be as a result of improved techniques using, for example, the Virtual Point Transformation to provide compatibility. Regarding hybrids of different modelling approaches, we saw in 10.21 how the component-based framework could be extended to include statistical representations of short wavelength systems. This could potentially be taken further, for example to include other modelling approaches such as energy methods.

Speculating further, there is the possibility of *real time* and *time domain* VAVPs. A VAVP could be considered *real-time* if the time taken to predict the forward response at a target location is less than the time of flight from the excited interface to that target location. The prediction would then be available before the response had actually occurred. Unlikely as this sounds, a not dissimilar process is already employed in feed-forward active control, particularly to cancel road noise in vehicles. After all, if the response due to an upstream excitation can be predicted before it actually occurs, then a control signal can in principle be designed to cancel that response. As such, real-time VAVPs could form the first stage of improved active control systems. This also has parallels with the concept of *digital-twins* where continual real-time monitoring and optimisation of model parameters is envisaged as part of the trends of the fourth industrial revolution.

Any frequency-domain analysis is theoretically restricted to linear, time-invariant systems, although slowly varying FRFs can potentially be handled by adapting properties and recalculating. For faster varying FRFs it would be necessary to solve the systems in the time-domain. The same can be said for non-linearity. Potentially then, recasting the entire VAVP framework in the time-domain could open up many new possibilities.

11.3 EPILOGUE

In the last forty years things have moved on significantly. At the time of writing we have reached something of a milestone in that the simulated sound and vibration of complex industrial products is now good enough, at best, to be indistinguishable from recordings of the real thing. While initial construction of a VAVP might be a lengthy and complex process, judging by the continued take-up of the ideas in industry, this effort is seen as a worthwhile investment to enjoy the advantages it brings.

Among these benefits is the fact that components can be combined, modified, interchanged and 'switched on', just like a real physical prototypes. This allows us to test what-if scenarios with great *speed* and *convenience*. Being based on independent properties, the process of building a VAVP also has the potential to *clarify* the contractual responsibilities of component suppliers and OEMs who assemble the components. The component data, unlike physical components, can then be sent electronically anywhere in the world more or less instantaneously. Moreover, the use of independent quantities provide a basis for improved *target setting*. This links into *numerical modelling* since, on the same basis, if target quantities are clear then the outputs required from a modelled component are also clear and may be shared without revealing full details of the model. Arguably though, the greatest benefit is that the process of breaking a machine down into its vibro-acoustic components and reassembling them virtually provides *insight* and understanding that is so valuable for engineers. Who knows where this will take them in the next forty years.

Index